KT-549-643

BARRY'S INTRODUCTION TO CONSTRUCTION OF BUILDINGS

Third edition

Stephen Emmitt
Professor of Architectural Practice
University of Bath
and Visiting Professor in Innovation Sciences at Halmstad University, Sweden

and

Christopher A. Gorse
Professor of Construction & Project Management
Leeds Metropolitan University

WILEY Blackwell

Bromley Libraries

30128 80268 860 9

This edition first published 2014
Third edition © 2014 by John Wiley & Sons, Ltd

Registered office: John Wiley & Sons, Ltd, The Atrium, Southern Gate, Chichester, West Sussex, PO19 8SQ, UK

Editorial offices: 9600 Garsington Road, Oxford, OX4 2DQ, UK
The Atrium, Southern Gate, Chichester, West Sussex, PO19 8SQ, UK

For details of our global editorial offices, for customer services and for information about how to apply for permission to reuse the copyright material in this book please see our website at www.wiley.com/wiley-blackwell.

The right of the author to be identified as the author of this work has been asserted in accordance with the UK Copyright, Designs and Patents Act 1988.

All rights reserved. No part of this publication may be reproduced, stored in a retrieval system, or transmitted, in any form or by any means, electronic, mechanical, photocopying, recording or otherwise, except as permitted by the UK Copyright, Designs and Patents Act 1988, without the prior permission of the publisher.

Designations used by companies to distinguish their products are often claimed as trademarks. All brand names and product names used in this book are trade names, service marks, trademarks or registered trademarks of their respective owners. The publisher is not associated with any product or vendor mentioned in this book. This publication is designed to provide accurate and authoritative information in regard to the subject matter covered. It is sold on the understanding that the publisher is not engaged in rendering professional services. If professional advice or other expert assistance is required, the services of a competent professional should be sought.

Library of Congress Cataloging-in-Publication Data
Emmitt, Stephen.
  Barry's introduction to construction of buildings / Stephen Emmitt, Christopher A. Gorse. – Third edition.
      1 online resource.
  Includes bibliographical references and index.
  Description based on print version record and CIP data provided by publisher; resource not viewed.
  ISBN 978-1-118-85654-3 (ePub) – ISBN 978-1-118-85668-0 (Adobe PDF) – ISBN 978-1-118-25542-1 (pbk.)   1. Building.   I. Gorse, Christopher A.   II. Title.   III. Title: Introduction to construction of buildings.
  TH146
  690–dc23

                              2014002855

A catalogue record for this book is available from the British Library.

Wiley also publishes its books in a variety of electronic formats. Some content that appears in print may not be available in electronic books.

Cover image courtesy of the authors and iStock Photo
Cover design by Andy Meaden

Set in 10/12 pt Minion Pro Regular by Toppan Best-set Premedia Limited
Printed and bound in Malaysia by Vivar Printing Sdn Bhd

2   2015

Contents

Preface

Robin Barry's *The Construction of Buildings* was first published in 1958 and quickly became an established source of information for students of building design and construction. When we became involved the initial task was to embark on a major redesign and updating exercise, taking five volumes of the work and distilling it into two comprehensive books: *Barry's Introduction to Construction of Buildings* and *Barry's Advanced Construction of Buildings*. The intention was, and still is, that the 'introduction' volume deals with topics normally taught in the first year of a student's studies and the 'advanced' volume addressed topics usually covered in the second year. Our philosophy remains the same: to provide informative and engaging material that will help students of building design and construction understand the fundamentals of how we construct sustainable buildings. Once again we have tried to do this in a way that represents exceptional value to the reader.

In the third editions of the books, we have made the books easier to navigate by repositioning some of the material. For example, all of the content relating to services has been included in the *Introduction* volume, which has allowed some space to better address indoor climate (Chapters 11 and 12). We have also added a new chapter that deals with heat loss and heat loss calculations (Chapter 13). *Barry's Advanced Construction of Building* contains a new chapter on obsolescence and revitalisation of our building stock (Chapter 11). Combined the two books cover the entire life cycle of buildings underpinned by an environmentally sustainable ethos.

Application of fundamental principles of building will be coloured by prevailing national and international legislation and, to a certain extent, local traditions of building: we feel that this is how it should be, rather than slavishly copying the details and information in the *Barry* books. Although we have continued to make reference to the Building Regulations for England and Wales to help explain some of the issues, we have tried to remain faithful to Robin Barry's original philosophy and describe building elements in relation to their function. Thus the contents are applicable to readers, regardless of specific location.

Stephen Emmitt
Christopher A. Gorse

Acknowledgements

Over the years our students have continued to be the inspiration for writing books and they deserve our heartfelt credit for helping to keep our feet firmly on the ground by asking the 'why' and 'how' questions. Feedback from our readers and reviewers also helps us to keep the *Barry* series relevant and topical. It would, of course, be an impossible task to write this book without the support and assistance of our colleagues in academia and constant interaction with industry, for which we are extremely grateful. We would like to mention and thank Mike Armstrong (Shepherd Group), Joanne Bridges (Yorkon), Mikkel Kragh (Dow Corning), Shaun Long (Rossi Long Consulting), Karen Makin (Roger Bullivant), Jennifer Muston (Rockwool B.V. / Rockpanel Group), Gordon Throup (Big Sky Contracting) and Paul Wilson (Interserve). We would also like to thank the numerous other individuals and organisations, many of whom have been very generous with their time, allowing access for photography and giving valuable advice. We trust a 'global' acknowledgement of our gratitude will go some way to acknowledge their collective help.

About the Companion Website

This book's companion website is at *www.wiley.com/go/barrysintroduction* and offers invaluable resources for students and lecturers:

1 Introduction

The aim of this short introductory chapter is to highlight some of the factors that determine how buildings are constructed and also to provide some context to the chapters that follow. Related issues are dealt with in the introduction to *Barry's Advanced Construction of Buildings*. A brief overview of the function and performance of buildings leads into a discussion about environmental factors and sustainable building. This is followed by a description of the general principles of construction, concluding with some comments on legislation, sources of information and making informed choices.

1.1 The function and performance of buildings

Buildings are constructed, altered, upgraded, restored or demolished for a variety of reasons. Whether the aim is simply to provide more space or to make a financial gain from speculative development, all building projects need to fulfil a function and meet set performance criteria, no matter how fundamental or sophisticated the client's requirements may be. Buildings, regardless of function, will have an impact (either positive or negative) on the environment throughout their entire life. Environmental impact will be influenced by many factors, such as the responsible sourcing and manufacturing of environmentally sustainable materials and building products; the decisions taken during the construction process; the actions of owners and occupants through a long period of use, reuse, alteration and repair; through to deconstruction at the end of the building's useful life. At this 'final' stage many materials and components can be recovered and reused, or recycled for use in new building products, helping to reduce the amount of material sent to landfill and improving the environmental impact of buildings. Environmental performance of buildings is an important consideration for all building projects, be they new build or work to existing structures.

Function

The primary function of a building is to provide shelter from our weather, a container for living, working and playing in. The principal functional requirements include:

❑ Shelter
❑ Security

Barry's Introduction to Construction of Buildings, Third Edition. Stephen Emmitt and Christopher A. Gorse.
© 2014 John Wiley & Sons, Ltd. Published 2014 by John Wiley & Sons, Ltd.

❑   Safety (and comfort)
❑   Ease of use and operation (functionality)
❑   Ease of maintenance, periodic repair and replacement/upgrading
❑   Adaptability and durability
❑   Ability to reuse and recycle materials and components at a future date

The overall goal is to achieve these functions in an economical, safe and timely fashion using the most appropriate resources available and with minimal negative impact on the environment. These primary functional requirements are explored in the chapters that follow in relation to specific building elements.

Performance

The performance of the building will be determined by a number of interrelated factors set by the client, legislation and society. Clients' performance requirements will vary from project to project. However, the main considerations are likely to be:

❑   Space, determined by a figure for floor area and/or volume (and related to anticipated use)
❑   Thermal and acoustic performance (the quality of the indoor climate)
❑   Design life and service life of the building and specific building elements
❑   Cost of construction, cost in use and cost of demolition/deconstruction and recycling
❑   Quality of the finished building (functionality and durability)
❑   Appearance of the finished building
❑   Environmental impact

Other specific performance criteria will relate to the use of the building, for example the provision of special work surfaces for catering establishments. Legislative performance requirements are set out in building codes and regulations (see 'Building Control and Building Regulations'). Specific performance requirements, for example the thermal insulation of walls and fire protection of doors, must be met or bettered in the proposed construction method.

Quality

Function and performance will influence the quality of the building. The quality of the completed building, as well as the process that brings it about, will also be determined by the quality of thought behind the design process, the quality of the materials and products specified, and the quality of the work undertaken. There are a number of different quality issues:

(1) Quality control is a managerial tool that ensures both work and products conform to predetermined performance specifications. Getting the performance specification right is an important step in getting the required quality, be it for an individual component or the whole building.

(2) Quality assurance is a managerial system that ensures quality service to predetermined parameters. The ethos of total quality management aims at continual improvement and greater integration through a focus on client satisfaction. Manufacturers, contractors and professional consultants use this.

(3) Quality of the finished artefact will be determined by a number of variables constant for all projects, namely, the:

- ○ Interaction and characteristics of the participants engaged in design, manufacture and assembly
- ○ Effectiveness of the briefing process
- ○ Effectiveness of the design decision-making process and resultant information
- ○ Effectiveness of the assembly process
- ○ Effectiveness of communications
- ○ Time constraints
- ○ Financial constraints
- ○ Manner in which users perceive their built environment

The required quality of materials and workmanship will be set out in the written specification. Good quality materials and good quality work tend to carry a higher initial cost than lower quality alternatives; however, the overall feel of the building and its long-term durability may be considerably improved: we tend to get what we pay for. When making decisions about the materials and components to be used it is important to consider the whole life cost of the materials, not just their initial capital cost and the cost of labour to assemble the materials.

Economics

The building site and the structures constructed on the land are economic assets. In addition to the cost of the land there are three interrelated costs to consider. The first is the initial cost, the cost of designing and erecting the building. This is usually the primary and sometimes the only concern of clients and developers. It covers professional fees and associated costs involved in land acquisition and permissions, the capital cost of materials and components and the labour costs associated with carrying out the work.

The second cost to consider is the cost of the building in use, i.e. the costs associated with routine maintenance and replacement and the costs associated with heating and servicing the building over its life. These costs can be reduced by sensitive design and detailing, for example designing a building to use zero energy and to be easy to maintain will carry significant cost benefits over the longer term (not to mention benefits to the environment). All materials and components have a specified design life and should also have a specified service life. Designers and contractors need to be aware of these factors before starting work, thus helping to reduce defects and maintenance requirements before construction commences.

The third cost is the cost of materials recovery at the end of the life of the building, i.e. the cost of demolition, recycling and disposal. All three areas of cost associated with building should be considered within a whole life cost model, from which decisions can be made about the type of materials and components to be used and the manner in which they are to be assembled (and subsequently disassembled). This links with issues concerning main-tenance, repair renovation and recycling.

1.2 Environmental factors

There is an extensive literature concerning the environmental impact of building materials, products and components, construction activities and the use (and misuse) of buildings during their lifetime. We know that we must do more to respect our planet and build in a way that has a positive impact on our environment. From a construction perspective consideration should be given to the method of construction, maintenance and repair, future adaptability of the structure and the recycling of materials as and when the building is demolished or substantially remodelled. This is particularly important at the detailing and specification stage when materials and components are selected. There are many ways in which we can improve the relationship between our artificial environment and our natural one. For example, detailing buildings so as to reduce unnecessary waste during production not only helps to reduce landfill, it also saves time and money. Similarly, detailing and constructing a building in such a way as to make it easy to disassemble at the end of its service life will enable precious components and materials to be recovered with minimal damage, and hence minimal waste.

Climate change

There continues to be considerable speculation as to the future impact of climate change. In the UK the general consensus is that the average temperature will continue to rise, as will the amount of rainfall and the average wind speed. The message from the weather forecasters is wetter, warmer and windier. This has given rise to a number of concerns about the suitability of the existing building stock and also to the technologies being employed for the erection of new buildings. How, for example, do these predicted changes impact on the way in which we detail the external fabric of buildings? Are existing Codes, Standards and building practices adequate? The general consensus is that we should adopt a cautious approach, although we would urge against over-detailing and over-specifying, which may be wasteful.

Some concern has been expressed about new buildings, especially homes that are built from lightweight materials, such as timber-framed, steel-framed, modular and other light-weight construction systems. The fear is that with an expected increase in temperatures the internal temperature of lightweight construction may become too high during the summer, thus necessitating air conditioning (increased energy demands) and/or better shading and natural ventilation. Buildings constructed of heavy walls, with small windows and sun shading devices (e.g. shutters, verandas) are less susceptible to temperature fluctuation. However, there are plenty of places around the world that have a warmer climate than the UK and where lightweight construction is used successfully. The answer to the problem is not so much about the type of construction used, rather the manner in which the building is designed to respond better to its immediate environment (e.g. verandas and shading devices).

Passive design includes the selection of energy-efficient building materials so that there is very little or no need to provide renewable technologies. This is sometimes referred to as 'fabric first'. A good example is *Passivhaus* (Passive House), which effectively eliminates the need for space heating through a highly insulated building fabric. Taking this concept a little further the *Activhaus* (Active House) concept aims to design and construct a building that generates more energy than it uses. Buildings that are constructed using straw bales and rammed earth also adopt the fabric first philosophy to eliminate the need for space heating. Level 6 of the Code for Sustainable Homes equates to a property with no net CO_2 emissions.

Environmental impact

There are a wide variety of approaches to the construction of buildings, and with increased attention focussed on ecologically friendly construction a number of different approaches are possible. Some have their roots in vernacular architecture and others in technological advancement, although most approaches combine features present in both old and new construction techniques. Strategies adopted can include, for example, the reuse of salvaged components and recycled materials from redundant buildings, designing buildings that may be disassembled with minimal damage to the components used, buildings that are designed to decompose after a predetermined time frame, incorporation of renewal energy sources, and so on. Care is required as many of these methods are largely untried (or the techniques have been forgotten) and it will take some time before we can really know for certain how they perform in situ. We do, however, urge all readers to consider the impact on the environment of their preferred construction method by adopting a whole life approach to the design, construction use and reuse/recycling of buildings (see also *Barry's Advanced Construction of Buildings*).

Energy efficiency and environmental performance

The environmental performance of buildings has long been a cause for concern, but it is an area in which it is difficult for the building owner to get reliable information. Designers and builders must make a greater effort to provide buildings with:

- ❏ Lower running costs
- ❏ Enhanced air quality and natural daylight
- ❏ Use of low-allergy materials
- ❏ Use of environmentally friendly materials
- ❏ Water efficiency (and recycling) measures
- ❏ Ease of adaption and alteration
- ❏ Future proofing (easy upgrading of energy-efficient technologies)

If these (and related) factors are addressed at the conceptual and detailed design phases then the initial cost of the construction is likely to be similar to a project that is less energy efficient and less environmentally friendly. Add to this the considerable cost savings over the life of the building and it is difficult to understand why buildings are still being constructed with such scant regard for the whole life performance of the constructed works.

Our existing building stock is a little more problematic, simply because it may be a challenge to make improvements to the building fabric and services to improve their low carbon credentials. With an estimated 27 million older homes needing to be upgraded to meet energy efficiency targets, the challenge is substantial. The ability to upgrade the building fabric and retrofit appropriate technologies is addressed in Chapter 11 of *Barry's Advanced Construction of Buildings.*

1.3 General principles of construction

Whatever approach taken to the design and erection of our buildings there are a number of fundamental principles that hold true. The building has to resist gravity and hence remain safe throughout its design life and substantial advice is provided in regulations and standards. Every building is composed of some common elements:

❑ Foundations (see Chapters 2 and 3)
❑ Floors (see Chapter 4)
❑ Walls (see Chapter 5)
❑ Roof (see Chapter 6)
❑ Windows and doors (see Chapters 7 and 8)
❑ Stairs and ramps (see Chapter 9)
❑ Surface finishes (see Chapter 10)
❑ Services (see Chapters 11 and 12)

It is vital for the success of the building project and the use of the constructed building that an integrated approach is adopted. It is impossible to consider the choice of, for example, a window without considering its interaction with the wall in which it is to be positioned and fixed, maintained and eventually replaced. It follows that the window should exhibit the same, or very similar, thermal and acoustic performance characteristics as the wall. The same argument holds for all building services, which should be integrated with the building structure and fabric in such a way as to make access for routine maintenance, repair and upgrading a safe and straightforward event which does not cause any damage.

It is common to classify construction methods as either loadbearing or framed construction.

Loadbearing construction

Masonry loadbearing construction is well established in the British building sector and despite a move towards more prefabrication, loadbearing construction tends to be the preferred option for many house builders and small commercial buildings (see Chapter 5). There is a heavy reliance on the skills of the site workers and on wet trades, e.g. bricklaying, plastering, and so on. Quality control is highly dependent upon the labour used and the quality of the supervision on site.

In a typical loadbearing cavity wall construction the main loads are transferred to the foundations via the internal loadbearing wall. The external skin serves to provide weather protection and aesthetic quality. Primarily 'wet' construction techniques are employed.

Framed construction

Framed construction has a long pedigree in the UK, starting with the framed construction of low-rise buildings from timber and followed by early experiments with iron and reinforced concrete frames. Subsequent development of technologies and advances in production have resulted in three main materials being used for low-rise developments: timber, steel and concrete (see also *Barry's Advanced Construction of Buildings* for additional information). Framed construction is better suited to prefabrication and off-site manufacturing than masonry loadbearing construction. Dry techniques are used and quality control is easier because the production process is repetitive and a large amount of the work is carried out in a carefully controlled environment. Site operations are concerned with the correct placement and connection of individual component parts in a safe and timely manner.

In a typical framed cavity wall construction the main loads are transferred to the foundations via the structural frame. The external skin serves to provide weather protection and aesthetic quality. It is common practice in most of the UK to clad timber- and steel-framed buildings with brickwork; thus from external appearances it might be impossible to determine whether the construction is framed or loadbearing.

Design and constructability

The functional and performance requirements will inform the design process, from the initial concepts right through to the completion of the detailed designs and production of the information (drawings, schedules and specifications) from which the building will be constructed. The design of the junction between different materials, i.e. the solution for how different parts are assembled, is crucial in helping to meet the performance and functional requirements of the overall building. Good design and detailing will help the contractor and subcontractors to assemble the building safely and economically. Good design and detailing, combined with good workmanship, will contribute to the durability and ease of use of the building over its life.

The manner in which materials are joined together will be determined by their material properties, shapes and sizes available, type of joint required, construction method (e.g. framed or loadbearing) and the safe sequence of assembly (and anticipated disassembly). Interfaces between materials and components can be quite complex and will be specific to particular materials and components, although in simple terms the following methods are used widely to join separate parts, either in isolation or in combination.

❏   Gravity. The simple placing of materials so that they stay in place due to their mass (e.g. stone on stone) or shape (e.g. interlocking roof tiles) is common, although it tends to be used in conjunction with an adhesive joint or mechanical fixing. Masonry is usually laid in mortar in loadbearing construction and roof tiles need to be clipped in position at regular intervals.

❑   Screws and bolts. Screws and bolts perform a similar function to each other, in joining two (or more) materials together by a mechanical fixing. Screws are widely used for joining timber, with the thread of the screw drawing the timber components together through the act of screwing through one piece of the material into the other. Bolts tend to be used for joining two pieces of metal and are (usually) placed in pre-drilled holes. A nut is threaded onto the end of the bolt and the bolt and nut are tightened to hold the materials together. The advantage of screws and bolts is that it is relatively straightforward to unscrew the screw or undue the bolt with minimum damage to the materials. Both the screw and the bolt can be reused. This is helpful for routine maintenance and inspection and also for recovering materials and reusing them at a future date.

❑   Nails. Nails are driven through the first material into the second using a hammer or a nail gun, with the materials held together by friction. This is a common method of joining two materials together, although it is difficult to withdraw the nail without causing damage to the materials and the nail.

❑   Adhesives, glues and welds. A wide variety of materials are employed to stick or bond one material to another. These include lime and cement mortars, chemical adhesives, glues and welds. Unless the bond is designed to be comparatively weak in comparison with the materials being joined together, e.g. lime mortar in brickwork, it will be very difficult to disassemble the construction without damaging the materials.

❑   Mastics. Mastics are primarily used to fill a joint. These 'flexible' filling materials are designed to allow movement between adjacent materials and to prevent the penetration of rain and wind through the joint. Mastic materials are usually forced into the joint. Ease of removal of the mastic will be determined by the material properties of the mastic and the shape of the joint.

Constructability (or buildability) is an approach to building design and construction that seeks to eliminate non-productive work on site, make the production process simpler and provide the opportunity for more efficient site management and safer working. Thus, designing and detailing for constructability requires an understanding of how components are manufactured off site, as well as how the building is to be assembled (the sequence of work packages) on the site. The core message of constructability is more simplicity (of joints between materials), greater standardisation (to avoid unnecessary cutting on site and hence reduce material waste) and better communication between designer, manufacturer and builder. These three principles also relate to the eventual disassembly of the building at some date in the future when materials and building products will be recovered, reused and recycled. An ethical approach should be taken to the sustainable sourcing of all building materials and products, which means that those making the decisions about which materials and products to specify and purchase must understand the supply chain and seek assurances about the provenance of each and every item. Some of the practical considerations are concerned with:

❑   Timescale
❑   Availability of labour and materials (supply chain logistics)
❑   Sequence of construction and tolerances (constructability)
❑   Reduction of waste (materials, labour, time and energy)

❏ Temporary protection from the weather
❏ Integration of structure, fabric and services
❏ Maintenance and replacement
❏ Disassembly and recycling strategies

Prefabrication and off-site production

In recent years the emphasis has been firmly on prefabrication and off-site production. This is, of course, only one of many different approaches and is usually more suited to repeat building types than one-off projects. However, the range of prefabricated units is expanding and considerable improvements in product quality and health and safety may be made through the use of prefabricated components and proprietary systems. This has tended to move the skills away from the building site into the controlled environment of the factory. Site operations become limited to the lifting, positioning and fixing of components into the correct position, and emphasis is on delivery of components to site 'just in time' and the specification of the correct tolerances to allow operations to be conducted safely. As the technologies improve and the number of off-site manufacturers grow the choice for designers and contractors is becoming much wider. Prefabrication and off-site production is described in Chapter 8 of *Barry's Advanced Construction of Buildings*.

An alternative approach

Conventional construction methods rely on a plentiful supply of resources, many of which have started to become less plentiful and hence more expensive. Alternative approaches (and attitudes) to construction, in the philosophy and use of materials and energy, seek to minimise environmental impact through sensitive design and specification. The mantra is:

❏ Reduce
❏ Reuse
❏ Recycle
❏ Revitalise

The energy expended in the extraction, working and transportation of materials to the site and the total resources used during construction should be included in the calculation of the structure's efficiency. Integration of resource-friendly concepts into the design and construction processes can significantly reduce the environmental impact of the con-structed works. Similarly, the occupants' habits and environmental ideals will affect the operating efficiency of the building. Adopting a less mechanised (and hence less conven-tional) approach to construction may be seen as a step in the wrong direction by some, but, for many, natural materials and labour intensive methods provide a realistic alternative. Primary drivers behind a non-conventional approach to construction may be one, or more likely, a combination of the following factors:

❏ Lower initial construction costs – affordability
❏ Energy efficiency – low heating (and cooling) costs throughout the life of the building

❏   Use of local materials
❏   Use of local (semi-skilled) labour, community involvement or self-build
❏   Cultural compatibility with the local environment
❏   Simplicity of design
❏   Easy to adapt as needs change
❏   Comfort
❏   Implementation of environmental ideals and principles
❏   Ease of disassembly and materials recovery at a future date

The following underlying issues need a little more explanation.

Cost of labour

Labour costs comprise a substantial part of the initial cost of most building projects. One way of mitigating labour costs is to employ a method of construction that is quick and efficient, although these tend to carry a high cost that is associated with the technologies and machinery required to manufacture and erect the building. Another approach is to engage in some form of self-build or self-help scheme, assuming that the self-builder has the time to invest in the project and has, or can readily acquire, the necessary skills to implement a quality product. For example, straw bale construction and rammed earth structures are attractive to owner-builders (self-builders) because of the cheap cost of the raw materials and the large savings in labour costs to be made by providing their own labour. Also new innovations, such as hollow polystyrene interlocking blocks (fitting together much like Lego bricks) that are filled with concrete to make a structural wall with low thermal conductivity, are attractive to the do-it-yourself builder. Experienced labour may, however, still be required for the foundation work, roof framing, electrical wiring and plumbing. Where possible the labour should be sourced locally, thus helping to stimulate the local economy.

Cost of materials

Compared with manufactured materials, the initial cost of materials for some of the non-conventional approaches may be considerably cheaper, although the increased use of manual labour may well offset this saving if some or all of the labour has to be paid for at the market rate. In the majority of cases there will be considerable life cycle cost benefits for the entire structure. Similarly, by using simple construction techniques the ease and hence cost of maintenance, repair and replacement should also be better than more conventional approaches. Adopting a passive design philosophy may help to reduce some of the services provision and need for integration; for example, passive ventilation instead of mechanical. Materials should be sourced locally, preferably from renewable resources.

Genius loci – the importance of site

The importance of the site and the manner in which the building is positioned on, or within, the ground becomes even more critical with some of the alternative approaches. Many of these materials are more sensitive to damage from moisture than conventional building products, and they may be considerably less durable unless competently detailed and constructed. Site sensitivity is a crucial factor in ensuring a durable and trouble-free building. The proposed site of the building must be carefully analysed in terms of the microclimate, soil type, position of water table, etc. Then (and only then) should a decision be taken as to the most appropriate materials and construction techniques to employ. For example, some sites may be better suited to earth sheltered construction than straw bale construction and vice versa. In some cases a more traditional approach may be a better option once the data gleaned from a thorough site analysis have been collected and analysed. Readers with a strong desire to build using a particular material, for example, straw bales, must first find an appropriate site.

1.4 Regulations and approvals

A number of approvals need to be in place before building work commences. The two main consents required in the UK are from the appropriate town planning authority and building control. Specific conditions relating to town planning consent and building regulation approval will be influenced by the physical characteristics of the site and its immediate surroundings.

Planning consent

Issues concerning local town planning approval are outside the scope of this book; however, it is important to recognise that (with a few exceptions) planning approval must be applied for and have been granted before any construction or demolition work commences. The legislation concerning the right to develop, alter and/or demolish buildings is extensive and professional advice should be sought before applying for the appropriate approvals. The process of obtaining approval can be very time consuming (preparing the necessary information for submission, allowing time for consultation and decisions, etc.) and conditions attached to the approval may affect the construction process (e.g. restricted times of working, conditions on materials to be used, etc.). Sometimes the application may be unsuccessful, leading to an appeal or a submission of a revised proposal. Planning consent will permit development; it does not deal with how the building is to be constructed safely; this is dealt with by building control.

Building control and building regulations

During the last 50 or so years there has been a considerable increase in building control legislation, which initially was the province of local authorities through building by-laws and later replaced by national building regulations. In the UK building control is governed by three differing, though broadly similar, sets of legislation, for England and Wales, Northern Ireland and Scotland, respectively. Building regulations aim to ensure the health and safety of people in and around buildings by setting functional requirements for the design and construction of buildings. The regulations also aim to promote energy efficiency of buildings and sustainable construction and to contribute to the needs of people with disabilities.

In England and Wales the Building Act 1984 and Building Regulations (1985) set out functional requirements for buildings and health and safety requirements that may be met through the practical guidance given in the Approved Documents; these in turn refer to British Standards and Codes of Practice. In Northern Ireland construction is covered by the Building Regulations (Northern Ireland) 2000 with Approved Documents, similar to those for England and Wales.

In Scotland the Building (Scotland) Act 2003 has replaced the old prescriptive standards with performance standards. The Act is a response to European harmonisation of standards and their use in Scotland as required under the Construction Products Directive (CPD). The Act has two objectives: to allow greater flexibility for designers in meeting minimum performance standards and to ensure greater consistency across the country. There are separate Guidance Documents for domestic and non-domestic buildings, both of which are divided into six subject areas that match the essential requirements of the CPD, namely, structure, fire, environment, safety, noise and energy.

The Approved/Guidance Documents give practical guidance to meeting the require-ments, but there is no obligation to adopt any particular solution in the documents if the stated functional requirements can be met in some other way. The stated aim of the current regulations is to allow freedom of choice of building form and construction so long as the stated (minimum) performance requirements are satisfied. In practice the likelihood is that the majority of designers will accept the guidance given in the Approved/Guidance Docu-ments as if the guidance were prescriptive. This is the easier and quicker approach to construction, rather than proposing some other form of construction that would involve calculation and reference to a bewildering array of British Standards, Codes and Agrément Certificates.

Although the guidance in this book is meant to be as broad as possible, when we make specific reference to regulations we have, both for consistency and to avoid confusion, worked to the Approved Documents for England and Wales. However, the principles for readers in Northern Ireland and Scotland are broadly similar, differences in detailing build-ings arising from a combination of legislation, response to local climate and building tradi-tion. Full details of the Acts as well as the Approved Documents and Guidance Documents are available online (http://www.planningportal.gov.uk).

Robust details

One development linked to the Approved Documents (for England and Wales) is the pub-lication of detailed drawings that show compliance with the Approved Documents, known as robust details. The aim of the robust details is to assist the house-building industry in achieving the performance standards published in the Approved Documents, especially the requirements relating to Part E and Part L. Robust details are intended to reduce risks and problems that can arise as a result of building to higher performance standards and, in the spirit of the Approved Documents, are intended as guidance. Robust details are pre-tested to higher standards than those required by the Approved Documents, and they should:

❏ Be practical to build on site
❏ Be tolerant to variations in standards of workmanship
❏ Exceed the current performance standards

Further details are available from the government department Communities and Local Government (see Appendices) and via manufacturers' homepages.

Associated legislation

In addition to the legislation set down in the Building Regulations, other legislations affect the way in which buildings are designed and built. For example, health and safety and fire safety legislation is covered in a number of documents that sit alongside the Approved/ Guidance Documents. The European Union is particularly active in promoting consistent standards across the Union through a series of directives. For example the European Union Council Directive 89/106/EEC (1988), the CPD, requires all construction products to satisfy the Essential Requirements, which deal with health and safety issues, namely, mechanical resistance and stability; safety in case of fire; hygiene, health and environment; safety in use; protection against noise; and energy economy and heat retention. Over 600 CEN standards have been mandated under the CPD.

Codes and guidance

Regulations and guidance can also apply to specific types of buildings, e.g. housing. _The Code for Sustainable Homes (2008)_ was introduced in England in 2007 to provide guidance to house builders in meeting the requirements of the _Energy Performance of Buildings Directive_.

1.5 Making choices and sources of information

The design and construction of buildings is concerned with making choices. Decisions have to be made about the design of the building and its details, which necessitates the selection of materials and components to realise the design intention and aspirations of the client. At the construction stage decisions have to be made about what mechanical plant to use, how best to sequence the work so that operations are conducted safely and efficiently, and what to do when an unexpected problem occurs. During the life of the building decisions will need to be made about how best to replace damaged or worn components and how to upgrade buildings to improve their functionality and performance. Then at the end of the building's useful life decisions will need to be taken about how to deconstruct the building safely and economically while also maximising the reuse of materials, products and components. The contents of this book and _Barry's Advanced Construction of Buildings_ are designed to assist with that decision-making exercise.

Sources of information

Construction is essentially a process of assembly where products are chosen from manufacturers' catalogues and/or from the builders' merchants and are put together using a raft of different fixing techniques. Each new project brings with it a new set of challenges and a fresh search for information to answer specific problems. Some of the main sources of information for readers working in Britain are:

- ❑ Building Regulations and Codes
- ❑ British (BS), European (EN) and International (ISO) standards
- ❑ British Board of Agrément (BBA) certificates

❑   Building Research Establishment (BRE) publications
❑   Trade association publications
❑   Manufacturers' technical literature
❑   Compendia of technical literature
❑   Technical articles and guides in professional journals
❑   Building information centres
❑   Organisational knowledge (codified in 'standard' details and specifications)

Readers should make use of the information and details provided by manufacturers but should avoid making a choice based on information from one source. Explore at least three different suppliers to compare functional and performance requirements, including costs and availability – then compare this with current legislation and standards – thus making an informed decision based on achieving the best value within given parameters.

A note of caution

Before proceeding any further it is necessary to make an important observation about the contents of this book and *Barry's Advanced Construction of Buildings*. The principles and details illustrated are meant as a guide to the construction of buildings. Details should not be copied without thinking about what is really going on. This also applies to details given in guidance documents. Readers should be asking questions such as: How is the building to be assembled, maintained and disassembled safely and efficiently? Is the detail in question entirely suitable for the task in hand? We make this point because approaches to detailing and to construction vary from region to region (e.g. a building located in a wet and sheltered area of the UK may benefit from a pitched roof with a large overhang, but a similar building in a dry and exposed part of the country may benefit from a pitched roof with clipped eaves or even a flat roof). Building practices and regulations also differ from country to country (e.g. differences between the regulations relating to England and Wales and those relating to Scotland) and so it is impossible to cover every eventuality for every reader. Instead we would urge readers to engage in some critical thinking, analyse the details and seek out alternative approaches where necessary. Buildings exist within a local context and they should be detailed to be in harmony with nature, i.e. they should respond to their *genius loci* in a positive and sustainable way. The other point we must make is that regulations and building practices will change over time, therefore this book should carry a 'best before date'; as the regulations change some of the details will need to be revised. On a more positive note, the details will be a useful source of reference for readers involved in refurbishment and conservation projects at a future date. Our advice is to work closely with consultants, manufacturers and suppliers with the aim of applying their specialist knowledge to the benefit of the construction process, thinking critically and making informed decisions.

2 Site Analysis and Set-Up

The physical characteristics of the site and its immediate environment will influence the decisions to be made about a building's design and construction. The information gathered from a thorough site analysis is, therefore, a vital exercise and must be completed before any design or construction work commences. This information can then be collated in a performance appraisal document from which informed decisions about the best way to proceed may be made.

2.1 Function of the site analysis

Prior to any construction operations, the client/developer will want to know whether it is economically viable to build on the proposed site. Because the nature and condition of the ground and soil below the surface of the site are an unknown quantity they pose a considerable risk to the construction project, with the potential to cause delays and additional costs, both of which can be substantial. Inadequate soil investigation and hence inappropriate foundation design can lead to structural problems at a later date, which is a problem for the building owners and users as well as the building insurers. Many projects are built on brownfield sites (land that has been previously built on and used) and it may be necessary to seal, stabilise or remove any contaminated ground, toxic waste or other dangerous substances before commencing the main construction works. The extent of contamination must be established before any work commences on the site.

The main purpose of site analysis is to identify and hence reduce the risks associated with the development by recording site features and soil characteristics, helping to determine the design and cost of suitable foundations and structure. A thorough site analysis is an essential first step that will assist development, design and construction decisions. The site analysis helps:

❏ The client to assess whether the project is viable (best done in consultation with professional advisors)
❏ The client, designer, structural engineer and contractor to locate the best position for the building, avoiding or accommodating identified problems where possible, while making the best possible use of physical features and environmental conditions

Barry's Introduction to Construction of Buildings, Third Edition. Stephen Emmitt and Christopher A. Gorse.
© 2014 John Wiley & Sons, Ltd. Published 2014 by John Wiley & Sons, Ltd.

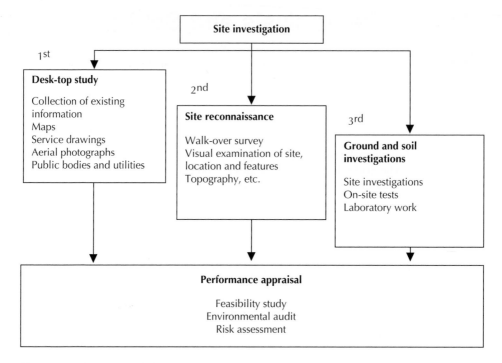

Figure 2.1 Sequence of site analysis.

❑ The engineers to design the most suitable foundation system
❑ The mechanical and electrical consultants to design the service provision
❑ The designers and contractor to ensure that safe construction methods are used
❑ The environmental consultants to identify the most suitable way of dealing with any contaminants and problem materials, e.g. remediation works, material reuse, on-site treatment and disposal options to licensed tips.

Sequence of activities

The site analysis comprises three interrelated research activities (Figure 2.1): the desk-top study, the site reconnaissance, and the ground and soil investigations. The soil investigation may also involve laboratory tests on liquids and gases found in soil samples. The order in which these activities are carried out will depend to a large extent on the nature of the development and the timescales involved. The preferred sequence of overlapping activities is shown in Figure 2.1 and described in Sections 2.2–2.4. This should result in a comprehensive appraisal document, as described in Section 2.5.

2.2 The 'desk-top' study

The 'desk-top study' is a vital element in any site analysis exercise. The study involves the collection of all documents and materials that can be obtained without having to visit the site. There is a considerable amount of information available from local and national authorities, museums, private companies and research groups (see also Appendix A). The client or previous owners may also have relevant information to hand.

Although the different site investigation operations often overlap, care should be taken not to commence with expensive ground exploration and soil tests before the desk-top study is completed. This is partly to avoid unnecessary work and expenses – for example, information from the desk-top study may reveal that recent site and soil investigations are available – and partly a health and safety issue since the approximate location of services and potential hazards must be known before carrying out any physical investigations. Early and thorough research may also show that the site is unsuitable for development, or that special measures are required before proceeding further.

Information required

Ownership(s) and legal boundaries

The client should provide, via a legal representative, information pertaining to the exact location of the site boundaries and responsibilities for maintaining them. These will need to be checked against a measured survey and any areas of uncertainty checked by the legal representative. Other issues to be determined by the client's legal representative include:

❏ Rights of way
❏ Rights of light
❏ Rights of support (for adjoining properties)
❏ Legal easements
❏ Ownership of land (essential where parcels of land are being assembled to make a larger site)
❏ Rights of tenants, etc.

Ground conditions

As a first step it is usual to collect information on soil and subsoil conditions from the county and local authority. This includes knowledge from maps, geological surveys, aerial photography and works for buildings and services adjacent to the site, which may in itself give an adequate guide to subsoil conditions. Geological maps from the British Geological Survey, information from local geological societies, Ordnance Survey maps, mining, river and coastal information may also be useful.

Services

All suppliers of services should be contacted to confirm the position of pipes and cables and the nature of the existing supply, i.e. its capacity. This includes gas, mains water, sewage and surface drainage pipes as well as electricity, broadband and telephone cables. It is usual for a representative of the supplier to visit the site to identify their equipment (which might be in a different location to that shown on plans).

Contaminated land and methane

Previous use of land can give some clues as to the likely contaminants to be found and the local authority may have records that can help in this regard. However, extensive soil testing should be carried out to ascertain the nature and extent of any contamination (see also Chapter 3). Methane is associated with landfill sites, and local knowledge about the position of old tips/landfill can prove useful at an early stage.

Radon

In certain areas of the UK, radon gas occurs naturally within the underlying ground and poses a health threat to the inhabitants of buildings. Local authorities provide advice on the likely level of contamination and will advise on the extent of protection required to prevent the radon gas from entering the building.

Mining activity

The UK has a long and varied history of mining. Coal mining is the most common and advice on mining records and coverage can be accessed via The Coal Authority; however, certain geographical areas may have specific mining issues, e.g. salt mining in areas of Cheshire and tin mining in Cornwall. The position of mines and mine shafts, and their size, depth and condition may affect the positioning of buildings on a site. Work to make old mine workings safe may add significant costs to the development, hence the need for thorough research before finalising site layout or commencing any work on site.

Flooding

Damage to property and possessions and the associated disruption to businesses and family life has become a serious concern in recent years as the frequency of flooding has increased. Factors include heavy rainfall, buildings sited on, or too close to, flood plains and inadequate maintenance of rivers, watercourses and surface water drainage. Thorough checks should be made about previous flooding of the site (if any), proximity to flood plains and any special requirements suggested or required by the various authorities and insurers.

Typical sources of information

Information will always be site-specific; however, the following list of information sources serves as a general guide (see also Appendix A):

❑ Ordnance Survey – detailed maps in many different formats are available from Ordnance Survey, Romsey Road, Southampton S016 4GU (http://www.ordnancesurvey.co.uk)
❑ Historical maps (http://www.old-maps.co.uk) and libraries local to the site
❑ Geological maps – the British Geological Survey is the national repository for geosciences data in the UK. Information provided includes maps, records and materials, including borehole cores and specimens from across the UK. Address: London Information Office, British Geological Survey, Earth Galleries, Natural History Museum, Exhibition Road, London SW7 2DE (http://www.bgs.ac.uk)
❑ Hydrogeological maps – soil reports and publication lists are obtainable from soil survey and Land Research Centre, Cranfield University, Silsoe, Bedford MK 45 4DT
❑ Meteorological information – monthly and annual reports are available on air temperature, wind speed, rainfall and sunshine. Such information is useful when designing

the building and for scheduling construction operations. Statistics on averages and extremes are also available. The Met Office, Room JG6, Johnson House, London Road, Bracknell, Berks RG12 2SY (http://www.metoffice.gov.uk)

❑ Hydrological information – surface water run-off data are collected by water authorities, private water undertakings and local authorities

❑ Site history:
  ○ Previous owners and developers
  ○ Site surveys and drawings used for previous development
  ○ Records held by Building Control
  ○ Local newspaper archives
  ○ Records held by the local planning authority

❑ Gas supplier – location of gas mains

❑ Electricity supplier – location of electricity cables

❑ Electricity generating board – mains electricity cables

❑ Water suppliers – water supply mains

❑ Mains sewers

❑ Local authority – local sewers

❑ Telecommunications authority – telephone and optical cables

❑ Rail authority – railways

❑ British Water Board – canals

❑ Coal Authority (http://www.coal.decc.gov.uk) – mining activity

❑ Aerial photographs – there are many collections of aerial photographs dating back over many decades. A directory of organisations and agencies that hold aerial photographs can be obtained from Publications Department, Aslib, The Association for Information Management, Information House, 20–24 Old Street, London EC1V 9AP

2.3 Site reconnaissance

Written approvals from the client and/or the property owners must be in place and a thorough risk assessment exercise must be carried out before entering the site, especially before any invasive investigations are carried out (which may require separate written permission). Obvious considerations are related to trespass and criminal damage, although the prime concern must be for the safety of those doing the investigations. The majority of sites will have been used previously (and may still be in use) and might contain buildings that are structurally unsound (which may be redundant or still in use despite their condition). Specialists should be appointed to establish the condition and safety of existing structures and whether or not asbestos is present. Figure 2.2 provides an overview of the type of information that can be collected during a site reconnaissance.

The visual inspection of the site

A visit to the site and its surroundings should always be made to record everything relevant to the proposed development. The site reconnaissance is often referred to as the visual inspection or the 'walkover'. From experience we have found that two pairs of eyes (or

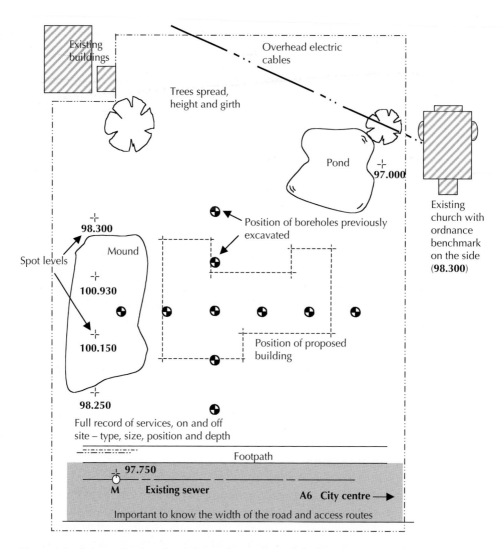

Figure 2.2   Schematic showing information collected during site reconnaissance.

more) are always better than one and so the visual inspection should be undertaken by at least two, and preferably three, people, e.g. the architect, engineer and contractor, with each taking their own notes but discussing features as they come across them. Careful observation should be made of the nature of the subsoil, vegetation, evidence of marshy ground, signs of groundwater and flooding, irregularities in topography, ground erosion and ditches and flat ground near streams and rivers where there may be soft alluvial soil. A record should be made of the foundations of old buildings on the site. Cracks and other signs of movement in adjacent buildings should be noted. When undertaking site reconnaissance on contaminated land, ensure as far as possible that all hazards have been identified and that correct safety procedures are followed. In preparation for the site reconnaissance, all of the maps and records should be assembled so that any differences or omissions found when walking over and observing the site can be recorded. A visual inspection of physical site boundaries should be made and compared with any legal documents that show boundaries.

British Standard procedure for walkover surveys

When conducting a walkover survey, the British Standard for site investigations (BS 5930:1999) suggests that the surveyor should:

- ❏ Traverse the whole area on foot (if possible and safe to do so)
- ❏ Establish the proposed location of work on plans
- ❏ Identify and record any differences on the plans and maps
- ❏ Record details of existing services, trees, structures, buildings and obstructions
- ❏ Check access and determine capability of sustaining heavy construction traffic
- ❏ Record water levels, fluctuations in levels, direction of flow and flow rate
- ❏ Identify adjacent property and the likelihood of it being affected by proposed works
- ❏ Identify any previous or current activities that may have led to contamination
- ❏ Record mine or quarry workings, old structures and other features
- ❏ Record obvious features that pose immediate hazard to public health and safety or the environment
- ❏ Record any areas of discoloured soil, evidence of gas production or underground combustion

During a site reconnaissance the following ground information and features should be noted (BS 5930:1999):

- ❏ Record surface features on site and on adjacent land, note the following:
  - ○ Type and variability of surface conditions
  - ○ Compare land and topography with previous records; check for fill, erosions and cuttings.
  - ○ Any steps in the surface that may indicate geological faults. Steps in mining areas may be the result of subsidence. Other evidence of subsidence caused by mining may include compression or tensile damage to structures and roads, structures out of plumb and interference with the line of drainage patterns.
  - ○ Mounds and hummocks in relatively flat country often indicate former glacial conditions, e.g. glacial gravel.
  - ○ Where the ground is terraced and broken on hill slopes this may be due to landslips; small steps and inclined tree trunks may be evidence of creep and ground movement.
  - ○ Crater-type holes in chalk or limestone usually indicate swallow holes that have been filled with a soft material.
  - ○ Low-lying flat areas in hill country may be the site of a previous lake and can indicate the presence of soft silts and peat
- ❏ Record details of ground conditions in quarries and cuttings.

- ❏ Record groundwater levels (these are often different from streams, ponds and lakes).
- ❏ Identify the position of wells and springs.
- ❏ Note the nature of vegetation in relation to soil type and wetness of soil. Unusual green patches, reeds, rushes, willow trees and poplars usually indicate wet ground conditions.
- ❏ Investigate structures in the vicinity of areas having a settlement history.

Identification and physical location of services

Before undertaking any digging, e.g. for trail holes, it is necessary to clearly identify the nature of the services on the site and their actual position. Unfortunately, the majority of the plans provided by the service providers only give an approximate location of their pipes and cables; therefore, some detective work is required on site. The first task is to identify all inspection covers and the nature of the service, and to compare their positions with those on the drawings. Handheld sonic and magnetic detecting devices to help locate the position of services are available from most plant hire firms. Exact position and depth can be established by carefully hand digging trial pits to expose pipes, cables and conduits. The service providers will also be keen to establish exact positions in an attempt to prevent damage to their pipes and cables. The organisations responsible for particular services should be invited to the site to help to establish the exact position, size and capacity of their supply and to resolve any uncertainty.

In order to develop the site to its full potential services may need to be re-routed, if the appropriate authority will give permission. This is usually an expensive option, which could threaten the viability of the project. Alternatively, the proposed position of the building may need to be adjusted to enable the project to proceed without undue disruption to major service routes.

Surveys

Measured survey

A land surveyor will conduct a topographical survey to establish the physical boundaries, existing features and variations in ground level. Most land surveyors have a standard list of features to be established during the land survey, although it is not uncommon to direct the land surveyors to particular areas so that they record all the necessary features on and immediately adjacent to the site. The survey will be provided as a digital file to import directly into computer aided design (CAD)/building information modelling (BIM) software.

Condition survey

Condition surveys are used to record the physical condition of boundaries and adjoining property as well as the buildings on the site to be protected, refurbished or altered. Before commencing any work that is likely to result in vibration (e.g. demolition, excavations, piling, heavy construction traffic, and so on) it is important to undertake a full condition survey of surrounding and adjoining properties and structures. This serves as a record for any subsequent claims for damage and also serves as a good source of design information, for example, indicating how particular materials have weathered. Detailed drawings and written descriptions of the property should be supported with photographic evidence.

Photographic and video surveys

Photographic and video surveys are useful tools to prompt one's memory when back in the office. They also provide a record of the original condition of the site and adjoining land/property in case of any dispute or claim for damage. Photographic surveys should be conducted in a systematic and thorough manner, with the position of the photographer and direction of view noted on a site plan to avoid any future confusion.

2.4 Soil investigations

Details of the subsoil should include soil type, consistency or strength, soil structure, moisture conditions and the presence of roots. From the nature of the subsoil the bearing capacity, seasonal volume changes and other possible ground movements are assumed. To determine the nature of the subsoil below foundation level it is necessary either to excavate trial pits some depth below the assumed foundation level or to bore in the base of the trial hole to withdraw samples.

The common methods of obtaining samples include:

❑   Trial pits
❑   Boreholes – cable percussive boreholes and rotary drilled borehole
❑   Window sampling and dynamic probe testing

These methods are explained in further detail later in this chapter.

When proposing work to existing buildings (e.g. adding another floor) it will be necessary to expose the existing foundation in a number of places to check if it was built as detailed on drawings (if available) and also to check the subsoil below the foundation. Whichever system is adopted will depend on economy, the proposed building works and the nature of the subsoil. Trial pits or boreholes should be sufficient in number to determine the nature of the subsoil over and around the site of the building and should be at most, say, 30 m apart.

Ground movements that may cause settlement are:

❑   Compression of the soil by the load of the building
❑   Seasonal volume changes in the soil
❑   Mass movement in unstable areas such as made up ground and mining areas where there may be considerable settlement
❑   Ground made unstable by adjacent excavations or by de-watering, for example, due to an adjacent road cutting.

Foundation design and subsoil examination

To select a foundation from tables, or to design a foundation, it is necessary to calculate the loads on the foundation and to determine the nature of the subsoil, its bearing capacity, likely behaviour under seasonal and groundwater level changes, and the possibility of ground movement. Where the nature of the subsoil is known from geological surveys, adjacent building work, trial pits or borings and the loads on foundations are small, as for single domestic buildings, it is generally sufficient to excavate for foundations and confirm, from the exposed subsoil in the trenches, that the ground is as anticipated. Table 2.1 provides a guide on the allowable bearing pressure of different classifications of ground.

Where loads may be more substantial or there is little evidence available on the nature of the ground, boreholes and trial pits will need to be excavated. The depth of exploration will be related to the proposed foundation and loads imposed on the subsoil.

Under pad foundations there is a significant pressure on the subsoil to a depth and breadth of about one-and-a-half times the width of the foundation; the depth of pressure is slightly greater for strip foundations (Table 2.2 and Figure 2.4). If, at any point where the foundation exerts this bulb of pressure, the soil-bearing capacity is less than the load exerted by the foundation, then appreciable settlement of the foundation can occur and damage the building (Figure 2.3). It is important, therefore, to know or ascertain the nature of the subsoil both at the level of the foundation and for some depth below.

Table 2.1   Allowable bearing pressure of soil and ground

Soil and ground classification	Bearing capacity (kN/m²)
Rocks	
Strong sandstone	4000
Schists	3000
Strong shale	2000
Granular soils	
Dense sand and gravel	>600
Medium dense gravel	200–600
Loose sand and gravel	<200
Compact sand	>300
Loose sand	<100
Cohesive soils	
Stiff boulder clay	300–600
Stiff clay	150–300
Firm clay	75–150
Soft clay and silt	<75

Adapted from BS 8004 (1986).

Table 2.2   Depth of effective bulb pressure

Foundation type	Effective bulb pressure
Strip	3 × the width of foundation
Pad	1.5 × the width of foundation
Raft	1.4 × the width of foundation

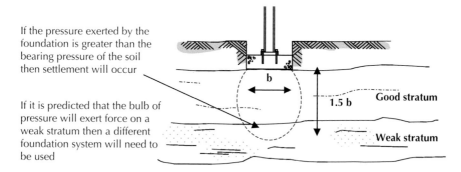

If the pressure exerted by the foundation is greater than the bearing pressure of the soil then settlement will occur

If it is predicted that the bulb of pressure will exert force on a weak stratum then a different foundation system will need to be used

b

1.5 b   Good stratum

Weak stratum

Figure 2.3 Example of foundation exerting pressure on weak stratum: may lead to settlement.

Figure 2.4 shows typical forces exerted on the substructure at a depth of one-and-a-half times the width of the foundation. The loads placed on a pad foundation are generally distributed over a greater surface area, reducing the load per unit area. Strip foundations transfer the loads from the wall directly to the ground. Although strip foundations do spread the load over a slightly greater surface area they exert continual pressure along the ground (Figure 2.4).

Care should be taken where foundations are positioned close to each other, or a new foundation is placed next to an existing structure. Each foundation exerts a force on the ground that is distributed at an angle of approximately 45°. Where the pressure exerted on the ground by the foundations overlaps, the soil has to bear the force of both foundations (Figure 2.5). If the force exerted by both foundations is greater than the bearing capacity of the soil, the ground will fail and the foundation may settle.

Trial pits

To make an examination of the subsoil trial pits and/or boreholes are excavated. Trial pits are usually excavated by a mechanical excavator, or in some cases by hand tools, to a depth of 3–4 m. The nature of the subsoil is determined by examination of the sides of the excavations. Soil and rocks can be examined in situ on the faces of the excavated pit, and samples taken for further laboratory tests. The trial pit also provides an indication of the ease of dig (or excavation), trench stability and groundwater conditions. For exploration of shallow depths (up to 3 m) this is usually more economical than boreholes. The pits are usually rectangular, being approximately 1.2 × 1.2 m in plan. The pits should be excavated in the vicinity of the proposed structure; if the pit is located under a proposed foundation particular attention needs to be given to the material used to backfill the hole. In such situations material should be of sufficient strength and well compacted. As each trial pit is excavated and inspected a report should be made of the inspection. Typical information contained in the inspection, the trial pit log, is shown in Figure 2.6.

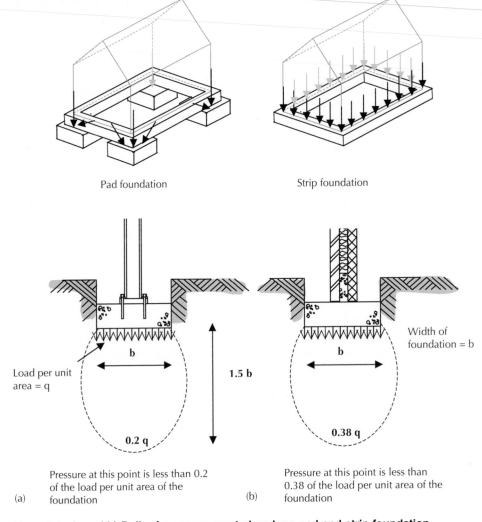

Pad foundation

Strip foundation

Load per unit area = q

b

1.5 b

0.2 q

Width of foundation = b

b

0.38 q

(a) Pressure at this point is less than 0.2 of the load per unit area of the foundation

(b) Pressure at this point is less than 0.38 of the load per unit area of the foundation

Figure 2.4 (a and b) Bulb of pressure exerted under a pad and strip foundation.

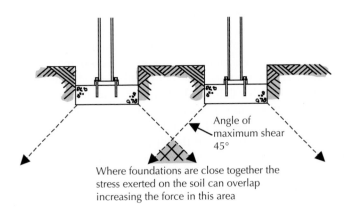

Angle of maximum shear 45°

Where foundations are close together the stress exerted on the soil can overlap increasing the force in this area

Figure 2.5 Combined stress exerted on subsoil from foundations in close proximity.

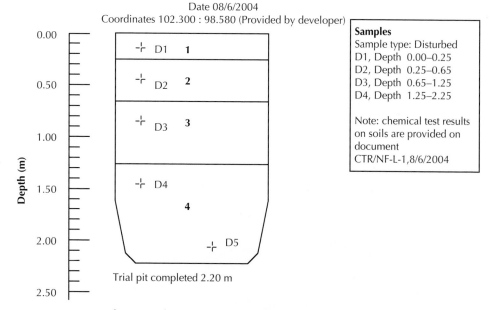

Sketch of face of trial pit
Location: Nicholls Farm – Leeds
No 1. – Face A
Date 08/6/2004
Coordinates 102.300 : 98.580 (Provided by developer)

Samples
Sample type: Disturbed
D1, Depth 0.00–0.25
D2, Depth 0.25–0.65
D3, Depth 0.65–1.25
D4, Depth 1.25–2.25

Note: chemical test results
on soils are provided on
document
CTR/NF-L-1,8/6/2004

Trial pit completed 2.20 m

Width

Strata
(1)   Topsoil and grass 220 mm
(2)   Made ground: firm to stiff orange brown sandy clay with a little gravel of limestone
(3)   Made ground: soft to firm dark grey clay with occasional rootlets
(4)   Medium dense dark grey clay

Notes:
(1)   Groundwater: no groundwater encountered
(2)   Variability of faces: all faces were similar
(3)   Stability of faces: all faces were stable
(4)   Weather: overcast

Figure 2.6   Trial pit log.

Boreholes

'Borehole' is often used as a generic term that represents the various methods used to excavate and extract disturbed and undisturbed soil samples. All samples taken from boreholes should be sealed as soon as possible to minimise any loss of moisture before testing. Some of the most commonly used methods are as follows.

Auger boring – rotary boring methods

These holes are usually made by hand or powered auger into the ground. Auger holes are typically 75–150 mm in diameter. Short helical augers are used and disturbed samples of soil are collected as they are brought to the surface. Such methods are not widely used as they do not allow the soil to be examined in situ and are not capable of penetrating to the depth of boreholes. Rotary drilling is used where the boreholes are being cut into very dense gravel or bedrock. Samples or bedrocks are recovered in seamless plastic tubes, are logged by an engineer, and then taken for laboratory testing (Photograph 2.1).

Photograph 2.1   Rotary drilling rig.

Photograph 2.2   Tracked percussive sampler.

Window samplers

A window sampler is a steel tube about 1 m long with a hole cut into the wall of the tube allowing the disturbed sample to be viewed or soil samples taken from the tube. The tube is driven into the ground using a lightweight percussion hammer and extracted with the aid of jacks. A range of tube diameters are available. In practice the large tubes are driven in first and removed leaving a hole for smaller tubes to be inserted and driven in further. Samples can be obtained down to a depth of 8 m.

Window sampling can be carried out using either handheld pneumatic samplers or tracked percussive samplers (Photograph 2.2). The samples are retrieved in seamless plastic tubes. A qualified engineer logs the samples (Photograph 2.3). Window sampling is suited to sites with restricted access, where disturbance is to be kept to a minimum and contamination investigation. The percussive samplers are also normally capable of doing penetrometer testing. Penetrometer testing is a continuous soil test procedure which enables the relative density or strength of the ground to be determined. Further information can be found at the Structural Soils web site, http://www.soils.co.uk.

Percussion boring

Boreholes are normally made using light percussion equipment. The 150–200 mm diameter weighted hollow tube is dropped into the hole so that the soil becomes lodged within the tube. The tube is then lifted to the surface and the sample removed. In clay soils the method relies on the cohesive properties of the soil to hold it in the tube (clay cutter). In granular soils a hollow tube with a flap over its base is used (shell or bailer). A single day's drilling can excavate up to 15 m (typically 7–15 m).

The materials that are collected during the drilling and excavation can be retained as disturbed samples; however, in cohesive soils a 100 mm diameter tube can be dropped to

Photograph 2.3   Soil samples ready for logging.

the bottom of the hole to collect an undisturbed sample (undisturbed sampling is generally confined to cohesive soils). The tube is then taken back to the laboratory for further tests. Standard penetration tests (SPTs) and vane tests can be carried out in the borehole as the drilling and excavation process proceeds.

Photographic evidence from boreholes
In very stiff clays and fissured rock it is possible to take photographs of the strata using a remote-controlled borehole camera. If it is clear that a borehole has entered a void, photographic evidence may determine the nature of the void, e.g. mine shaft, sewer, cave, and so on.

Depth and location of exploratory investigation

The information obtained from boreholes and trial pits will be used to design suitable foundation systems. Boreholes must penetrate through all unsuitable deposits such as unconsolidated fill, degradable material, peat, organic silts and very soft compressible clay. Once a suitable bearing stratum is reached, the depth of the borehole can be calculated. For low-rise buildings the depth of the borehole can be calculated using the following formulae (BRE, 1995).

Isolated pad or raft foundations
Where the structure rests on an isolated pad or raft foundation (Figure 2.7) the following formula is used to calculate the depth of borehole required:

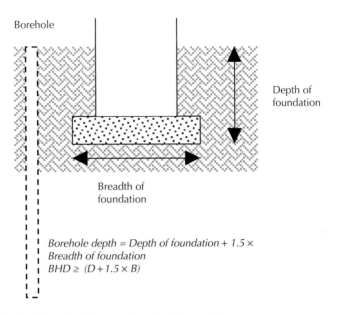

Borehole

Depth of
foundation

Breadth of
foundation

Borehole depth = Depth of foundation + 1.5 ×
Breadth of foundation
BHD ≥ (D + 1.5 × B)

Figure 2.7   Depth of borehole for pad and raft foundations.

$$\text{Borehole depth} = \text{Depth of foundation} + 1.5 \times \text{Breadth of foundation}$$

$$\text{BHD} \geq (\text{D} + 1.5 \times \text{B})$$

Closely spaced strip or pad foundations
Where the structure rests on closely spaced strip or pad foundations (Figure 2.8) the following formula is used to calculate the depth of borehole required:

$$\text{Borehole depth} = \text{Depth of foundation} + 1.5 \times (2 \times \text{Space between foundations} + \text{Breadth of foundation})$$

$$\text{BHD} \geq \text{D} + 1.5(2\text{S} + \text{B})$$

Structure on friction piles
Where friction piles are used (Figure 2.9), the calculation is:

$$\text{Borehole depth} = \text{Depth of pile foundation} + 3 \times \text{Breadth}$$

$$\text{BHD} \geq (\text{D} + 3\text{B})$$

BRE (1995) procedure for site investigations in low-rise buildings suggests that boreholes should be drilled in straight lines to facilitate section drawings (Figure 2.10).

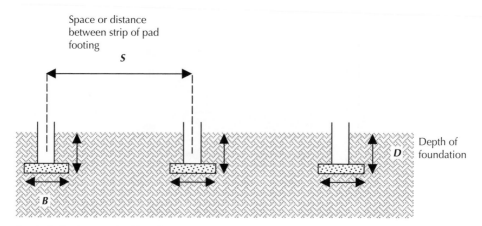

Space or distance between strip of pad footing

S

B

D — Depth of foundation

Borehole depth = Depth of foundation + 1.5 × (2 × Space between foundations + Breadth of foundation)
$BHD \geq D + 1.5(2S + B)$
Used where S is less than 5 times B

Figure 2.8   Depth of borehole for closely spaced strip or pad foundation.

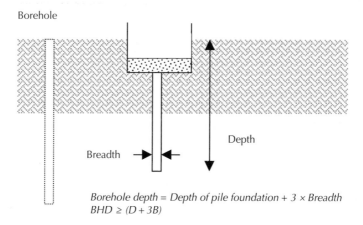

Borehole

Depth

Breadth

Borehole depth = Depth of pile foundation + 3 × Breadth
$BHD \geq (D + 3B)$

Figure 2.9   Depth of borehole for friction piles.

Ground and soil tests

There are a wide variety of on-site and laboratory tests that can be used to establish the characteristic ground and soil conditions. The extent of soil investigation will be based on the nature of the building and characteristics of the site. More detailed site and laboratory studies will provide more information, reducing the risks inherent in building on unknown ground. Laboratory and on-site tests that can be used include the following.

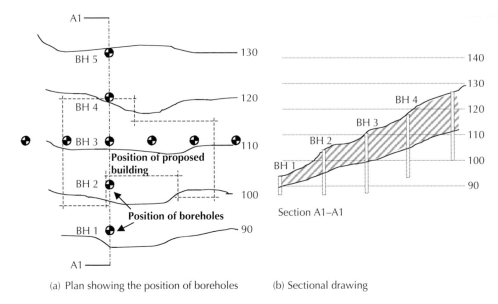

(a) Plan showing the position of boreholes   (b) Sectional drawing

Figure 2.10   (a and b) Plan and section showing the position of boreholes.

On-site test

☐ Plate load test
☐ Vane shear test
☐ California bearing ratio (CBR) test
☐ Dry density/moisture relationship
☐ SPT
☐ Lightweight dynamic penetrometers
☐ Cone penetration tests (CPT)
☐ Methane/oxygen/carbon dioxide/barometric pressure test

Laboratory work

☐ Triaxial compression tests
☐ Liquid and plastic limit tests
☐ Sieve analysis – particle size and distribution
☐ Moisture content
☐ pH value tests

Further information on-site and soil investigation is included in *Barry's Advanced Construction of Buildings.*

2.5 The performance appraisal

When the information has been collected from the desk-top study, the site reconnaissance and the ground and soil investigations, it must be brought together in the form of a final report, the performance appraisal.

Structure of the report

The main headings should address the following.

Feasibility study

Is the proposed development feasible? Are alterations required to improve constructability given that more information is available than at the start of the project? How does the information gathered affect the client requirements as stated in the development brief?

Environmental audit

Environmental impact assessments and environmental impact statements can be prepared and recommendations considered. How does this information affect the proposed development?

Risk assessment

Given that all building activity carries some degree of risk, it is important that risk assessments are carried out. Decisions can be made to eliminate, reduce or accept the risk associated with particular work packages.

Conclusions and recommendations

Clear and concise conclusions based on the information gathered need to be made and any areas requiring further detail/research clearly stated. All recommendations need to be related to the client brief and supported with accurate costing where possible, thus enabling informed decisions to be taken.

2.6 Site set-up and security

Site planning is essential to ensure the safe and efficient flow of work, especially for sites with limited space and/or those located in busy areas such as city centres. Although this is covered under the literature on construction management a brief summary of the main issues to be considered is presented here because it may influence decisions about how best to design and construct the building.

Site set-up

Once legal possession of the site has been obtained by the contractor it is possible to prepare the area ready for construction.

Access

Safe access to the site is required for personnel, construction plant and delivery vehicles during construction. Firefighting equipment must also have clear access in case of a fire or emergency occurring during the works. Temporary vehicular access may be allowed in consultation with the appropriate Highways, Police and Town Planning departments. Safe access for large cranes and wide, heavy and/or long loads will also need careful planning.

Storage and waste

Materials and plant must be stored so as to protect them from the weather, from damage from site operations and from theft. This applies equally to materials stored on site for a short period and those stored for a longer period. Space for the construction and subsequent protection from damage of sample panels is a related consideration. The reduction of waste on site is linked to good site management and to good detailing. When detailing a building, attention should be given to reducing the amount of cutting and hence waste generated on the site (which is expensive to dispose of).

Services and accommodation

The well-being of construction workers and visitors to the site is an important factor to consider, and space must be provided for office space and comfort facilities. These are usually provided in specialist prefabricated site units that are hired by the contractor for the duration of the project. On highly constrained sites these units are stacked vertically to save space. Access to services will also be required, including electricity (for heating and power), water and foul drainage.

Security and safety

Site security is required to protect materials and plant from theft and malicious damage, and to stop members of the public from inadvertently wandering on to the site and hence endangering their safety. Perimeter fencing must be secure and all access points constantly monitored. In addition to fencing and physical barriers to unauthorised entry, many contractors also employ security firms to provide additional protection at night and at weekends. Monitoring of materials entering and leaving the site is also required to prevent pilfering. Legal requirements, tools and guidance on how to manage health, safety and welfare can be found on the Health and Safety Executive web site (http://www.hse.gov.uk).

Statutory requirements

All work must comply with the relevant Building Regulations, Town Planning consent (conditions may relate to the protection of trees and boundaries, archaeological digs, restricted working times, etc.) and other statutory requirements relating to the construction works such as topical health and safety legislation [including the Construction Design and Management (CDM) Regulations 2007]. Contractors should also be working to the most appropriate guidelines, such as the Considerate Contractors scheme. House builders may also need to consult the most recent edition of the *NHBC Standards*.

Levelling and setting out

One of the first tasks to undertake is to establish the positions of the buildings and the various floor/datum levels. This requires careful surveying and measuring – and checking – to make sure everything is where it is intended to be. The land must then be taken down to workable levels. For simple housing developments the ground floor level is usually positioned slightly higher than the existing ground level to reduce the need for deep excavation and to avoid problems with groundwater penetration.

Building and grid lines should be set out and before any work can commence a temporary benchmark should be established. Temporary benchmarks will provide a fixed level on the site for the duration of the works. All other levels for footings, floor levels, road levels, etc., can be determined by working out the difference in levels from the temporary benchmark.

In order to establish the on-site temporary benchmark a fixed reference point off-site needs to be found, i.e. Ordnance Survey record benchmarks that have been established on existing structures. Benchmarks can often be found on churches, bridges and other large structures (and should be shown on the land surveyor's drawings). Once the level has been found it can be transferred to site. Site-based temporary benchmarks should be placed on a sturdy structure in positions where they are not liable to be knocked or disturbed. The benchmark should be recorded on a site plan for reference during construction.

Reference points on the ground also need to be determined to position the building and associated works correctly. North- and east-based coordinates are used to position grid lines; these may be obtained using a global positioning system (GPS) or using two reference points with known coordinates. Alternatively, the building or grid lines can be simply set out off existing structures.

Reduced level dig

Once a level has been established and grid lines or building lines marked out the reduced level dig can commence. Generally, areas of the site where major excavation will take place are reduced. Under buildings the reduced level dig is excavated to the underside level of the floor construction; this is the largest volume of excavation that should take place. Reducing the level of the site where buildings are positioned helps to provide a level site for ease of work. Foundations that go deeper than the underside of the floor construction are excavated independently from this level.

Temporary weather protection

Workers, building materials and building work will need to be protected from the vagaries of the weather. Materials delivered to site will require temporary weather protection until they are placed in the correct position. Similarly, work will need to be protected as it proceeds, e.g. brickwork requires protection from rain and frost. Getting the building weathertight as quickly as possible is a major help to protecting workers from wind, rain and sun. Alternatively, temporary weather protection over the entire building will provide protection to all work activities.

Cold weather working

As a general rule, work should not proceed when the air temperature is below or likely to fall below 2°C. Frozen material must not be used. The temperature should be monitored and recorded at regular intervals. A maximum and minimum thermometer, sited in the shade, will indicate whether the temperature is falling or rising. Weather forecasting services, such as that provided by the Meteorological Office, can help in planning work over a slightly longer period. Work must be planned and adjusted to mitigate the effects of cold weather. Strong winds can reduce the temperature of mortar and concrete (and site personnel) more quickly than still conditions; therefore, the wind chill factor must be considered in the planning of work. Topographical features, such as hollows in the ground or sites overshadowed by high trees or buildings, will have their own microclimate, often warming up slower than sites that benefit from the sun's rays.

Stored materials must be protected from cold weather. Covers can be placed over materials to protect them from frost, ice, rain and snow and should be in place at the end of each working day. Covers should also be used to protect newly laid masonry and concrete. During prolonged cold weather heaters can be used to protect materials from frost damage.

Concrete should not be placed in foundation trenches or oversite when the ground is frozen because ground movement will damage the work when the temperature rises and the ground thaws. All formwork, reinforcement and other surfaces that come into contact with fresh concrete must be free of frost, ice and snow. Ready-mixed concrete must be a minimum of 50°C when delivered (a requirement of BS 5328). Site-mixed concrete should not be used if the air temperature is below 2°C. However, if the aggregate temperature is above 2°C and free of frost and snow, and the water for mixing is heated (but not over 60°C), the cement is not heated, the ground is not frozen and the cast work can be protected against frost, then work can proceed. At low temperatures curing times will be extended, depending on the severity of the conditions, by up to 6 days.

Masonry should not be laid when the temperature is below or is likely to fall below 2°C unless heating is provided and the materials have not been damaged by frost or are frozen. New masonry must be protected against frost damage with covers and heating in severe weather.

The use of admixtures should be used sparingly and always in accordance with the manufacturers' recommendations. Plasticisers provide improved frost resistance to mature mortar and concrete. Accelerators may help the mortar or concrete to set before the temperature falls.

3 Groundwork and Foundations

The foundation of a building is that part of walls, piers and columns in direct contact with, and transmitting loads to, the ground. The building foundation is sometimes referred to as the artificial foundation, and the ground on which it bears as the natural foundation. Early buildings were founded on rock or firm ground; it was not until the beginning of the twentieth century that concrete was increasingly used as a foundation base for walls. With the introduction of local and then national building regulations, standard forms of concrete foundations have become accepted practice in the UK, along with more rigorous investigation of the nature and bearing capacity of soils and bedrock. This chapter provides an introduction to groundwork and foundations, which is developed further in *Barry's Advanced Construction of Buildings* (Chapter 3).

3.1 Functional requirements

The primary functional requirement of a foundation is strength and stability.

Strength and stability

The combined, dead, imposed and wind loads on a building must be transmitted to the ground safely, without causing deflection or deformation of the building or movement of the ground that would impair the stability of the building and/or neighbouring structures. Foundations should also be designed and constructed to resist any movements of the subsoil.

Foundations should be designed so that any settlement is both limited and uniform under the whole of the building. Some settlement of a building on a soil foundation is inevitable. As the building is erected the loads placed on the foundation increase and the soil is compressed. This settlement should be limited to avoid damage to service pipes and drains connected to the building. Bearing capacities for various rocks and soils are assumed and these capacities should not be exceeded in the design of the foundation to limit settlement.

Barry's Introduction to Construction of Buildings, Third Edition. Stephen Emmitt and Christopher A. Gorse.
© 2014 John Wiley & Sons, Ltd. Published 2014 by John Wiley & Sons, Ltd.

3.2 Bedrock and soil types

Ground is the general term for the Earth's surface, which varies in composition within the two main groups: rocks and soils. Rocks include hard, strongly cemented deposits such as granite, and soils, the loose un-cemented deposits such as clay. Under the imposed load of a building, rocks suffer negligible compression and soils measurable compression. The size and depth of a foundation is determined by the structure and size of the building it supports and also the nature and bearing capacity of the ground supporting it.

Rocks

Rocks may be divided into three broad groups as igneous, sedimentary and metamorphic.

Igneous rocks
Igneous rocks, such as granite, dolerite and basalt, are those formed by the fusion of minerals under great heat and pressure. Beds of strong igneous rock occur just below or at the surface of the ground. Because of the density and strength of these rocks it would be sufficient to raise walls directly off the rock surface. For convenience it is usual to cast a bed of concrete on the roughly levelled rock surface as a level surface on which to build. The concrete bed need be no wider than the wall thickness it supports.

Sedimentary rocks
Sedimentary rocks, such as limestone and sandstone, are those formed gradually over thousands of years by the settlement of particles of calcium carbonate or sand to the bottom of bodies of water, where the successive layers of deposit have been compacted as beds of rock by the weight of water above. Because of the irregular and varied deposit of the sediment, these rocks were formed in layers or laminae. In dense rock beds the layers are strongly compacted and in others the layers are weakly compacted and may vary in the nature of the layers and so have poor compressive strength. Because of the layered nature of these rocks the material should be laid as a building stone with the layers at right angles to the loads.

Metamorphic rocks
Metamorphic rocks, such as slates and schists, are those changed from igneous, sedimentary rocks or from soils into metamorphic rocks by pressure or heat or both. These rocks vary from dense slates in which the layers of the material are barely visible to schists in which the layers of various minerals are clearly visible and may readily split into thin plates. Because of the mode of the formation of these rocks the layers or planes rarely lie horizontally in the ground and so generally provide an unsatisfactory or poor foundation.

Soil

Soil is the general term for the upper layer of the Earth's surface that consists of various combinations of particles of disintegrated rock, such as gravel, sand or clay, with some organic remains of decayed vegetation generally close to the surface.

Topsoil

The surface layer of most of the low-lying land in the UK that is most suited to building consists of a mixture of loosely compacted particles of sand, clay and an accumulation of decaying vegetation. This layer of topsoil, which is about 100–300 mm deep, is sometimes referred to as vegetable topsoil. It is loosely compacted, supports growing plant life and is unsatisfactory as a foundation because of its poor bearing capacity. It should be stripped from the site and retained for landscaping around the site.

Subsoil

Subsoil is the general term for soil below the topsoil. It is unusual for a subsoil to consist of gravel, sand or clay by itself. The majority of subsoils are mixes of various soils. Gravel, sand and clay may be combined in a variety of proportions. To make a broad assumption of the behaviour of a particular soil under the load on foundations, it is convenient to group soils such as gravel, sand and clay by reference to the size and nature of the particles. The three broad groups are coarse-grained non-cohesive, fine-grained cohesive and organic. The nature and behaviour under the load on foundations of the soils in each group are similar.

Coarse-grained non-cohesive soils

Soils that are composed mainly of, or combinations of, sand and gravel consist of largely siliceous, unaltered products of rock weathering. They have no plasticity and tend to lack cohesion, especially when dry. Under pressure of the loads on foundations the soils in this group compress and consolidate rapidly by some rearrangement of the coarse particles and the expulsion of water. A foundation on coarse-grained non-cohesive soils settles rapidly by consolidation of the soil as the building is erected, so that there is no further settlement once the building is completed.

Gravel consists of particles of a natural coarse-grained deposit of rock fragments and finer sand. Many of the particles are larger than 2 mm. Sand is the natural sediment of granular, mainly siliceous, products of rock weathering. Particles are smaller than 2 mm, are visible to the naked eye and the smallest size is 0.06 mm. Sand is gritty, has no real plasticity and can easily be powdered by hand when dry. Dense, compact gravel and sand requires a pick to excavate for foundation trenches.

Fine-grained cohesive soils

Fine-grained cohesive soils, such as clays, are a natural deposit of the finest siliceous and aluminous products of rock weathering. Clay is smooth and greasy to the touch, shows high plasticity, dries slowly and shrinks appreciably on drying. Under the pressure of the load on foundations clay soils are very gradually compressed by the expulsion of water through the very many fine capillary paths, so that buildings settle gradually during building work and this settlement may continue for some years after the building is completed. The initial and subsequent small settlement by compression during and after building on clay subsoils will generally be uniform under most small buildings; thus no damage is caused to the structure and its connected services.

3.3 Ground movement

Approved Document A states that the building shall be constructed so that ground movement caused by swelling, shrinkage or freezing of the subsoil, or landslip or subsidence, which can be reasonably foreseen, will not impair the stability of the building.

Volume change

Firm, compact, shrinkable clays suffer appreciable vertical and horizontal shrinkage on drying, and expansion on wetting, due to seasonal changes. Seasonal volume changes under grass extend to about 1 m below the surface in Great Britain and up to depths of 4 m or more below large trees. The extent of volume changes, particularly in firm clay soils, depends on seasonal variations and the proximity of trees and shrubs. The greater the seasonable variation, the greater the volume change. The more vigorous the growth of shrubs and trees in firm clay soils, the greater the depth below surface the volume change will occur. As a rough guide it is recommended that buildings on shallow foundations should not be closer to a single tree than the height of the tree at maturity, and one-and-a-half times the height at maturity of groups of trees, to reduce the risk of damage to buildings by seasonal volume changes in clay subsoils.

When shrubs and trees are removed to clear a site for building on firm clay subsoils there will, for some years after the clearance, be ground recovery as the clay gradually recovers moisture previously taken by the shrubs and trees. The design and depth of foundations of buildings must allow for this gradual expansion to limit damage by differential settlement. Similarly, if vigorous shrub or tree growth is stopped by removal, or started by planting, near to a building on firm clay subsoil with foundations at a shallow depth, it is likely that gradual expansion or contraction of the soil will cause damage to the building by differential movement. Significant seasonal volume change, due to deep-rooted vegetation, may be pronounced during periods of drought and heavy continuous rainfall.

The most economical and effective foundation for low-rise buildings on shrinkable clays close to deep-rooted vegetation is a traditional system of short-bored piles and ground beams (Figure 3.1c) or the precast piles and beams such as those shown in Figure 3.2 and Photograph 3.1. The piles should be taken down to a depth below which vegetation roots will not cause significant volume changes in the subsoil. Single deep-rooted vegetation such as shrubs and trees as close as their mature height to buildings, and groups of shrubs and trees one-and-a-half times their mature height to buildings, can affect foundations on shrinkable clay subsoils. When planting new trees and shrubs close to buildings proprietary root limiting products should be used (sometimes termed 'root protectors') to help prevent the spread of tree roots and hence provide some protection against damage to foundations. Expert advice should be sought from a tree surgeon.

Frost heave

Where the water table is high, i.e. near to the surface, soils, such as silts, chalk, fine gritty sands and some lean clays, may expand when frozen. This expansion is due to crystals of ice forming and expanding in the soil and so causing frost heave. In the UK, groundwater near the surface rarely freezes at depths of more than 0.5 m, but in exposed positions on

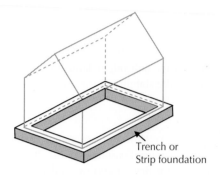

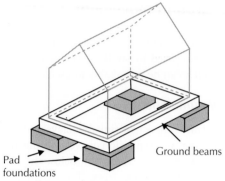

Trench or
Strip foundation

Pad
foundations

Ground beams

(a) Strip foundations

Used to transfer long continuous loads
(such as walls). The width and depth of the
foundation will depend on the nature of
ground and building loads. Where the
ground conditions are good strip
foundations are the preferred choice to
support walls.

(b) Pad foundations

More commonly used under point loads
such as columns but can be used under
ground beams to transfer continuous loads
(such as walls). The width and depth of
each pad foundation will depend on the soil
conditions and building loads.

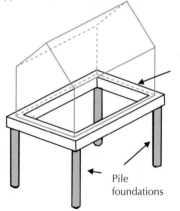

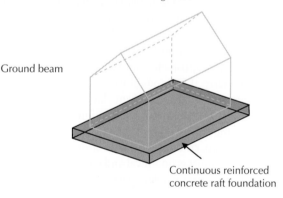

Ground beam

Pile
foundations

Continuous reinforced
concrete raft foundation

(c) Pile foundations

The pile foundation takes the load of the
building through made up ground or weak
soil to loadbearing strata. The ground
beams transfer the building loads to the
piles.

(d) Raft foundations

Reinforced concrete raft foundations
spread the load over the whole building
area reducing the load per unit area. Raft
foundations are used where building loads
are high or ground conditions are poor.

Figure 3.1   Simple schematic of foundation types.

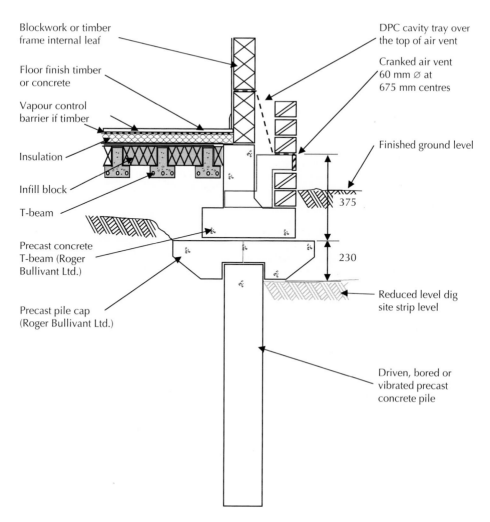

Blockwork or timber frame internal leaf

Floor finish timber or concrete

Vapour control barrier if timber

Insulation

Infill block

T-beam

Precast concrete T-beam (Roger Bullivant Ltd.)

Precast pile cap (Roger Bullivant Ltd.)

DPC cavity tray over the top of air vent

Cranked air vent 60 mm ∅ at 675 mm centres

Finished ground level

375

230

Reduced level dig site strip level

Driven, bored or vibrated precast concrete pile

Figure 3.2 Precast concrete foundation system (adapted from http://www.roger-bullivant .co.uk).

open ground during frost it may freeze up to a depth of 1 m. For unheated buildings and heated buildings with insulated ground floors, a foundation depth of 450 mm is generally sufficient against the possibility of damage by ground movement due to frost heave.

Made up ground

Areas of made up ground are often used for buildings as the demand for new buildings increases. Because of the varied nature of the materials used to fill depressions in the ground and hence raise ground levels, and the uncertainty of the bearing capacity of

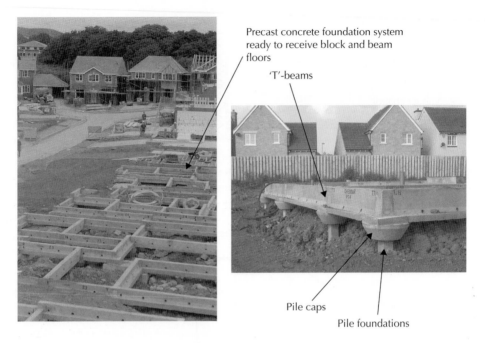

Precast concrete foundation system ready to receive block and beam floors

'T'-beams

Pile caps

Pile foundations

Photograph 3.1   Precast concrete foundations systems (http://www.roger-bullivant.co.uk).

the fill, conventional foundations (Figure 3.1a and b) may be unsatisfactory. Ground investigation is required to establish the most suitable foundation design. The bearing ground may be some distance below the surface level of the made up ground and to excavate for conventional strip foundations would be uneconomic. A solution is to use pile foundation to support reinforced concrete ground beams on which walls are raised, as illustrated in Figure 3.1c.

Unstable ground

There are extensive areas of ground where mining and excavations for coal, chalk, sand and gravel may have made the ground unstable. Thus the ground and building foundations may be subject to periodic, unpredictable subsidence. Specialist advice should always be sought to establish the extent of previous workings and the most appropriate method of designing foundations for such situations.

Where it is known that the ground may be unstable and there is no ready means of predicting the possibility of mass movement of the subsoil and it is expedient to build, a solution is to use some form of reinforced concrete raft under the whole of the building, as illustrated in Figure 3.1d. The concrete raft, which is cast on or just below the surface, is designed to spread the load of the building over the whole of the underside of the raft so that, in a sense, the raft floats on the surface. Alternatively, if there is good loadbearing strata, some metres below the surface, pile foundations can be used.

Precast pile and beam foundation systems

Planning requirements that restrict the removal of trees and encourage building on brown-field sites have resulted in increased use of pile foundations for domestic developments. Precast concrete foundations are often used to overcome difficult ground conditions, problems caused by vegetation or where the speed of construction is important. 'Fast track' pile foundations systems have been developed that are suitable for both timber-framed and traditional masonry construction (Figure 3.2). The general trend towards mechanised installation of foundation systems helps to remove the risks associated with working in deep excavations, workers coming into contact with wet concrete and manual handling of heavy materials.

Precast piles can be installed by a top-driven hydraulic hammer. Once the correct resistance is achieved, the piles can be cropped to the correct level so that the pile caps and T-beams can be quickly installed. Other types of pile may be used, according to ground conditions. These could be driven steel, continuous flight auger (CFA) or a unique displacement auger pile (CHD), described in *Barry's Advanced Construction of Buildings*.

3.4 Foundation construction

The move to closely regulated systems has resulted in the use of foundations that are less prone to problems than some earlier methods of construction. One negative effect of this standardised approach is that some foundations may be over-designed for the loads they carry, either to avoid any possibility of foundation failure or simply through the application of inappropriate foundations for a particular building type. For example, it is not uncommon for timber-framed buildings to be built off foundations designed for heavier loadbearing masonry construction, essentially a lack of thought resulting in wasted materials and unnecessary expense.

There are a number of familiar approaches to foundation construction, from strip foundations, piles and rafts as described further (Figure 3.3), all of which are constructed of concrete. It should be noted that pile foundations can also be constructed from timber, steel and compacted stone, as well as unreinforced and reinforced concrete.

Concrete

Concrete is the name given to a mixture of particles of sand and gravel, the aggregate, bound together with cement, the matrix. Fine aggregate is natural sand, which has been washed and sieved to remove particles larger than 5 mm. Coarse aggregate is gravel that has been crushed, washed and sieved so that the particles vary from 5 to 50 mm in size. The fine and coarse aggregates are delivered separately. By combining them in the correct proportions, a concrete with very few voids or spaces in it can be made that produces a strong concrete.

The cement most used is ordinary Portland cement. It is manufactured by heating a mixture of finely powdered clay and limestone with water to a temperature of about 1200°C, at which the lime and clay fuse to form a clinker. This clinker is ground with the addition of a little gypsum to a fine powder of cement. Cement powder reacts with water and its

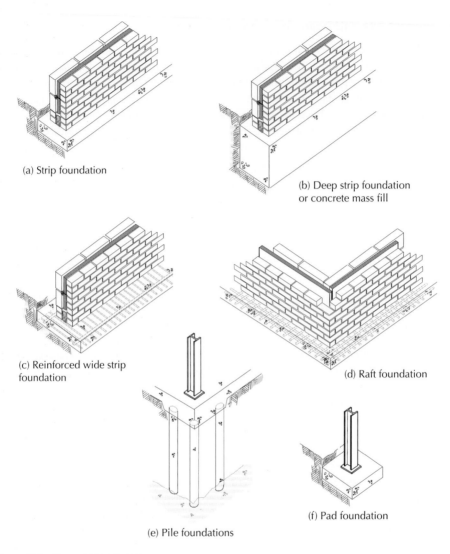

(a) Strip foundation

(b) Deep strip foundation
or concrete mass fill

(c) Reinforced wide strip
foundation

(d) Raft foundation

(e) Pile foundations

(f) Pad foundation

Figure 3.3   Foundation types.

composition gradually changes; the particles of cement bind together and adhere strongly to materials with which they are mixed. Cement hardens gradually after it is mixed with water. Some 30 minutes to an hour after mixing with water the cement is no longer plastic and it is said that the initial set has occurred. About 10 hours after mixing with water, the cement has solidified and it increasingly hardens to a dense solid mass after 7 days.

Water–cement ratio
Materials used for making concrete are mixed with water for two reasons: first, to cause the reaction between cement and water, which results in the cement acting as a binding

agent; second, to make the material sufficiently plastic to be easily placed in position. The ratio of water to cement used in concrete affects its ultimate strength. If too little water is used the concrete will be so stiff that it cannot be compacted, and if too much water is used the concrete will not develop full strength. Very little water is required to ensure that a full chemical reaction takes place within the concrete mix. Excess water will not be used and it will leave very small voids within the concrete as it evaporates. The water added must be sufficient to allow the chemical reaction to take place and enable the concrete to be worked (poured or vibrated). The amount of water required to make concrete sufficiently plastic (workable) depends on the position in which the concrete is to be placed. Plasticisers are added to the concrete mixture and enable the concrete to be more workable and fluid without increasing the quantities of water used. Using plasticisers keeps the strength of the concrete high without increasing the quantity of cement.

Concrete mixes

The materials used in reinforced concrete are commonly weighed and mixed in large concrete mixers. It is not economical for builders to employ expensive concrete mixing machinery for small buildings, and the concrete for foundations and floors is usually delivered to site ready mixed, except for small batches that are mixed by hand or in a portable mixer.

British Standard 5328: specifying concrete, including ready-mixed concrete, gives a range of mixes. One range of concrete mixes in the Standard, ordinary prescribed mixes, is suited to general building work such as foundations and floors. These prescribed mixes should be used in place of the nominal volume mixes such as 1:3:6 cement, fine and coarse aggregate by volume, which have been used in the past. The prescribed mixes, specified by dry weight of aggregate, used with 100 kg of cement, provide a more accurate method of measuring the proportion of cement to aggregate and, as they are measured against the dry weight of aggregate, allow for close control of the water content and therefore the strength of the concrete. Table 3.1 equates the old nominal volumetric mixes of cement and aggregate with the prescribed mixes and indicates the uses for these mixes.

Prescribed mixes are designated by the letters and numbers C7.5P, C10P, C15P, C20P, C25P and C30P; the letter C stands for 'compressive', the letter P for 'prescribed' and the number indicates the 28 day characteristic cube crushing strength in newtons per square millimetre (N/mm^2) that the concrete is expected to attain. Newton is a unit of force named after Sir Isaac Newton. The prescribed mix specifies the proportions of the mix to give an indication of the strength of the concrete sufficient for most building purposes, other than designed reinforced concrete work.

Table 3.1   Concrete mixes

Nominal volume mix	BS 5328 Standard mixes	Uses
1:8 all-in	ST1	Foundations
1:3:6		
1:3:6	ST2	Site concrete
1:2:4	ST3	
1:1½:3	ST4	Site concrete reinforced

Ready-mixed concrete

Ready-mixed concrete plants are able to supply to all but the most isolated building sites. These plants prepare carefully controlled concrete mixes, which are delivered to site by lorries on which the concrete is churned to delay setting. Standard lorries, which hold the concrete mixers, deliver up to $6\,m^3$ loads; however, some plants do have $8\,m^3$ lorries, and smaller lorries of $4\,m^3$ and less for occasional small loads.

To order ready-mixed concrete it is necessary to specify the prescribed mix, e.g. C10P, the cement, type and size of aggregate and workability. The workability required will depend on the situation where the concrete is to be used. Very stiff mixes (low workability) are used to bed kerb stones, concrete flags, road gullies and slabs. If highly workable and fluid mixes were used in these situations the precast concrete units would simply sink through the concrete. However, when a stiff mix is used, the precast or stone units can be easily bedded at the correct level. If concrete needs to be poured into deep formwork, which has considerable reinforcement, the concrete needs to be sufficiently fluid to pass through the gaps in the reinforcement to the bottom of the formwork.

Soluble sulphates

There are water soluble sulphates in some soils, such as plastic clay, which react with ordinary cement and in time will weaken concrete. Sulphate-resistant cements are more resistant to the destructive action of sulphates. Sulphate-resisting Portland cement has a reduced content of aluminates that combine with soluble sulphates in some soils and is used for concrete in contact with those soils.

Strip foundations

Strip foundations consist of a continuous strip, usually of steel-reinforced concrete, formed centrally under loadbearing walls (Figure 3.4). This continuous strip serves as a level base

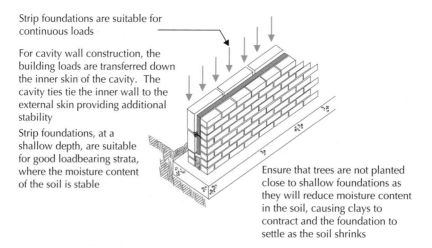

Strip foundations are suitable for continuous loads

For cavity wall construction, the building loads are transferred down the inner skin of the cavity. The cavity ties tie the inner wall to the external skin providing additional stability

Strip foundations, at a shallow depth, are suitable for good loadbearing strata, where the moisture content of the soil is stable

Ensure that trees are not planted close to shallow foundations as they will reduce moisture content in the soil, causing clays to contract and the foundation to settle as the soil shrinks

Figure 3.4   Strip foundation.

on which the wall is built. The width of the foundation is that necessary to spread the load exerted on the foundations to an area of subsoil which is capable of supporting the load without undue compaction. The bearing capacity of the soil should be greater than the loads imposed by the building's foundation.

The continuous strip of concrete is spread in trenches that have been excavated down to an undisturbed level of compact soil. The strip of concrete may well need to be no wider than the thickness of the wall. In practice the concrete strip will generally be wider than the thickness of the wall for the convenience of covering the whole width of the trench and to provide a wide enough level base for bricklaying below ground. A continuous strip foundation of concrete is the most economic form of foundation for small buildings on compact soils (Photograph 3.2).

The width of a concrete strip foundation depends on the bearing capacity of the subsoil and the load on the foundations: the greater the bearing capacity of the subsoil, the less the width of the foundation and vice versa. Table 3.2 (from Approved Document A) sets out the recommended minimum width of concrete strip foundations related to six specified categories of subsoil and calculated total loads on foundations.

The dimensions given in Table 3.2 are indicative of what might be acceptable in the conditions specified rather than absolutes to be accepted regardless of the conditions prevailing on individual sites. Figure 3.5a and b illustrates the important dimensions.

The strip foundation for a cavity external wall and a solid internal, loadbearing wall (illustrated in Figure 3.4) would be similar to the width recommended in the Approved Documents for a firm clay subsoil when the load on the foundations is no more than 50 kN/linear metre. In practice the linear load on the foundation of a house would be appreciably less than 50 kN/linear metre and the foundation may well be made wider than the minimum requirement for the convenience of filling a wider trench with concrete, due to the width of the excavator's bucket, or for the convenience of laying brick below ground, allowing adequate working space.

If the thickness of a concrete strip foundation (without steel reinforcement) were appreciably less than its projection each side of a wall, the concrete might fail through the weight of the wall causing a 45° shear crack as illustrated in Figure 3.6. If this occurred the bearing surface of the foundation on the ground would be reduced to less than that necessary for stability. In Figure 3.6, where the load is placed directly on top of the concrete foundation and compressed between the ground, the foundation remains stable. However, the concrete outside the 45° angle (where compressive forces are distributed) experiences some tension as the reaction of the ground attempts to lever the foundation upwards; thus the foundation would fail.

Stepping strip foundations

When strip foundations are used on sloping sites it may be necessary to step the foundation (Figure 3.7 and Photograph 3.3). In order to step the foundation, the full thickness of the upper foundation should overlap twice the height of the step (thickness of upper foundation), or 300 mm, whichever is greater. Consideration should also be given to the brickwork and blockwork that will be built on top of the foundation. The brickwork and blockwork should tie in at the step to avoid the need for cutting bricks and blocks and also to avoid the possibility of reducing the stability of the wall, as illustrated in Figure 3.8.

The trenches for the foundations are excavated.

A continuous membrane is then laid over the whole area of the building. Traditionally the membrane would be used to prevent the penetration of damp into the building; nowadays such barriers are often used to prevent gases as well as moisture from entering into the dwelling. The membrane or damp-proof membrane (dpm) acts as a barrier separating the building from the ground.

The reinforcement for the trench foundation (ground beam) is then positioned in line and at the correct level. The main trenches and beams will be positioned under loadbearing walls.

Spacers are used to hold the reinforcement off the ground. This allows concrete to go underneath the reinforcement ensuring adequate cover all around the reinforcement.

Ensuring the steel reinforcement is correctly positioned within the concrete will mean that they can bond together. The matrix formed between the reinforcement and concrete allows the foundation to deal with both the compressive and tensile forces. If the steel did not bond with the concrete the forces would not be able to transfer from one material to the other and the foundation would fail.

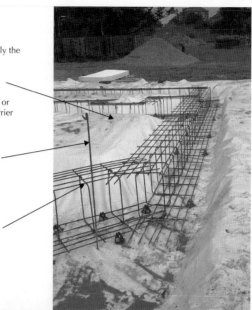

The trench foundations are poured up to the underside of the floor slab. The internal blockwork wall can then be built up to floor level; this will act as edge support when the concrete floor is poured.

Part of the ground beam reinforcement remains uncovered during this stage. This will eventually be cast into the floor forming a complete reinforced concrete ground beam.

Light steel mesh reinforcement will be placed over the whole of the floor slab. This will prevent the concrete slab from cracking if the ground between the ground beams settles.

Finally, the floor slab is cast and levelled off and the external cavity wall is built up to damp-proof course (dpc) level.

Photograph 3.2   Trenches and ground beams (http://www.leedsmet.ac.uk/teaching/vsite).

Table 3.2 Minimum width of strip foundations

| Types of subsoil | Condition of subsoil | Field test applicable | Total load of loadbearing walling – not more than (kN/linear metre) | | | | | |
			20	30	40	50	60	70
			Minimum width of strip foundation (mm)					
Rock	Not inferior to sandstone, limestone or firm chalk	Requires at least a mechanically operated pick for excavation	Equal to the width of the wall					
Gravel	Compact	Requires pick for excavation						
Sand	Compact	Wooden peg 50 mm square cross-section hard to drive in beyond 150 mm	250	300	400	500	600	650
Clay	Stiff	Cannot be moulded with the fingers and requires pick for removal	250	300	400	500	600	650
Sandy clay	Stiff							
Clay	Firm	Can be moulded by substantial pressure with fingers, can be excavated with a spade	300	350	450	600	750	850
Sandy clay	Firm							
Sand	Loose	Can be excavated with a spade. Wooden peg 50 mm × 50 mm cross-section easily driven	400	600	When the loads exceed 30 kN/linear metre in such weak soil conditions, the widths of strip foundations are not specified in the Building Regulations. Alternative types of foundation would be more suitable.			
Silty sand	Loose							
Clayey sand	Loose							
Silt	Soft	Fairly easy to mould with fingers. Easy to excavate	450	650				
Clay	Soft							
Sandy clay	Soft							
Silty clay	Soft							
Silt	Very soft	Extrudes between fingers when squeezed	600	850				
Clay	Very soft							
Sandy clay	Very soft							
Silty clay	Very soft							

Source: Approved Document A (2000).

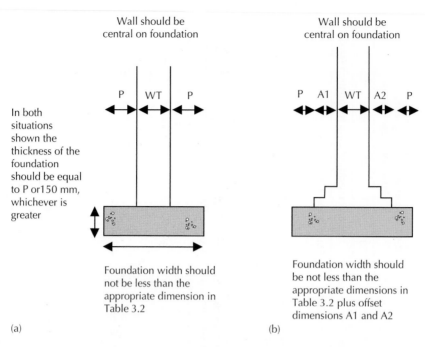

In both situations shown the thickness of the foundation should be equal to P or150 mm, whichever is greater

Wall should be central on foundation

P   WT   P

Foundation width should not be less than the appropriate dimension in Table 3.2

(a)

Wall should be central on foundation

P   A1   WT   A2   P

Foundation width should be not less than the appropriate dimensions in Table 3.2 plus offset dimensions A1 and A2

(b)

Figure 3.5   (a and b) Foundation dimensions (adapted from Approved Document A).

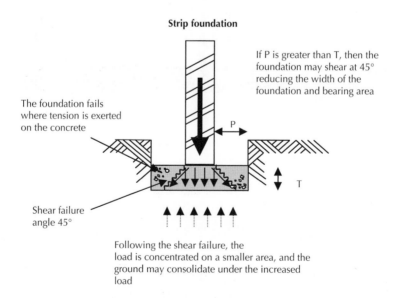

Strip foundation

If P is greater than T, then the foundation may shear at 45° reducing the width of the foundation and bearing area

The foundation fails where tension is exerted on the concrete

P

T

Shear failure angle 45°

Following the shear failure, the load is concentrated on a smaller area, and the ground may consolidate under the increased load

Figure 3.6   Shear failure in a strip foundation.

Minimum overlap L = twice
height of step or 300 mm
whichever is greater

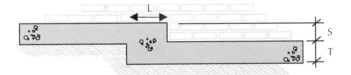

The step (S) should not
be greater than the
foundation depth (T)

Figure 3.7 Stepped strip foundations.

Photograph 3.3 Stepped strip foundation.

Modular heights
Brick 65 + 10 mm
Blocks 215 + 10 mm
Blocks laid flat 100 + 10 mm
Blocks laid flat 150 + 10 mm

Walling selected so that
the courses tie in with
the step

Different heights can
be achieved by
selecting a combination
of bricks and blocks,
e.g. to accommodate a
step of 300 mm a
course of bricks and a
course of blocks can be
used

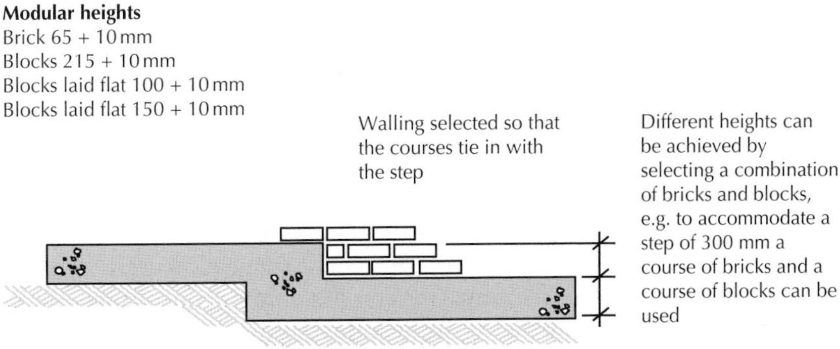

Figure 3.8 Brick coursing to stepped foundations.

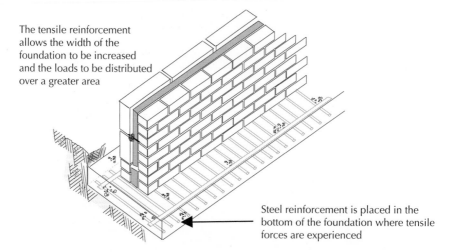

The tensile reinforcement allows the width of the foundation to be increased and the loads to be distributed over a greater area

Steel reinforcement is placed in the bottom of the foundation where tensile forces are experienced

Figure 3.9   Steel-reinforced wide strip foundation.

Wide strip foundation

Distributing the load over a larger area reduces the load per unit area on the ground. Strip foundations on subsoils with poor bearing capacity, such as soft sandy clays, may need to be considerably wider than traditional (narrow) strip foundations. However, to keep increasing the width and depth of the concrete ensuring that the foundation does not shear makes the process uneconomical. The alternative is to form a strip of steel-reinforced concrete, illustrated in Figure 3.9.

Concrete is strong in compression but is weak in tension. The effect of the downward pressure of the wall on the middle of the foundation and the opposing force of the ground spread across the base of the foundation attempts to bend the foundation upwards; this places the top of the foundation in compression and the base of the foundation in tension.

These opposing pressures will tend to cause the shear cracking illustrated in Figure 3.6 and Figure 3.10. To add tensile properties to the foundation, steel reinforcing bars are cast in the lower edge where tension will occur (Figure 3.11). There has to be a sufficient cover of concrete below the steel reinforcing rods to ensure a good bond between the concrete and steel and to protect the steel from corrosion. Together the steel and concrete make up a composite material that can resist both tensile and compressive forces.

Narrow strip (trench fill or deep strip) foundation

Stiff clay subsoils have good bearing strength but are subject to seasonal volume change. Because of seasonal changes and the withdrawal of moisture by deep-rooted vegetation (such as shrubs and trees) it is practice to adopt a foundation depth of at least 0.9 m to provide a stable foundation. Although the base of a deep strip foundation will go to a depth where the clay soil is unaffected by seasonal changes in moisture content, the soil at the external face of the foundation will still expand and contract as it becomes saturated and

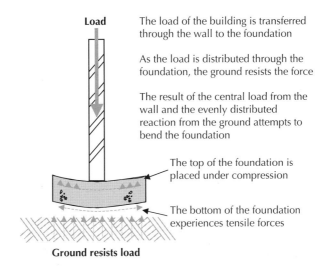

Load

The load of the building is transferred through the wall to the foundation

As the load is distributed through the foundation, the ground resists the force

The result of the central load from the wall and the evenly distributed reaction from the ground attempts to bend the foundation

The top of the foundation is placed under compression

The bottom of the foundation experiences tensile forces

Ground resists load

Figure 3.10   Distribution of forces in strip foundations.

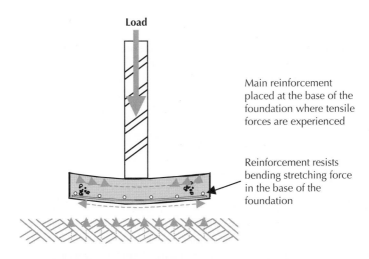

Load

Main reinforcement placed at the base of the foundation where tensile forces are experienced

Reinforcement resists bending stretching force in the base of the foundation

Figure 3.11   Accommodating tensile forces within strip foundations.

dries out. To prevent lateral pressure being applied to the side of the foundation, a 50 mm thick compressible sheet material may be used (Figure 3.12).

Because of the good bearing capacity of the clay, the foundation may need to be a little wider than the thickness of the wall to be supported. Trenches are usually formed by using mechanical excavators; thus the width of the trench is determined by the width of the excavator bucket available, which should not be less than the minimum required width of foundation. The trench is filled with concrete as soon as possible so that the sides of the

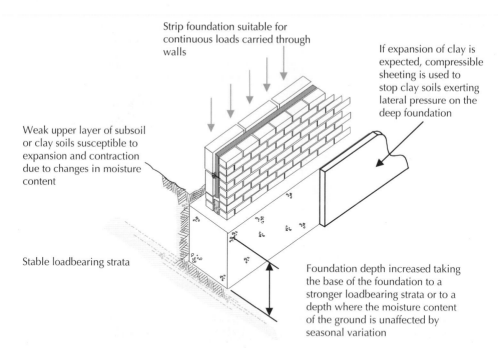

Strip foundation suitable for continuous loads carried through walls

If expansion of clay is expected, compressible sheeting is used to stop clay soils exerting lateral pressure on the deep foundation

Weak upper layer of subsoil or clay soils susceptible to expansion and contraction due to changes in moisture content

Stable loadbearing strata

Foundation depth increased taking the base of the foundation to a stronger loadbearing strata or to a depth where the moisture content of the ground is unaffected by seasonal variation

Figure 3.12   Deep strip or mass fill strip foundation.

trench do not fall in and the exposed clay bed does not dry out and shrink (as illustrated in Figure 3.13a). Where the trench is particularly wide and it is uneconomical to fill the whole trench with concrete it is common practice to use trench blocks. The trench blocks are laid flat and are often the full width of the walling above them (as illustrated in Figure 3.13b and Photograph 3.4).

Short-bored pile foundations

Where the subsoil is of firm, shrinkable clay, which is subject to volume change due to deep-rooted vegetation for some depth below the surface and where the subsoil is of soft or uncertain bearing capacity for a few metres below the surface, it may be economic to use a system of short-bored piles as a foundation. Piles are concrete columns, which are either precast and driven (hammered) into the ground or cast in holes that are augured (drilled) into the ground down to a level of a firm, stable, stratum of subsoil (Figure 3.14).

Piles that are excavated to a depth of less than 4 m below the surface are termed short bore, which refers to the comparatively short length of the piles as compared with the much longer piles used for larger buildings. Short-bored piles are generally from 2 to 4 m long and from 250 to 350 mm in diameter.

Holes are augered into the ground by machine. An auger is a form of drill comprising a rotating shaft with cutting blades that screw into the ground. The soil is either withdrawn and lifted to the surface as the shaft rotates or, with small augers, once the auger has cut

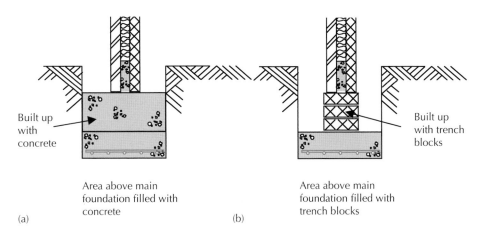

Built up
with
concrete

Built up
with trench
blocks

Area above main
foundation filled with
concrete

Area above main
foundation filled with
trench blocks

(a)   (b)

Figure 3.13   (a and b) Use of trench fill and trench blocks in strip foundations.

Photograph 3.4   Trench blocks – used to build wall up to ground level.

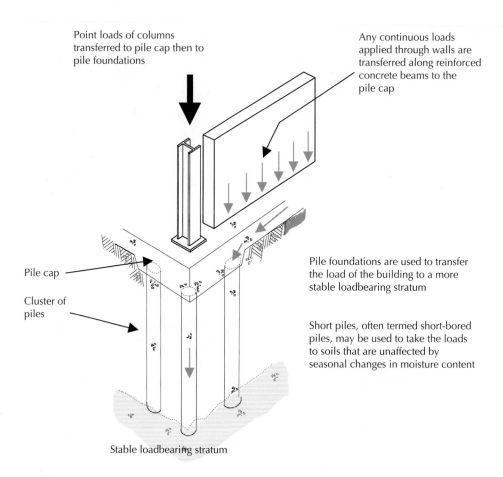

Point loads of columns
transferred to pile cap then to
pile foundations

Any continuous loads
applied through walls are
transferred along reinforced
concrete beams to the
pile cap

Pile cap

Cluster of
piles

Pile foundations are used to transfer
the load of the building to a more
stable loadbearing stratum

Short piles, often termed short-bored
piles, may be used to take the loads
to soils that are unaffected by
seasonal changes in moisture content

Stable loadbearing stratum

Figure 3.14   Pile foundations.

into the ground, it is withdrawn and the soil is removed from the blades. The advantage of this system of augered holes is that samples of the subsoil are withdrawn, from which the nature and bearing capacity of the subsoil may be assessed. The piles may be formed of concrete or, more usually, a light steel cage of reinforcement is lowered into the hole and concrete poured or pumped into the hole and compacted to form a pile foundation. The piles are cast underneath the corners and intersection of loadbearing walls and at regular intervals between to reduce the span and depth of the reinforced ground beam, which transfers the wall and building loads to the foundation. A reinforced concrete ground beam is then cast over the piles as illustrated in Figure 3.15. The spacing of the piles depends on the loads to be supported and on economic sections of ground beam (see Figure 3.15).

Pad foundations

Pad foundations can be used to carry point loads. They can also be designed so that the loads of the walls and the buildings are transferred through ground beams that rest on the

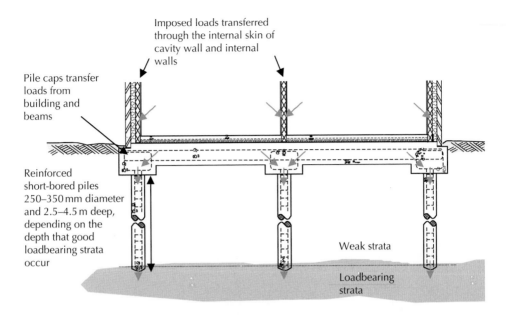

Imposed loads transferred through the internal skin of cavity wall and internal walls

Pile caps transfer loads from building and beams

Reinforced short-bored piles 250–350 mm diameter and 2.5–4.5 m deep, depending on the depth that good loadbearing strata occur

Weak strata

Loadbearing strata

Figure 3.15   Section illustrating pile foundations.

pad foundations. Pad foundations transfer the loads to a lower level where soil of sufficient loadbearing strata exists (Figure 3.16). The width of a pad foundation can be increased to distribute the loads over a greater area, thus reducing the pressure on the ground. Photograph 3.5 shows the excavation sequence of pad foundations and ground beams.

On made up ground and ground with poor bearing capacity where a firm, natural bed of strata, for example, gravel or sand, is a few metres below the surface, it may be economic to excavate for isolated piers of brick or concrete to support the load of buildings. The piers will be built at the intersection of the walls and under the more heavily loaded sections of wall, such as that between windows up the height of the building.

Pits are excavated down to the necessary level and the sides of the excavation are temporarily supported. Isolated pads of concrete are then cast in the bottom of the pits. Brick piers or reinforced concrete piers are built or cast on the pad foundations up to the underside of the reinforced concrete beams that support walls as illustrated in Figure 3.17. The ground beams or foundation beams may be just below or at ground level, the walls being raised off the beams.

The advantage of this system of foundation is that pockets of tipped stone or brick and concrete rubble that would obstruct bored piling may be removed as the pits are excavated and that the nature of the subsoil may be examined as the pits are dug to select a level of sound subsoil. This advantage may well be justification for this labour-intensive form of construction.

Raft foundations

A raft foundation consists of a raft of reinforced concrete under the whole of a building. Raft foundations may be used for buildings on compressible ground such as very soft clay,

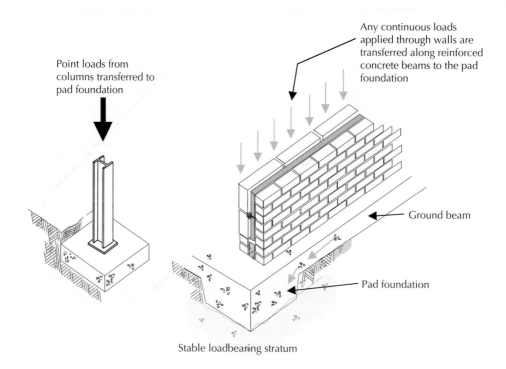

Point loads from columns transferred to pad foundation

Any continuous loads applied through walls are transferred along reinforced concrete beams to the pad foundation

Ground beam

Pad foundation

Stable loadbearing stratum

Figure 3.16   Pad foundations.

alluvial deposits and compressible fill material where strip, pad or pile foundations would not provide a stable foundation without excessive excavation. The reinforced concrete raft is designed to transmit the load of the building and distribute the load over the whole area under the raft, reducing the load per unit area placed on the ground (Figure 3.18). Distributing the loads in this way causes little, if any, appreciable settlement. The two types of raft foundation commonly used are the flat raft and the wide toe raft.

The flat slab raft foundation may be used under small buildings such as bungalows and two storey houses where the comparatively small loads on the foundations can be spread safely and economically under the raft. The concrete raft is of uniform thickness and reinforced top and bottom against both upward and downward bending. The construction sequence is as follows:

(1) Vegetable topsoil is removed.
(2) A blinding layer of concrete 50 mm thick is spread and levelled to provide a level base so that the steel reinforcement cage can be constructed.
(3) Where the raft is not cast directly against the ground, formwork may be required to contain any concrete up-stands.
(4) Once the reinforcement is correctly spaced and tied together in position, the concrete can be poured, vibrated and levelled.

(a) The foundation is marked out and excavated to the correct level. The machine operator attempts to keep the sides of the excavation as true and square as possible

(b) This pad foundation has been excavated down to good loadbearing strata. The bottom of the foundation is clean and ready for inspection by the building control officer

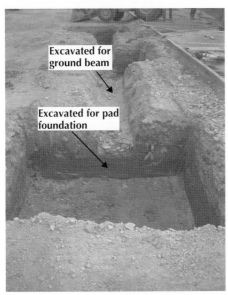

(c) Once one pad foundation is complete, the ground beam, which spans between the pad foundations, is excavated, then the next pad foundation is dug out

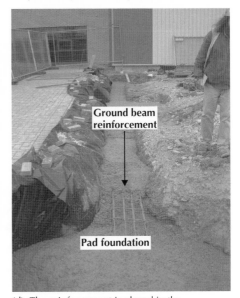

(d) The reinforcement is placed in the ground beam foundation, positioned correctly and then the concrete is poured to the correct level. The concrete should be vibrated to remove air bubbles

Photograph 3.5   Excavation and casting of pad foundation and ground beam.

Once the foundations have set the building can be constructed on top of them

The columns are a point load and are positioned and bolted to the pad foundation

The walls, which span between the columns, rest on the ground beam. The loads from the walls are transferred via the ground beam to the pad foundation

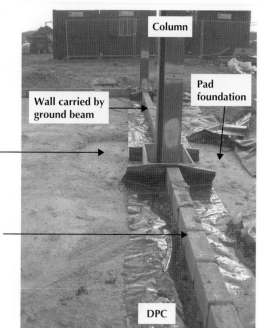

(e) The loads from the columns (point loads) and walls (distributed loads) are transferred to the foundations

Photograph 3.5 *(Continued)*

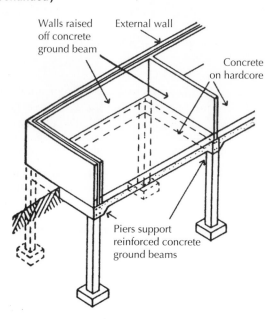

Figure 3.17 Pad foundations supported on beams and piles.

The load is distributed over the whole area of the building

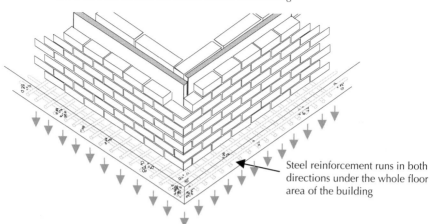

Steel reinforcement runs in both directions under the whole floor area of the building

Figure 3.18   Raft foundation.

(5) A waterproof membrane can be positioned either underneath the structural concrete or on top of it, beneath the insulation. Traditionally the damp-proof membrane (dpm) was placed on top of the blinding; it is now more common for the dpm to sit on top of the insulation (providing the insulation is impermeable). When the dpm is positioned above the insulation it not only prevents groundwater penetration but also reduces the possibility of interstitial condensation forming (see Chapter 4).

(6) Rigid insulation boards are placed on top of the structural concrete.

(7) Finally a 40 mm sand/cement screed finish is spread and levelled on top of the raft.

When the reinforced concrete raft has dried and developed sufficient strength the walls are raised as illustrated in Figure 3.19. The concrete raft is usually at least 150 mm thick.

In areas subject to mining subsidence the flat slab is cast on a bed of fine granular material, 150 mm thick, so that the raft is not keyed to the ground and is therefore unaffected by horizontal ground strains. Where the ground has poor compressibility and the loads on the foundations would require a thick, uneconomic flat slab, it is usual to cast the raft as a wide toe raft foundation. The raft is cast with a reinforced concrete stiffening edge beam, from which a reinforced concrete toe extends as a base for the external leaf of a cavity wall, as shown in Figure 3.20 and Figure 3.21.

Raft foundation on a sloping site

On sites where the slope of the ground is such that there is an appreciable fall in the surface across the width or length of a building, and a raft foundation is to be used, it is necessary either to cut into the surface or to provide additional fill under the building, or a combination of both, to provide a level base for the raft.

A 40–50 mm sand/cement screed provides the finished floor surface. Edge insulation is used to prevent cold bridging

Damp-proof membrane covers the whole area of the foundation; all joints should be properly sealed (taped)

Rigid insulation board (impermeable) is placed on top of the structural concrete

Once the reinforcement and any necessary formwork is in place the concrete is poured

Using steel or concrete spacers, the main reinforcement is correctly positioned, ready for incorporation within the concrete slab

Thin (40–50 mm) layer of concrete is spread over the ground, providing a level platform for the reinforcement cage to be constructed

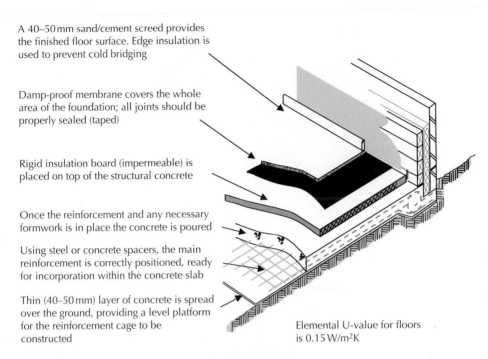

Elemental U-value for floors is 0.15 W/m²K

Figure 3.19   Construction of a concrete raft foundation.

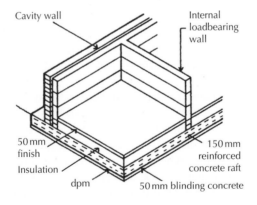

Cavity wall

Internal loadbearing wall

50 mm finish

Insulation

dpm

150 mm reinforced concrete raft

50 mm blinding concrete

Figure 3.20   Flat slab raft.

Where the slope is shallow and the design and use of the building allows, a stepped raft may be used down the slope, as illustrated in Figure 3.22. A stepped, wide toe, reinforced concrete raft is formed with the step or steps made at the point of a loadbearing internal wall or at a division wall between compartments. Drains are positioned to remove surface water running down the slope that might otherwise be trapped against steps and promote dampness in the building.

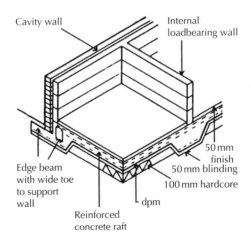

Figure 3.21   Edge beam raft.

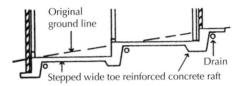

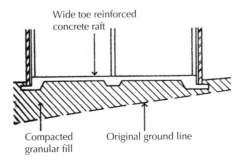

Figure 3.22   Raft on sloping site.

The level raft illustrated in Figure 3.22 is cast on imported granular fill that is spread, consolidated and levelled as a base for the raft. The disadvantage of a level raft on a sloping site is the cost of the additional granular fill: the advantage is a level bed of uniform consistency under the raft. As an alternative the system of cut and fill may be used to reduce the volume of imported fill.

Raft foundations are usually formed on ground of soft subsoil or made up ground where the bearing capacity is low or uncertain, to minimise settlement. There is some possibility of there being some slight movement of the ground under the building which would fracture drains and other service pipes entering the building through the raft. Service pipes rising through the raft should run through collars, cast in the concrete, which will allow some movement of the raft without fracturing the service pipes.

Foundations on sloping sites

The natural surface of ground is rarely level. On sloping sites an initial decision to be made is whether the ground floor is to be above ground at the highest point or partly sunk below ground as illustrated in Figure 3.22. Where the ground floor is to be at or just above ground level at the highest point, it is necessary to import some dry fill material such as hardcore to raise the level of the oversite concrete and floor. This fill will be placed, spread and consolidated up to the external wall once it has been built.

The consolidated fill will impose some horizontal pressure on the wall. To make sure that the stability of the wall is adequate to withstand this lateral pressure it is recommended practice that the thickness of the wall should be at least a quarter of the height of the fill bearing on it as illustrated in Figure 3.23a and b. The thickness of a cavity wall is taken as the combined thickness of the two leaves unless the cavity is filled with concrete when the overall thickness is taken.

To reduce the amount of fill necessary under solid floors on sloping sites a system of cut and fill may be used as illustrated in Figure 3.24. The disadvantage of this arrangement is that the ground floor is below ground level at the highest point, and it is necessary to form an excavated dry area to collect and drain surface water that would otherwise run up to

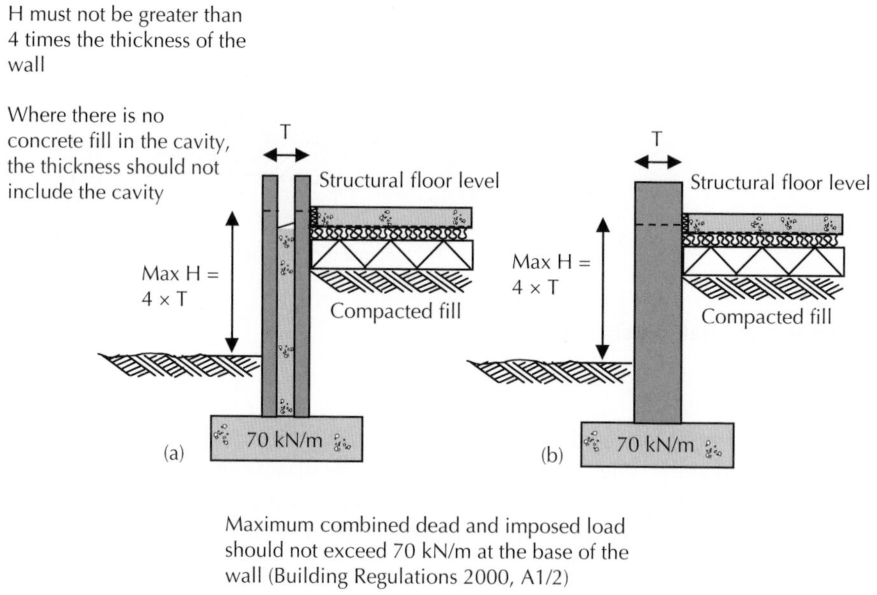

H must not be greater than 4 times the thickness of the wall

Where there is no concrete fill in the cavity, the thickness should not include the cavity

Structural floor level

Max H = 4 × T

Compacted fill

(a) 70 kN/m

Structural floor level

Max H = 4 × T

Compacted fill

(b) 70 kN/m

Maximum combined dead and imposed load should not exceed 70 kN/m at the base of the wall (Building Regulations 2000, A1/2)

Figure 3.23 Depth of internal hardcore fill and external wall thickness.

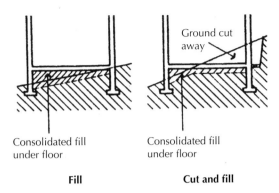

Fill

Cut and fill

Figure 3.24   Fill and cut and fill.

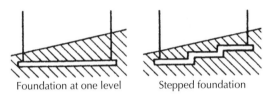

Foundation at one level

Stepped foundation

Figure 3.25   Foundation on sloping site.

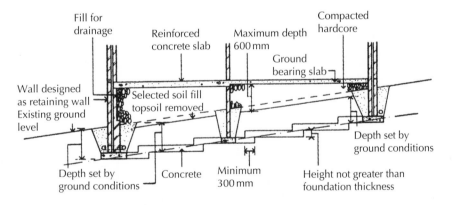

Figure 3.26   Stepped foundation.

the wall and cause problems of dampness. To reduce the depth of excavation and the foundations a stepped foundation is used as illustrated in Figure 3.25.

Figure 3.26 is an illustration of the stepped foundation for a small building on a sloping site where the subsoil is reasonably compact near the surface and will not be affected by volume changes. The foundation is stepped up the slope to minimise excavation and walling below ground. The foundation is stepped so that each step is no higher than the thickness of the concrete foundation and the foundation at the higher level overlaps the lower foundation by at least 300 mm.

The loadbearing walls are raised and the foundation trenches around the walls backfilled with selected soil from the excavation. The concrete oversite and solid ground floor may be cast on granular fill no more than 600 mm deep or cast or placed as a suspended reinforced concrete slab. The drains shown at the back of the trench fill are laid to collect and drain water to the sides of the building.

Alternative approaches

With a growing number of alternative approaches to construction inspired by more sustainable architecture and also advances in prefabrication, some alternative approaches to traditional concrete foundations are being used.

Framed building resting on gabions is one such approach (see Figure 3.27). The idea here is that when the building has exceeded its life the structure and the foundation can be easily recovered, recycled and reused with little loss of resources. Placing the gabions on the ground, rather than in the ground, can help to reduce the health and safety risks associated with excavation for the foundations. Uses of such technology tend to be limited to small, often temporary, buildings. Structural calculations will be required to demonstrate compliance with the Building Regulations.

Thermally insulated foundations

With the move towards zero energy buildings, it is becoming common to ensure that a thermal break extends around the whole building (Figure 3.28). Efforts are also required

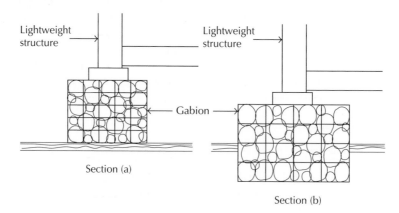

Figure 3.27   Gabion foundation, resting on (a) or in (b) the ground.

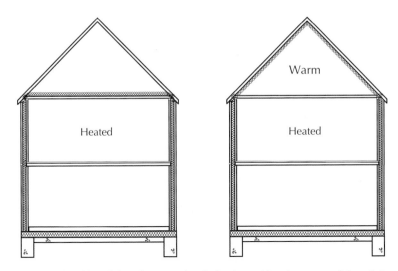

(a) Thermal break cold roof: foundations isolated   (b) Thermal break warm roof: foundations

Figure 3.28   Continuous thermal break above foundations.

to reduce the thermal bridging at junctions and connections with structural elements. The thermal components of the wall are extended down to meet the insulation materials across the floor and around the foundation (Figure 3.29).

For domestic and low rise non-domestic buildings it is possible to transfer comparatively small loads through insulating materials. It is also becoming more common to insulate under structural floor slabs and below foundations using rigid insulating materials that have structural properties. The insulation acts with the foundation to form a loadbearing structure. The insulation can sustain small forces placed on it during the construction phase, such as the load of the concrete foundation while it is setting and curing. When the concrete has set the building loads are transferred through the ground beams, pads and piles. The insulation retains its shape and form and resists forces imposed by surrounding materials and ground, but does not take the full load of the building. These foundations systems are capable of achieving U-values of around $0.1\,\text{W/m}^2\text{K}$.

Figure 3.30 and Figure 3.31 illustrate a ground beam and oversite concrete floor isolated from the ground with a structural thermal barrier which can form part of a Passivhaus building. Where loads are particularly high, such as that imposed by a column on a pad foundation, it is still possible to reduce the thermal bridging by insulating on top of the pad foundation and also insulating around the structural steel column, as illustrated in Figure 3.30. This can lead to U-values of around $0.05\,\text{W/m}^2\text{K}$. Where the fabric is so well insulated heat losses are considered negligible, and these details are suitable for use within nearly zero carbon building systems. Passivhaus principles work on the basis that the fabric U-values are less than $0.15\,\text{W/m}^2\text{K}$.

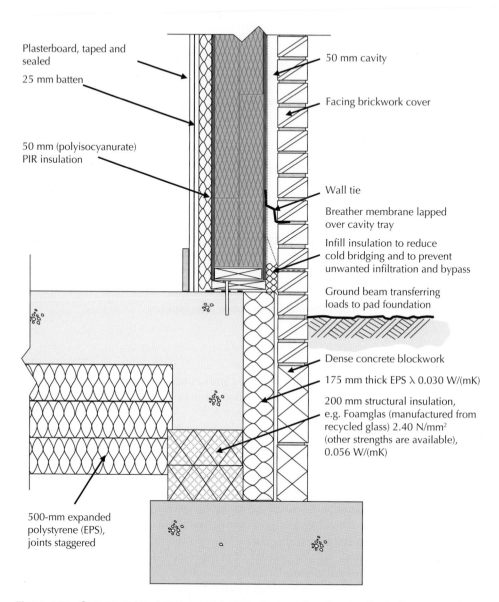

Plasterboard, taped and sealed

25 mm batten

50 mm (polyisocyanurate) PIR insulation

500-mm expanded polystyrene (EPS), joints staggered

50 mm cavity

Facing brickwork cover

Wall tie

Breather membrane lapped over cavity tray

Infill insulation to reduce cold bridging and to prevent unwanted infiltration and bypass

Ground beam transferring loads to pad foundation

Dense concrete blockwork

175 mm thick EPS λ 0.030 W/(mK)

200 mm structural insulation, e.g. Foamglas (manufactured from recycled glass) 2.40 N/mm^2 (other strengths are available), 0.056 W/(mK)

Figure 3.29   Structural insulation used to transfer construction loads until concrete has cured.

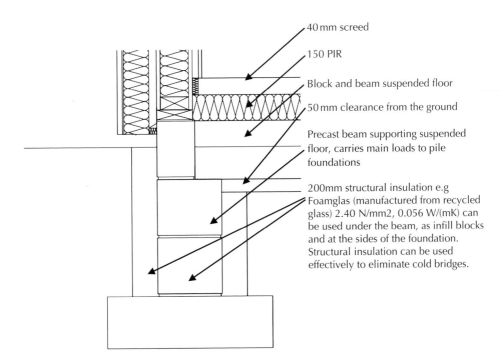

40 mm screed

150 PIR

Block and beam suspended floor

50 mm clearance from the ground

Precast beam supporting suspended floor, carries main loads to pile foundations

200mm structural insulation e.g Foamglas (manufactured from recycled glass) 2.40 N/mm2, 0.056 W/(mK) can be used under the beam, as infill blocks and at the sides of the foundation. Structural insulation can be used effectively to eliminate cold bridges.

Figure 3.30 Structural insulation carrying the loads of the structurally insulated panels (SIPs).

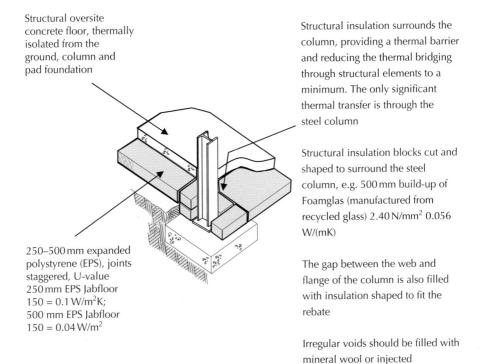

Structural oversite concrete floor, thermally isolated from the ground, column and pad foundation

Structural insulation surrounds the column, providing a thermal barrier and reducing the thermal bridging through structural elements to a minimum. The only significant thermal transfer is through the steel column

Structural insulation blocks cut and shaped to surround the steel column, e.g. 500 mm build-up of Foamglas (manufactured from recycled glass) 2.40 N/mm^2 0.056 W/(mK)

250–500 mm expanded polystyrene (EPS), joints staggered, U-value 250 mm EPS Jabfloor 150 = 0.1 W/m^{2}K; 500 mm EPS Jabfloor 150 = 0.04 W/m^2

The gap between the web and flange of the column is also filled with insulation shaped to fit the rebate

Irregular voids should be filled with mineral wool or injected polystyrene beads to reduce bypass

Figure 3.31 Thermal break around column on a pad foundation (to Passivhaus standard).

3.5 Site preparation and drainage

Turf and vegetable topsoil should be removed from the ground to be covered by a building, to a depth sufficient to prevent later growth. Tree and bush roots, which might encourage later growth and any pockets of soft compressible material, which might affect the stability of the building, should also be removed. The reasons for removing this vegetable soil are to prevent plants, shrubs or trees from attempting to grow; this would exert pressure on the concrete and crack it. Also once covered over, the vegetation contained in the soil will decay, causing voids to form below the concrete. The depth of vegetable topsoil varies and on some sites it may be necessary to remove 300 mm or more.

Contaminants

Approved Document C contains a list of contaminants in, or on, the ground to be covered by a building, that may be a danger to health or safety. Contaminants can be identified from planning records or local knowledge of previous uses. Sites that are likely to contain contaminants include:

- Asbestos works
- Chemical or gas works
- Coal carbonisation plants and ancillary by-products
- Industries making or using wood preservatives
- Landfill sites
- Waste disposals sites
- Metal works
- Munitions factories
- Nuclear installations
- Oil stores
- Paper printing works
- Railway land
- Scrap yards
- Sewage works
- Tanneries

Approved Document C also recommends action necessary if any contaminants are discovered, as described in Table 3.3.

Site drainage

Surface water (storm water) is the term used for rainwater and melted snow that falls on the surface of the ground, including open ground such as fields, paved areas and roofs. Rainwater that falls on paved areas and roofs generally drains to surface water (storm water) drains and then to soakaways or mains drainage. Paved areas are usually laid to falls to channels and gullies that drain to surface water drains. Rainwater falling on natural open ground will in part lie on the surface of impermeable soils, evaporate to air, run off to streams and rivers, and soak into permeable soils as groundwater.

Table 3.3   Possible signs of contaminants and actions

Signs of possible contaminants	Possible contaminant	Relevant action
No sign of vegetation, or poor or unnatural growth	Metals Metal compounds	None
No sign of vegetation, or poor or unnatural growth	Organic compounds Gases	Removal
Surface colour and contour or materials may be unusual indicating waste	Metals Metal compounds	None
	Oily and tarry wastes	Removal, filling or sealing
	Asbestos (loose)	Filling or sealing
	Other mineral fibres	None
	Organic compounds including phenols	Removal or filling
	Combustible material including coal and coke dust	Removal or filling
	Refuse and waste	Total removal or seek specialist advice
Fumes and odours may indicate organic chemicals at very low concentrations	Flammable explosive and asphyxiating gases including methane and carbon dioxide	Removal
	Corrosive liquids	Removal, filling or sealing
	Faecal animal and vegetable matter (biologically active)	Removal or filling
Drums and containers (whether full or empty)	Various	Removal with all contaminated ground

Adapted from Approved Document C2, table 2.

Groundwater is that water held in soils at and below the water table (which is the depth at which there is free water below the surface). The level of the water table will vary seasonally, being closest to the surface during rainy seasons and deeper during dry seasons when most evaporation to air occurs.

In Approved Document C there is a requirement for 'adequate subsoil drainage', to avoid passage of ground moisture to the inside of a building or to avoid damage to the fabric of the building. Subsoil drainage is also required where the water table can rise to within 0.25 m of the lowest floor, where the water table is high in dry weather and where the site of the building is surrounded by higher ground.

Subsoil drains

Subsoil drains are used to improve the run-off of surface water and the drainage of groundwater to maintain the water table at some depth below the surface to:

❏   Improve the stability of the ground
❏   Avoid surface flooding
❏   Alleviate or avoid dampness in basements
❏   Reduce humidity in the immediate vicinity of buildings

Subsoil drains are either open jointed or jointed, porous or perforated pipes of clayware, concrete, pitch fibre or plastic. The pipes are laid in trenches to follow the fall of the ground, generally with branch drains discharging to a ditch, stream or drain. On impervious subsoils, such as clay, it may be necessary to form a system of drains to improve the run-off of surface water and drain subsoil to prevent flooding. Some of the drain systems used are natural, herringbone, grid, fan and moat or cut-off.

Natural system
Natural system, which is commonly used for field drains, uses the natural contours of the ground to improve run-off of surface groundwater to spine drains in natural valleys that fall towards ditches or streams. The drains are laid in irregular patterns to follow the natural contours as illustrated in Figure 3.32a.

Herringbone system
Herringbone system, illustrated in Figure 3.32b, fairly regular runs of drains connect to spine drains that connect to a ditch or main drain. This system is suited to shallow, mainly one-way, slopes that fall naturally towards a ditch or main drain and can be laid to a reasonably regular pattern to provide a broad area of drainage.

Grid system
Grid system is an alternative to the herringbone system for draining one-way slopes where branch drains are fed by short branches that fall towards a ditch or main drain, as illustrated in Figure 3.33a. This system may be preferred to the herringbone system, where the run-off is moderate, because there are fewer drain connections that may become blocked.

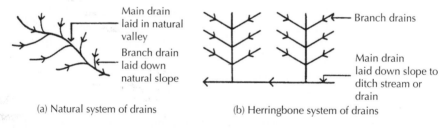

Main drain laid in natural valley
Branch drain laid down natural slope
Branch drains
Main drain laid down slope to ditch stream or drain

(a) Natural system of drains   (b) Herringbone system of drains

Figure 3.32   (a) Natural system. (b) Herringbone system.

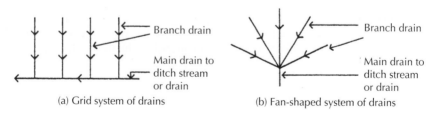

Branch drain
Main drain to ditch stream or drain
Branch drain
Main drain to ditch stream or drain

(a) Grid system of drains   (b) Fan-shaped system of drains

Figure 3.33   (a) Grid system. (b) Fan system.

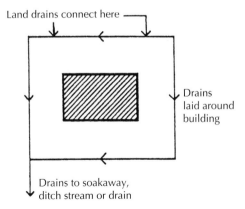

Land drains connect here

Drains laid around building

Drains to soakaway, ditch stream or drain

Figure 3.34   Moat or cut-off system.

Fan system

A fan-shaped layout of short branches, illustrated in Figure 3.33b, drains to spine drains that fan towards a soakaway, ditch or drain on narrow sites. A similar system is also used to drain the partially purified outflow from a septic tank to an area of subsoil where further purification will be affected.

On sloping building sites on impervious soil where an existing system of land drains is already laid and where a new system is laid to prevent flooding, a moat or cut-off system is used around the new building to isolate it from general land drains, as illustrated in Figure 3.34.

The moat or cut-off system of drains is laid some distance from and around the new building to drain the ground between it and the new building and to carry water from the diverted land drains down the slope of the site. The moat drains should be clear of paved areas around the house.

Laying drains

Groundwater (land) drains are laid in trenches at depths of 0.6 and 0.9 m in heavy soils and 0.9–1.2 m in light soils. The nominal bore of the pipes is usually 75 and 100 mm for main drains and 65 or 75 mm for branches. The drainpipes are laid in the bed of the trench and surrounded with coarse gravel (without any fine gravel). A filtering material (traditionally this was inverted turf or straw) is then placed on top of the gravel; this allows the water to percolate through to the drain without allowing fine material to pass through and block up the surrounding gravel. Excavated material is backfilled into the drain trench up to the natural ground level. The drain trench bottom may be shaped to take and contain the pipe or finished with a flat bed as illustrated in Figure 3.35, depending on the nature of the subsoil and convenience in using a shaping tool.

Where drains are laid to collect mainly surface water the trenches are filled with clinker, gravel or broken rubble to drain water either to a drain or without a drain as illustrated in

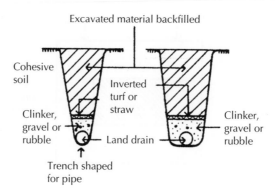

Figure 3.35   Land drains.

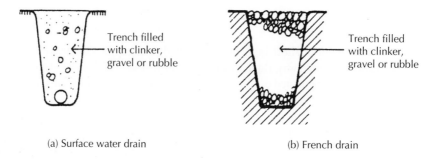

(a) Surface water drain   (b) French drain

Figure 3.36   (a) Surface water drain. (b) French drain.

Figure 3.36 in the form known as a French drain. Whichever is used will depend on the anticipated volume of water and the economy of dispensing with drainpipes.

Support for foundation trenches

The trenches to be dug for the foundations of walls may be excavated by hand for single small buildings, but where, for example, several houses are being built at the same time it is often economical to use mechanical trench diggers. If the trenches are of any significant depth it may be necessary to fix temporary timber supports to stop the sides of the trench from falling in. The nature of the soil being excavated mainly determines the depth of trench for which timber supports to the sides should be used. Working in trenches is dangerous and all health and safety guidance must be complied with in terms of providing adequate support for trenches and hence safe working conditions.

Soft granular soils readily crumble and the sides of trenches in such soil may have to be supported for the full depth of the trench. The sides of trenches in clay soil do not usually require support until a depth of approximately 1.5 m, particularly in dry weather. In wet weather, if the bottom of the trench in clay soil gets filled with water, the water may wash out the clay from the sides at the bottom of the trench and the whole of the sides above may collapse.

The purpose of temporary timbering supports to trenches is to uphold the sides of the excavation as necessary to avoid collapse of the sides, which may endanger the lives of those working in the trench, and to avoid the wasteful labour of constantly clearing falling earth from the trench bottoms.

The material most used for temporary support for the sides of excavations for strip foundations is rough sawn timber. The timbers used are square section struts, across the width of the trench, supporting open poling boards, close poling boards and wailing or poling boards and sheeting.

Whichever system of timbering is used there should be as few struts, i.e. horizontal members, fixed across the width of the trench as possible as these obstruct ease of working in the trench. Struts should be cut to fit tightly between poling or wailing boards and secured in position so that they are not easily knocked out of place. For excavations more than 1.5 m deep in compact clay soils it is generally sufficient to use a comparatively open timbering system as the sides of clay will not readily fall in unless very wet or supporting heavy nearby loads. A traditional system of struts between poling boards spaced at about 1.8 m intervals as illustrated in Figure 3.37 will usually suffice.

Where the soil is soft, such as soft clay or sand, it will be necessary to use more closely spaced poling boards to prevent the sides of the trench between the struts from falling in. To support the poling boards horizontal wailings are strutted across the trench, as illustrated in Figure 3.38.

Traditional timber support systems are still used. They are particularly useful in built-up areas where services regularly interrupt the path of, and cross, the trenches. Photograph 3.5 and Photograph 3.7 show a close boarded trench system, and Photograph 3.8 shows how the traditional system is used to accommodate services within the trench and also those that cross the trench.

For trenches in dry granular or loose soil it may be necessary to use sheeting to the whole of the sides of trenches. Rough timber sheeting boards are fixed along the length and up

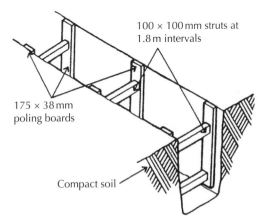

100 × 100 mm struts at 1.8 m intervals

175 × 38 mm poling boards

Compact soil

Figure 3.37   Open system: struts and poling boards.

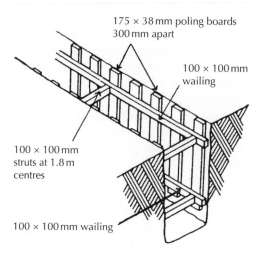

Figure 3.38 **Struts, wailing and poling boards.**

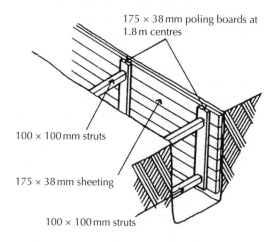

Figure 3.39 **Struts, poling boards and sheeting.**

the sides of the trench to which poling boards are strutted, as illustrated in Figure 3.39 and in Photograph 3.6, Photograph 3.7 and Photograph 3.8. The three basic arrangements of timber supports for trenches are indicative of some common systems used and the sizes given are those that might be used.

Although the traditional method of timbering is still used to provide temporary trench support it is more common to use steel shoring systems such as steel walers, adjustable vertical shores, adjustable props, trench sheets and trench boxes as illustrated in Figure

Photograph 3.6 Close boarded trench support.

3.40, Figure 3.41, Figure 3.42, Figure 3.43 and Figure 3.44 and in Photograph 3.9, Photograph 3.10, Photograph 3.11, Photograph 3.12 and Photograph 3.13.

When timbering is used for temporary support it is often used with adjustable steel struts instead of timber struts. Trench boxes, such as those shown in Photograph 3.9 and Photograph 3.10, are often used because they provide a safe area in which to work. Trench boxes, trench sheet and adjustable struts are often used because they can be inserted at ground level, without the need to enter the excavation. The hydraulic shores can be braced with by connecting a pump (Photograph 3.13) so that the struts and supporting frame can be expanded bracing the sheets in position.

Trench boxes cannot be used where services cross the excavation. Because trench support has to be chosen to suit particular site conditions it is common to find a combination of systems in more complex situations.

Photograph 3.7 Traditional trench support with edge protection.

Photograph 3.8 Traditional trench support with poling boards adjusted to accommodate services.

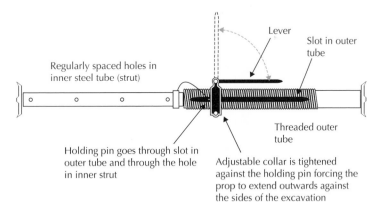

| Lever |
| Slot in outer tube |

Regularly spaced holes in inner steel tube (strut)

Holding pin goes through slot in outer tube and through the hole in inner strut

Threaded outer tube

Adjustable collar is tightened against the holding pin forcing the prop to extend outwards against the sides of the excavation

Standard sizes of strut available

Extension in metres	
Closed	**Open**
0.495	0.737
0.705	1.137
1.029	1.740

Figure 3.40 Adjustable steel strut.

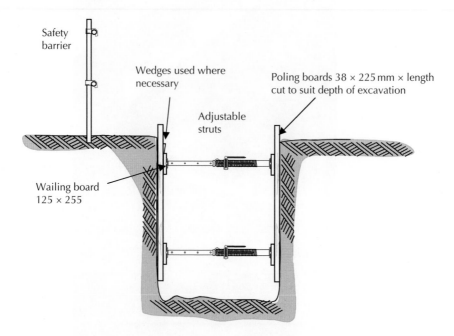

Figure 3.41 Open or closed timbering using adjustable steel strut.

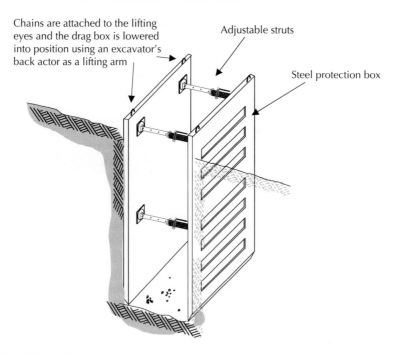

Figure 3.42 Trench box.

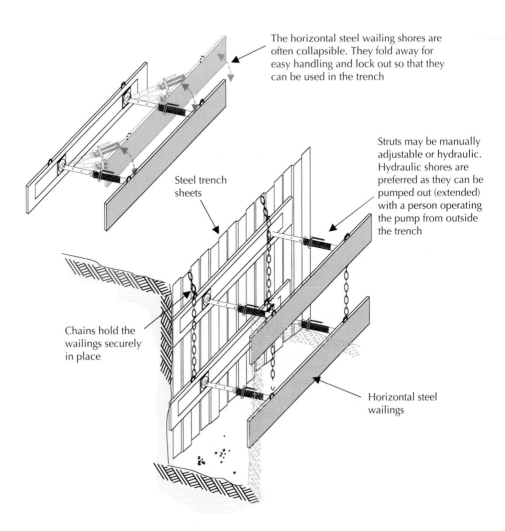

The horizontal steel wailing shores are often collapsible. They fold away for easy handling and lock out so that they can be used in the trench

Struts may be manually adjustable or hydraulic. Hydraulic shores are preferred as they can be pumped out (extended) with a person operating the pump from outside the trench

Steel trench sheets

Chains hold the wailings securely in place

Horizontal steel wailings

Figure 3.43   Shoring: horizontal steel wailing.

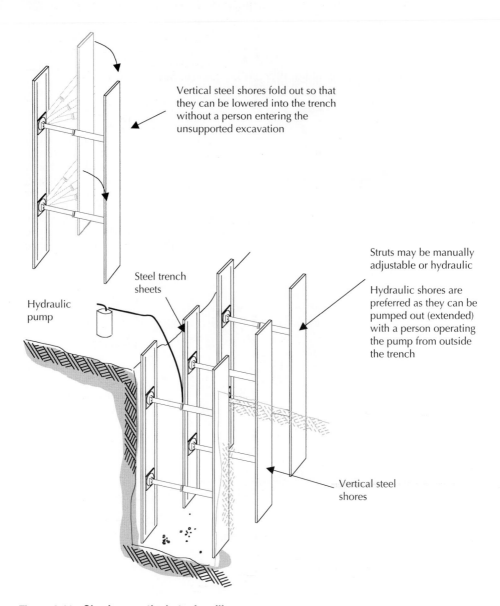

Vertical steel shores fold out so that they can be lowered into the trench without a person entering the unsupported excavation

Struts may be manually adjustable or hydraulic

Hydraulic shores are preferred as they can be pumped out (extended) with a person operating the pump from outside the trench

Steel trench sheets

Hydraulic pump

Vertical steel shores

Figure 3.44   Shoring: vertical steel wailing.

Photograph 3.9   Trench box.

Photograph 3.10   Inspection chamber shores and trench box.

Photograph 3.11   Adjustable shore and trench sheets.

Photograph 3.12   Shores hung from the top of the trench sheets and secured in place using a hydraulic pump.

Photograph 3.13   Hydraulic pump.

4 Floors

Concrete and timber are the two materials most used for the construction of ground and upper floors, the choice of one over another determined largely by the span required and the required performance in terms of fire safety and the resistance to the passage of heat and sound. Ground and upper floors constructed of concrete or timber are described in this chapter. Figure 4.1 provides examples of some of the more common types of floor construction.

4.1 Functional requirements

The functional requirements of a floor are:

- Strength and stability
- Resistance to weather and ground moisture
- Durability and freedom from maintenance
- Fire safety – resisting spread and passage of fire
- Fire safety – providing stable support for occupants to evacuate
- Resistance to passage of heat
- Resistance to the passage of sound

Strength

The strength of a floor depends on the characteristics of the materials used for the structure of the floor, such as timber, steel or concrete. The floor structure must be strong enough to support safely the dead load of the floor and its finishes, fixtures, partitions and services, and the anticipated imposed loads of the occupants and their movable furniture and equipment. BS 6399: Part 1 is the Code of Practice for dead and imposed loads for buildings. Where imposed loads are small, as in single-family domestic buildings of not more than three storeys, a timber floor construction is usual. The lightweight timber floor structure is adequate for the small loads over small spans. Precast concrete block and beam flooring offers an economical and quick alternative to timber floors. For larger imposed loads and wider spans a reinforced concrete floor is used, both for strength in support and also for resistance to fire. Approved Document A includes tables of recommended sizes

Barry's Introduction to Construction of Buildings, Third Edition. Stephen Emmitt and Christopher A. Gorse.
© 2014 John Wiley & Sons, Ltd. Published 2014 by John Wiley & Sons, Ltd.

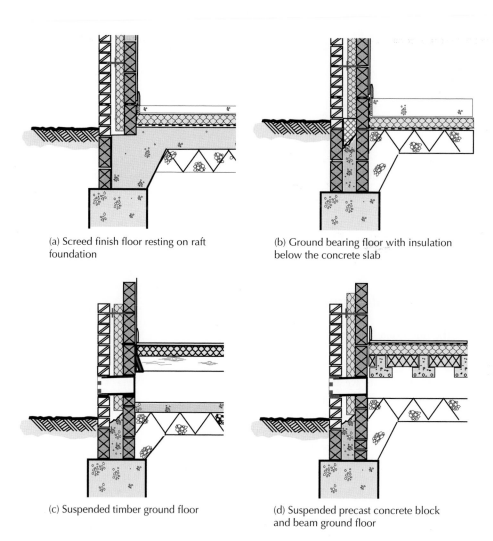

(a) Screed finish floor resting on raft foundation

(b) Ground bearing floor with insulation below the concrete slab

(c) Suspended timber ground floor

(d) Suspended precast concrete block and beam ground floor

Figure 4.1   Ground floor types.

and spacing for softwood timber floor joists of two strength classes, for various dead loads and spans.

Stability

A floor is designed and constructed to serve as a horizontal surface to support people and their furniture, equipment or machinery. The floor should have adequate stiffness to remain stable and horizontal under the dead load of the floor structure and such partitions and other fixtures it supports and the anticipated static and live loads it is designed to support. The floor structure should also support and accommodate services either in its depth, or below or above the floor, without affecting its stability.

Solid ground and basement floors are often built off the ground from which they derive support. The stability of such floors depends, therefore, on the characteristics of the concrete under them. Upper or suspended floors are supported by walls or beams and should have adequate stiffness to minimise deflection under load. Under load a floor will deflect and bend, and this deflection or bending should be limited to avoid cracking of rigid finishes such as plasterboards, which are attached to the ceiling directly below the floor.

Resistance to weather and ground moisture

The requirements of the Building Regulations for the resistance of the passage of moisture through ground floors to the inside of buildings are described as follows.

Durability and freedom from maintenance

All floors should be durable for the expected life of the building and require little maintenance or repair. The durability and freedom from maintenance of floor finishes will depend on the nature of the materials used and the wear to which they are subject.

Fire safety

Suspended upper floors should be so constructed as to provide resistance to fire for a period adequate for the escape of the occupants from the building. The notional periods of resistance to fire, from ½ to 4 hours, depending on the size and use of the building, are set out in the Building Regulations.

Resistance to the passage of heat

A floor should provide resistance to transfer of heat where there is normally a significant air temperature difference on the opposite sides of the floor. This would include any building which was heated but would not include some external buildings, such as garages.

Resistance to the passage of sound

Upper floors that separate dwellings (party floors), or separate noisy from quiet activities, should act as a barrier to the transmission of sound. The comparatively low mass of a timber floor will transmit airborne sound more readily than a high mass concrete floor, so that party floors between dwellings, for example, are generally constructed of concrete. The hard surfaces of the floor and ceiling of both timber and concrete floors will not appreciably absorb airborne sound, which will be reflected. The sound absorption of a floor can be improved by introducing a soft absorbent material such as carpet or felt. Absorbent tiles and finishes can also be used on walls and ceilings to reduce echo effects caused by sound reflecting and bouncing back off hard surfaces.

When upgrading existing buildings, sound insulation can be improved by filling between timber joists and sealing voids, or by constructing a floating floor, or suspended floor, both of which introduce breaks between the floor surface and the structural floor. Where cavities exist between the floor surface and the structural floor, sound insulating materials can be introduced to add further sound reductions. Floating floors prevents the impact sound from foot traffic transferring to the main structure. Specially designed acoustic-based resilient floor coverings that reduce impact sound are also available.

4.2 Ground supported concrete slab

A typical concrete ground supported slab is shown in Figure 4.2. It is general practice first to build the external and internal loadbearing walls from the concrete foundation up to the level of the damp-proof course (dpc). The hardcore bed and the concrete slab are then spread and levelled within the area created by the walls. If the hardcore is spread and consolidated over the whole area of the ground floor, into and around excavations used to construct foundations and where soft ground has been removed, there should be very little settlement of the ground that supports the floor slab. The hardcore should be thoroughly consolidated using a vibrating plate or roller. Settlement cracking in ground floor slabs is often due to inadequate hardcore bed, poor filling of excavation for trenches or ground movement due to moisture changes (swelling or shrinkage of clay soils). Where appreciable settlement is anticipated it is best to reinforce the slab and build it into walls as a suspended reinforced concrete slab. The gap below the floor will allow clay soils to swell and contract without causing damage to the structural floor.

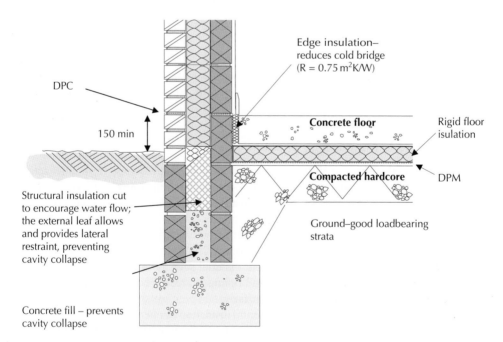

Figure 4.2 Concrete ground bearing floor construction with structural edge insulation in the cavity to prevent cold bridge.

Hardcore

Hardcore is the name given to the infill of materials such as crushed and graded bricks, stone or concrete, which are hard and do not readily absorb water or deteriorate. Hardcore should be spread until it is level, then compacted using a mechanical vibrating plate or roller. The hardcore provides a stable bed for the concrete slab; this hardcore bed is usually 100–300 mm deep. The materials used for hardcore should be chemically inert, not appreciably affected by water and should be free from water-soluble sulphates. A method of testing materials for soluble sulphate is described in *Building Research Station (BRS) Digest 174.* The materials used for hardcore are:

❏ Brick or tile rubble. Clean, hard broken brick or tile crushed and graded. Bricks should be free of plaster and wood. On wet sites the bricks should not contain appreciable amounts of soluble sulphate.
❏ Concrete rubble. Clean, broken, well-graded concrete is a good material for hardcore. The concrete should be free from plaster, wood or other building materials.
❏ Gravel and crushed hard rock. Clean, well-graded gravel and crushed hard rock are both excellent but somewhat expensive materials for hardcore.
❏ Chalk. Broken chalk is a good material for hardcore providing it is protected from expansion due to frost. Once the site concrete is laid it is unlikely to be affected by frost.
❏ Road planings. When roads are resurfaced the top of the road is planed off. The planings provide a very good strong hardcore, which binds together when compacted.

Blinding layer

It was common for hardcore to have a blinding layer of dry concrete or sand, 50 mm deep, placed on it before the concrete was laid. The purpose of this was to prevent the wet concrete running down between the lumps of broken brick or stone, and to form a smooth bed for the damp-proof membrane (dpm). Now that hardcores are well graded with a mixture of fine and course material, when the hardcore is adequately compacted the surface finish is relatively smooth and level; thus a blinding layer is unnecessary. Where a reinforced cage is to be used within the concrete slab it is useful to have a level surface from which to work. Thus blinding may be used so that the reinforcement cage can be easily constructed on the level surface, and properly spaced and positioned off the concrete blinding.

Damp-proof membrane

A requirement of the Building Regulations is that floors shall adequately resist the passage of moisture to the inside of the building. Concrete is permeable to moisture; therefore, it is necessary to use a dpm under, in or on top of ground supported concrete floor slabs as an effective barrier to moisture rising from the ground (Figure 4.3). The membrane should be continuous with the dpc in walls to prevent moisture rising between the edges of the concrete slab and walls. A dpm should be impermeable to water in either liquid or vapour form and be tough enough to withstand possible damage during the laying of screeds,

Without a vapour barrier it is possible that interstitial condensation may form on top of the damp-proof membrane

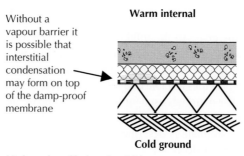

Warm internal

Cold ground

(a) dpm placed below the rigid impermeable insulation

Concrete floor slab

dpm

Insulation

Hardcore

Formation level (ground)

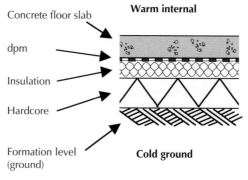

Warm internal

Cold ground

(b) dpm placed above the rigid impermeable insulation

Timber floor finish

Vapour control layer

Rigid insulation

Concrete floor slab

dpm

Hardcore

Formation level (ground)

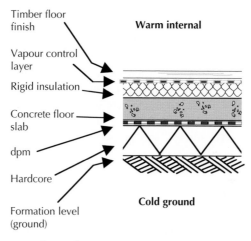

Warm internal

Cold ground

(c) Insulation above concrete with timber finish dpm and vapour control layer required

Figure 4.3 (a, b and c) Sections showing the alternative position of the dpm in solid concrete ground floors.

concrete or floor finishes. The position of the dpm will vary in accordance with the proposed method of construction.

dpm below concrete

The membrane is spread on a blinding layer or directly onto the hardcore if well graded and smooth. The edges of the membrane are turned up the faces of the external and internal walls so that it may unite and overlap the dpc in the wall. The membrane should be spread with some care to ensure that it is not punctured and that the edge upstands are kept in place as the concrete is subsequently laid.

The advantage of a dpm under the concrete is that it will be protected from damage during subsequent building operations. The concrete remains dry and protected from impurities. However, unless the concrete is likely to suffer chemical attack, the dampness caused by groundwater is not a cause for concern and the dpm can just as easily be placed on top of the concrete. When rigid impermeable insulation is used the dpm should be placed on the warm side (internal side) of the insulation, reducing the possibility of interstitial condensation (Figure 4.3a). Although concrete does have a level of natural resistance to the passage of moisture poor workmanship may mean that water vapour can penetrate the structure. Where under floor heating is used the membrane should be under the concrete (Figure 4.3b).

If the insulation is placed on top of the concrete and a timber finish is applied to the surface a vapour control layer must be used (Figure 4.3c). If a vapour barrier is not used, warm moist air will penetrate through the timber finish to the cold side of the insulation and form condensation on top of the dpm. If the insulation is permeable then the dpm must go below the insulation and a vapour barrier may be used on top of the insulation.

Surface dpm

Floor finishes such as pitch mastic and mastic asphalt that are impermeable to water can serve as a combined dpm and floor finish. These floor finishes should be laid to overlap the dpc in the wall to seal the joint between the concrete and the wall. Where hot soft bitumen or coal-tar pitch are used as an adhesive for woodblock floor finishes, the continuous layer of the impervious adhesive can serve as a waterproof membrane.

dpm below a floor screed

An alternative method is to place the dpm between the floor screed and the thermal insulation, as illustrated in Figure 4.4. At the junction of wall and floor the membrane should overlap the dpc in the wall. An advantage of positioning the dpm above the insulation is that it can be used to secure the upstand edge insulation in place while concrete is being placed. If the dpm is laid below the insulation it is necessary to spread a separating layer over the insulation to prevent wet screed running into the joints between the insulation boards. The separating layer should be building paper or 500 gauge polythene sheet. To avoid damage to the insulation layer and the dpm it is necessary to take care in tipping, spreading and compacting wet concrete or screed. Scaffold boards should be used for barrowing and tipping concrete.

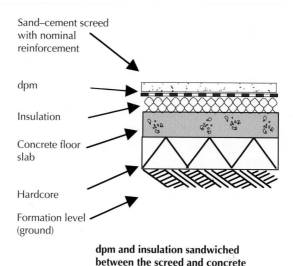

Sand–cement screed with nominal reinforcement

dpm

Insulation

Concrete floor slab

Hardcore

Formation level (ground)

dpm and insulation sandwiched between the screed and concrete

Figure 4.4   Position of dpm in solid ground floors.

Materials used for the dpm

The materials used as dpm must be impermeable to water both in liquid and vapour form and also be sufficiently robust to withstand accidental (puncture) damage to the membrane during building operations.

Polythene and polyethylene sheet

Polythene or polyethylene sheet is commonly used as a dpm with oversite concrete for all but severe conditions of dampness. These sheets should be chemically inert and not be affected by alkalis and acids present in the subsoil. It is recommended that the sheet should be at least 0.25 mm (250 μm) thick (1200 gauge). Sheets are also available in 300 and 500 μm thicknesses and are supplied in rolls 4 m wide by 25 m long. The sheets are spread over the blinding and lapped 150 mm at joints and continued across surrounding walls, under the dpc for the thickness of the wall. Where site conditions are reasonably dry and clean, the overlap joints between the sheets are sealed with mastic or mastic tape between the overlapping sheets and the joint completed with a polythene jointing tape as illustrated in Figure 4.5.

For this lapped joint to be successful the sheets must be dry and clean or the jointing tape will not adhere to the surface of the sheets. Where site conditions are too wet to use mastic and tape, the joint is made by welting the overlapping sheets with a double welted fold as illustrated in Figure 4.6, and this fold is kept in place by weighing it down with bricks or securing it with tape until the screed or concrete has been placed. The sheet should be used so that there are only joints one way as it is impractical to form a welt at junctions of joints.

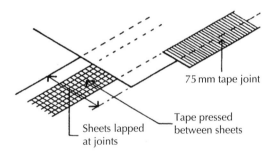

Figure 4.5   Jointing laps in polythene sheet.

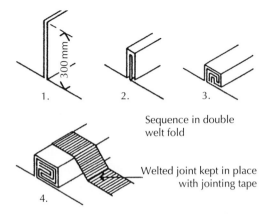

Figure 4.6   Double welted fold joint in polythene sheet.

Where the level of the dpm is below that of the dpc in walls, it is necessary to turn it up against walls so that it can overlap the dpc or be turned over as dpc as illustrated in Figure 4.7.

To keep the sheet in place as an upstand to walls it is necessary to keep it in place with bricks or blocks laid on the sheet against walls until the concrete has been placed and the bricks or blocks removed as the concrete is run up the wall. At the internal angle of walls a cut is made in the upstand sheet to facilitate making an overlap of sheet at corners.

Penetrations for service pipes will need to be sealed using proprietary Top Hat units and sealed using double-sided sealing tape in accordance with the manufacturer's instructions. Manufacturers also provide guidance on repairing accidental punctures to the dpm. All membranes must be covered (e.g. with a screed) as soon as possible after laying to prevent accidental damage and to protect the membrane from the damaging effect of sunlight.

Recycled content membranes
Increasingly the manufacturers of damp-proofing materials are offering products manufactured entirely from post-use waste. Reprocessed and recycled polyethylene dpms are

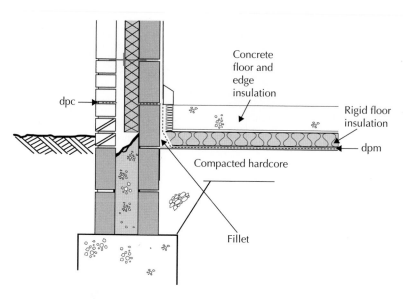

Figure 4.7   **Dpm turn-up.**

available that comply with BS EN 13967:2004 'Flexible sheets for waterproofing' and have similar or better performance compared with traditional dpms.

Hot pitch or bitumen

A continuous layer of hot applied coal-tar pitch or soft bitumen is poured on the surface and spread to a thickness of not less than 3 mm. In dry weather a concrete binding layer is ready for the membrane 3 days after placing. The surface of the concrete should be brushed to remove dust and primed with a solution of coal-tar pitch or bitumen solution or emulsion. Properly applied pitch or bitumen layers serve as an effective dpm both horizontally and spread up inside wall faces to unite with dpcs in walls.

Bitumen solution, bitumen/rubber emulsion or tar/rubber emulsion

These cold applied solutions are brushed on to the surface of concrete in two or three coats to a finished thickness of not less than 2.5 mm, allowing each coat to dry and harden before the next is applied. Bitumen-based solvent-free emulsion liquid dpms are suitable for horizontal and vertical application given the material's adhesive, elastic and waterproofing properties.

Bitumen sheet

Sheets of bitumen with hessian, fibre or mineral fibre base are spread on the concrete oversite or on a blinding of stiff concrete below the concrete, in a single layer with the joints between adjacent sheets lapped 75 mm. The joints are then sealed with a gas torch, which melts the bitumen in the overlap of the sheets sufficient to bond them together. Alternatively the lap is made with hot bitumen spread between the overlap of the sheets, which are then pressed together to make a damp-proof joint. The bonded sheets may be carried across adjacent walls as a dpc, or up against the walls and then across as dpc where the membrane and dpc are at different levels.

The polythene or polyester film and self-adhesive rubber/bitumen compound sheets, described in Chapter 3 of *Barry's Advanced Construction of Buildings* under 'Tanking', can also be used as dpms, with the purpose cut, shaped cloaks and gussets for upstand edges and angles. This type of membrane is particularly useful where the membrane is below the level of the dpc in walls. Bitumen sheets, which may be damaged on building sites, should be covered by the screed or concrete slab as soon as possible to avoid damage.

Mastic asphalt or pitch mastic

These materials are spread hot and finished to a thickness of at least 12.5 mm. This expensive dpm is used where there is appreciable water pressure under the floor and as 'tanking' to basements as described in *Barry's Advanced Construction of Buildings*.

4.3 Suspended concrete floor slabs

Suspended concrete slabs or block and beam floors are used where the ground:

- ❏ Slopes
- ❏ Has poor or uncertain bearing capacity
- ❏ Is liable to volume change (swells and shrinks)

In such situations it may be wise to form the ground floor as a suspended reinforced concrete slab or to use a block and beam floor, supported by external and internal loadbearing walls, which are independent of the ground. Suspended concrete floors can be constructed using:

- ❏ Precast reinforced concrete planks or slabs
- ❏ Block and beam floor systems or
- ❏ In situ reinforced concrete slabs

All of these are described later for upper floors.

T-beam method

It is common practice to construct raised concrete floors using the concrete block and inverted T-beam method (Photograph 4.1; also see Section 4.6, 'Reinforced Concrete Upper Floors'). Brick or concrete block sleeper walls are built off the ground slab to support the concrete 'T'-beams. The precast inverted T-beams are located on the perimeter walls and internal sleeper walls; concrete infill blocks are then inserted between the beams. Spacing between sleeper walls is dictated by an economical span for the inverted T-beams. A concrete topping or screed is spread and levelled over the precast concrete units. Where there is a likelihood of an accumulation of gas building up in the space below the floor, the space should be at least 150 mm clear and cross-ventilation should be provided.

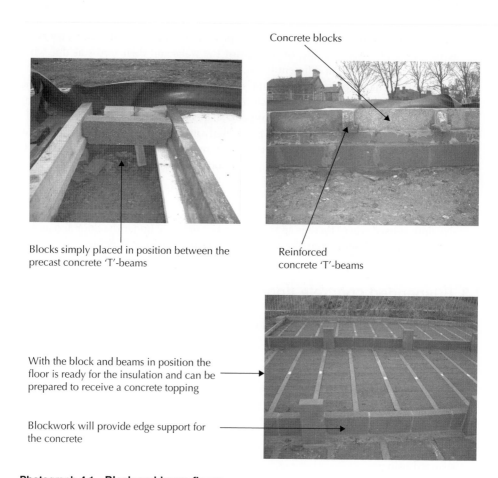

Concrete blocks

Blocks simply placed in position between the precast concrete 'T'-beams

Reinforced concrete 'T'-beams

With the block and beams in position the floor is ready for the insulation and can be prepared to receive a concrete topping

Blockwork will provide edge support for the concrete

Photograph 4.1   Block and beam floors.

Solid reinforced concrete beams usually shaped like an inverted T in section are precast to the required length. The depth of the beams is from 130 to 250 mm. The beams are made in lengths of up to 6 m. The T-beams are reinforced with mild steel reinforcing bars to provide adequate support for the deadweight of the floor and anticipated dead and live loads.

Precast lightweight concrete infill blocks are made to fit between and bear on the T-beams. Some of the blocks are hollow (for lightness), although it is now more common to use solid blocks. It is possible to use rigid insulation between the beams instead of concrete blocks (Figure 4.8 and Photograph 4.2).

Floor surface

The term floor finish is generally used to describe the material or materials that are applied to a floor surface as a finished surface, such as tiles (see Chapter 10). It is important to

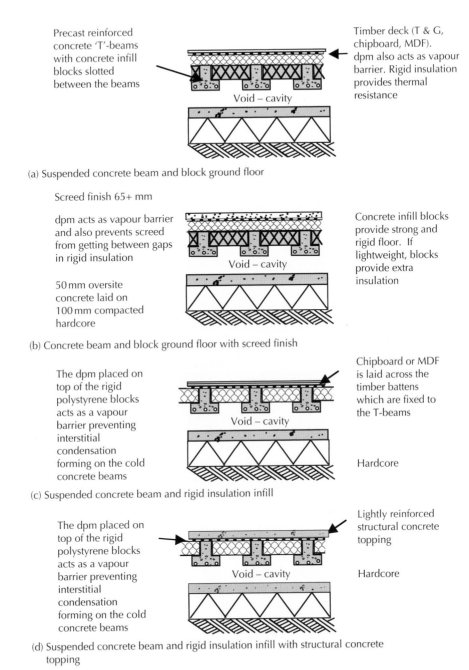

Precast reinforced concrete 'T'-beams with concrete infill blocks slotted between the beams

Timber deck (T & G, chipboard, MDF). dpm also acts as vapour barrier. Rigid insulation provides thermal resistance

Void – cavity

(a) Suspended concrete beam and block ground floor

Screed finish 65+ mm

dpm acts as vapour barrier and also prevents screed from getting between gaps in rigid insulation

Concrete infill blocks provide strong and rigid floor. If lightweight, blocks provide extra insulation

Void – cavity

50 mm oversite concrete laid on 100 mm compacted hardcore

(b) Concrete beam and block ground floor with screed finish

The dpm placed on top of the rigid polystyrene blocks acts as a vapour barrier preventing interstitial condensation forming on the cold concrete beams

Chipboard or MDF is laid across the timber battens which are fixed to the T-beams

Void – cavity

Hardcore

(c) Suspended concrete beam and rigid insulation infill

The dpm placed on top of the rigid polystyrene blocks acts as a vapour barrier preventing interstitial condensation forming on the cold concrete beams

Lightly reinforced structural concrete topping

Void – cavity

Hardcore

(d) Suspended concrete beam and rigid insulation infill with structural concrete topping

Figure 4.8 Alternative arrangements for block and beam floors.

Precast concrete beams rest on internal blockwork skin

Rigid insulation infill panel is supported by the concrete beams

The insulation panel is specially designed so that it laps under the concrete beam preventing a cold bridge

Once the concrete 'T'-beams are in position the lightweight insulation panels are quickly positioned

Once the concrete beams and insulation panels are in position a layer of concrete can then be applied over the whole floor

Photograph 4.2   Precast concrete 'T'-beams and insulation infill panels.

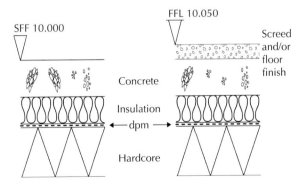

Figure 4.9   Structural and finished floor levels.

distinguish clearly between the level of the structural floor finish (SFF) and the finished floor level (FFL) to avoid confusion on site (Figure 4.9).

For sheds, workshops, stores and garages, the SFF of the concrete is sometimes used as the finished floor surface to save the cost of an applied floor finish. Unless finished and sealed correctly, concrete floors have poor resistance to wear and in a short time the surface of the concrete 'dusts'. Being a coarse-grained material, concrete cannot be washed clean, and if it becomes stained the stains are permanent.

Extensive areas of concrete floor may be levelled and finished by power floating. Concrete floors provide a satisfactory base for the thicker floor finishes such as mastic asphalt, tiles and woodblocks. For the thin finishes such as plastic, linoleum, rubber sheet and tile, the more precisely level, smooth surface of a screeded base is necessary. However, with the increasing precision of concrete laying machines that are guided and levelled by lasers, floors can be produced that are level within a tolerance of $\pm 1.5\,\text{mm}$. The use of concrete laying machines is restricted by access through the building to the floor area and obstructions within the structure.

Floor screeds

The purpose of a floor screed is to provide a level surface to which a floor finish can be applied. The word screed is used to describe the wet sand–cement mix that is first laid across the length and width of the floor. If screed finishes are to be used over large areas it may be necessary to lay the screed in strips or bays. The screed strips or bays are carefully levelled in both directions to set out a precise level finish. The main bulk of the mix (which is semi-dry) is then spread and levelled between the screeds.

Generally screed mixes have just sufficient water for hydration; the screed mixes do not have the wet workable properties of concrete typically used in floor construction. Too much water will cause the screen to slump. The screed has a composition such that when it is formed to the level required it remains in position. This is not the same for self-levelling screeds or compounds which are much more fluid in composition and are poured in thin layers onto the floor, lightly floated and then left to level out.

The thickness of the screed and the mix of cement and sand depend on the surface on which the screed is laid. The cement-rich mix used in a screed will shrink as it dries; thus the thinner the screed, the more rapidly it will dry and the more it will shrink and crack.

On the majority of building sites the concrete ground and upper floors are cast and roughly levelled as a working platform for subsequent building operations. To avoid damage to screeded surfaces that will serve as a finished floor surface or as a level base or substrate to applied floor finishes it is usual to lay a screed after the concrete floor has dried and hardened.

Bonded screeds

Where it is practical to lay a screed on a concrete base within 3 hours of placing the concrete it will bond strongly to the concrete. The screed will also dry slowly with the concrete so that drying shrinkage and cracking of the screed relative to that of the concrete will be minimised. For this monolithic construction of screed a thickness of 12 mm of screed will suffice (Figure 4.10).

Semi-bonded screeds

A screed laid on a concrete base that has set and hardened should be at least 40 mm thick. To provide a good bond between the screed and the concrete, the surface of the concrete should be hacked by mechanical means, cleaned and dampened and then covered by a thin grout, or wet mix, of water and cement before the screed is laid. With a good bond to the concrete base a separate screed at least 40 mm thick will dry sufficiently slowly to avoid serious shrinkage cracking (Figure 4.11).

Independent or unbonded screeds

Where a screed is laid on an impermeable dpm there will be no bond between the screed and the concrete base so that drying shrinkage of the screed is unrestrained. So that the screed does not dry too rapidly and suffer shrinkage cracking, the screed in this unbonded construction should be at least 50 mm thick, reinforced with light mesh (Figure 4.12).

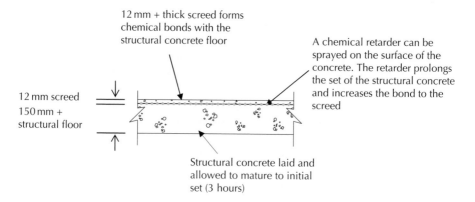

12 mm + thick screed forms chemical bonds with the structural concrete floor

A chemical retarder can be sprayed on the surface of the concrete. The retarder prolongs the set of the structural concrete and increases the bond to the screed

12 mm screed
150 mm +
structural floor

Structural concrete laid and allowed to mature to initial set (3 hours)

Figure 4.10   Bonded screed.

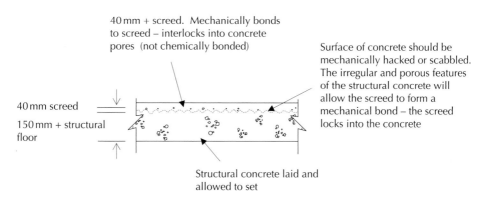

40 mm + screed. Mechanically bonds
to screed – interlocks into concrete
pores (not chemically bonded)

Surface of concrete should be
mechanically hacked or scabbled.
The irregular and porous features
of the structural concrete will
allow the screed to form a
mechanical bond – the screed
locks into the concrete

40 mm screed

150 mm + structural
floor

Structural concrete laid and
allowed to set

Figure 4.11   Semi-bonded screed.

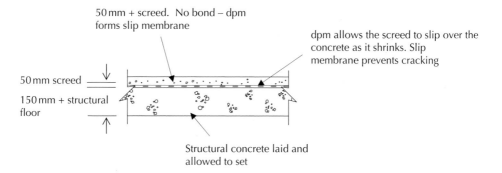

50 mm + screed. No bond – dpm
forms slip membrane

dpm allows the screed to slip over the
concrete as it shrinks. Slip
membrane prevents cracking

50 mm screed

150 mm + structural
floor

Structural concrete laid and
allowed to set

Figure 4.12   Unbonded screed.

Floating floor – screed

A screed laid on a layer of compressible thermal or sound insulating material should be at least 65 mm thick for domestic buildings and 75 mm for other buildings, if this floating construction is not to crack due to drying shrinkage and the deflection under loads on the floor (Figure 4.13).

For screeds up to 40 mm thick, a mix of cement and clean sand in the proportions by weight of 1 : 3 to 1 : 4½ is used. The lower the proportion of cement to sand, the less the drying shrinkage. For screeds over 40 mm thick a mix of fine concrete is often used in the proportions of 1 : 1½ : 3 of cement, fine aggregate and coarse aggregate with a maximum of 10 mm for the coarse aggregate.

Screed should be mixed with just sufficient water for workability. The material is spread over the surface of the base to the required level and then it is finished with a plastic or steel float. The screed should be cured, that is, allowed to dry out slowly over the course of several days. Curing prevents the water needed for hydration from evaporating out of the screed. The water within the wet concrete mix can be held in place by covering the

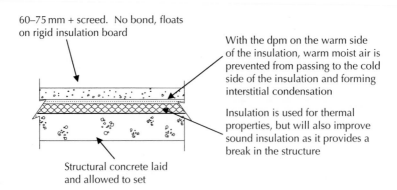

60–75 mm + screed. No bond, floats on rigid insulation board

With the dpm on the warm side of the insulation, warm moist air is prevented from passing to the cold side of the insulation and forming interstitial condensation

Insulation is used for thermal properties, but will also improve sound insulation as it provides a break in the structure

Structural concrete laid and allowed to set

Figure 4.13   Floating floor – screed.

screed with polythene sheeting, damp hessian or a liquid chemical curing agent (which is sprayed on to the surface of the screed). Premixed cement screed materials are available and can be delivered to site with polymer or steel fibre for reinforcement.

4.4 Suspended timber ground floors

A suspended (or raised) timber ground floor is constructed as a timber platform of boards nailed across timber joists bearing on ½ brick (B) thick sleeper walls. The raised timber floor is formed inside the external walls and internal walls, as illustrated in Figure 4.14.

Sleeper walls

Sleeper walls are ½ B thick and built directly off the site concrete up to 1.8 m apart (Figure 4.14). These sleeper walls are generally built at least three courses of bricks high and sometimes as much as 600 mm high. The walls are built honeycombed to allow free circulation of air below the floor, the holes in the wall being ½ B wide by 65 mm deep, as illustrated in Figure 4.15. The guidance in Approved Document C requires a space of at least 75 mm from the top of the concrete to the underside of a wall plate and at least 150 mm to the underside of the floor joists.

Ventilation

The space below the floor is usually ventilated by inserting air bricks, made of terracotta or plastic, in the walls below the floor so that air from outside the building can circulate at all times under the floor. This is to prevent stagnant air and the possibility of dry rot from developing under the floor. Airbrick dimensions are 215 × 65, 215 × 140 or 215 × 215 mm. These bricks are built into external and internal walls for each floor to provide 1500 mm of ventilation for each metre run of wall. The bricks are built in just above ground level and below the floor, as illustrated in Figure 4.14 and Figure 4.15.

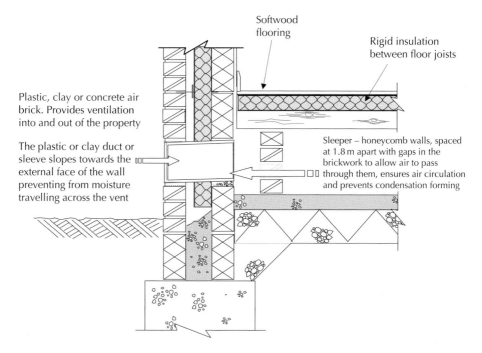

Softwood
flooring

Rigid insulation
between floor joists

Plastic, clay or concrete air
brick. Provides ventilation
into and out of the property

The plastic or clay duct or
sleeve slopes towards the
external face of the wall
preventing from moisture
travelling across the vent

Sleeper – honeycomb walls, spaced
at 1.8 m apart with gaps in the
brickwork to allow air to pass
through them, ensures air circulation
and prevents condensation forming

Figure 4.14   Section through a suspended timber floor.

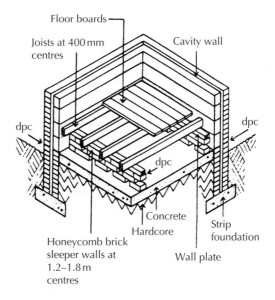

Floor boards

Joists at 400 mm
centres

Cavity wall

dpc

dpc

dpc

Concrete

Hardcore

Strip
foundation

Honeycomb brick
sleeper walls at
1.2–1.8 m
centres

Wall plate

Figure 4.15   Sleeper walls, joists and boards.

Wall plate

A wall plate is a continuous length of softwood timber that rests along the length of a wall. The function of a wall plate for timber joists is twofold. It forms a firm level surface on which the timber floor or roof joists can bear and to which the floor joists can be nailed. The wall plate spreads the point loads from joists uniformly along the length of the wall below. The plate makes it considerably easier to fix floor and ceiling joists to the wall. A wall plate is usually a 100 × 75 mm timber and is laid on its widest face so that there is a 100 mm surface width on which the timber joists bear. A dpc should be laid on top of the sleeper walls under the wall plate to prevent any moisture rising through sleeper walls to the timber floor.

Floor joists

Floor joists are rectangular sections of sawn softwood timber from 38 to 75 mm thick and from 75 to 225 m deep, spaced from 400 to 600 mm apart. The minimum bearing for floor and roof joists is 35 mm (Approved Document A). The span of a joist is the distance measured along its length between walls that support it. The sleeper walls built to support the joists are usually 1.8 m apart or less; thus the span of the joists is 1.8 m or less. The best method of supporting the ends of the joists at external walls and at internal brick partitions is to build a honeycombed sleeper wall some 50 mm away from loadbearing walls to carry the ends of the joists, as illustrated in Figure 4.15. The sleeper wall is built away from the main wall to allow air to circulate through the holes in the honeycomb of the sleeper wall. The ends of the joists are positioned so that they are 50 mm clear of the inside face of the wall.

From a calculation of the dead and imposed loads on the floor the most economical size and spacing of joists can be selected from the tables in Approved Document A and from this the spacing of the sleeper walls to support the joists can be found. Similarly the thickness of the floorboards to be used will determine the spacing of the joists; the thicker the board, the greater the spacing of the joists. Timber floorboards, chipboard, medium-density fibreboard (MDF) or plywood boards are laid across the joists and are then screwed or nailed to them to form a firm, level floor surface. Screwing the boards to the joists tends to be more time consuming compared to nailing, but it allows ease of access, e.g. to electrical wiring and pipes positioned in the floor void.

Floor surface

Floorboards for timber floors are usually 16, 19, 21 or 28 mm thick and 65, 90, 113 or 137 mm wide and up to about 5 m in length. The common way of cutting boards is with a projecting tongue on one edge and a groove on the opposite edge of each board. The tongued and grooved (T & G) boards are laid across the floor joists and cramped together. The boards, as they are cramped up, are screwed or nailed to the joists with two screws or nails to each board bearing on each joist. The joint between the end of one board and the end of another is described as the heading joint. The heading joints in floorboards should always be staggered in some regular manner. Obviously the heading joint ends of boards must be cut so that the ends of both boards rest on a joist to which the ends are nailed. A usual method of staggering heading joints is illustrated in Figure 4.16. T & G boards

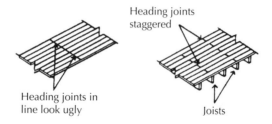

Figure 4.16   Heading joints.

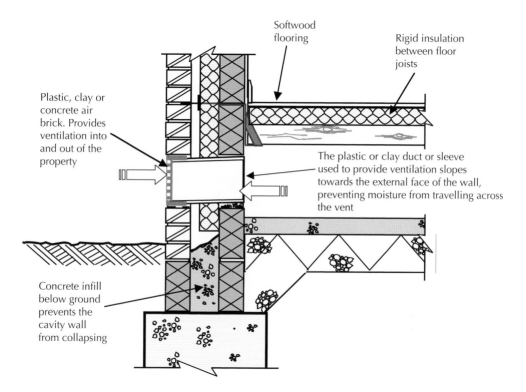

Figure 4.17   Section through wall and floor junction, with airbrick and under-floor ventilation.

provide an attractive floor finish, which can be finished with stain or varnish to improve durability and appearance.

The recommendation in Approved Document A is that T & G boards screwed or nailed to joists spaced at up to 500 mm should be at least 16 mm finished thickness and at wider spacing up to 600 mm, 19 mm finished thickness. Boards of compressed wood chips, chipboards, are commonly used today as a substitute for T & G boards. The use of large T & G chipboard minimises joints and allows for more rapid construction. The boards are nailed or screwed to the timber joists. These boards will need to be covered with an applied floor finish, such as carpet or vinyl sheet, to provide an attractive finish (Figure 4.17).

Thermal insulation

To meet the requirements of the Building Regulations for resistance to heat transfer through ground floors it may be necessary to insulate suspended timber ground floors. The most practical way of insulating a suspended timber ground floor is to fix mineral wool roll, mat or quilt or semi-rigid slabs between the joists. Rolls or quilt of loosely felted glass fibre or mineral wool are supported by a mesh of plastic that is draped over the joists and stapled in position to support the insulation. Semi-rigid slabs or batts of fibreglass or mineral wool are supported between the joists by nails or battens of wood nailed to the sides of the joists.

4.5 Resistance to the passage of heat

Approved Document L includes provision for the insulation of ground floors. To reduce heat losses through thermal bridges around the edges of solid floors and so minimise problems of condensation and mould growth, edge insulation should be used, particularly where the wall insulation is not carried down below the ground floor slab. Edge insulation is formed as a vertical strip between the edge of the slab and the wall as illustrated in Figure 4.18. The depth (width) of the strips of edge insulation is selected so that the whole edge of the concrete floor and screed is insulated.

Achieving insulation values in different types of floor construction

The only practical way of improving the insulation of a solid ground floor to the required U-value is to add a layer of rigid insulation board with a high thermal resistance to the

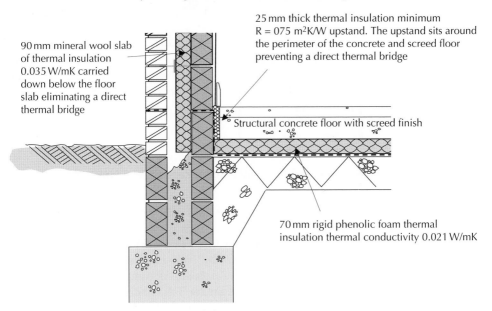

90 mm mineral wool slab of thermal insulation 0.035 W/mK carried down below the floor slab eliminating a direct thermal bridge

25 mm thick thermal insulation minimum R = 075 m²K/W upstand. The upstand sits around the perimeter of the concrete and screed floor preventing a direct thermal bridge

Structural concrete floor with screed finish

70 mm rigid phenolic foam thermal insulation thermal conductivity 0.021 W/mK

Figure 4.18 Edge insulation under structural concrete and cavity insulation carried down to reduce the cold bridge.

floor. The layer of insulation may be laid below a chipboard or plywood panel floor finish or below a timber boarded finish or below the screed finish to a floor or under the concrete floor slab. With insulation under the screed or slab it is important that the density of the insulation board is sufficient to support the load of the floor itself and imposed loads on the floor. A density of at least $16\,kg/m^3$ is recommended for domestic buildings.

The advantage of laying the insulation below the floor slab is that the high-density slab, which warms and cools slowly (slow thermal response) in response to changes in temperature of the constant low-output heating systems, will not lose heat to the ground. With the insulation layer and the dpm below the concrete floor slab it is necessary to continue the dpm and insulation up vertically around the edges of the slab to unite with the dpc in walls as illustrated in Figure 4.19. Figure 4.20 shows insulation arrangements to suit different floor types.

Where the wall insulation is in the cavity, or on the inside face of the wall, it is necessary to avoid a cold bridge across the foundation wall and the edges of the slab. This is achieved by fitting insulation around the edges of the slab or by continuing the insulation down inside the cavity, as illustrated in Figure 4.21.

Materials for under floor insulation

Any material used as an insulation layer to a solid, ground supported floor must be sufficiently strong and rigid to support the weight of the floor or the weight of the screed and floor loads without undue compression and deformation. To meet this requirement one of the rigid board or slab insulants is used. The thickness of the insulation is determined by the nature of the material from which it is made and the construction of the floor, to provide the required U-value.

Some insulants absorb moisture more readily than others and may be affected by ground contaminants. Where the insulation layer is below the concrete floor slab, with the dpm above the insulation, an insulant with low moisture absorption characteristics should be used. The materials commonly used for floor insulation are mineral wool slabs, extruded polystyrene, cellular glass and rigid polyurethane foam boards.

4.6 Reinforced concrete upper floors

Reinforced concrete floors have a better resistance to damage by fire and can safely support greater superimposed loads than timber floors of similar depth. The resistance to fire, required by building regulations for most offices, large blocks of flats, factories and public buildings, is greater than can be obtained with a timber upper floor; therefore, some form of reinforced concrete floor has to be used. The types of reinforced concrete floor that are used for small buildings are self-centring T-beams and infill blocks, hollow beams and monolithic in situ cast floors. The word centring is used to describe the temporary platform on which in situ cast concrete floors are constructed and supported until the concrete has sufficient strength to be self-supporting. The term self-centring is used to define those precast concrete floor systems that require no temporary support.

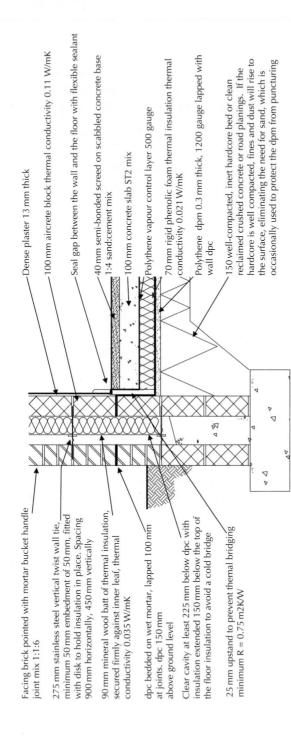

Dense plaster 13 mm thick

100 mm aircrete block thermal conductivity 0.11 W/mK

Seal gap between the wall and the floor with flexible sealant

40 mm semi-bonded screed on scabbled concrete base 1:4 sand:cement mix

100 mm concrete slab ST2 mix

Polythene vapour control layer 500 gauge

70 mm rigid phenolic foam thermal insulation thermal conductivity 0.021 W/mK

Polythene dpm 0.3 mm thick, 1200 gauge lapped with wall dpc

150 well-compacted, inert hardcore bed or clean reclaimed crushed concrete or road planings. If the hardcore is well compacted, fines and dust will rise to the surface, eliminating the need for sand, which is occasionally used to protect the dpm from puncturing

Facing brick pointed with mortar bucket handle joint mix 1:1:6

275 mm stainless steel vertical twist wall tie, minimum 50 mm embedment of 50 mm, fitted with disk to hold insulation in place. Spacing 900 mm horizontally, 450 mm vertically

90 mm mineral wool batt of thermal insulation, secured firmly against inner leaf, thermal conductivity 0.035 W/mK

dpc bedded on wet mortar, lapped 100 mm at joints. dpc 150 mm above ground level

Clear cavity at least 225 mm below dpc with insulation extended 150 mm below the top of the floor insulation to avoid a cold bridge

25 mm upstand to prevent thermal bridging minimum R = 0.75 m2K/W

Figure 4.19   Working detail: edge insulation used to reduce cold bridge, dpm carried up the wall and overlapped with internal dpc (courtesy of J. Bradley).

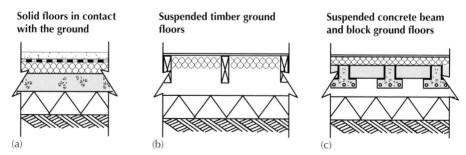

Solid floors in contact with the ground

Suspended timber ground floors

Suspended concrete beam and block ground floors

(a)   (b)   (c)

Note: when insulation is placed between floor joists, the thermal conductivity of the joists must be taken into account in the U-value calculations.

Figure 4.20   (a, b and c) Insulation within different types of floor construction.

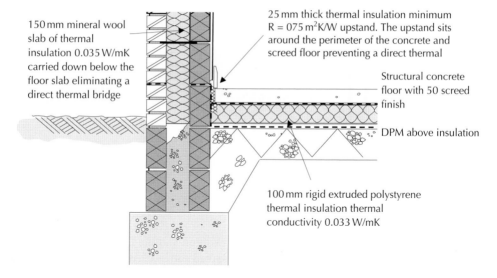

150 mm mineral wool slab of thermal insulation 0.035 W/mK carried down below the floor slab eliminating a direct thermal bridge

25 mm thick thermal insulation minimum R = 075 m²K/W upstand. The upstand sits around the perimeter of the concrete and screed floor preventing a direct thermal

Structural concrete floor with 50 screed finish

DPM above insulation

100 mm rigid extruded polystyrene thermal insulation thermal conductivity 0.033 W/mK

Figure 4.21   dpm under insulation and screed.

Precast 'T'-beam and infill block floor

This type of reinforced concrete floor is used for comparatively small spans and loads (previously described in Section 4.3). The advantage of this system is that two workers can safely handle the units without the need for lifting gear. For floors that need a greater bearing capacity, the block and beam floors can be finished with a concrete topping. The structural topping ties the blocks and beams together, making a composite floor. The T-beams are placed at 270 mm centres with their ends built into walls or bearing on beams

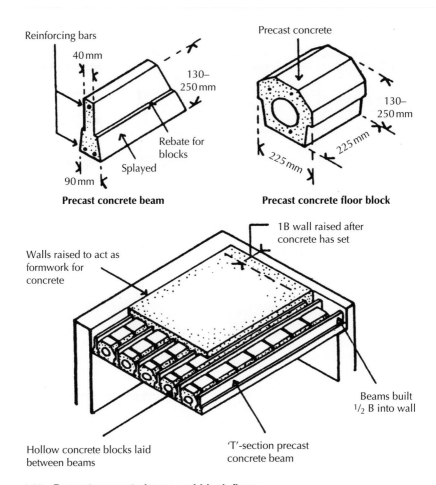

Figure 4.22 Precast concrete beam and block floor.

of at least 90 mm. The blocks are placed in position and the floor completed with a layer of structural concrete topping, 50 mm thick spread and levelled ready for a screed or power floated finish as illustrated in Figure 4.22. The purpose of the constructional concrete topping is to spread the loads on the floor over the blocks and beams. The underside, or soffit, of the floor is covered with plaster or will provide support for a suspended ceiling. This comparatively cheap floor system provides reasonable resistance to airborne sound and resistance to fire.

Hollow beam floor units

Hollow, reinforced concrete beams are precast around inflatable formers to produce the hollow cross section. The beams are rectangular in section with the steel reinforcement cast in the lower angles of the beam. The top of the beams are indented to provide a key for the concrete topping, as illustrated in Figure 4.23 and Photograph 4.3.

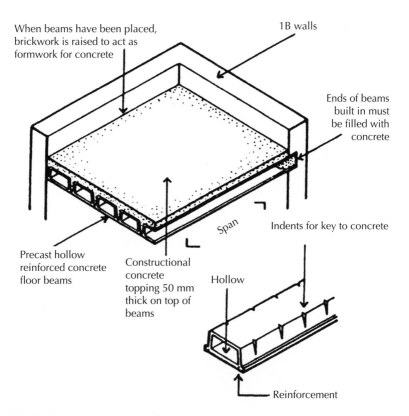

Figure 4.23   Hollow concrete beam floor.

The beams are usually 355 mm wide, from 130 to 205 mm deep and up to 6 m long. The depth of the beam depends on the superimposed loads and the span. Because of their length and weight, lifting gear is necessary to raise and lower the beams into place, as illustrated in Photograph 4.3b. The beams are placed side by side with their ends bearing ½ B on or into brick loadbearing walls or on to steel beams. If the ends of the beams are built into walls the ends should be solidly filled with concrete because a hollow beam is not strong enough to bear the weight of heavy brickwork. The walls of the beams are made thin so that they are light in weight for transporting and hoisting into position. The thin walls of the beams are not strong enough to carry the direct weight of furniture. So that point loads are transferred and distributed, a layer of concrete usually 50 mm thick is levelled over the beams. The concrete is termed 'constructional concrete topping'; it is an integral part of this floor system, spread and levelled on top of the beams.

Reinforced concrete and clay block floor

The resistance to damage by fire of a reinforced concrete floor depends on the protection, or cover, of concrete underneath the steel reinforcement. Under the action of heat, concrete

(a)   Concrete floor beams

(b)   Each beam is hoisted into position and placed on the
supporting walls or beams

Photograph 4.3   Hollow concrete floor beams.

is liable to expand and come away from its reinforcement. If, instead of concrete, pieces of
burnt clay tile are cast into the floor beneath the reinforcing bars, the floor has a better
resistance to fire than it would have with a similar thickness of concrete.

The particular advantage of this type of floor is its good resistance to damage by fire,
and it is sometimes termed 'fire-resisting reinforced concrete floor'. To keep the deadweight
of the floor as low as possible, compatible with strength, it is constructed of in situ reinforced concrete beams with hollow TC infilling blocks cast in between the beams. A typical
TC block is shown in Figure 4.24. This type of floor has to be given temporary support
with timber or steel temporary support (centring). The TC blocks and the reinforcement

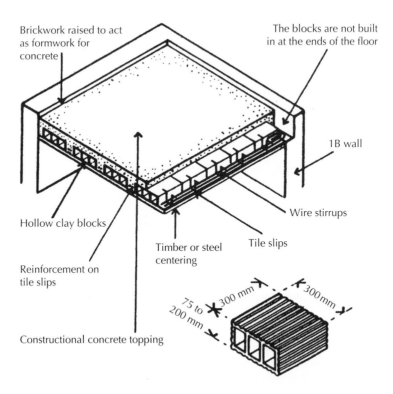

Figure 4.24   Terracotta block floor.

are set out on the temporary support, and pieces (slips) of clay tile are placed underneath the reinforcing bars. Concrete is then placed and compacted between the TC blocks and spread 50 mm thick over the top of the blocks.

The floor is built into walls ½ B thick as shown. This type of floor can span up to 5 m and the depth of the blocks, the depth of the finished floor, and the size and number of reinforcing bars depend on the superimposed loads and span. This type of floor is rarely used in developed countries because it is labour intensive. Considerable labour is involved in placing the hollow TC pots, reinforcement and temporary support. The flooring system is suited to those countries where hollow clay blocks are extensively used for infill walls to reinforced concrete frame buildings.

Monolithic reinforced concrete floor

A monolithic reinforced concrete floor is one unbroken solid mass, between 100 and 300 mm thick, of in situ reinforced concrete. To support the concrete while it is still wet and plastic, and for 7 days after it has been placed, temporary support (formwork or centring) has to be used. This takes the form of rough timber marine plywood boarding or steel sheets, supported on timber or steel beams and posts. The steel reinforcement is laid out on top of the support and raised 20 mm or more above the formwork by means of small

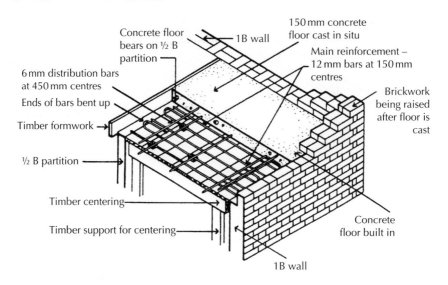

Figure 4.25   Monolithic reinforced concrete floor.

concrete blocks, wire chairs or plastic spacers (called spacers), which are tied to the reinforcing bars with wire. The wet concrete is then placed and spread over the reinforcement and the formwork. It is compacted and levelled off. Figure 4.25 illustrates a single monolithic concrete floor with part of the concrete taken away to show reinforcement and timber centring.

Steel reinforcing bars are cast into the underside of the floor with 20 mm or more concrete cover below them to prevent the steel rusting and to give it protection in case of fire. The thicker the concrete cover to reinforcement, the greater the resistance of the floor to fire. Structural floors are reinforced with a combination of high tensile steel and mild steel reinforcement. The high tensile steel is usually positioned at the bottom of the concrete floor to withstand any tensile forces.

Floors are usually designed to span in one direction, although they can span in two directions, especially if supported by walls on all sides. From the loads and the span the required thickness of concrete can be determined and then the cross-sectional area of steel reinforcement determined. The main reinforcement usually consists of 12 mm diameter high tensile steel rods spaced from 150 to 225 mm apart, and these span across the floor between walls supporting the floor. The diameter will increase as the loads imposed and the span of the floor increases. Mild steel rods, 6 mm in diameter, are wired across the main reinforcement spaced 450–900 mm apart and are called distribution rods or bars. These rods are tied to the main reinforcement with wire and keep the main reinforcing rods correctly spaced while the concrete is being placed. The main purpose of these mild steel rods is to assist in distributing point loads on the floor, ensuring that the forces are uniformly distributed over the mass of the concrete.

Because the formwork required to give temporary support to a monolithic concrete floor tends to obstruct and hence delay building operations, 'self-centring' concrete floors (concrete floors with permanent formwork) are often used for multi-storey buildings. Monolithic concrete floors are used for heavily loaded and specially designed construction and for stairs, ramps and small spans.

Steel 'rib-deck' concrete floors

Profiled cold rolled steel decking is often used as permanent formwork. Profiled steel is placed between supporting walls or beams, reinforcement is positioned and concrete is poured into place. Once the concrete has set the profiled steel formwork becomes mechanically bonded to the concrete forming a matrix floor. The steel formwork also acts as reinforcement to concrete floor. This type of floor construction has become one of the principal floor systems for multi-storey steel-framed buildings, as described in *Barry's Advanced Construction of Buildings*. Once the steel frame is in position the permanent steel rib deck can be quickly tacked to the steel, the reinforcement positioned and the concrete poured. Rib-deck flooring is not usually used with masonry wall construction, timber frames or concrete frames.

Functional requirements specific to concrete floors

Fire safety
The resistance to fire of a reinforced concrete floor depends on the thickness of concrete cover to steel reinforcement, as the expansion of the steel under heat will tend to cause the floor to crack and ultimately give way. Also, if steel is exposed to the direct heat of the fire it will rapidly lose its strength. Approved Document B specifies the minimum concrete cover required between the external face of the concrete and the reinforcement.

Resistance to the passage of heat
A reinforced concrete upper floor that is exposed to outside air and one that separates a heated from an unheated space has to be insulated against excessive transfer of heat by a layer of insulating material. The insulation is usually laid on the top of the floor under a screed or boarded platform floor surface.

Resistance to the passage of sound
The mass of a concrete floor will provide some appreciable resistance to the transfer of airborne sound. Sound energy is absorbed in the mass of dense concrete. Where it is necessary to provide resistance to impact sound a form of floating floor surface may be necessary. Floating floors separate the surface floor from the structure, which reduces the ability of the sound vibrations to travel from the floor finish to the structural materials.

4.7 Timber upper floors

Floor joists

Strength and stability
A timber floor is supported by softwood timber joists, usually between 38 and 75 mm thick and 75–235 mm deep. The required depth of joists depends on the dead and imposed loads

and the span. The spacing of the joists is usually 400, 450 or 600 mm measured from the centre of one joist to the next. Tables in Approved Document A set out the required size of timber joists for given spans and two strength classes, with given spacing of joists for various loads for single-family dwellings of up to three storeys. To economise in the use of timber, the floor joists of upper floors usually span the least width of rooms, from external walls to internal loadbearing walls. The maximum economical span for timber joists is between 3.6 and 4.0 m.

Double floors – spanning timber floor joists between steel beams

Where the span of a timber floor is greater than the lengths of timber commercially available, or where such lengths become uneconomical, it is convenient to use a steel or timber beam to provide intermediate support for timber joists. This combination of beam and the joists is described as a double floor. Steel beams are generally used because of their small section.

The supporting steel beam may be fixed under the joists or wholly or partly hidden in the depth of the floor. To provide a fixing for the ends of the joists, timber plates are bolted to the bottom flange of the beam and the ends of the joists are scribed (shaped) to fit into the joist over the plates to which they are nailed. To provide a fixing for floorboards timber bearers are nailed to the sides of joists across the supporting steel beam, as illustrated in Figure 4.26. The ends of the supporting steel beam are built into loadbearing walls and bedded on a pad stone to spread the load along the wall. Where the supporting steel beam projects below the ceiling it is cased in plasterboard, which will also act as fire protection.

Strutting between joists

To maintain joists in the vertical position in which they were initially fixed, timber strutting is used. Herringbone strutting consists of short lengths of softwood timber about 50 × 38 mm nailed between the joists, as illustrated in Figure 4.27. As the struts are nailed between the joists they tend to spread and secure the joists in an upright position. To

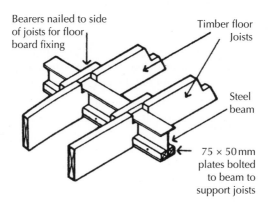

Bearers nailed to side of joists for floor board fixing

Timber floor Joists

Steel beam

75 × 50 mm plates bolted to beam to support joists

Figure 4.26   Double floor.

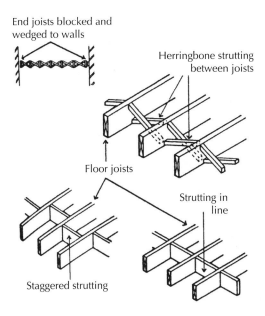

Figure 4.27   Strutting between joists.

provide rigid strutting between walls, wedges are fixed between the joists and walls at both ends of the strutting.

The recommendation in Approved Document A is that joists which span less than 2.5 m do not require strutting; those that span from 2.5 to 4.5 m require one row of struts at mid-span and those with more than 4.5 m span require two rows of struts spaced one-third of the span.

As an alternative to herringbone strutting, solid strutting may be used. This consists of short lengths of timber, of the same section as the joist, which are nailed between the joists either in line or staggered, as in Figure 4.27. This is not usually as effective as the herringbone system, because unless the short solid lengths are cut very accurately to fit the sides of the joists they do not firmly strut between the joists.

End support for floor joists

For stability, the end of floor joists must have adequate support from walls or beams. Timber joists built into the inner skin of a cavity wall must not project into the cavity or across separating or compartment walls where they may encourage the spread of fire. It may be wise to treat the ends of the joists with a preservative against the possibility of decay due to moisture penetration. It is common in such situations for the joists to bear directly on the brick or block wall with plastic spacers under each joist to level the joists (Figure 4.28). As an alternative to building in the ends of timber joists, joist hangers are used. Galvanised, pressed steel joist hangers are made with straps for building into horizontal courses and a stirrup to support a joist end (Figure 4.29 and Photograph 4.4). The joist hangers are built into horizontal brick or blockwork as walls are raised and the joists fitted

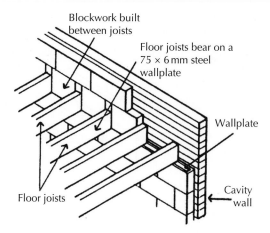

Figure 4.28   Joists ends built into cavity wall.

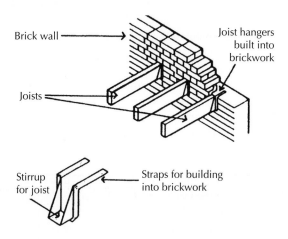

Figure 4.29   Galvanised pressed steel joist hanger.

and levelled later or the joists, with the hangers nailed to their ends, are given temporary support as the brick or blockwork is raised and the hangers are built into horizontal courses. The advantage of joist hangers is that joist ends are not exposed to possible damp, and there is no need for cutting brick or block to fit around joists.

Lateral restraint for walls
To provide lateral support to walls by floors, Approved Document A recommends the use of straps or joist hangers to provide lateral support for walls at each storey floor level above ground to transfer lateral forces on walls, such as wind, to floors.

(a) Galvanised steel joist hangers

(b) Hangers spaced and positioned along the party wall ready to receive joists

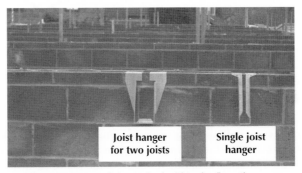

(c) Where extra strength is required within the floor (for example, around stair openings) the joist may be doubled up

(d) Joists fixed into steel joist hanger. Horizontal strutting holds the joists firmly in place

(e) Joist and joist hangers positioned around floor opening

Photograph 4.4   Joist hangers.

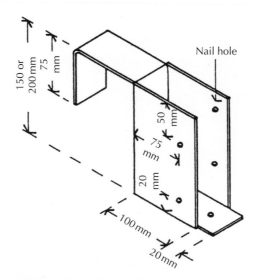

Figure 4.30 Galvanised steel restraint joist hanger.

Lateral support is required to any external compartment or separating wall longer than 3 m at every floor, roof and wall junction and any internal loadbearing wall, not being a compartment or separating wall, of any length at the top of each storey and roof.

Walls should be strapped to floors at intervals of not more than 2 m with 30 × 5 mm straps (see Chapter 5). Straps are not required in the longitudinal direction of joists in houses of not more than two storeys, if the joists are at no more than 1.2 m centres and where the joists are supported by restraint type joist hangers, illustrated in Figure 4.30, at not more than 2 m centres.

Notches and holes

Notches and holes are often cut into timber joists so that electric cables, water and gas pipes can pass through. So that the notches and holes do not seriously weaken the strength of the floor, limitations of size are given as practical guidance by the Building Regulations. Notches should be no deeper than 0.125 times the depth of a joist and cut no closer to the support than 0.07 of the span, nor further away than 0.25 times the span. Holes should be of no greater diameter than 0.25 the depth of joist, drilled on a neutral axis and not less than two diameters apart (centre to centre) and located between 0.25 and 0.4 times the span from the support, as illustrated in Figure 4.31.

Floor surface

The surface of a timber upper floor is the same as that described for a timber ground floor.

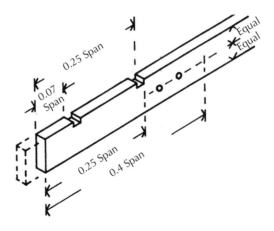

Figure 4.31   Notches and holes in timber joists.

Functional requirements specific to timber upper floors

Fire safety

Structural floors of dwelling houses of two or three storeys are required to have a minimum period of fire resistance of half an hour. Floors that meet the half an hour fire resistance include:

❑   Floor with T & G boards or sheets of plywood or chipboard at least 15 mm thick, joists at least 37 mm wide and a ceiling of 12.5 mm plasterboard with 5 mm neat gypsum plaster finish
❑   Floor with 21 mm thick T & G boards, sheets of plywood or chipboard on joists at least 37 mm wide with a ceiling of 12.5 mm plasterboard with joints taped and filled

Resistance to the passage of heat

Timber upper floors that are exposed to outside air, such as floors over carports, have to be insulated against heat transfer to meet the requirements of the Building Regulations. The dwelling's target Fabric Energy Standard (FEES) for exposed floors is 0.13 W/m² K (with a maximum allowable U-value 0.25 W/m² K allowing design flexibility). Insulation is usually placed between the joists, and depending upon the depth of the floor joists additional rigid insulation may be placed over the joists to help achieve the U-value. To avoid cold bridges the insulation must extend right across the floor in both directions up to the surrounding walls.

Resistance to the passage of sound

A boarded timber floor with a rigid plasterboard ceiling affords poor resistance to the transmission of sound. To reduce the transmission of airborne sound it is necessary to increase the mass of a floor to restrict the flow of energy through it. To reduce the transmission of impact sound it is necessary to provide some soft material, such as a carpet, between the cause of the impact and the hard surface.

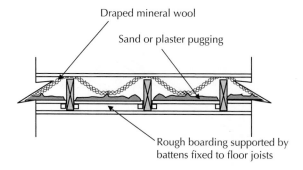

(a) Draped mineral wool and sand or plaster pugging on rough board between floor joists

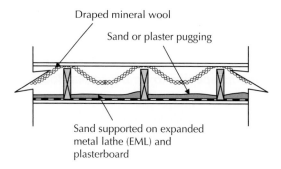

(b) Draped mineral wool over joists with sand pugging on EML and plasterboard

Figure 4.32   Traditional sound insulation for timber floors.

The traditional method of insulating timber floors against sound was to spread a layer of plaster or sand, termed 'pugging', on rough boarding fixed between the joists or on expanded metal lath and plaster, as illustrated in Figure 4.32. Pugging is effective in reducing the transmission of airborne sound but has little effect in deadening impact sound. To deaden impact sound it is necessary to lay some resilient material on the floor surface or between the floor surface material and the structural floor timbers. The combination of a resilient layer under the floor surface and pugging between joists will affect appreciable reduction of impact and airborne sound. Where the floorboards are nailed directly to joists, through the resilient layer, as illustrated in Figure 4.32, much of the impact sound deadening effect is lost. Approved Document E includes specifications for the construction of concrete and timber floors.

For concrete floors the resistance to airborne sound depends mainly on the mass of the concrete. Resistance to impact sound may be provided by a soft covering of carpet or other resilient material. Alternatively, the resistance to airborne sound can be provided mainly by the mass of the concrete floor and partly by a floating top layer. The top layer consists of a platform of T & G boards or chipboard nailed to timber battens that are laid on a

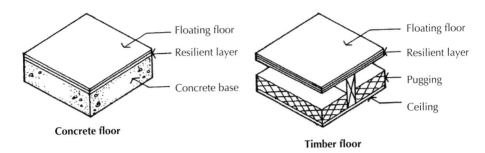

Figure 4.33   Sound insulation of floors.

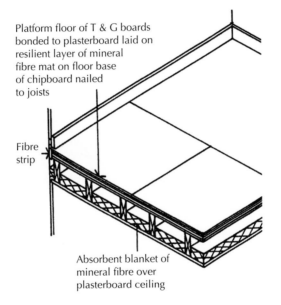

Figure 4.34   Platform floor.

13 mm thickness of resilient material as indicated in Figure 4.33, with the edges of the resilient layer turned up around the edges of the floating top layer. This floating layer is laid loose on the concrete base. The carpet or other finish can be laid on top.

For timber floors pugging is placed or fixed between the joists for resistance to airborne sound, and a floating floor is used to resist impact sound and to some extent resist airborne sound. This platform floor is constructed as a floor of 18 mm thick T & G boards or chipboard with all joints glued. The boarded platform is spot bonded to a base of 19 mm thick plasterboard. The boarded platform is laid on a 25 mm thick layer of resilient mineral fibre on a floor base of 12 mm thick boarding or chipboard nailed to the joints, as illustrated in Figure 4.34. The ceiling comprises two layers of plasterboard, with joints staggered, to a

Typical methods of improving acoustic and fire performance in compartment and separate floors

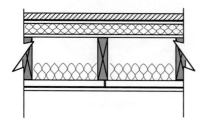

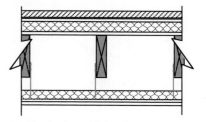

(a) Mineral wool and composite plaster board
21 mm softwood floor boarding
15 mm Gyproc sound block (composite board)
Rigid acoustic insulation on top of existing floor
(mineral wool)
100 mm acoustic wool between joists
15 mm Gyproc sound block below existing
ceiling

(b) Ceiling suspended from floor
21 mm softwood floor boarding
15 mm Gyproc sound block (composite board)
Rigid acoustic insulation on top of existing
floor (mineral wool)
80 mm acoustic wool on hung ceiling provides
continuous sound and fire resistance
Two layers of 15 mm Gyproc sound block
suspended from floor joists

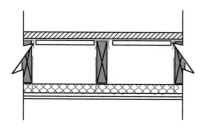

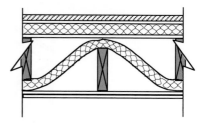

(c) Ceiling hung suspended from floor with
resilient channels placed on joists and sound
resisting planks between joist
21 mm softwood floor boarding
19 mm sound block planks between joists
(plaster-based composite board)
100 mm acoustic wool on hung ceiling
provides continuous sound and fire resistance
Two layers of 15 mm Gyproc sound block
suspended from floor joists

(d) Separate ceiling joists or sub-floor isolates
floor from ceiling
21 mm softwood floor boarding
15 mm Gyproc sound block (composite board)
Rigid acoustic insulation on top of existing
floor (mineral wool).
100 mm acoustic wool between joists lapped
over and under joists
Two layers of 15 mm Gyproc sound block
below existing ceiling

All tapered joints are taped and filled. Screw fixing is recommended where
the composite plaster boards are fixed direct to timber joists, rather than nail
fixing, as this minimises the risks of fixing defects

Figure 4.35   (a, b, c and d) Timber compartment floors (adapted from http://www.british-gypsum.com).

finished thickness of 30 mm on which 100 mm of absorbent mineral fibre is laid. To provide a level floor surface for platform floors it is necessary to fix the floorboards by nailing them to battens or by fixing the boards to a firm level base of plasterboard or similar material, so that the boards can be cramped together. The boards are bonded to the plasterboard base with strips or pads of adhesive to keep them flat, as the tongues in the edges of the boards are cramped up into the grooves of adjacent boards to produce a level floor finish.

To minimise flanking transmission of sound from the floor surface to the surrounding walls, a strip of resilient fibre material is fixed between the edges of platform, floors and surrounding solid walls, and a gap of at least 3 mm is left between the skirting and floating floors. To limit the transmission of airborne sound through gaps in the construction it is important to seal all gaps at junctions of wall and ceiling finishes and to avoid or seal breaks in the floor around service pipes. Mineral wool and other resilient materials that are specifically manufactured for their acoustic properties can be used under the floorboards and as absorbent pugging between the floor joists.

Party floors

The function of a party floor is the same as that of a party wall. In residential units it may be necessary to make more substantial separations between floors resisting the passage of fire and sound. Plasterboard and mineral wool can be used to improve the acoustic performance and fire resistance. In the Building Regulations a distinction is made between floors that act as sound barriers, referred to as separating floors, and floors that resist the passage of fire, referred to as compartment floors. Floors may be required to resist the passage of fire and sound and thus are both separating floors and compartment floors (Figure 4.35).

5 Walls

A wall is a continuous, usually vertical structure, which is thin relative to its length and height. The prime function of an external wall is to provide shelter against wind, rain and the daily and seasonal variations of outside temperature normal to its location, for reasonable indoor comfort. To provide adequate shelter a wall should have sufficient strength and stability to be self-supporting and also to support roofs and upper floors. Internal walls divide space into smaller areas, rooms and compartments. To differentiate the structural requirements of those walls that carry the loads from roofs and upper floors in addition to their own weight from those that are free-standing and carry only their own weight, the terms loadbearing and non-loadbearing are used. The majority of walls for single-, double- and triple-storey buildings are constructed with loadbearing masonry walls or are framed from timber, steel or concrete. The type of wall used will generally depend on the availability of materials and labour, economic factors and the design approach. This chapter describes the main principles of wall construction. The focus is mainly on loadbearing walls and timber-framed buildings familiar to domestic and small-scale developments. Details of steel and concrete frames, cladding and prefabricated systems are described in *Barry's Advanced Construction of Buildings*.

5.1 Functional requirements

The function of a wall is to enclose and protect a building or to divide space within a building. A wide variety of materials are used to construct walls, ranging from the familiar stone, brick and block, timber, concrete, glass and steel, through to the less common straw bale and earth construction and various hybrid systems. Regardless of the materials used, the commonly accepted functional requirements of a wall are:

❑ Strength and stability
❑ Resistance to weather and ground moisture
❑ Durability and freedom from maintenance
❑ Fire safety
❑ Resistance to the passage of heat
❑ Resistance to airborne and impact sound
❑ Security

Barry's Introduction to Construction of Buildings, Third Edition. Stephen Emmitt and Christopher A. Gorse.
© 2014 John Wiley & Sons, Ltd. Published 2014 by John Wiley & Sons, Ltd.

❑   Aesthetics
❑   Strength and stability

The strength of the materials used in wall construction is determined by the strength of a material in resisting compressive and tensile stress and the way in which the materials are put together. The usual method of determining the compressive and tensile strengths of a material is to subject samples of the material to tests to assess the ultimate compressive and tensile stresses at which the material fails in compression and in tension. From these tests the safe working strengths of materials in compression and in tension are set. The safe working strength of a material is considerably less than the ultimate strength, to provide a safety factor against variations in the strength of materials and their behaviour under stress. The characteristic working strengths of materials, to an extent, determine their use in the construction of buildings.

The moderate compressive and tensile strength of timber members has long been used to construct a frame of walls, floors and roofs for houses. Steel and concrete are used principally for their considerable strength as the structural frame members of large buildings. The compressive strength of brick and stone combined with the durability, fire resistance and appearance of the materials makes then an attractive choice. In the majority of small buildings, such as houses, the compressive strength of brick and stone is rarely fully utilised because the functional requirements of stability and exclusion of weather dictate a thickness of wall in excess of that required for strength alone.

Stability of a wall may be affected by foundation movement, eccentric loading, lateral forces (wind) and expansion due to changes in temperature and moisture. Eccentric loads (those not acting on the centre of the wall), such as from floors and roofs, and lateral forces, such as wind, tend to deform and overturn walls. The greater the eccentricity of the loads and the greater the lateral forces, the greater the tendency of a wall to deform, bow out of the vertical and lose stability. To prevent loss of stability, due to deformation under load, Building Regulations and codes set limits to the height or thickness ratios (slenderness ratios) to provide reasonable stiffness against loss of stability (the Building Regulations can be downloaded from the Government planning portal, http://www.planningportal.gov.uk). To provide stiffness against deformation under load, lateral (horizontal) restraint is provided by walls and roofs that are tied to the wall, and by intersecting walls and piers that are bonded or tied to the wall along its length. Irregular profile walls have greater stiffness against deformation than straight walls because of the buttressing effect of the angle of the walls, as illustrated in Figure 5.1. The more pronounced the chevron, zigzag, offset or serpentine of the wall, the stiffer it will be. Similarly the diaphragm and fin walls, described in *Barry's Advanced Construction of Buildings*, are stiffened against overturning and loss of stability by the cross ribs or diaphragms built across the wide cavity to diaphragm walls and the fins or piers that are built and bonded to straight walls in the fin wall construction.

Resistance to weather and ground moisture

A requirement of the Building Regulations is that walls should adequately resist the passage of moisture to the inside of the building. Moisture includes water vapour and liquid water. Moisture may penetrate a wall by absorption of water from the ground that is in contact

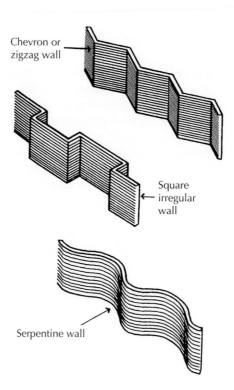

Figure 5.1   Irregular profile walls.

with the foundations or through rain and snow falling on the wall. Impermeable materials are used to form damp-proof courses (dpcs) and damp-proof membranes (dpms) to prevent water rising in floors and walls. The function of the dpm is described more fully in Section 5.2. The ability of a wall to resist the passage of water to its inside face depends on its exposure to wind-driven rain and the construction of the wall.

The exposure of a wall is determined by its location and the extent to which it is protected by surrounding higher ground, or sheltered by surrounding buildings or trees, from rain driven by the prevailing winds. In Great Britain the prevailing westerly winds from the Atlantic Ocean cause more severe exposure to driving rain along the west coast of the country than do the cooler and drier easterly winds on the east coast. British Standard 5628: Part 3 defines five categories of exposure as: very severe, moderate/severe, sheltered/moderate, sheltered and very sheltered.

Maps are available that provide an indication of the degree of exposure around the UK.

The Met Office has the most comprehensive meteorology library and should be used to obtain such information:

❑   http://www.metoffice.gov.uk

Local knowledge and specific physical site characteristics (e.g. detailing and weathering of neighbouring buildings) are also valuable indicators of exposure. The cavity wall has been

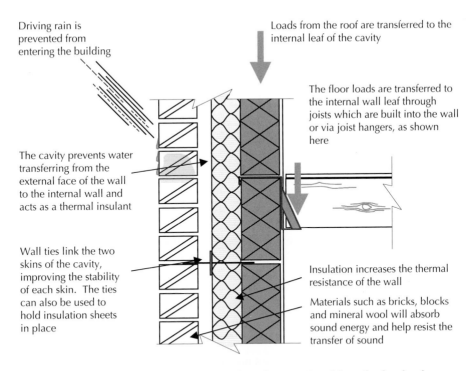

Driving rain is prevented from entering the building

Loads from the roof are transferred to the internal leaf of the cavity

The floor loads are transferred to the internal wall leaf through joists which are built into the wall or via joist hangers, as shown here

The cavity prevents water transferring from the external face of the wall to the internal wall and acts as a thermal insulant

Wall ties link the two skins of the cavity, improving the stability of each skin. The ties can also be used to hold insulation sheets in place

Insulation increases the thermal resistance of the wall

Materials such as bricks, blocks and mineral wool will absorb sound energy and help resist the transfer of sound

Figure 5.2   Cavity wall: resisting the external environment and transferring loads.

particularly successful in separating the internal and external environments, transferring loads and resisting rain and wind penetration (Figure 5.2).

Durability and freedom from maintenance

The durability of a wall is indicated by the frequency and extent of the work necessary to maintain minimum functional requirements and an acceptable appearance. Where there are agreed minimum functional requirements such as exclusion of rain and thermal properties, the durability of different walls may be compared through the cost of maintenance over a number of years.

Fire safety

Walls (combined with doors and windows) are an important element in providing fire protection. The two principal considerations are the structural integrity of the wall in a fire and the surface spread of flame, which is determined by the materials used.

Specifying a minimum period of fire resistance for the elements of the structure may restrict premature failure of the structural stability of a building in a fire. An element of structure is defined as part of a structural frame, a loadbearing wall and a floor. The requirements are that the elements should resist collapse for a minimum period of time in which the occupants may escape in the event of fire. Periods of fire resistance vary from 30 minutes for dwellings with a top floor not more than 5 m above ground to 120 minutes for an industrial building, without sprinklers, whose top floor is not more than 30 m above ground.

Fire may spread over the surface of materials that encourage the spread of flame across their surfaces. In Approved Document B is a classification of the performance of linings relative to surface spread of flame over walls and ceilings and limitations in the use of thermoplastic materials used in roof lights and lighting diffusers.

Resistance to the passage of heat

The building interior is heated by the transfer of heat from heaters and radiators to air (conduction), the circulation of heated air (convection) and the radiation of energy from heaters and radiators to surrounding colder surfaces (radiation). This internal heat is transferred through colder enclosing walls, roofs and floors by conduction, convection and radiation to colder outside air. (See Chapter 13 for more information on heat loss and heat loss calculations.) Walls need to provide adequate thermal resistance to heat loss, which is usually achieved through the mass and thickness of the wall and/or the inclusion of thermal insulation material(s) to the wall assembly.

Resistance to the passage of sound

Sound is transmitted as airborne sound and impact sound. Airborne sound is generated as cyclical disturbances of air from, for example, a radio, which radiate from the source of the sound with diminishing intensity with distance from the source (Figure 5.3). The vibrations in the air caused by the sound source will set up vibrations in enclosing walls and floors, which will cause vibrations of air on the opposite side of walls and floors.

Impact sound is caused by contact with a surface, for example, the slamming of a door or footsteps on a floor which set up vibrations in walls and floors that in turn cause vibrations of air around them that are heard as sound (Figure 5.3). The most effective insulation against airborne sound is a dense barrier such as a solid wall, which absorbs the energy of the airborne sound waves. The heavier and more dense the material of the wall, the more effective it is in reducing sound. The Building Regulations require walls and floors to provide reasonable resistance to airborne sound between dwellings and between machine rooms, tank rooms, refuse chutes and habitable rooms (see Section 5.15 'Internal and Party Walls').

The more dense the material, the more readily it will transmit impact sound. A knock on a part of a rigid concrete frame may be heard some considerable distance away. Insulation against impact sound will therefore consist of some absorbent material that will act to cushion the impact, such as a carpet on a floor, or serve to interrupt the path of the sound, as, for example, the absorbent pads under a floating floor. Noise generated in a room may be reflected from the walls and ceilings and build up to an uncomfortable intensity inside the room, particularly where the wall and ceiling surfaces are hard and smooth. To prevent the build-up of reflected sound some absorbent material should be applied to walls and ceilings, such as acoustic tiles or curtains, to absorb the energy of the sound waves.

Breaks in the structure, such as the cavity, prevent structure-borne sound from progressing. The sound energy must change back to airborne sound to pass through the cavity

The dense interconnected materials allow the sound energy to travel through the structure

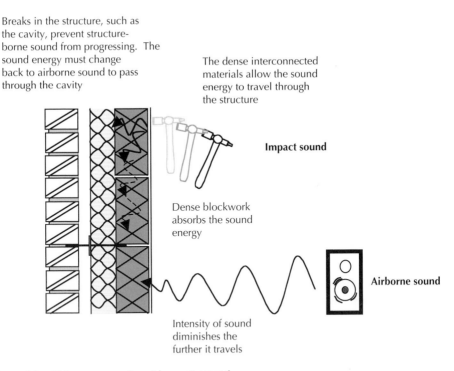

Impact sound

Dense blockwork absorbs the sound energy

Airborne sound

Intensity of sound diminishes the further it travels

Figure 5.3   Airborne sound and impact sound.

Security

Walls, in conjunction with doors and windows, help to provide a secure enclosure. In domestic properties unauthorised entry to property usually takes place through doors or windows. In commercial premises, especially where buildings contain goods that are subject to theft, it is not unusual for forced entry to occur through walls and roofs. Therefore the resistance of the wall to, for example, ram raiding may be a primary function of the wall.

Aesthetics

Walls are important visually. They are a vital component in the design of a building, making a major contribution to the character of the building. Choice of materials will be dependant upon satisfying the functional and performance requirements listed earlier and on satisfying the client's and designer's aesthetic goals, other considerations being the walling materials used on neighbouring buildings and any requirements of the local planning authority. The desired appearance of the walls may also influence whether loadbearing or framed construction is chosen.

5.2 Damp-proof courses (dpcs)

The function of a dpc is to act as a barrier to the passage of moisture or water between the parts separated by the dpc. The movement of moisture or water may be upwards in the foundation of walls and ground floors, downwards in parapets and chimneys, or horizontal, where a cavity wall is closed at the jambs of openings.

There should be a continuous horizontal dpc above ground in walls whose foundations are in contact with the ground, to prevent moisture from the ground rising through the foundation to the wall above ground, which otherwise would make wall surfaces damp and damage wall finishes. The dpc should be continuous for the whole length and thickness of the wall and be at least 150 mm above finished ground level to avoid the possibility of a build-up of material against the wall acting as a bridge for moisture from the ground as illustrated in Figure 5.4.

Flexible dpcs

Lead

Lead is an effective barrier to moisture and water. It is liable to corrosion in contact with freshly laid lime or cement mortar and should be protected by a coating of bitumen or

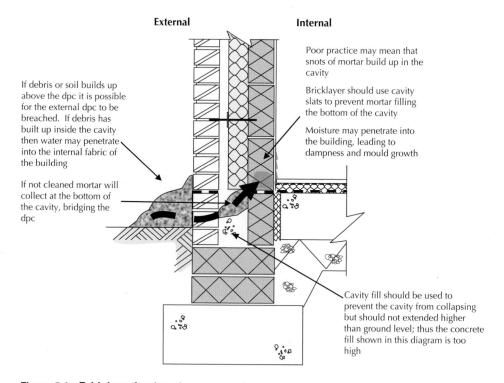

Figure 5.4   Bridging of a dpc above ground.

bitumen paint applied to the mortar surface and both surfaces of the lead. Lead is durable and flexible and can suffer distortion due to moderate settlement in walls without damage. It is an expensive material and is little used today other than for ashlar stonework or as a shaped dpc in chimneys. Cheaper 'lead substitute' materials, which have very similar properties to lead, are widely available. A lead dpc should weigh not less than 19.5 kg/m^2 (Code No. 4, 1.8 mm thick). Lead should be laid in rolls across the full thickness of a solid wall and each leaf of a cavity wall. The lead should be lapped at joints along the length of the wall and at all intersections, with the lap being at least 100 mm or the width of the dpc.

Copper

Copper is an effective barrier to moisture and water, is flexible, has high tensile strength and can suffer distortion due to moderate settlement in a wall without damage; it is an expensive material and is rarely used today. Copper as a dpc should be annealed, at least 0.25 mm thick and have a nominal weight of 2.28 kg/m^2.

Bitumen dpc

Bitumen dpcs are reasonably flexible and can withstand distortion due to moderate settlement in walls without damage. Bitumen dpcs, which are made in rolls to suit the thickness of walls, are bedded on a level bed of mortar and lapped at least 100 mm or the width of the dpc at running joints and intersections. Bitumen is economical, flexible, reasonably durable and convenient to lay. Bitumen is used in conjunction with other materials such as hessian, fibre, rubber or polymers to make a durable and workable dpc. The combination of a mortar bed, bitumen dpc and the mortar bed over the dpc for brickwork makes a comparatively deep mortar joint.

Polyethylene sheet and thermoplastic polymetric products

Polyethylene dpcs have largely been replaced by thermoplastic polymetric products, although most on-site staff still refer to the flexible dpcs as polythene. The flexible polymer dpcs are the most cost-effective and are most commonly used for both horizontal and vertical applications in domestic construction. Most of the dpcs are manufactured from reprocessed materials and supplied in 30 m long rolls and widths to suit standard brick and block walls (Photograph 5.1). The polymer roll is a clean and easy to handle material that remains flexible and workable at low temperatures. It can withstand distortion due to moderate settlement in a wall without damage and is an effective barrier against moisture. The sheets should have an embossed surface to increase mechanical slip resistance, reducing the possibility of the brickwork moving on the dpc. It is laid on an even bed of mortar and lapped at least the width of the dpc at running joints and intersections. Being a thin sheet material makes a thinner mortar joint than bitumen dpc and is sometimes preferred for that reason. Its disadvantage as a dpc is that it is fairly readily damaged by sharp particles in mortar or the coarse edges of brick. All laps and joints should be fully lapped by at least 100 mm and sealed using a jointing tape in accordance with the manufacturer's instructions.

Around door openings and windows where the internal skin of blockwork needs to be returned and the cavity closed, a vertical dpc is installed to prevent the passage of water (Photograph 5.2). Cavity trays are often formed out of polythene and are often used as

Photograph 5.1   Polythene dpc rolls.

Photograph 5.2   Vertical polythene dpc.

Photograph 5.3   Cavity tray made from polythene.

dpcs. Where the cavity is bridged such as above window and door openings a cavity tray may be used to ensure water is guided away from the internal skin (Photograph 5.3).

Polymer-based sheets
Polymer-based sheets, such as pitch polymer and co-polymer thermoplastic, are thinner than bitumen sheets and are used where the thicker bitumen dpc mortar joint would be unsightly. This dpc material, which has its laps sealed with adhesive, may be punctured by sharp particles and edges. Co-polymer thermoplastic dpcs have better tensile strength and higher tear and puncture resistance compared with pitch polymer dpcs.

Ethylene Propylene Diene Monomer (EPDM) rubber sheets
EPDM rubber sheet materials are also available.

Semi-rigid dpcs

Mastic asphalt

Mastic asphalt, spread hot in one coat to a thickness of 13 mm, forms a semi-rigid dpc, impervious to moisture and water. Moderate settlement in a wall may well cause a crack in the asphalt through which moisture or water may penetrate. It is an expensive form of dpc, which shows on the face of walls as a thick joint, and it is rarely used.

Rigid dpcs

Slate

Thin slates are sufficiently impermeable to water to serve as an effective dpc in solid walls of brick or stone. Slates are laid on a bed of mortar in two courses, with staggered joints as illustrated in Figure 5.5. Because of the small units of slate and the joints being staggered this dpc can remain reasonably effective where moderate settlement occurs. To be effective the edges of the slates should be exposed on a wall face and not be covered, which makes a deep joint and a strong architectural statement. This form of construction is rarely used today but is seen in existing buildings, and knowledge of the technology is of use when renovating existing buildings.

Brick dpcs

Two or three courses of dense engineering bricks may form an effective dpc. The effectiveness of the dpc is determined more by the permeability of the joining material than the bricks. This is rarely used these days other than as an architectural feature at the base of the building (and usually in conjunction with a modern flexible dpc).

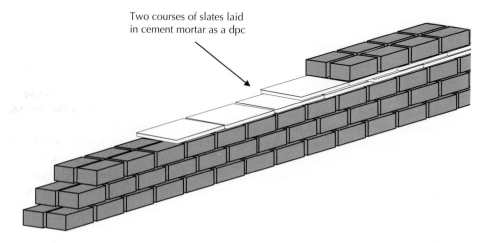

Two courses of slates laid in cement mortar as a dpc

Figure 5.5   Traditional slate dpc.

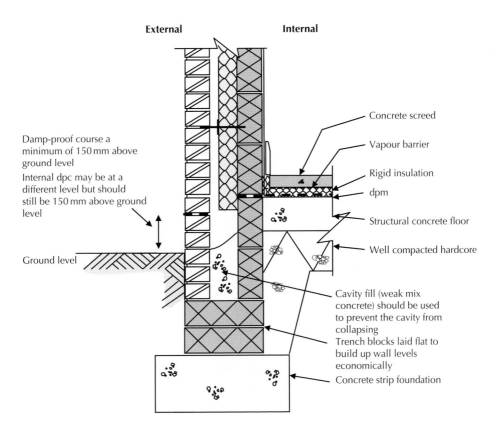

External   **Internal**

Concrete screed

Vapour barrier

Damp-proof course a
minimum of 150 mm above
ground level

Rigid insulation

Internal dpc may be at a
different level but should
still be 150 mm above ground
level

dpm

Structural concrete floor

Well compacted hardcore

Ground level

Cavity fill (weak mix
concrete) should be used
to prevent the cavity from
collapsing

Trench blocks laid flat to
build up wall levels
economically

Concrete strip foundation

Figure 5.6   dpc at different levels.

dpcs in cavity walls

A requirement of the Building Regulations is that the cavity should be carried down at least 150 mm below the level of the lowest dpc (Figure 5.6, Figure 5.7a and b, and Figure 5.8). A dpc in an external wall should ideally be at the same level as the dpm for the convenience of overlapping the two materials to make a damp-proof joint. Where the dpcs in both leaves of a cavity wall are at least 150 mm above outside ground level and the floor level is at, or just above, ground level, it is necessary to dress the dpm up the wall and into the level of the dpc. Figure 5.7a and b and Figure 5.8 show various methods of dressing the dpc to meet the dpm. This is a laborious operation, which makes it difficult to make a moisture tight joint at angles and intersections. The solution is to lay the dpc in the inner leaf of the cavity wall, level with the dpm in the floor, as illustrated in Figure 5.6.

Where the level of the foundation is near the surface, as with trench fill systems, it may be convenient to build two or more courses of solid blockwork up to ground level on which the cavity wall is raised, as illustrated in Figure 5.6 and Figure 5.7a. As little vegetable topsoil has been removed, the floor level finishes some way above ground and the dpm in the floor

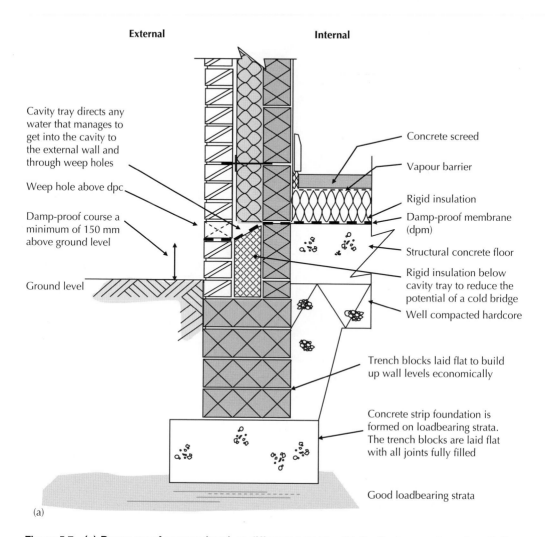

External **Internal**

Cavity tray directs any water that manages to get into the cavity to the external wall and through weep holes

Weep hole above dpc

Damp-proof course a minimum of 150 mm above ground level

Ground level

Concrete screed

Vapour barrier

Rigid insulation

Damp-proof membrane (dpm)

Structural concrete floor

Rigid insulation below cavity tray to reduce the potential of a cold bridge

Well compacted hardcore

Trench blocks laid flat to build up wall levels economically

Concrete strip foundation is formed on loadbearing strata. The trench blocks are laid flat with all joints fully filled

Good loadbearing strata

(a)

Figure 5.7   (a) Damp-proof course (tray) at different heights. (b) Cavity tray on top of partially filled cavity.

can be united with the dpc at the same level. The cavity insulation is taken down to the base of the cavity to continue wall insulation down to serve in part as edge insulation to the floor construction (Figure 5.7a).

It is accepted practice to finish the cavity in external walling at the level of the dpc, at least 150 mm above ground. The wall is built as a solid wall up to 150 mm below the dpc, as illustrated in Figure 5.6, or the cavity below ground is filled with concrete to

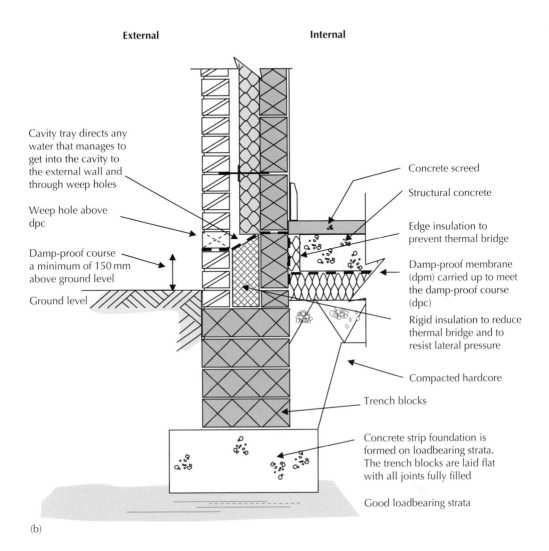

External   **Internal**

Cavity tray directs any
water that manages to
get into the cavity to
the external wall and
through weep holes

Concrete screed

Structural concrete

Weep hole above
dpc

Edge insulation to
prevent thermal bridge

Damp-proof course
a minimum of 150 mm
above ground level

Damp-proof membrane
(dpm) carried up to meet
the damp-proof course
(dpc)

Ground level

Rigid insulation to reduce
thermal bridge and to
resist lateral pressure

Compacted hardcore

Trench blocks

Concrete strip foundation is
formed on loadbearing strata.
The trench blocks are laid flat
with all joints fully filled

Good loadbearing strata

(b)

Figure 5.7 (Continued)

resist the ground pressure (Figure 5.7a and b and Figure 5.8). With this arrangement the
requirements of the Building Regulations recommend the use of a cavity tray at the
bottom of the cavity. Various arrangements of cavity trays are shown in Figure 5.7a and
b and Figure 5.8. This tray takes the form of a sheet of a flexible, impermeable material
such as one of the flexible dpc materials, which is laid across the cavity from a level
higher in the inner leaf so that it falls towards the outer leaf to catch and drain any snow
or moisture that might enter the cavity. The cavity thus acts as both tray and dpc to the
cavity wall leaves.

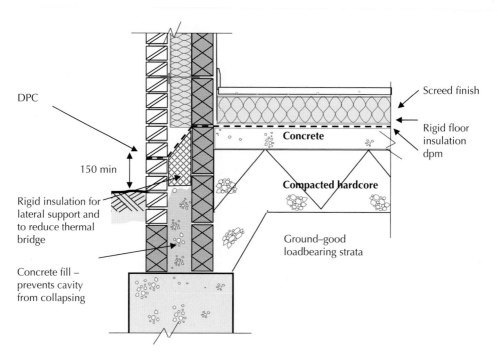

DPC

150 min

Rigid insulation for
lateral support and
to reduce thermal
bridge

Concrete fill –
prevents cavity
from collapsing

Screed finish

Rigid floor
insulation
dpm

Concrete

Compacted hardcore

Ground–good
loadbearing strata

Figure 5.8   DPC – cavity tray with rigid insulation to reduce thermal bridging.

5.3 Stone

Because natural stone is expensive it is mainly used as a facing material bonded or fixed
to a backing of brickwork or concrete. Many of the larger civic and commercial buildings
are faced with natural stone because of its durability, texture, colour and sense of perma-
nence. Natural stone is also used as the outer leaf of cavity walls for houses (both loadbearing and framed construction) in areas where local quarries can supply stone at reasonable
cost. Much of the time-consuming, and therefore expensive, labour associated with cutting,
shaping and finishing natural stone has been reduced by the use of power tools. However,
natural stone is an expensive material and various artificial stone products have been
developed as cheaper alternatives.

Natural stone

The natural stones used in building may be classified by reference to their origin as (1)
igneous, such as granite and basalt; (2) sedimentary, such as limestones and sandstones;
and (3) metamorphic, such as slates and marble. Natural stone has been used in the construction of buildings because it was thought that any hard, natural stone would resist the
action of wind and rain for centuries. Many natural stones have proved to be extremely

durable for a hundred or more years; however, there have been some failures due to a poor selection of the material, poor workmanship and/or poor detailing. Variation within the stone may cause localised weathering problems on exposure to rain and chemical pollutants in the air.

Some natural stones are comparatively soft and moist when first quarried but will gradually harden. Building stones should be seasoned (allowed to harden) for periods of up to a few years, depending on their size and type. Natural stones that are stratified, limestone and sandstone, must be used so that they lie on their natural bed to support compressive stress.

Reconstituted (reconstructed) stone

Reconstructed stone is made from an aggregate of crushed stone, cement and water. The stone is crushed so that the maximum size of the particles is 6 mm and it is mixed with cement in the proportions of one part cement to three or four parts of stone. Portland cement, white cement or coloured cement may be used to simulate the colour of a natural stone as closely as possible. A comparatively dry mix of cement, crushed stone and water is prepared and cast in moulds. The mix is thoroughly consolidated inside the moulds by vibrating and is left to harden in the moulds for at least 24 hours. The stones are then taken out of the moulds and allowed to harden gradually for 28 days.

Well-made reconstructed stone has much the same texture and colour as the natural stone from which it is made and can be cut, carved and dressed just like natural stone. It is not stratified, is free from flaws and is sometimes a better material than the natural stone from which it is made. Moulded cast stones made by repetitive casting can often be produced more cheaply than similar natural stones that have to be cut and shaped.

A cheaper form of cast stone is made with a core of ordinary concrete, faced with an aggregate of crushed natural stone and cement. This material should more properly be called cast concrete. The core is made from clean gravel, sand and cement and the facing is made from crushed stone and cement to resemble the texture and colour of a natural stone. The crushed stone, cement and water is first spread in the base of the mould to a thickness of about 25 mm, the core concrete is added and the mix consolidated. If the stone is to be exposed on two or more faces the natural stone mix is spread up the sides and the bottom of the mould. This type of cast stone cannot be carved as it has only a thin surface of natural looking stone.

Functional requirements of stone

Strength and stability

The strength of sound building stone lies in its very considerable compressive strength. The ultimate or failing stress of stone used for walling is about $300–100 \, \text{N/mm}^3$ for granite, $195–27 \, \text{N/mm}^3$ for sandstone and $42–16 \, \text{N/mm}^3$ for limestone. The use of stone as a facing material makes little use of the inherent compressive strength of the material. The strength and stability of a stone wall is affected by the same limitations that apply to walls of brick or block.

Resistance to weather and ground moisture

To prevent moisture rising from the ground through foundation walls it is necessary to form a continuous horizontal dpc some 150 mm above ground level. Historically this was achieved using dense stone, such as granite, that does not readily absorb moisture, or a thin sheet lead dpc; today a thin strip of reinforced plastic is used. The resistance to the penetration of wind-driven rain was not generally a consideration in the construction of solid masonry walls. The very considerable thickness of masonry walls of traditional large buildings was such that little, if any, rain penetrated to the inside face.

With the use of stone largely as a facing material it is necessary to construct walls faced with stone as cavity walling with a brick or block inner leaf separated by a cavity, as illustrated in Figure 5.9. The outer leaf illustrated is built with natural stone blocks bonded to a brick backing, with full width stones in every other course and the stones finished on face in ashlar masonry. This is an expensive form of construction because of the considerable labour costs in preparing the stones. As alternatives, the outer leaf of small buildings may be constructed with stone blocks for the full thickness of the outer leaf or, with larger buildings, the outer leaf may be constructed of brick to which a facing of stone slabs is fixed. Ashlar walling is constructed of blocks of stone that have been very accurately cut and finished square to specified dimensions so that the blocks can be laid, bedded and bonded with comparatively thin mortar joints, as illustrated in Figure 5.9. The very considerable labour involved in cutting and finishing individual stones is such that this type of walling is very expensive. Ashlar walling has been used for the larger, more permanent buildings and is now used principally as a facing material.

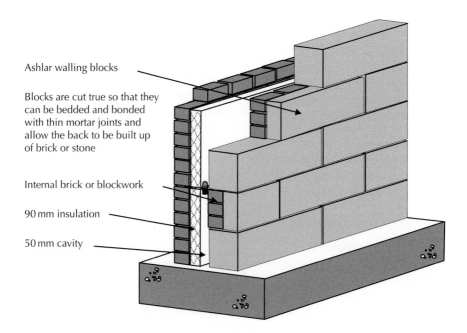

Ashlar walling blocks

Blocks are cut true so that they can be bedded and bonded with thin mortar joints and allow the back to be built up of brick or stone

Internal brick or blockwork

90 mm insulation

50 mm cavity

Figure 5.9   Cavity wall faced with ashlared stone and brick backing.

Durability and freedom from maintenance

Sound natural stone is highly durable as a walling material and will have a useful life of very many years in buildings that are adequately maintained. Granite, for example, is resistant to all usual weathering agents, including highly polluted atmospheres, and will maintain a high natural polished surface for a hundred years or more. Hard sandstones are very durable and inert to weathering agents but can stain over time due to the coarse-grained texture of the material, which retains dirt particles. The surface of sandstone may be cleaned from time to time to remove dirt stains by abrasive blasting with grit or chemical processes and thorough washing.

Sound limestone, sensibly selected and carefully laid, is durable for the anticipated life of the majority of buildings. In time the surface weathers by a gradual change of colour over many years. This is commonly held to be an advantage from the point of view of appearance. Limestones are soluble in rainwater that contains carbon dioxide so that the surface of a limestone wall is to an extent self-cleansing when freely washed by rain, while protected parts of the wall will collect and retain dirt. This effect gives the familiar black and white appearance of limestone masonry. The surface of limestone walls may be cleaned by washing with a water spray or by steam and brushing to remove dirt encrustations and the surface brought back to something near its original appearance.

Fire resistance

Natural stone is incombustible and will not support or encourage the spread of flame. The requirements of Part B of Schedule 1 to the Building Regulations for structural stability and integrity and for concealed spaces apply to walls of stone as they do for walls of block or brick masonry.

Resistance to the passage of heat

The natural stones used for walling are poor insulators against the transfer of heat and will contribute little to the thermal resistance in a wall. It is necessary to use some material with a low U-value as cavity insulation in walls faced with stone in the same way that insulation is used in cavity walls of brick or blockwork.

Resistance to the passage of sound

Because natural building stone is dense it has good resistance to the transmission of airborne sound, but will provide a ready path for impact sound. Dense material will absorb sound energy and the sound energy should diminish quickly as it attempts to travel through the dense material.

5.4 Stone masonry walls

Openings to stone walls

Masonry over door and window openings may be supported by flat stone lintels or by segmental or semi-circular arches.

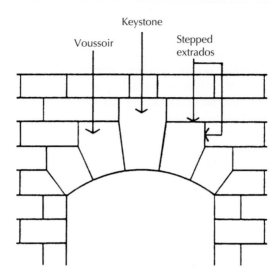

Figure 5.10 Segmental stone arch.

Lintels

A stone lintel for small openings of up to about a metre wide can be formed of one whole stone with its ends built into jambs and its depth corresponding to one or more stone courses. The poor tensile strength of stone limits the span of single stone lintels unless they are to be disproportionately deep. Over openings wider than about a metre it is usual to form lintels with three or five stones cut in the form of a flat arch. The stones are cut so that the joints between the ends of stones radiate from a common centre so that the centre, or key stone, is wedge shaped, as illustrated in Figure 5.10. The stones are cut so that the lower face of each stone occupies a third or a fifth of the width of the opening.

To prevent the key stone sinking due to settlement and so breaking the line of the soffit, it is usual to cut half depth joggles in the ends of the key stone to fit to rebates cut in the other stones. The joggles and rebates may be cut the full thickness of each stone and show on the face of the lintel, or more usually the joggles and rebates are cut on the inner half of the thickness of stones as secret joggles, which do not show on the face, as illustrated in Figure 5.11. The depth of the lintel corresponds to a course height, with the ends of the lintel built in at jambs as end bearing. Stone lintels are used over both ashlar and rubble walling. The use of lintels is limited to comparatively small openings due to the tendency of the stones to sink out of horizontal alignment. Photograph 5.4 provides an example of a flat arch. For wider openings some form of arch is used.

Arches

A stone arch consists of stones specially cut so that the joints between stones radiate from a common centre; the soffit is arched and the stones bond in with the surrounding walling. The individual stones of the arch are termed 'voussoirs', the arched soffit the 'intrados' and the upper profile of the arch stones the 'extrados' (see Figure 5.10) The voussoirs of the

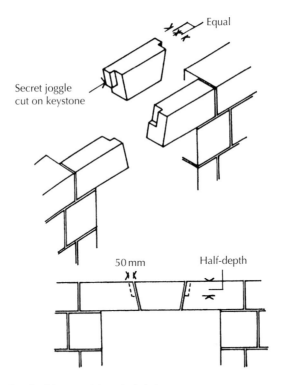

Equal

Secret joggle
cut on keystone

50 mm

Half-depth

Figure 5.11 Stone lintel with secret joggle joints.

Photograph 5.4 Flat arch with segmental stones and keystone.

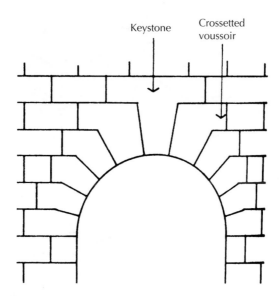

Figure 5.12   Semi-circular stone arch with crossetted voussoirs.

segmental arch are cut with steps that correspond in height with stone courses, to which the stepped extrados is bonded. The stones of an arch are cut so that there are an uneven number of voussoirs with a centre or key stone. The key stone is the last stone to be put in place as a key to the completion and the stability of the arch, hence the term key stone.

Crossetted arch

The semi-circular arch, illustrated in Figure 5.12, is formed with stones that are cut to bond into the surrounding walling to form stepped extrados and also to bond horizontally into the surrounding stones. The stones, voussoirs, are said to be crossetted, or crossed. The stones radiate from a common centre with an odd number of stones arranged around the half circle so that there is a central key stone. The extent that the crossetted top of each stone extends into the surrounding masonry is not necessarily dictated by the stretcher bond of the main walling. Any degree of bond into the main wall may be chosen and the bond of masonry adjusted accordingly. The effect of the crossetted voussoir is to emphasise the arch as structurally separate from the coursed masonry and is chosen for that effect. There is some waste involved in cutting stones to achieve this effect.

Ashlar masonry joints

Ashlar stones may be finished with smooth faces and bedded with thin joints, or the stones may have their exposed edges cut to form a channelled or V joint to emphasise the shape of each stone and to give the wall a heavier, more permanent appearance. The ashlar stones of the lower floor of large buildings are often finished with channelled or V joints and the wall above with plain ashlar masonry to give the base of the wall an appearance of strength. Ashlar masonry finished with channelled or V joints is said to be rusticated. A channelled

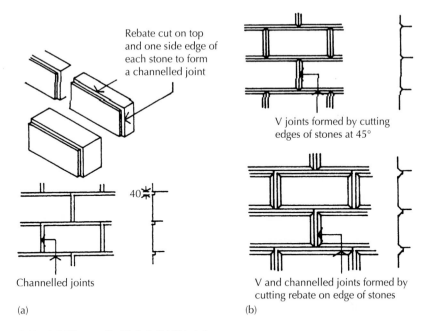

Figure 5.13   (a) Channelled joint. (b) V joint.

joint (rebated joint) is formed by cutting a rebate on the top and one side edge of each stone so that when the stones are laid, a channel rebate appears around each stone, as illustrated in Figure 5.13a. The rebate is cut on the top edge of each stone so that when the stones are laid rainwater, which may run into the horizontal joint, will not penetrate the mortar joint. A V joint (chamfered joint) is formed by cutting all four edges of stones with a chamfer so that when they are laid a V groove appears on face, as illustrated in Figure 5.13b. Often the edges of stones are cut with both V and channelled joints to give greater emphasis to each stone.

Tooled finish

Plain ashlar stones are usually finished with flat faces to form plain ashlar facing. The stones may also be finished with their exposed faces tooled to show the texture of the stone. Some of the tooled finishes used with masonry are illustrated in Figure 5.14. It is the harder stones such as granite and hard sandstone that are more commonly finished with rock face, pitched face, reticulated or vermiculated faces. The softer, fine-grained stones are usually finished as plain ashlar.

Cornice and parapet walls

It is common practice to raise masonry walls above the levels of the eaves of a roof, as a parapet. The purpose of the parapet is partly to obscure the roof and also to provide a depth of wall over the top of the upper windows for the sake of appearance in the proportion of

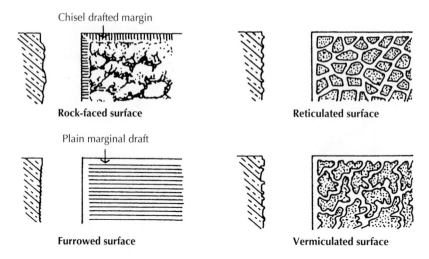

Chisel drafted margin

Rock-faced surface

Reticulated surface

Plain marginal draft

Furrowed surface

Vermiculated surface

Figure 5.14   Tooled finishes.

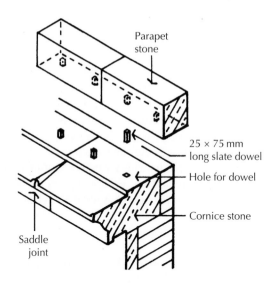

Parapet stone

25 × 75 mm long slate dowel

Hole for dowel

Cornice stone

Saddle joint

Figure 5.15   Cornice and parapet.

the building as a whole. In order to provide a decorative termination to the wall, a course of projecting moulded stones is formed. This projecting stone course is termed a cornice and it is generally formed of one or more courses of stone below the top of the parapet (Figure 5.15). This projection provides some protection to the wall below.

The parapet wall usually consists of two or three courses of stones capped with coping stones bedded on a dpc of sheet metal. The parapet is usually at least 1 B thick or of such thickness that its height above the roof is limited by the requirements of the Building Regulations. The parapet may be built of solid stone or stones bonded to a brick backing. The cornice is constructed of stones of about the same depth as the stones in the wall below, cut so that they project and are moulded for appearance sake. Because the stones project, their top surface is weathered (slopes out) to throw water off.

Saddle joint

The projecting, weathered top surface of coping stones is exposed and rain running off it will in time saturate the mortar in the vertical joints between the stones. To prevent rain soaking into these joints it is usual to cut the stones to form a saddle joint as illustrated in Figure 5.15. The exposed top surface of the stones has to be cut to slope out (weathering), and when this cutting is executed a projecting quarter circle of stone is left on the ends of each stone. When the stones are laid, the projections on the ends of adjacent stones form a protruding semi-circular saddle joint, which causes rain to run off away from the joints.

Weathering to cornices

Because cornices are exposed and liable to saturation by rain and possible damage by frost, it is good practice to cover the exposed top surface of cornice stones cut from limestone or sandstone with sheet metal. The sheet metal covering is particularly useful in urban areas where airborne pollutants may gradually erode stone. Sheet lead is preferred as a non-ferrous covering because of its ductility, which facilitates shaping, and its impermeability. Sheets of lead, Code No. 5, are cut and shaped for the profile of the top of the cornice and are laid with welted (folded) joints at 2 m intervals along the length of the cornice. The purpose of these comparatively closely spaced joints is to accommodate the inevitable thermal expansion and contraction of the lead sheet. The top edge of the lead is dressed up some 75 mm against the parapet as an upstand, and turned into a raglet (groove) cut in the parapet stones and wedged in place with lead wedges. The joint is then pointed with mortar. The bottom edge of the lead sheets is dressed (shaped) around the outer face of the stones and welted (folded). To prevent the lower edge of the lead sheet weathering being blow up in high winds, 40 mm wide strips of lead are screwed to lead plugs set in holes in the stone at 750 mm intervals and are folded into the welted edge of the lead, as illustrated in Figure 5.16. Where cornice stones are to be protected with sheet lead weathering there is no purpose in cutting saddle joints.

Cement joggle

Cornice stones project and one or more stones might in time settle slightly so that the decorative line of the mouldings cut on them would be broken and so ruin the appearance of the cornice. To prevent this possibility, shallow V-shaped grooves are cut in the ends of each stone so that when the stones are put together these matching V grooves form a square hole into which cement grout is run. When the cement hardens it forms a joggle, which locks the stones in their correct position.

Dowels

To maintain parapet stones in their correct position in a wall, slate dowels are used. The stones in a parapet are not kept in position by the weight of walling above and these stones

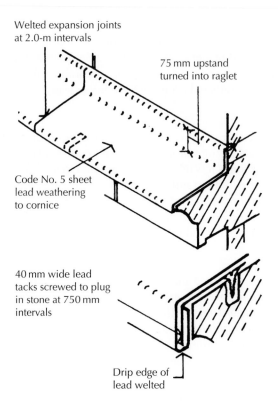

Welted expansion joints
at 2.0-m intervals

75 mm upstand
turned into raglet

Code No. 5 sheet
lead weathering
to cornice

40 mm wide lead
tacks screwed to plug
in stone at 750 mm
intervals

Drip edge of
lead welted

Figure 5.16   Lead weathering to cornice.

are, therefore, usually fixed with slate dowels. These dowels consist of square pins of slate that are fitted to holes cut in adjacent stones, as illustrated in Figure 5.15.

Cramps

Coping stones are bedded on top of a parapet wall as a protection against water soaking down into the wall below. There is a possibility that the coping (capping) stones may suffer some slight movement and cracks will develop in the joints between them. Rain may then saturate the parapet wall below and frost action may contribute to some movement and eventual damage. To keep coping stones in place a system of cramps is used. Either slate or non-ferrous metal is used to cramp the stones together. A short length of slate, shaped with dovetail ends, is set in cement grout (cement and water) in dovetail grooves in the ends of adjacent stones, as illustrated in Figure 5.17a. As an alternative, a gunmetal cramp is set in a groove and mortise in the end of each stone and bedded in cement mortar, as illustrated in Figure 5.17b. For coping stones cut from limestone or sandstone a sheet metal weathering is sometimes dressed over coping stones. The weathering of lead is welted and tacked in position over the stones.

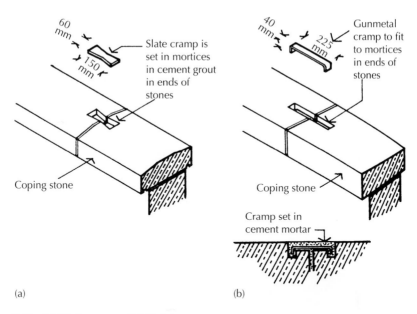

Figure 5.17   (a) Slate cramp. (b) Metal cramp.

Types of walling

Rubble walling

Rubble walling has been extensively used for agricultural buildings in towns and villages in those parts of the country where a local source of stone was readily available. The various forms of rubble walling may be classified as random rubble and squared rubble.

Random rubble

Uncoursed random rubble stones of all shapes and sizes are laid in mortar, as illustrated in Figure 5.18a and Photograph 5.5. No attempt is made to select and lay stones in horizontal courses. There is some degree of selection to avoid excessively wide mortar joints and also to bond stones by laying some longer stones both along the face and into the thickness of the wall, so that there is a bond stone in each square metre of walling. At quoins, angles and around openings, selected stones or shaped stones are laid to form roughly square angles.

Random rubble brought to course is similar to random rubble uncoursed except that the stones are selected and laid so that the walling is roughly levelled in horizontal courses at vertical intervals of from 600 to 900 mm, as illustrated in Figure 5.18b and Photograph 5.6. As with uncoursed rubble, transverse and longitudinal bond stones are used.

Squared rubble

Squared rubble uncoursed is laid with stones that come roughly square from the quarry in a variety of sizes. The stones are selected at random, are roughly squared with a walling

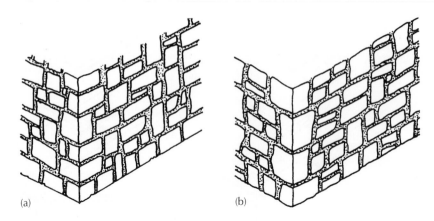

Figure 5.18   Random rubble uncoursed (a) and coursed (b).

Photograph 5.5   Random rubble uncoursed.

hammer and laid without courses, as illustrated in Figure 5.19a. As with random rubble, both transverse and longitudinal bond stones are laid at intervals. Snecked rubble is a term for squared rubble in which a number of small squared stones, snecks, are laid to break up long continuous vertical joints. Snecked rubble is often difficult to distinguish from squared random rubble.

Photograph 5.6   Squared random rubble coursed.

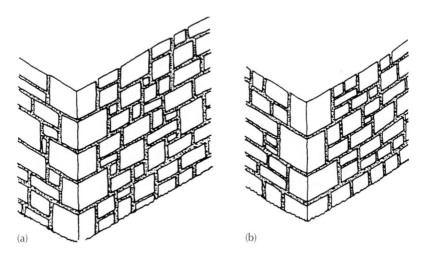

(a)   (b)

Figure 5.19   (a) Squared rubble uncoursed. (b) Squared rubble coursed.

Photograph 5.7   Stone coursed.

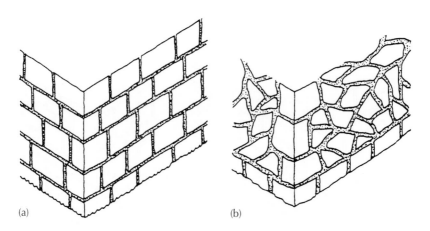

(a)   (b)

Figure 5.20   (a) Squared rubble coursed. (b) Polygonal walling.

Squared rubble brought to course is constructed from roughly square stone rubble, selected and squared so that the work is brought to courses every 300–900 mm intervals, as illustrated in Figure 5.19b and Photograph 5.7. Squared rubble course is built with stones that are roughly squared so that the stones in each course are roughly the same height and the courses vary in height, as illustrated in Figure 5.20a and Photograph 5.8. The face of

Photograph 5.8   Squared rubble coursed.

the stones may be roughly dressed to give a rock-faced appearance or dressed smooth to give a more formal appearance.

Polygonal walling

Stones that are taken from a quarry where the stone is hard, have no pronounced laminations and come in irregular shapes that can be laid as polygonal walling. The stones are selected and roughly dressed to fit when laid, to an irregular pattern, with no attempt at regular courses or vertical joints. At corners and as a base, roughly square-edged stones are used, as illustrated in Figure 5.20b. Photograph 5.9 shows a polygonal wall with the edges squared up by cornet stones. Photograph 5.10 shows a polygonal wall that is uncoursed.

Flint walling

Flint walling is traditional to East Anglia and the south and south-east of England. Both field and shore flints are used. The flints used for walling are up to 300 m in length and from 75 to 250 mm in width and thickness. The flints or cobbles may be used whole or split to show the heart of the flint, and also knapped or snapped so that they show a roughly square face. Flint walling is built with a dressing of stone or brick at angles and in horizontal lacing courses that level the wall at intervals. Figure 5.21a is an illustration of whole flints laid without courses in brick dressing to angles and as lacing, and Figure 5.21b is an illustration of knapped flints laid to courses in stone dressing. Photograph 5.11 shows a whole

Photograph 5.9   Polygonal walling.

Photograph 5.10   Polygonal walling uncoursed.

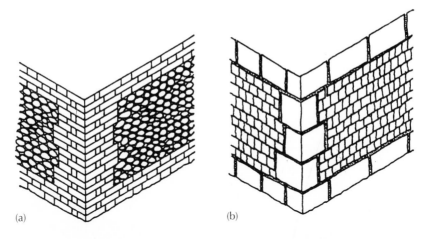

Figure 5.21   (a) Whole flint wall. (b) Knapped flint wall.

Photograph 5.11   Whole flint wall with brick stretcher band.

Photograph 5.12   Random knapped flint uncoursed.

flint wall with brick stretcher band and Photograph 5.12 shows a random knapped flint uncoursed wall.

5.5 Bricks and brickwork

The word 'brick' is used to describe a small block of burnt clay of such size that it can be conveniently held in one hand and it is slightly longer than twice its width. The great majority of bricks in use today are made from clay, although bricks can also be made from sand and lime or concrete. Glass bricks and blocks are also widely available. In the UK the standard brick size is $215 \times 102.5 \times 65$ mm, as illustrated in Figure 5.22. With a 10 mm mortar joint the working size becomes $225 \times 112.5 \times 75$ mm. Bricks may be manufactured to other shapes and sizes and are usually known as 'specials'.

Types of brick

Clay bricks
Clay differs quite widely in composition from place to place and the clay dug from one part of a field may differ from that dug from another part of the same field. Clay is ground in mills, mixed with water to make it plastic and then moulded, either by hand or machine, to the shape and size of a brick. Bricks that are shaped and pressed by hand in a sanded

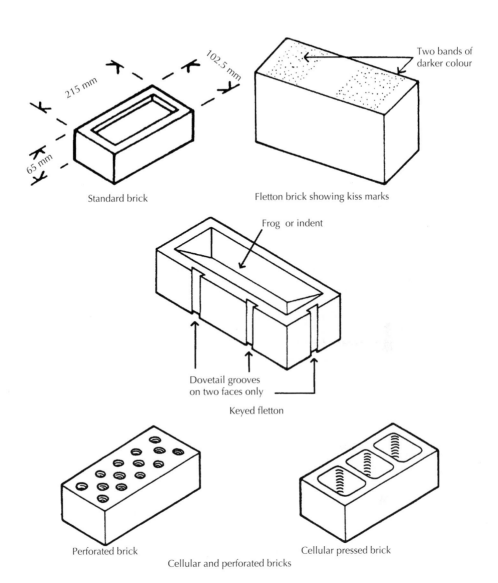

Standard brick

Fletton brick showing kiss marks

Two bands of darker colour

Frog or indent

Dovetail grooves on two faces only

Keyed fletton

Perforated brick

Cellular pressed brick

Cellular and perforated bricks

Figure 5.22   Types of brick.

wood mould and then dried and fired have a sandy texture, are irregular in shape and colour, and are used as facing bricks due to the variety of their shapes, colour and texture. Machine made bricks are either hydraulically pressed in steel moulds or extruded as a continuous band of clay. The continuous band of clay, the section of which is the length and width of a brick, is cut into bricks by a wire frame. Bricks made this way are called 'wire cuts'. The moulded brick is baked to dry out the water and is burnt at a high temperature so that part of the clay fuses the whole mass of the brick into a hard durable unit. Because there is wide variation in the composition of the clays suitable for brick making, and because it is possible to burn bricks over quite a wide range of temperatures sufficient to fuse the material into a durable mass, a large variety of bricks is available.

Calcium silicate bricks (sand–lime)

Calcium silicate bricks are generally known as sand–lime bricks. The bricks are made from a carefully controlled mixture of clean sand and hydrated lime which is mixed together with water, moulded to brick shape and then hardened in a steam oven. The resulting bricks are very uniform in shape and colour and are normally a dull white. Coloured sand–lime bricks are made by adding a colouring matter during manufacture. The material from which they are made can be carefully selected and accurately proportioned to ensure a uniform hardness, shape and durability that are quite impossible with the clay used for most bricks.

Concrete bricks

Concrete bricks are manufactured in the same size as clay bricks. They tend to be more consistent in shape, size and colour than clay bricks and come in a variety of colours and finishes. Appearance and properties vary between manufacturers, although the concrete brick does have a different appearance from clay bricks, which extends the choice available.

Brick classifications

Bricks may be classified in accordance with their uses as commons, facing and engineering bricks, although they should be specified according to performance requirements, i.e. strength, thermal performance, water absorption, and so on.

Commons

These are bricks that are sufficiently hard to carry the loads normally supported by brickwork safely, but because they have a dull texture or are a poor colour they are used for situations where they will not be exposed to view.

Facings

This includes any brick that is sufficiently hard burnt to carry normal loads. Manufacturers offer an extensive range of colours and textures from which to choose.

Engineering bricks

These are bricks that have been made from selected clay and carefully burnt so that the finished brick is very solid and hard and is capable of safely carrying much heavier loads than other types of brick. These bricks are mainly used for walls carrying exceptionally heavy loads, for brick piers and general engineering works, especially for works below ground.

Special bricks

A range of special bricks are made for specific uses in fairface brickwork. These bricks are made from fine clays to control and reduce shrinkage deformation during firing.

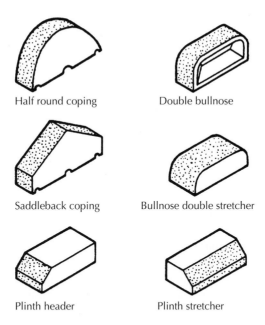

Half round coping   Double bullnose

Saddleback coping   Bullnose double stretcher

Plinth header   Plinth stretcher

Figure 5.23   Special bricks.

Figure 5.23 is an illustration of some typical (standard) specials. The two coping bricks, the half round and the saddleback, are for use as coping to a one brick thick parapet wall. The bricks are some 50 mm wider than the thickness of a one brick wall so that they overhang the wall each side to shed rainwater: the grooves on the underside of the hangover form a drip edge. The two bullnose specials are made as a capping or coping for one brick thick walls. They are of the same length as the thickness of the wall on which they are laid, to provide protection and a flush finish to the wall. The plinth bricks are used to provide and cap a thickening to the base of walls, a plinth. A plinth at the base of a wall gives some definition to the base of a wall as opposed to the wall being built flush out of the ground. A plinth may be formed by the junction of a solid $1^1/_2$ brick thick wall built from the foundation up to a cavity wall.

Special specials

Manufacturers also make purpose-made specials, known as 'special specials' to suit particular design requirements. Additional time will be needed to design and manufacture the special specials and this will need to be considered in the programming of the construction works. These bricks tend to be expensive because they are made to order and are often made in small quantities. However, used as a design feature, they can prove to add considerable value to brick walls.

Physical properties of bricks

When specifying bricks, certain properties need to be defined that relate to durability, form and appearance. A performance specification will need to consider the following parameters: size and type, compressive strength, frost resistance, soluble salt content and visual appearance. This information is provided by brick manufacturers. It is common practice to specify facing bricks by name (manufacturer and brick name), which will automatically confirm the brick's characteristics, thus determining the exact appearance (colour and texture) required.

Compressive strength

The compressive strength of bricks is found by crushing 12 of them individually until they fail or crumble. The average compressive strength of the brick is stated as newton per millimeter of surface area required to ultimately crush the brick. The crushing resistance varies from about $3.5\,N/mm^2$ for soft facing bricks up to $140\,N/mm^2$ for engineering bricks.

Water absorption and suction

The amount of water a brick will absorb is a guide to its density and therefore its strength in resisting crushing. The level of water absorption is most critical for bricks to be used below dpc level or for dpcs. Absorption rates vary between 1% and 35%. Bricks with high suction rates absorb water rapidly from the mortar and hence they are more difficult to reposition as the work proceeds than bricks with medium to low suction rates.

Thermal and moisture movement

All building materials move as a result of the expansion and contraction caused by temperature or moisture changes. The amount of movement depends on the materials and conditions. Allowance must be made for movement in masonry walls through careful positioning of control joints (Table 5.1). Brick manufacturers offer guidance based on current standards. Photograph 5.13 shows a movement joint in a brick and flint clad wall. Expansion in long walls without adequate control joints may result in the bricks at the end of the wall oversailing the dpc (Figure 5.24) or the corner of walls cracking (Figure 5.25). Where building or different structures meet it is always advisable to have control or movement joints. Photograph 5.14 shows a movement joint at the junction between two properties.

For clay bricks it is recommended that the joint should be capable of accommodating 10 mm movement. Because materials will be used to construct and fill the joint this may mean that the joint is between 12 and 20 mm wide. Calcium silicate bricks suffer from contraction rather than expansion. Where joints are used to accommodate shrinkage, the wall should be structurally sound when the joint is at its widest (when the wall has shrunk to its full extent). Control joints should be located where there is lateral support (Figure 5.26).

Table 5.1 Thermal and moisture movement joints in brickwork

Brick type	Type of movement	Recommended distance between joints (mm)
Clay fired bricks	Expansion	12
Calcium silicate bricks	Shrinkage	7.5–9
Concrete blocks or bricks	Shrinkage	6

Photograph 5.13   Movement joint.

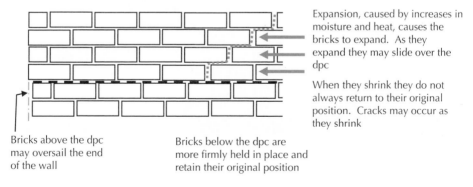

Expansion, caused by increases in moisture and heat, causes the bricks to expand. As they expand they may slide over the dpc

When they shrink they do not always return to their original position. Cracks may occur as they shrink

Bricks above the dpc may oversail the end of the wall

Bricks below the dpc are more firmly held in place and retain their original position

Figure 5.24   Elevation of a wall: expansion of brickwork causing oversailing.

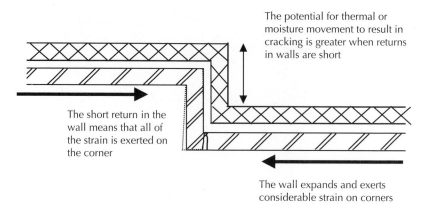

The potential for thermal or moisture movement to result in cracking is greater when returns in walls are short

The short return in the wall means that all of the strain is exerted on the corner

The wall expands and exerts considerable strain on corners

Figure 5.25   Plan of a wall junction: expansion and shrinkage at the corner joint.

Photograph 5.14   Brickwork movement joint and fire stop at a party wall junction.

Extra ties at 300 mm centres should be used each side of the joint. Wire and plastic wall ties can accommodate differential movement between the skins of cavity walls by bending and flexing. In some situations it may be necessary to place a wall tie directly across the movement joint. Specially designed sleeves can be used with straight wall ties: these allow for movement along the length of the ties but resist movement in other directions (Figure 5.27).

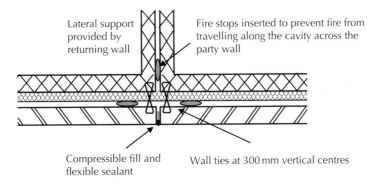

Lateral support provided by returning wall

Fire stops inserted to prevent fire from travelling along the cavity across the party wall

Compressible fill and flexible sealant

Wall ties at 300 mm vertical centres

Figure 5.26   Lateral stability across control joints.

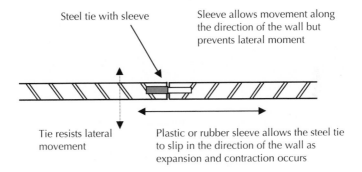

Steel tie with sleeve

Sleeve allows movement along the direction of the wall but prevents lateral moment

Tie resists lateral movement

Plastic or rubber sleeve allows the steel tie to slip in the direction of the wall as expansion and contraction occurs

Figure 5.27   Lateral stability across control joints provided by ties with sleeves.

Movement due to differential settlement

Where it is anticipated that differential settlement may occur due to different ground conditions, foundation designs or building loads, control joints should be used to allow separate parts of the building to move without causing cracking to the facing or structural elements. In long buildings or terraced housing it may be necessary to separate the buildings and incorporate control joints that allow for differential settlement while retaining the appearance of the building.

Appearance

There is a large range of bricks from which to choose and choice is often made on personal preference for the colour and texture of the brick, assuming it meets the set performance requirements. Choice is also influenced by that used on neighbouring buildings, local town planning authority requirements and the existing brick when working on repair and alteration works.

Frost resistance

If bricks are saturated and the water within them freezes (expanding as it freezes up to $-4°C$) it is likely that small cracks will develop, causing the face of the brickwork to break

Photograph 5.15   Face of brickwork removed by frost attack.

away (Photograph 5.15). Care is required in exposed areas, for example, parapet walls, chimney stacks, around dpcs, in relation to the selection of bricks and the detailing of the respective junctions to ensure durability. With increased emphasis on higher thermal standards and thermally upgrading existing buildings, the additional insulation results in less heat passing through the wall. During the winter months this means that the masonry, which is directly exposed to the external environment, receives less heat energy from the interior of the building and as such is more susceptible to damage from frost. Hence the durability of the masonry needs careful consideration, especially in exposed locations.

Efflorescence

Clay bricks contain soluble salts that migrate, in solution in water, to the surface of brickwork as water evaporates to outside air. These salts will collect on the face of brickwork as an efflorescence of white crystals that appear in irregular, unsightly patches. This efflorescence of white salts is most pronounced in parapet walls, chimneys and below dpcs where brickwork is most liable to saturation. The concentration of salts depends on the soluble salt content of the bricks and the degree and persistence of saturation of brickwork. The efflorescence of white salts on the surface is unsightly and usually causes no damage. In time the salts may be washed from surfaces by rain. Heavy concentration of salts can cause spalling and powdering of the surface of bricks, particularly those with smooth faces. This effect is sometimes described as crypto-efflorescence. The salts trapped behind the smooth face of bricks expand when wetted by rain and cause the face of the bricks to crumble and disintegrate. Efflorescence may also be caused by absorption of soluble salts from a cement-rich mortar or from the ground. There is no way of preventing the absorption of soluble salts from the ground by brickwork below the horizontal dpc level, although the effect can be reduced considerably by the use of dense bricks below the dpc. Soluble salts can migrate through mortar and natural stone in the same way as a brick wall. Photograph 5.16 shows

(a)

(b)

Photograph 5.16   (a) A combination of water saturation and freeze thaw cycles has caused the joints to spall and salts within the mortar to migrate to the surface. (b) Efflorescent salt crystals are clearly visible at the surface of the wall.

efflorescence on a solid stone wall. Photograph 5.16a shows how the mortar has spalled off the surface as the efflorescent crystals have expanded. In Photograph 5.16b the efflorescence can be clearly seen around the joints.

Sulphate attack on mortars and renderings
When brickwork is persistently wet, as in foundations, retaining walls, parapets and chimneys, sulphates in bricks and mortar may in time crystallise and expand and cause mortar and renderings to disintegrate. To minimise this effect bricks with low sulphate content should be used.

5.6 Bonding bricks

When building a wall it is usual to lay bricks in regular, horizontal courses so that each brick bears on two bricks below. The bricks are said to be 'bonded' since they bind together by being laid across each other along the length of the wall, as illustrated in Figure 5.28. The advantage of bonding is that the wall acts as a whole so that the load of a beam carried by the topmost brick is spread to the two bricks below it, then to the three below that and so on down to the base or foundation course of bricks. The failure of one poor quality brick such as 'A' in a wall and a slight settlement under part of the foundation such as 'B' and 'C' in Figure 5.28 will not affect the strength and stability of the whole wall as the load carried by the weak brick and the two foundation bricks is transferred to the adjacent bricks. Because of the bond, window and door openings may be formed in a wall, the load of the wall above the opening being transferred to the brickwork on each side of the openings by an arch or lintel. The effect of bonding is to stiffen a wall along its length and also to some small extent against lateral pressure, such as wind.

Types of bond

Stretcher bond
The four faces of a brick, which may be exposed in fairface brickwork, are the two long stretcher faces and the two header faces illustrated in Figure 5.29 (also see Photograph

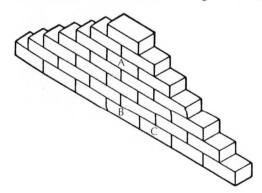

Figure 5.28   Bricks stacked pyramid fashion.

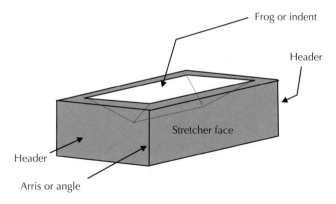

Figure 5.29   Brick faces.

Photograph 5.17   Aspects of a brick.

5.17). The face on which the brick is laid is the bed. Some bricks have an indent or frog formed in one of the bed faces. The purpose of the frog or indent is to assist in compressing the wet clay during moulding. The frog also serves as a reservoir of mortar onto which bricks in the course above may more easily be bedded.

Brick length can vary appreciably, especially those that are hand moulded and those made from plastic clays that will shrink differentially during firing. It was common practice to describe the thickness of a wall by the length of a brick, that is 1/2 B, 1 B, $1^1/_2$ B wall or a 2 B wall, rather than a precise dimension (the wall in Figure 5.30 and Figure 5.31 may be described as a $^1/_2$ B thick wall and the walls in Figure 5.32 are described as 1 B thick walls). This convention is used in this book; however, dimensions should always be specified to avoid any possible confusion.

The external leaf of a cavity wall is often built of brick. The most straightforward way of laying bricks in an outer leaf of a cavity wall is with the stretcher face of each brick showing externally. So that bricks are bonded along the length of the wall, they are laid with the

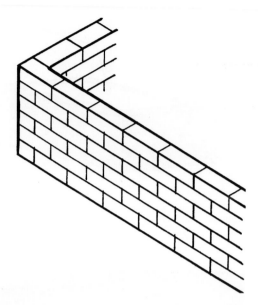

Figure 5.30 Stretcher bond.

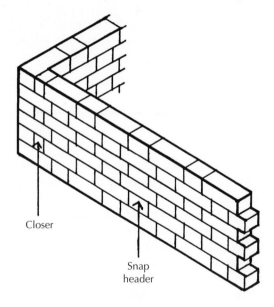

Closer

Snap
header

Figure 5.31 Flemish bond with snap headers.

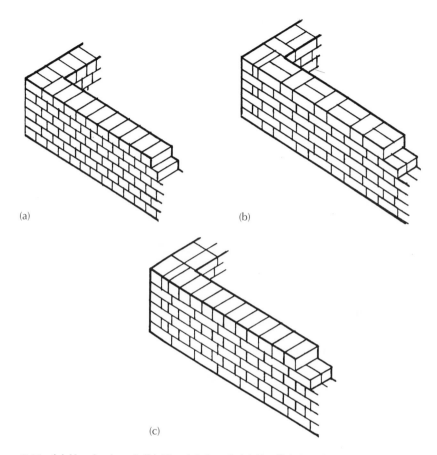

Figure 5.32   (a) Header bond. (b) Flemish bond. (c) English bond.

vertical joints between bricks lying directly under and over the centre of the bricks in the courses under and over. This is described as stretcher bond as illustrated in Figure 5.30 and Photograph 5.18.

At the intersection of two half brick walls at corners or angles and at the jambs, sides of openings, the bricks are laid so that a header face shows in every other course to complete the bond, as illustrated in Figure 5.30. The appearance of a wall laid in a stretcher bond may look somewhat monotonous because of the mass of stretcher faces showing. To provide some variety the wall may be built with snap headers so that a stretcher face and a header face show alternately in each course with the centre of the header face lying directly under and over the centre of the stretcher faces in courses below and above, as illustrated in Figure 5.31. This form of fake Flemish bond is achieved by the use of half bricks, hence the name 'snap header'. The combination and variety in colour and shape can add appreciably to the appearance of a wall. The additional labour and likely wastage of bricks add somewhat to the cost.

Photograph 5.18 Band of brick on end in a stretcher bond.

English and Flemish bond

A solid wall, 1 B and thicker, is bonded along its length and through its thickness. The two basic ways in which a solid brick wall may be bonded are with every brick showing a header face with each header face lying directly over two header faces below or with header faces centrally over a stretcher face in the course below. The bond in which only header faces show is termed 'heading' or 'header bond'. This bond is little used as the great number of vertical joints and header faces is generally considered unattractive. The bond in which header faces lie directly above and below a stretcher face is termed Flemish bond (Figure 5.32b). This bond is generally considered the most attractive bond for facing brickwork because of the variety of shades of colour between header and stretcher faces dispersed over the whole face of the walling.

English bond, illustrated in Figure 5.32c, avoids the repetition of header faces in each course by using alternate courses of header and stretcher faces with a header face lying directly over the centre of a stretcher face below. The colour of header faces may differ from the colour of stretcher faces. In English bond this difference is shown in successive horizontal courses (Photograph 5.19).

Bonding at angles and jambs

At the end of a wall at a stop end, at an angle or quoin and at jambs of openings the bonding of bricks has to be finished up to a vertical angle. To complete the bond a brick ¼ B wide has to be used to close or complete the bond of the ¼ B overlap of face brickwork. A brick,

Photograph 5.19   English bond.

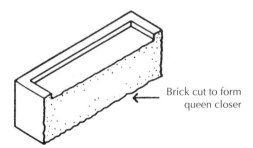

Brick cut to form
queen closer

Figure 5.33   Queen closer.

cut in half along its length, is used to close the bond at an angle. This cut brick is termed a 'queen closer', illustrated in Figure 5.33. If the narrow width queen closer were laid at the angle, it might be displaced during bricklaying. To avoid this possibility the closer is laid next to a header, as illustrated in Figure 5.34. The rule is that a closer is laid next to a quoin (corner) header.

There is often an appreciable difference in the length of facing bricks so that a solid wall 1 B thick may be difficult to finish as a wall 'fairface' on both sides. The word

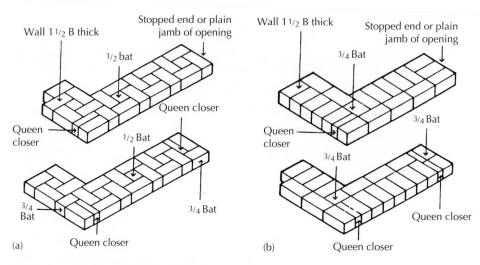

Figure 5.34   (a) Double Flemish bond. (b) English bond.

fairface describes a brick wall finished with a reasonably flat and level face for the sake of appearance. Where a 1 B wall is built with bricks of uneven length it may be necessary to select bricks of much the same length as headers and use longer bricks as stretchers. Walls $1\frac{1}{2}$ B thick may be used for substantial walling for larger buildings where the walling is finished fairface on both sides.

To complete the bond of a solid wall $1\frac{1}{2}$ B thick in double Flemish bond it is necessary to use cut half bricks in the thickness of the wall as illustrated in Figure 5.34a. At angles and stop ends of a wall, queen closers are laid next to quoin headers and a three quarter length cut brick is used. Cutting half-length ($\frac{1}{2}$ bats) and three quarter length bricks and closers is time consuming and wasteful. A $1\frac{1}{2}$ B thick wall, finished fairface on both sides and showing English bond on both sides, requires considerably less cutting of bricks to complete the bond, as illustrated in Figure 5.34b. It is only necessary to cut closers and three quarter length bricks to complete the bond at angles and stop ends.

Garden wall bond
Garden wall bonds are designed specifically to reduce the number of through headers to minimise the labour in selecting bricks of roughly the same length for use as headers. Usual garden wall bonds are three courses of stretchers to every one course of headers in an English garden wall bond and one header to every three stretchers in Flemish garden wall bond, as illustrated in Figure 5.35. The reduction in the number of through headers does to an extent weaken the through bond of the brickwork. Other combinations such as two or four stretchers to one header may be used. Tops of garden walls are finished with special coping bricks or a brick or stone capping or coping to provide weather protection to the top of the wall.

English garden wall bond   Flemish garden wall bond

Figure 5.35   Garden wall bonds.

5.7 Blocks and blockwork

Building blocks are wall units, larger in size than a brick, which can be handled by one person. Building blocks are made of concrete or clay.

Concrete blocks

These are used extensively for both loadbearing and non-loadbearing walls. A concrete block wall can be laid in less time than a similar brick wall. Lightweight aggregate concrete blocks have good insulating properties against transfer of heat and are used for the inner leaf of cavity walls. Concrete blocks may be used as a fairface external wall finish. Blocks are accurately moulded to uniform sizes and are made from aggregates to provide a variety of colours and textures.

Concrete blocks are manufactured from cement and either dense or lightweight aggregates as solid, cellular or hollow blocks as illustrated in Figure 5.36. A cellular block has one or more holes or cavities that do not pass wholly through the block and a hollow block is one in which the holes pass through the block. The thicker blocks are made with cavities or holes to reduce weight and drying shrinkage.

The most commonly used size of both dense and lightweight concrete blocks is 440 mm long × 215 mm high. The height of the block is chosen to coincide with three courses of brick for the convenience of building in wall ties and bonding to brickwork. The length of the block is chosen for laying in stretcher bond. For the leaves of cavity walls and internal loadbearing walls 100 mm thick blocks are used. For non-loadbearing partition walls 75 mm thick lightweight aggregate blocks are used. Either 440 × 215 mm or 390 × 190 mm blocks may be used. Concrete blocks may be specified by their minimum average compressive strength for:

❑   All blocks not less than 75 mm thick
❑   A maximum average transverse strength for blocks less than 75 mm thick, which are used for non-loadbearing partitions

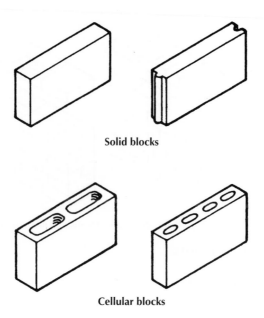

Solid blocks

Cellular blocks

Figure 5.36   Concrete blocks.

The usual compressive strengths for blocks are 2.8, 3.5, 5.0, 7.0. 10.0, 15.0, 20.0 and 35.0 N/mm². The compressive strength of blocks used for the walls of small buildings of up to three storeys, recommended in Approved Document A, is between 2.8 and 7 N/mm², depending on the loads carried. Concrete blocks may also be classified in accordance with the aggregate used in making the block and some common uses.

Dense aggregate blocks for general use

The blocks are made of cement, natural aggregate or blast-furnace slag. The usual mix is one part of cement to six or eight parts of aggregate by volume. These blocks are as heavy per cubic metre as bricks, they are not good thermal insulators and their strength in resisting crushing is less than that of most well-burnt bricks. The colour and texture of these blocks is far from attractive and they are usually covered with plaster or a coat of rendering. These blocks are used for internal and external loadbearing walls, including walls below ground.

Lightweight aggregate concrete blocks for general use in building

The blocks are made of cement and one of the following lightweight aggregates: granulated blast-furnace slag, foamed blast-furnace slag, expanded clay or shale, or well-burnt furnace clinker. The usual mix is one part cement to six or eight of aggregate by volume.

Lightweight aggregate concrete blocks are used primarily for internal non-loadbearing walls. These blocks, usually 75 or 100 mm thick, are made with the same lightweight aggregate as those in Class 2. These blocks are manufactured as solid, hollow or cellular depending largely on the thickness of the block. The thin blocks are solid and either square edged or with a tongue and groove in the short edges so that there is a mechanical bond between blocks to improve the stability of internal partitions. The poor structural stability may be improved by the use of storey height door linings, which are secured at floor and ceiling levels. Thin block internal partitions afford negligible acoustic insulation and poor support for fittings, such as bookshelves secured to them. The thicker blocks are either hollow or cellular to reduce weight and drying shrinkage.

As water dries out from precast concrete blocks, the shrinkage that occurs, particularly with lightweight blocks, may cause serious cracking of plaster and rendering applied to the surface of a wall built with them. Obviously, the wetter the blocks the more they will shrink. It is essential that these blocks be protected on building sites from saturation by rain both when they are stacked on site before use and while walls are being built.

Clay blocks

Hollow clay building blocks are made from selected brick clays that are press moulded and burnt. These hard, dense blocks are hollow to reduce shrinkage during firing and also to reduce their weight. They are grooved to provide a key for plaster, as illustrated in Figure 5.37. The standard block is 290 long × 215 mm high and 75, 100 or 150 mm thick. Clay blocks are comparatively lightweight, do not suffer moisture movement, and have good resistance to damage by fire and poor thermal insulating properties. In the UK, these blocks are mainly used for non-loadbearing partitions. They are extensively used in southern Europe as infill panel walls to framed buildings where the tradition is to render the external face of buildings on which the blocks provide a substantial mechanical key for rendering.

Bonding blocks

Blocks are made in various thicknesses to suit most wall requirements and are usually laid in stretcher bond. Thin blocks, used for non-loadbearing partitions, are laid in running stretcher bond with each block centred over and under blocks above and below. At return angles full blocks bond into the return wall in every other course, as illustrated in Figure 5.38. So as not to disturb the full width bonding of blocks at angles, for the sake of stability, a short length of cut block is used as closer and infill block. Thicker blocks are laid in off centre running bond with a three quarter length block at stop ends and sides of openings. The off centre bond is acceptable with thicker blocks as it avoids the use of cut blocks to complete the bond at angles.

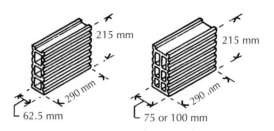

215 mm · 290 mm · 62.5 mm · 215 mm · 290 mm · 75 or 100 mm

Figure 5.37   Clay blocks.

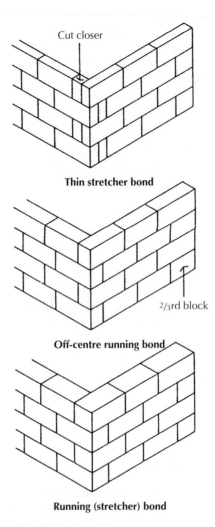

Figure 5.38   Bonding building blocks.

Thick blocks, whose length is twice their width, are laid in running (stretcher) bond as illustrated in Figure 5.38, and cut blocks are only necessary to complete the bond at stop ends and sides of opening. At the 'T' junctions of loadbearing concrete block walls it is sometimes considered good practice to butt the end face of the intersecting walls with a continuous vertical joint to accommodate shrinkage movements and to minimise cracking of plaster finishes. Where one intersecting wall serves as a buttress to the other, the butt joint should be reinforced by building in split end wall ties at each horizontal joint across the butt joint to bond the walls. Similarly, non-loadbearing block walls should be butt jointed at intersections and the joint reinforced with strips of expanded metal bedded in horizontal joints across the butt joint.

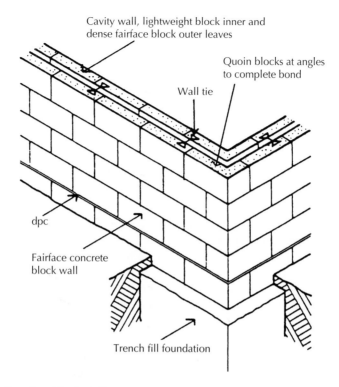

Cavity wall, lightweight block inner and
dense fairface block outer leaves

Quoin blocks at angles
to complete bond

Wall tie

dpc

Fairface concrete
block wall

Trench fill foundation

Figure 5.39   Bonding block walls.

Concrete block walls of specially produced blocks to be used as a fairface finish are bonded at angles to return walls with specially produced quoin blocks for the sake of appearance, as illustrated in Figure 5.39. The 'L'-shaped quoin blocks are made to continue the stretcher bond around the angle into the return walls. Quoin blocks are little used for other than fairface work as they are liable to damage in handling and use and add considerably to the cost of materials and labour.

5.8 Mortar

It is essential that brickwork and blockwork be laid in true horizontal courses, and the only way this can be done with bricks is to lay them on mortar. The basic requirements of a mortar are that it should harden to such an extent that it can carry the weight normally carried by bricks, without crushing, and that it should be sufficiently plastic when laid to take the varying sizes of bricks. It must have a porosity similar to that of the bricks and it must not deteriorate due to the weathering action of rain or frost.

Composition of mortar

Aggregate for mortar – soft and sharp sand

Sand is a natural material which is reasonably cheap and which, if mixed with water, can be made plastic, yet has very good strength in resisting crushing. Its grains are also virtually impervious to the action of rain and frost. Sand is termed either 'soft' or 'hard'. Sand that is not washed and that contains some clay in it feels soft and smooth when held in the hand, hence the term soft sand. Sand, which is clean, feels coarse in the hand, hence the term sharp. These are terms used by bricklayers. When soft sand ('builders' sand') is used, the mortar is very smooth and plastic and it is much easier to spread and to bed the bricks in than a mortar made of sharp sand.

Matrix for mortar

The material required to bind the grains of sand together into a solid mass is termed the matrix and two materials can be used for this purpose: either lime or cement. Lime, which mixes freely with water and sand, produces a material that is smooth, plastic and easily spread. Portland cement produces a hard dense material that has more than adequate strength for use as mortar and is largely unaffected by damp conditions. A mortar in which the advantages of lime and cement are combined is termed a 'compo' mortar.

Ready-mixed mortar

Ready-mixed mortars have come into use particularly on sites where extensive areas of brickwork are laid. A wide range of lime and sand, lime cement and sand and cement, and sand mixes is available. The sand may be selected to provide a colour and texture, or the mix may be pigmented for the same reason. Lime mortar is delivered to site ready to use within the day of delivery. Cement mix and cement lime mortar is delivered to site ready mixed with a retarding admixture. The retarding admixture is added to cement mix mortars to delay the initial set of cement. The initial set of ordinary Portland cement occurs some 30 minutes after the cement is mixed with water, so that an initial hardening occurs to assist in stiffening the material for use as rendering on vertical surfaces, for example. The advantages of ready-mixed mortar are consistency of the mix, the wide range of mixes available, saving in site labour costs and reduction of waste of material common with site mixing.

Types of mortar

Cement mortar

When fine cement powder is mixed with water a chemical action between water and cement takes place. At the completion of this reaction the nature of the cement has so changed that it binds itself very firmly to most materials. As the reaction takes place the excess water evaporates, leaving the cement and sand to harden gradually into a solid mass. The hardening of the mortar becomes noticeable some few hours after mixing and is complete in a few days. The usual mix of cement and sand for mortar is from one part cement to three or four parts sand to one part of cement to eight parts of sand by volume, mixed with just sufficient water to render the mixture plastic. A mortar of cement and sand is very durable and is often used for brickwork below ground level and brickwork exposed to weather above roof level such as parapet walls and chimney stacks. When used with some types of bricks it can cause efflorescence. Because cement mortar has greater compressive strength than required for most ordinary brickwork and because it is not very plastic by itself, it is sometimes mixed with lime and sand.

Lime mortar

Lime is manufactured by burning limestone or chalk and the result of this burning is an off-white lumpy material known as quicklime. When quicklime is mixed with water a chemical change occurs during which heat is generated, and the lime expands to about three times its former bulk. This change is gradual and takes some days to complete, and the quicklime afterwards is said to be slaked, i.e. it has no more thirst for water; more precisely the lime is said to be hydrated. Lime for building is delivered to site ready slaked and is termed 'hydrated lime'. When mixed with water, lime combines chemically with carbon dioxide in the air, and in undergoing this change it gradually hardens into a solid mass, which firmly binds the sand. The particular advantage of lime is that it is a cheap, reusable, readily available material that produces a plastic material ideal for bedding bricks. Its disadvantages are that it is a messy, laborious material to mix, and it gains strength slower than cement. Because it is to some extent soluble in water it will lose its adhesive property in persistently damp situations, crumble and eventually fall out of joints. Protected from damp, a lime mortar will serve as an effective mortar for the life of most buildings. A lime mortar is usually mixed with one part of lime to three parts of sand by volume. The mortar is plastic, easy to spread and hardens into a dense mass of good compressive strength.

Hydraulic lime

Hydraulic lime is made by burning a mixture of chalk or limestone that contains clay. Hydraulic lime is stronger than ordinary lime and will harden in wet conditions, hence the name. Ordinary Portland cement, made from similar materials and burnt at a higher temperature, has largely replaced hydraulic lime.

Proportions by volume

Mortar for general brickwork may be made from a mixture of cement, lime and sand in the proportions set out in Table 5.2. These mixtures combine the strength of cement with

Table 5.2   Mortar mixes

	Air-entrained mixes		
Mortar designation	Cement:lime:sand	Masonry cement:sand	Cement:sand with plasticiser
1	1:0 to $1/4$:3		
2	1:$1/2$:4 to $4^1/2$	1:$2^1/2$ to $3^1/2$	1:3 to 4
3	1:1:5 to 6	1:4 to 5	1:5 to 6
4	1:2:8 to 9	1:$5^1/2$ to $6^1/2$	1:7 to 8
5	1:3:10 to 12	1:$6^1/2$ to 7	1:8

Source: Taken from BS 5628: Part 3:1985 (table 15).

the plasticity of lime, have much the same porosity as most bricks and do not cause efflorescence on the face of the brickwork. The mixes set out in Table 5.2 are tabulated from rich mixes (1) to weak mixes (5). A rich mix of mortar is one in which there is a high proportion of matrix, that is, lime or cement or both, to sand as in the 1:3 mix, and a weak mix is one in which there is a low proportion of lime or cement to sand as in the mix 1:3:12. The richer the mix of mortar, the greater its compressive strength; and the weaker the mix, the greater the ability of the mortar to accommodate moisture or temperature movements. The general uses of the mortar mixes given in Table 5.2 are as mortar for brickwork or blockwork as follows:

- ❏   Mix 1: for cills, copings and retaining walls
- ❏   Mix 2: parapets and chimneys
- ❏   Mix 3: walls below dpc
- ❏   Mix 4: walls above dpc
- ❏   Mix 5: internal walls and lightweight block inner leaf of cavity

Mortar plasticisers

Liquids known as mortar plasticisers, when added to water, effervesce. If very small quantities are added to mortar, when it is mixed, the millions of minute bubbles that form surround the hard sharp particles of sand and so make the mortar plastic and easy to spread. The particular application of these mortar plasticisers is that if they are used with cement mortar they increase its plasticity and there is no need to use lime. It seems that the plasticisers do not adversely affect the hardness and durability of the mortar and they are commonly and successfully used for mortars.

Jointing and pointing

Jointing

Jointing is the word used to describe the finish of the mortar joints between bricks or blocks, in brickwork or blockwork that is finished fairface (not subsequently covered with plaster, rendering or other finish). Flush joints are generally made as a 'bagged' or a 'bagged in' joint. The joint is made by rubbing coarse sacking or a brush across the face of the brickwork to rub away all protruding mortar and hence leave a flush joint. This type of joint, illustrated in Figure 5.40, can most effectively be used on brickwork where the bricks are uniform in shape and comparatively smooth faced, where the mortar will not spread over the face of the brickwork.

A bucket handle joint is a concave, slightly recessed joint, illustrated in Figure 5.40. A bucket handle joint may be formed by a jointing tool with or without a wheel attachment to facilitate running the tool along uniformly deep joints. The advantage of the bucket handle joint is that the operation compacts the mortar into the joint and improves weather resistance to some extent. Flush and bucket handle joints are mainly used for jointing as the brickwork is raised, the joint being made after the mortar has hardened sufficiently, or 'gone off'.

The struck and recessed joints shown in Figure 5.40 are more laborious to make and therefore more expensive. The struck joint is made with a pointing trowel that is run along

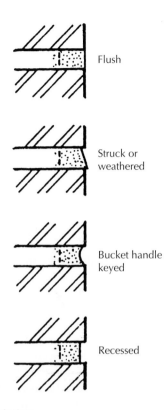

Figure 5.40   Jointing and pointing.

the joint either along the edges of uniformly shaped bricks or along a wood straight edge, where the bricks are irregular in shape or coarse textured, to form the splayed back joint. The recessed joint is similarly formed with a tool shaped for the purpose. Of the joints described, the struck joint is mainly used for pointing the joints in old brickwork and the recessed joint to emphasise the profile, colour and textures of bricks for appearance's sake to both new and old brickwork.

Pointing

Pointing is the operation of filling mortar joints with a mortar selected for colour and texture to either brickwork or blockwork. Mortar for pointing is a special mix of lime, cement and sand (or stone dust) chosen to produce a particular effect of colour and texture. The overall appearance of a fairface brick wall can be altered by the selection of mortar colour and type of pointing. The joints in new brickwork are raked out about 20 mm deep when the mortar has gone off sufficiently and before it has set hard and the joints are pointed as scaffolding is removed.

The mortar joints in old brickwork that was laid in lime mortar may in time crumble and be worn away by the action of wind and rain. The joints are raked out to a depth of about 20 mm and pointed with a mortar mix of cement, lime and sand that has roughly the same density as the brickwork. The operation of raking out joints is laborious and messy and the job of filling the joints with mortar for pointing is time consuming so that the cost of pointing old work is expensive. Pointing or re-pointing old brickwork is carried out both as protection for the old lime mortar to improve weather resistance and to improve the visual appearance of a wall.

5.9 Loadbearing brick and block walls

Functional requirements

Strength and stability
In Approved Document A there is practical guidance to meeting the requirements of the Building Regulations for the walls of small buildings of the following three types:

(1) Residential buildings of not more than three storeys
(2) Small single-storey non-residential buildings
(3) Small buildings forming annexes to residential buildings (including garages and outbuildings)

Height and width
The maximum height of residential buildings is given as 15 m from the lowest ground level to the highest point of any wall or roof in Approved Document A (Figure 5.41). Height is separately defined, for example, as from the base of a gable and external wall to half the height of the gable. The height of single-storey, non-residential buildings is given as 3 m from the ground to the top of the roof, which limits the guidance to very small buildings. The maximum height of an annexe is similarly given as 3 m, yet there is no definition of what is meant by annexe except that it includes garages and outbuildings. The least width

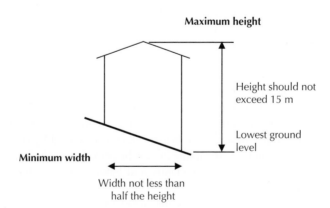

Figure 5.41   Maximum height and minimum width of walls for residential buildings (source: Approved Document A).

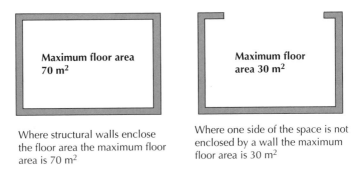

Figure 5.42   Structural walls and maximum floor area (source: Approved Document A).

of residential buildings is limited to not less than half the height. A diagram limits the dimensions of the wing of a residential building.

One further limitation is that no floor enclosed by structural walls on all sides should exceed 70 m² and a floor without a structural wall on one side, 30 m² (Figure 5.42). As the maximum allowable length of wall between buttressing walls, piers or chimneys is given as 12 m and the maximum span for floors as 6 m, the limitation is in effect a floor some 12 × 6 m on plan.

Strength

The guidance given in the Approved Document for walls of brick or block is based on compressive strengths of 5 N/mm² for bricks, and 2.8 N/mm² for blocks for walls up to two storeys in height, where the storey height is not more than 2.7 m and 7 N/mm² for bricks and blocks of walls of three-storey buildings where the storey height is greater than 2.7 m.

Stability

Thickness of walls

The general limitation of wall thickness given for stability is that solid walls of brick or block should be at least as thick as one-sixteenth of the storey height. This is a limiting slenderness ratio relating thickness of wall to height, measured between floors and floor and roof that provides lateral support and gives stability up the height of the wall. The minimum thickness of external, compartment and separating walls is given in a table in Approved Document A, relating thickness to height and length of wall as illustrated in Figure 5.43. Compartment walls are those that are formed to limit the spread of fire and separating walls (party walls), those that separate adjoining buildings, such as the walls between terraced houses. Cavity walls should have leaves at least 90 mm thick and the cavity at least 50 mm wide.

Internal loadbearing walls, except compartment and separating walls, should be half the thickness of external walls illustrated in Figure 5.43, minus 5 mm, except for the wall in the lowest storey of a three-storey building, which should be of the same thickness, or 140 mm, whichever is the greater.

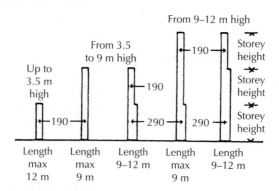

Figure 5.43   Minimum thickness of walls.

Table 5.3   Lateral support for walls

Wall type	Wall length	Lateral support required
Solid or cavity: external compartment separating	Any length	Roof lateral support by every roof forming a junction with the supported wall
	Greater than 3 m	Floor lateral support by every floor forming a junction with the supported wall
Internal loadbearing wall (not being a compartment or separating wall)	Any length	Roof or floor lateral support at the top of each storey

Source: Taken from Approved Document A (table 11). The Building Regulations.

Lateral support

For stability up the height of a wall lateral support is provided by floors and roofs as set out in Table 5.3. Walls that provide support for timber floors are given lateral support by 30 × 5 mm galvanised or stainless steel 'L' straps fixed to the side of floor joists at not more than 2 m centres for houses up to three storeys and 1.25 m centres for all storeys in all other buildings.

Lateral support from timber floors, where the joists run parallel to the wall, is provided by 30 × 5 mm galvanised iron on stainless steel strap anchors secured across at least three joists at not more than 2 m centres for houses up to three storeys and 1.25 m for all storeys in all other buildings. The 'L' straps are turned down a minimum of 100 mm on the cavity side of the inner leaf of cavity walls and into solid walls. Solid timber strutting is fixed between joists under the straps as illustrated in Figure 5.44. Solid floors of concrete provide lateral support for walls where the floor bears for a minimum of 90 mm in both solid and cavity walls.

To provide lateral support to gable end walls to roofs pitched at more than 15° a system of galvanised steel straps is used. Straps 30 × 5 mm are screwed to the underside of timber noggings fixed between three rafters, as illustrated in Figure 5.45, with timber packing

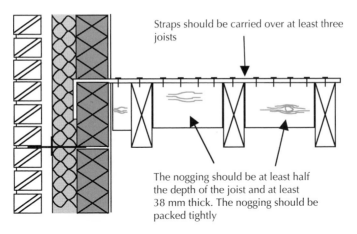

(a) Strap fixed to the top of joists

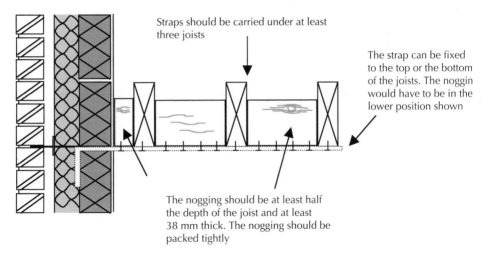

(b) Strap fixed below timber joists

Figure 5.44   Floors providing lateral restraint to walls.

pieces between the rafter next to the gable and the wall. The straps should be used at a maximum of 2 m centres and turned down against the cavity face of the inner leaf of a whole building block or down into a solid wall. To provide stability along the length and at the ends of loadbearing walls there should be walls, piers or chimneys bonded to the wall at intervals of not more than 12 m, to buttress and stabilise the wall.

The maximum spacing of buttressing walls, piers and chimneys is measured from the centre line of the supports as illustrated in Figure 5.46. The maximum length between piers and buttress walls is 3 m. The minimum length of a return buttressing wall should be equal

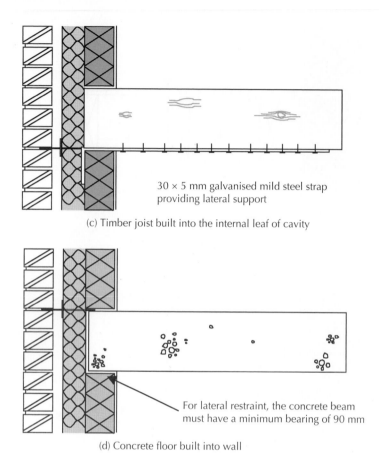

(c) Timber joist built into the internal leaf of cavity

30 × 5 mm galvanised mild steel strap
providing lateral support

For lateral restraint, the concrete beam
must have a minimum bearing of 90 mm

(d) Concrete floor built into wall

Figure 5.44   (Continued)

to one-sixth of the height of the supported wall. To be effective as buttresses to walls the return walls, piers and chimneys must be solidly bonded to the supported wall.

Where joists, floor intersection or rebates are cut into external walls it is necessary to ensure that that they do not provide a passage for air. To improve the energy efficiency of building the air permeability is controlled and reduced to a minimum. For example, in Passivhaus construction this is currently restricted to an airtightness of 0.6 air changes per hour at 50 Pa. To achieve airtight buildings the wall needs to be effectively sealed. It is common to parge masonry walls with a thin coat of mortar or plaster and to seal all junctions and interfaces. Parging was originally used to line chimneys or provide a waterproof cover to outer walls, but is now also used to improve airtightness.

Chases in walls (cuts and rebates in walls)
To limit the effect of chases cut into walls in reducing strength or stability, vertical chases should not be deeper than one-third of the thickness of solid walls or a leaf of a cavity wall

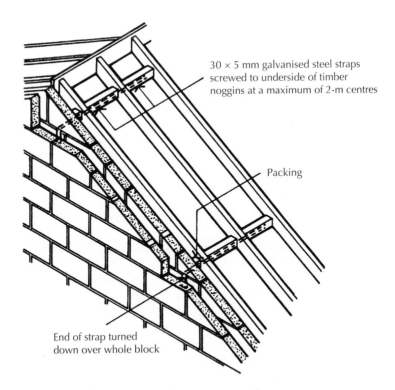

30 × 5 mm galvanised steel straps
screwed to underside of timber
noggins at a maximum of 2-m centres

Packing

End of strap turned
down over whole block

Figure 5.45   Lateral support to gable ends.

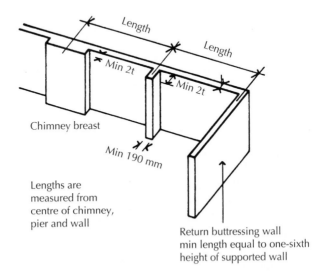

Length

Length

Min 2t

Min 2t

Chimney breast

Min 190 mm

Lengths are
measured from
centre of chimney,
pier and wall

Return buttressing wall
min length equal to one-sixth
height of supported wall

Figure 5.46   Length of walls.

and horizontal chases not deeper than one-sixteenth. A chase is a recess, cut or built into a wall, inside which small service pipes are run.

5.10 Solid wall construction

Up to the early part of the twentieth century, loadbearing walls were usually built as solid brickwork of adequate thickness to resist the penetration of rain to the inside face and to safely support the loads common to buildings both large and small. This has largely been replaced by cavity wall construction, although many buildings still exist that were built with solid walls.

Resistance to weather

A solid wall of brick will resist the penetration of rain to its inside face by absorbing rainwater that subsequently, in dry periods, evaporates to outside air. The penetration of rainwater into the thickness of a solid wall depends on the exposure of the wall to driving rain and the permeability of the bricks and mortar to water.

A solid 1 B thick wall may well be sufficiently thick to prevent the penetration of rainwater to its inside face in the sheltered positions common to urban areas on low-lying land. In positions of moderate exposure a solid wall $1^1/_2$ B thick will be effective in resisting the penetration of rainwater to its inside face. In exposed positions such as high ground and near the coast, a wall 2 B thick may be needed to resist penetration to inside faces, although a less thick wall protected with rendering or slate or tile hanging may be a more economical solution.

Rendering
The word rendering is used in the sense of rendering the surface of a brick or block wall smooth by the application of a wet mix of lime, cement and sand over the face of the wall. The rendering dries and hardens to a decorative protective coating that varies from dense and smooth to a coarse and open texture. Render improves the wall's resistance to rain penetration and alters its appearance. Materials and application of the various renders are described in Chapter 10.

Slate and tile hanging
In positions of very severe exposure to wind-driven rain, as on high open ground facing the prevailing wind and on the coast facing the open sea, it is necessary to protect both solid and cavity walls with an external cladding. Natural or manufactured slates and tiles can be used, hung on timber battens nailed to counter battens. Timber counter battens (50 × 25 mm) are nailed at 300 mm centres up the face of the wall to which timber slating or tiling battens are nailed at centres suited to the gauge (centres) necessary for double lap slates or tiles, as illustrated in Figure 5.47. As protection against decay, pressure impregnated softwood timber battens should be used, secured with non-ferrous fixings.

Where slate or tile hanging is used as cladding to a solid wall of buildings that are normally heated, then the thermal insulation can be fixed to the wall behind the counter

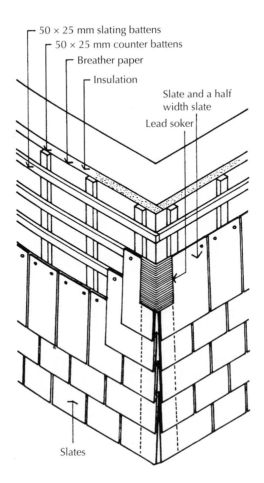

50 × 25 mm slating battens
50 × 25 mm counter battens
Breather paper
Insulation
Slate and a half
width slate
Lead soker
Slates

Figure 5.47 Slate hanging.

battens. Rigid insulation boards are fixed with a mechanically operated hammer gun that drives nails through both counter battens, a breather paper and the insulation boards into the wall. The continuous layer of breather paper that is fixed between the counter battens and the insulation is resistant to the penetration of water in liquid form but will allow water vapour to pass through it. Its purpose is to protect the outer surface of the insulation from cold air and any rain that might penetrate the hanging and to allow movement of vapour from within the structure to the external environment.

For vertically hung slating it is usual to use one of the smaller slates such as the 405 × 205 mm slate, which is headnailed to 50 × 25 mm battens and is less likely to be lifted and dislodged in high wind than longer slates would be. Each slate is nailed with non-ferrous nails to overlap two slates below, as illustrated in Figure 5.47, and double lapped by overlapping the head of slates two courses below.

At angles and the sides of openings a slate one and a half the width of slates is used to complete the overlap. This width of slate is specifically used to avoid the use of a half width

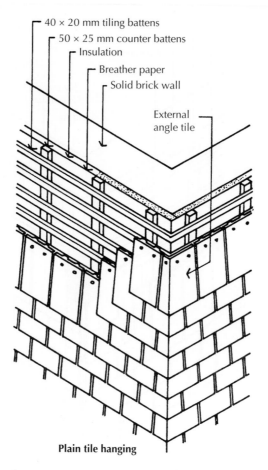

40 × 20 mm tiling battens
50 × 25 mm counter battens
Insulation
Breather paper
Solid brick wall
External angle tile

Plain tile hanging

Figure 5.48   Tile hanging.

slate that might easily be displaced in wind. Internal and external angles are weathered by lead soakers, which are hung over the head of slates, to overlap and make the joint weathertight. Slate hanging is fixed either to overlap or butt to the side of window and door frames with exposed edges of slates pointed with cement mortar or weathered with lead flashings. At lower edges of slate hanging a projection is formed on or in the wall face by means of blocks, battens or brick corbel courses on to which the lower courses of slates and tiles bell outwards slightly to throw water clear of the wall below.

Tile hanging is hung and nailed to 40 × 20 mm tiling battens fixed at centres to counter battens to suit the gauge of plain tiles. Each tile is hung to battens and also nailed, as security against wind, as illustrated in Figure 5.48. At internal and external angles special angle tiles may be used to continue the bond around the corner, as illustrated in Figure 5.48. As an alternative and also at the sides of openings, tile and a half width tiles may be used with lead soakers to angles and pointing to exposed edges or weathering to the sides of the openings.

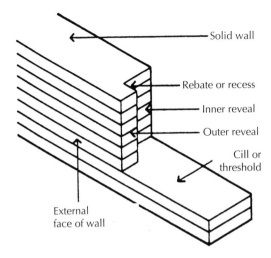

Figure 5.49   Rebated jamb.

Openings in solid walls

For the strength and stability of walling the size of openings in loadbearing walls is limited by regulations.

Jambs of openings

The jambs of openings for windows and doors in solid walls may be plain (square) or rebated. Plain or square jambs are used for small section window or door frames of steel and also for larger section frames where the whole of the external face of frames is to be exposed externally. The bonding of brickwork at square jambs is the same as for stop ends and angles with a closer next to a header in alternate courses to complete the bond. Rebated jambs are used to provide some protection to windows/door frames (Figure 5.49).

Rebated jambs

Historically rebated jambs have been used as a feature and to protect window and door fames. Figure 5.49 is a diagram of one rebated jamb and the terms commonly used. The thickness of brickwork that shows at the jamb of openings is described as the reveal. With rebated jambs there is an inner reveal and an outer reveal separated by the rebate. Figure 5.50 shows the brick arrangements for forming a rebated jamb in a Flemish bond. The amount of labour involved in creating such features makes them expensive and nowadays very rare.

Head of openings in solid walls

Solid brickwork over the head of openings has to be supported by either a lintel or an arch. The brickwork which the lintel or arch has to support is an isosceles triangle with 60° angles, formed by the bonding of bricks, as illustrated in Figure 5.51. The vertical joints between

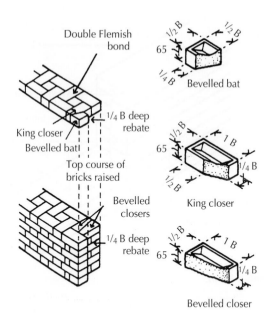

Figure 5.50   Bonding at rebated jambs.

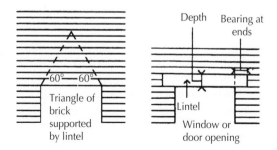

Figure 5.51   Head of openings.

bricks, which overlap ¼ B, form the triangle. In a bonded wall, if the solid brickwork inside the triangle were taken out, the load of the wall above the triangle would be transferred to the bricks of each side of the opening in what is termed 'the arching effect'.

Lintel (alternatively spelt lintol) is the name given to any single solid length of timber, stone, steel or concrete built in over an opening to support the wall above it, as shown in Figure 5.51. The ends of the lintel must be built into the brick or blockwork over the jambs to convey the weight carried by the lintel to the jambs. The area of wall on which the end of a lintel bears is termed its bearing at ends. The wider the opening, the more weight the lintel has to support and the greater its bearing at ends must be to transmit the load it

End of bar bent up   End of bar hooked

Figure 5.52   Ends of reinforcing rods.

carries to an area capable of supporting it. For convenience its depth is usually made a multiple of brick course height (75 mm), and the lintels are not usually less than 150 mm deep. Hardwood timber lintels have largely been replaced by concrete and steel lintels.

Reinforced concrete lintels

Lengths of steel rod (reinforcement) are cast into the bottom of concrete lintels to give them strength in resisting tensile or stretching forces. A minimum cover of concrete of 15 mm is necessary to avoid the possibility of corrosion and damage to the concrete around them. Reinforcing rods are usually of round section mild steel 10 or 12 mm diameter for lintels up to 1.8 m span. The ends of the rods should be bent up at 90° or hooked as illustrated in Figure 5.52. The purpose of bending up the ends is to ensure that when the lintel does bend, the rods do not lose their adhesion to the concrete around them. After being bent or hooked at the ends, the rods should be some 50 or 75 mm shorter than the lintel at each end.

Casting lintels

The word 'precast' indicates that a concrete lintel has been cast inside a mould, and has been allowed time to set and harden before it is built into the wall. The words 'in situ cast' indicate that a lintel is cast in position inside a timber mould fixed over the opening in walls. Precast lintels may be used for standard door and window openings, the advantage being that immediately the lintel is placed in position over the opening, brickwork can be raised on it, whereas the concrete in an in situ-cast lintel requires a timber mould or form-work and must be allowed to harden before brickwork can be raised on it. Lintels are cast in situ if a precast lintel would have been too heavy or cumbersome to have been hoisted and bedded in position. Precast lintels must be clearly marked to make certain that they are bedded with the steel reinforcement in its correct place, at the bottom of the lintel. Usually the letter 'T' or the word 'top' is cut into the top of the concrete lintel while it is still wet.

Prestressed concrete lintels

A prestressed lintel is made by casting concrete around high tensile, stretched wires, which are anchored to the concrete so that the concrete is compressed by the stress in the wires (see also *Barry's Advanced Construction of Buildings*). The load applied by the stressed wires, which compress the concrete, has to be overcome before the lintel will bend. Two types of prestressed concrete lintel are made: composite lintels and non-composite lintels.

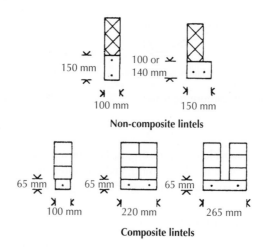

Figure 5.53   Prestressed lintels.

Composite lintels are stressed by wires at the centre of their depth and are used with brickwork, which acts as a composite part of the lintel in supporting loads. These comparatively thin precast lintels are built in over openings and brickwork is then built over them. Prestressed lintels over openings more than 1200 mm wide should be supported to avoid deflection until the mortar in the brickwork has set. When used to support blockwork the composite strength of these lintels is considerably less than when used with brickwork.

Non-composite prestressed lintels are made for use where there is insufficient brickwork over to act compositely with the lintel and also where there are heavy loads. These lintels are made to suit brick and block wall thicknesses, as illustrated in Figure 5.53.

Prestressed lintels may be used over openings in both internal and external solid walls. In external walls prestressed lintels are used where the wall is to be covered with rendering externally and for the inner leaf of cavity walls where the lintel will be covered with plaster. Precast reinforced concrete lintels may be exposed on the external face of both solid and cavity walling where the appearance of a concrete surface is desired.

Concrete boot lintels

The lintel is boot shaped in section with the toe part showing externally. The toe is usually made 65 mm deep. The main body of the lintel is hidden inside the wall and it is this part of the lintel which does most of the work of supporting the brickwork (Figure 5.54). Sometimes the detail is built so that the toe of the lintel finishes 25 or 40 mm back from the external face of the wall, as in Figure 5.54, to improve the visual appearance. The brickwork built on the toe of the lintel is usually ½ B thick for openings up to 1.8 m wide. The 65 mm deep toe, if reinforced as shown, is capable of safely carrying the two or three courses of ¹/₂ B thick brickwork over it. The brickwork above the top of the main part of the lintel bears mainly on it because the bricks are bonded. If the opening is wider than 1.8 m the main part of the lintel is sometimes made sufficiently thick to support most of the thickness

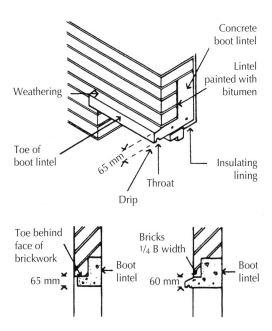

Figure 5.54   Boot lintels.

of the wall over, as in Figure 5.54. The bearing at ends where the boot lintel is bedded on the brick jambs should be of the same area as for ordinary lintels.

Although this section does not deal with cavity wall construction, Figure 5.55 illustrates how the concrete lintel used in solid wall construction evolved for use in cavity walls. Boot lintels (Figure 5.55c) should not be used in cavity wall construction because the solid concrete provides a thermal bridge across the cavity wall. Concrete lintels can be used in a cavity wall where two separate lintels are used to support each skin (Figure 5.55). The gap between the two skins of brickwork and blockwork can be filled with an insulated cavity closer. Further information on cavity wall construction can be found later in this section.

Pressed steel lintels

Galvanised pressed steel lintels are an alternative to concrete as a means of support to both loadbearing and non-loadbearing internal walls. Mild steel strip is pressed to shape, welded as necessary and galvanised. The steel lintels for support over door openings in loadbearing internal walls are usually in hollow box form, as illustrated in Figure 5.56. A range of lengths and sections is made to suit standard openings, wall thicknesses, course height for brickwork and adequate bearing at ends. For use over openings in loadbearing concrete block internal walls, it is usually necessary to cut blocks around the bearing ends of these shallow depth lintels. The exposed lintel faces are perforated to provide a key for plaster.

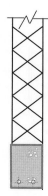

(a) Solid concrete lintel in a single skin wall; unsuitable for heated buildings

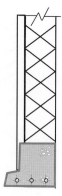

(b) Solid concrete boot lintel in a rendered block wall; unsuitable for modern heated buildings but found in existing buildings

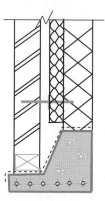

(c) Solid concrete boot lintel in cavity walls are unsuitable for heated buildings due to thermal bridging via the lintel

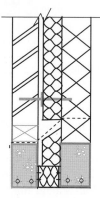

(d) By using two separate concrete lintels to support each leaf of masonry and providing a sealed cavity, the thermal bridging effect is reduced

Figure 5.55 **Evolution of the concrete lintel from solid wall to use in cavity construction.**

To support thin, non-loadbearing concrete blocks over narrow door openings in partition walls, a small range of corrugated, pressed steel lintels is made to suit block thickness. These shallow depth galvanized lintels are made to match the depth of horizontal mortar joints to avoid cutting of blocks. The corrugations provide adequate key for plaster run over the face of partitions and across the soffit of openings, as illustrated in Figure 5.57. These lintels act compositely with the blocks they support. To prevent sagging they should

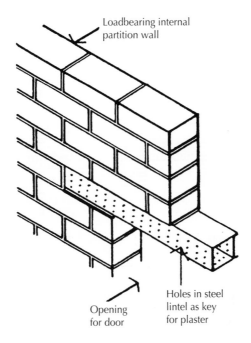

Figure 5.56   Steel lintels in internal walls.

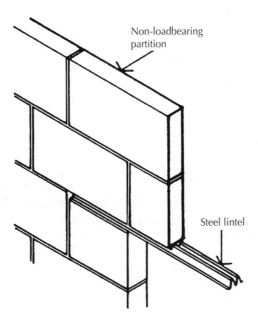

Figure 5.57   Corrugated steel lintel in internal wall.

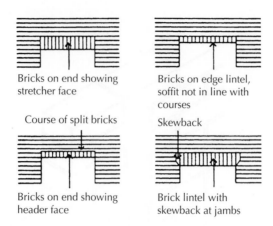

Bricks on end showing
stretcher face

Bricks on edge lintel,
soffit not in line with
courses

Course of split bricks

Skewback

Bricks on end showing
header face

Brick lintel with
skewback at jambs

Figure 5.58   Brick lintels.

be given temporary support at mid-span until the blocks above have been laid and the mortar hardened.

Brick lintels

A brick lintel may be formed as bricks on end, bricks on edge or coursed bricks laid horizontally over openings. Bricks laid in mortar give poor support to the wall above and usually need some form of additional support. A brick on-end lintel is generally known as a 'soldier arch' or 'brick on-end' arch (although the bricks are laid flat). The lintel is built with bricks laid on end with stretcher faces showing, as illustrated in Figure 5.58. For openings up to about 900 mm wide it was common to provide some support for soldier arches by building the lintel on the head of timber window and door frames. The wood frame served as temporary support as the bricks were laid, and support against sagging once the wall was built. A variation was to form skew back bricks at each end of the lintel with cut bricks so that the slanting surface bears on a skew brick in the jambs, as illustrated in Figure 5.58. The skew back does give some little extra stability against sagging.

Brick arches

An arch, which is an elegant and structurally efficient method of supporting brickwork, has for centuries been the preferred means of support for brickwork over the small openings for doors and windows and for arcades, viaducts and bridges.

Semi-circular arch

The most efficient method of supporting brickwork over an opening is by the use of a semi-circular arch, which transfers the load of the wall it supports most directly to the sides of the opening through the arch, Figure 5.59. A segmented arch, which takes the form of a segment (part) of a circle, is less efficient in that it transmits loads to the jambs by both vertical and outward thrust.

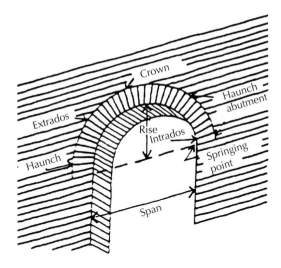

Figure 5.59   Semi-circular brick arch.

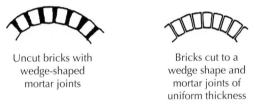

Uncut bricks with
wedge-shaped
mortar joints

Bricks cut to a
wedge shape and
mortar joints of
uniform thickness

Figure 5.60   Rough and axed arches.

Rough and axed arches

The two ways of constructing a curved brick arch are with bricks laid with wedge-shaped mortar joints or with wedge-shaped bricks with mortar joints of uniform thickness, as illustrated in Figure 5.60. An arch formed with uncut bricks and wedge-shaped mortar joints is termed a rough brick arch because the mortar joints are irregular and the finished effect is rough. In time the joints tend to crack and emphasise the rough appearance; thus rough archwork is rarely used for fairface work.

Arches in fairface brickwork are usually built with bricks cut to wedge shape with mortar joints of uniform width. Bricks are cut (either on or off site) to the required wedge shape by gradually chopping them to shape, hence the name 'axed bricks'. A template, or pattern, is cut from a sheet of zinc to the exact wedge shape to which the bricks are to be cut. The template is laid on the stretcher or header face of the brick as illustrated in Figure 5.61. Shallow cuts are made in the face of the brick each side of the template. These cuts are

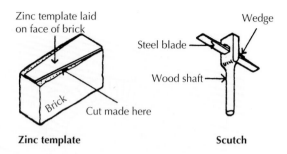

Figure 5.61   Axed brick.

made with a hacksaw blade or file and are to guide the bricklayer in cutting the brick. Then, holding the brick in one hand, the bricklayer gradually chops the brick to the required wedge shape. When the brick has been cut to a wedge shape, the rough, cut surfaces are roughly levelled with a coarse rasp. From the description this appears to be a laborious operation, but in fact the skilled bricklayer can axe a brick to a wedge shape in a few minutes. The axed wedged-shaped bricks are built to form the arch with uniform 10 mm mortar joints between the bricks.

Gauged brick arch

Gauged bricks are those that have been accurately prepared to a wedge shape so that they can be put together to form an arch with very thin joints between them. This does not improve the strength of the brick arch and is done for appearance. Hard burnt clay facing bricks cannot be cut to the accurate wedge shape required for this work because the bricks are too coarse grained. One type of brick used for gauged brickwork is called a rubber brick (or brick rubbers) because its composition is such that it can be rubbed down to an accurate shape on a flat stone. Rubbers are made from fine-grained sandy clays. They are moulded and then baked to harden them at a lower temperature than that at which clay bricks are burnt, the aim being to avoid fusion of the material of the bricks so that they can easily be cut and accurately rubbed to shape.

Sheet zinc templates are cut to the exact size of the wedge-shaped brick voussoirs. These templates are placed on the stretcher or header face of the brick to be cut and the brick is sawn to a wedge shape with a brick saw. Then they are carefully rubbed down by hand on a large flat stone until they are the exact wedge shape required as indicated by the template. The gauged rubber bricks are built to form the arch with joints between the bricks as thin as 1.5 mm thick. Mortar used between the gauged bricks is composed of either fine sand and cement and lime or lime and water, depending on the thickness of joint. The finished effect of accurately gauged red bricks with thin white joints between them is considered very attractive. Gauged bricks are used for flat camber arches.

Two-ring arch

Rough and axed bricks are used for both semi-circular and segmental arches and gauged brick for segmental and flat camber arches to avoid the more considerable cutting necessary

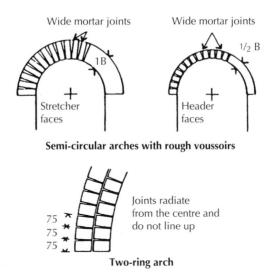

Semi-circular arches with rough voussoirs

Joints radiate from the centre and do not line up

Two-ring arch

Figure 5.62   Two-ring arch.

with semi-circular arches. Rough, axed or gauged bricks can be laid so that either their stretcher or their header face is exposed. Semi-circular arches are often formed with bricks showing header faces to avoid the excessively wedge-shaped bricks or joints that occur with stretcher faces showing. This is illustrated in Figure 5.62 by the comparison of two arches of similar span first with stretcher face showing and then with header face showing. If the span of the arch is of any considerable width, say, 1.8 m or more, it is often practice to build it with what is termed two or more rings of bricks, as illustrated in Figure 5.62. An advantage of two or more rings of bricks showing header faces is that the bricks bond into the thickness of the wall. Where the wall over the arch is more than 1 B thick it is practical to affect more bonding of arch bricks in walls or viaducts by employing alternate snap headers (half bricks) in the face of the arch.

Segmental arch

The curve of this arch is a segment, i.e. part of a circle, and designers of a building can choose any segment of a circle that they think suits their design. By trial and error over many years bricklayers have worked out methods of calculating a segment of a circle related to the span of this arch, which gives a pleasant looking shape, and which at the same time is capable of supporting the weight of brickwork over the arch. The recommended segment is such that the rise of the arch is 130 mm for every metre of span of the arch.

Centring

Temporary support for brick arches is necessary, and this is usually in the form of a timber framework to the profile of the underside of the arch, called centring. It is fixed and supported in the opening while the bricks of the arch are being built and the coursed brickwork over the arch laid. When the arch and brickwork are finished the centring is removed. A

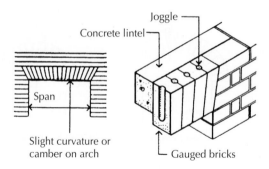

Figure 5.63   Flat gauged camber arch.

degree of both skill and labour is involved in arch building, in setting out the arch, cutting bricks for the arch and the abutment of coursed brickwork to the curved profile of the arch; thus an arched opening is more expensive than a plain lintel head.

Flat camber arch

A flat camber arch is not a true arch as it is not curved and might well be more correctly named flat brick lintel with voussoirs radiating from the centre, as illustrated in Figure 5.63. The bricks from which the arch is built may be either axed or gauged to the shape required so that the joints between the bricks radiated from a common centre and the widths of voussoirs measured horizontally along the top of the arch are the same. This width will be 65 mm or slightly less, so that there are an odd number of voussoirs; the centre is the key brick.

The centre from which the joints between the bricks radiate is usually determined either by making the skew or slating surface at the end of the arch 60° to the horizontal or by calculating the top of this skew line as lying 130 mm from the jamb for every metre of span. If the underside or soffit of this arch were made absolutely level it would appear to be sagging slightly at its centre. This is an optical illusion and it is corrected by forming a slight rise or camber on the soffit of the arch. This rise is usually calculated at 6 or 10 mm for every metre of span and the camber takes the form of a shallow curve. The camber is allowed for when cutting the bricks to shape. In walls built of hard coarse-grained facing bricks this arch is usually built of axed bricks. In walls built of softer, fine-grained facing bricks the arch is usually of gauged rubbers and is termed a flat gauged camber arch. This flat arch must be of such height on face that it bonds in with the brick course of the main walling. The voussoirs of this arch, particularly those at the extreme ends, are often longer overall than a normal brick and the voussoirs have to be formed with two bricks cut to shape.

Flat gauged camber arch

The bricks in this arch are jointed with lime and water, and the joints are usually 1.5 mm thick. Lime is soluble in water and does not adhere strongly to bricks, and in time the jointing material between the bricks may perish and the bricks might slip out of position.

To prevent this, joggles are formed between the bricks. These joggles take the form of semi-circular grooves cut in both bed faces of each brick, as shown in Figure 5.63, into which mortar is run.

Thermal insulation

To provide adequate insulation to a solid wall it is necessary to fix a layer of some light-weight insulating material to either the internal or exterior face of the wall.

Internal insulation

Internal insulation is used where solid walls have sufficient resistance to the penetration of rain, and/or alteration to the external appearance is not permitted or desirable. A disadvantage of internal insulation is that as the insulation is at, or close to, the internal surface, it will prevent the wall behind from acting as a heat store where constant, low-temperature, heating is used. There is also potential for interstitial condensation to form on the face of the wall; therefore a vapour barrier should be used on the surface of the insulation to ensure moisture is restricted from meeting the cold face of the wall. Photograph 5.20 shows 150 mm of polyisocyanurate (PIR) insulation fixed to the internally side of a solid 250 mm thick clay brick wall. The insulation on the plane elements provides an approximate U-value of $0.15 \, \text{W/m}^2 \text{K}$.

Photograph 5.20   150 mm of polyisocyanurate (PIR).

It is usual to cover the insulating layer with a lining of plasterboard or plaster so that the combined thickness of the inner lining and the wall have the required U-value. Internal linings for thermal insulation are either preformed laminated panels that combine a wall lining of plasterboard glued to an insulation board, or separate insulation material fixed to the wall and then covered with plasterboard or plaster. The method of fixing the lining to the inside wall surface depends on the surface to which it is applied.

Adhesive fixing

Adhesive fixing directly to the inside wall face is used for preformed laminate panels and for rigid insulation boards. Where the inside face of the wall is clean, dry, level and reasonably smooth, the panels or boards are secured with organic-based gap filling adhesive, applied in dabs and strips to the back of the boards or panels or to both the boards and wall. The panels or boards are then applied and pressed into position against the wall face and their position adjusted with a foot lifter.

Where the surface of the wall is uneven or rough the panels or boards are fixed with dabs or bonding, applied to both the wall surface and the back of the lining. Dabs are small areas of wet plaster bonding applied at intervals on the surface with a trowel, as a bedding and adhesive. The lining is applied and pressed into position against the wall. The wet dabs of bonding allow for irregularities in the wall surface and also serve as an adhesive. Some lining systems use secondary fixing in addition to adhesive; non-ferrous or plastic nails, or screws, driven or screwed through the insulation boards into the wall. In order to prevent air leakage around the edges of plasterboards applied using dot and dab, a ribbon (continuous strip) of plaster or fixing adhesive should be positioned around the perimeter of the plasterboard. Figure 5.64 is an illustration of laminated insulation panels fixed to the inside face of a solid wall.

Mechanical fixing

As an alternative to adhesive fixing, the insulating lining and the wall finish can be fixed to wood battens that are nailed to the wall with packing pieces as necessary, to form a level surface. The battens should be impregnated against rot and fixed with non-ferrous fixings. The insulating lining is fixed either between the battens or across the battens and an internal lining of plasterboard is then nailed to the battens, through the insulation. The thermal resistance of wood is less than that of most insulating materials. When the insulating material is fixed between the battens there will be thermal bridges through the battens, which may cause staining on wall faces.

Internal wall finish

An inner lining of plasterboard can be finished by taping and filling the joints or with a thin skim coat of plaster. A plaster finish of lightweight plaster and finishing coat is applied to the ready keyed surface of some insulating boards or to expanded metal lathing fixed to battens. Laminated panels of insulation, lined on one side with a plasterboard finish, are made specifically for the insulation of internal walls. The panels are fixed with adhesive or mechanical fixings to the inside face of the wall.

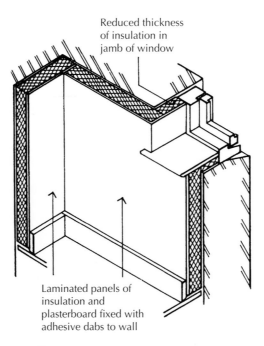

Reduced thickness
of insulation in
jamb of window

Laminated panels of
insulation and
plasterboard fixed with
adhesive dabs to wall

Figure 5.64   Internal insulation.

Vapour check (vapour barrier)

Moisture vapour pressure from warm moist air inside insulated buildings may find its way through internal linings and condense on cold outer faces. Where the condensate is absorbed by the insulation it will reduce the efficiency of the insulation, and where condensation saturates battens, they may rot. With insulation that is permeable to moisture vapour, a vapour check should be fixed on the room side of the insulation. A vapour barrier is one that completely stops the movement of vapour through it and a vapour check is one that substantially stops vapour. As it is difficult to make a complete seal across the whole surface of a wall including all overlaps of the barrier and at angles, it is in effect impossible to form a barrier and the term vapour check should more properly be used. Sheets of polythene with edges overlapped and taped together are commonly used as a vapour check.

External wall insulation

Insulating materials by themselves do not provide a satisfactory external finish to walls against rain penetration, nor do they provide an attractive finish; thus they are covered with a finish of cement rendering, paint or a cladding material such as tile, slate, profiled sheeting or weatherboarding. For rendered finishes, one of the inorganic insulants, rockwool or cellular glass in the form of rigid boards, is most suited. For cladding, one of the organic insulants is used because it is thinner for a comparable U-value.

Because the rendering is applied over a layer of insulation it will be subject to greater temperature fluctuations than it would be if it were applied directly to a wall, and so it is

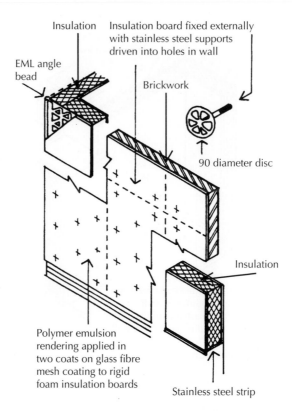

Figure 5.65   External insulation on expanded metal lath (EML).

more liable to crack. To minimise cracking due to temperature change and moisture movements, the rendering should be reinforced with a mesh securely fixed to the wall, and movement joints should be formed at not more than 6 m intervals. The use of a light coloured finish and rendering incorporating a polymer emulsion can also help to reduce cracking. As the overall thickness of the external insulation and rendering is too great to be returned into the reveals of existing openings it is usual to return the rendering by itself, or to fix some non-ferrous or plastic trim to mask the edge of the insulation and rendering. This will result in the reveals of openings acting as thermal bridges, which is not desirable. Figure 5.65 is an illustration of insulated rendering applied externally.

Slabs of compressed mineral wool are secured to the external face of the wall with stainless steel brackets, fixed to the wall to support and restrain the blocks that are arranged with either horizontal, bonded joints or vertical and horizontal continuous joints.

Flexible external wall insulation system

The system illustrated in Photograph 5.21a–e and Figure 5.66 show an external wall insulation system that can utilise PIR, expanded polystyrene (EPS), mineral wool or phonolic insulation materials. The ability to work with different insulants helps designers achieve

Photograph 5.21 (a) Existing building, with little thermal resistance and in need of renovation and aesthetic upgrade. (b) Thermal performance improved using external wall insulation applied, brick slips and render applied for aesthetics and to improve durability. The eaves of the building have been extended to provide adequate cover. (c) External wall insulation mechanically fixed to the wall. (d) Render applied to insulation. (e) Brick slips and render applied. (Images courtesy of Weber Saint Gobain, http://www.netweber.co.uk.)

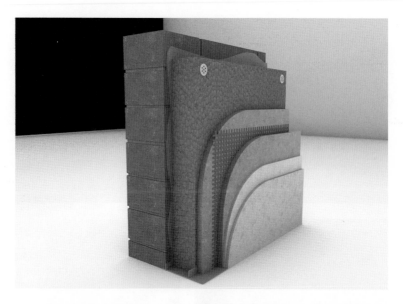

Figure 5.66   External wall insulation, with bond and mechanical fixing (image courtesy of Weber Saint Gobain, http://www.netweber.co.uk).

the desired aesthetics, durability, thermal performance and enables choice of systems to a specific budget. Applying the bonding layer over the whole area of the wall helps to seal the wall, improving airtightness and improving the bond between insulation and substrate. It is advisable to use both adhesive and mechanical bond where the quality of the substrate varies. See Photograph 5.21a–e.

As shown in Photograph 5.21c the insulation can be fixed by purpose-made plugs and mechanical fixings. These fixings reduce the thermal bridges that can occur when frames are used to attach the insulation.

Interlocking polystyrene building blocks

An alternative approach is to make the insulation an integral part of the wall construction. One example is the Styrostone system, which uses hollow interlocking polystyrene blocks (Figure 5.67 and Photograph 5.22a). Styrostone fits together much like children's toy construction bricks, Lego, with interlocking nodules on the top of the bricks and rebates along the bottom to receive the nodules (Photograph 5.22, Photograph 5.23, Photograph 5.24 and Photograph 5.25). The two skins of polystyrene that form the brick are tied together using plastic reinforcing spacers. The spacers ensure that the polystyrene forms retain their shape while the concrete fill is poured into the void. Ten polystyrene blocks can be laid on top of each other to form a 2.5 m high wall, after which the polystyrene chamber needs to be filled with concrete (Photograph 5.22b and c). Then, if necessary, the next lift of 10 blocks can be placed ready to receive the concrete. Special polystyrene blocks have also been made for use at corners and above window and doors (acting as lintels).

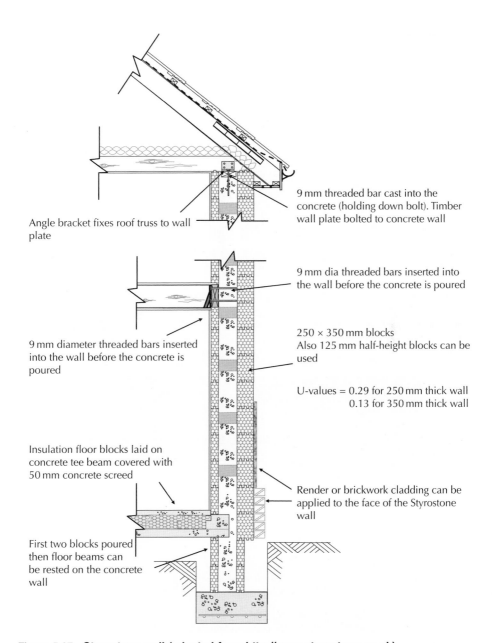

9 mm threaded bar cast into the concrete (holding down bolt). Timber wall plate bolted to concrete wall

Angle bracket fixes roof truss to wall plate

9 mm dia threaded bars inserted into the wall before the concrete is poured

9 mm diameter threaded bars inserted into the wall before the concrete is poured

250 × 350 mm blocks
Also 125 mm half-height blocks can be used

U-values = 0.29 for 250 mm thick wall
              0.13 for 350 mm thick wall

Insulation floor blocks laid on concrete tee beam covered with 50 mm concrete screed

Render or brickwork cladding can be applied to the face of the Styrostone wall

First two blocks poured then floor beams can be rested on the concrete wall

Figure 5.67   Styrostone wall (adapted from http://www.styrostone.co.uk).

Photograph 5.22   (a) Interlocking Styrostone block with spacers. (b) Blocks clipped together to form wall and permanent formwork. (c) Reinforcement inserted in the concrete to add lateral tensile strength. (http://www.styrostone.co.uk)

Photograph 5.23   (a) Styrostone walls and floor. (b) Support for floor positioned. (c) Styrostone floor – temporary props used until the concrete has sufficiently matured and set. (http://www.styrostone.co.uk)

Photograph 5.24   (a) Styrostone reinforced basement wall. (b) Styrostone basement wall. (http://www.styrostone.co.uk)

The polystyrene acts as a type of permanent or lost formwork, with the advantage of providing good thermal insulation. Reinforcement can be used to provide additional tensile strength where required (Photograph 5.22c). Due to the lightness of the blocks and the ease with which they are assembled, the structural wall is quickly assembled. The walls of a single-storey building can be fully constructed by three workers in just 1 day. The system has been tested and is capable of meeting current structural, thermal and acoustic requirements with considerable savings on many traditional forms of construction. U-values range from 0.29 for a 250 mm wall to 0.12 for a 350 mm wall. In Germany the method has been used to construct five-storey houses. To ensure that the dwellings fit in with the local vernacular, the Styrostone (polystyrene and concrete) structural membrane can be clad with render, stucco, stone or brick (Photograph 5.25). Specially designed brick slips have already been developed so that the Styrostone wall can be clad to give the appearance of brickwork,

Photograph 5.25 Polystyrene and infill concrete house – Styrostone block house ready for aesthetic cladding.

without the cost of a traditional masonry wall. The Styrostone blocks are of particular interest to the self-build market.

Recycled block construction

A similar system to that of Styrostone is the Durisol block construction (Photograph 5.26a–c). As an alternative to the polystyrene walling unit, the Durisol system uses special grade recycled wood, which is bonded together with a type of cement. Although made from wood, the system does not suffer from decay or attack from wood-boring beetles that would affect untreated timber. The blocks act as an insulating former, allowing concrete infill to give it the main structural strength. The building units have a range which includes 500 mm (length), 250 mm (high), 170 mm (thick) uninsulated blocks for use on non-loadbearing internal walls; 500 mm (length), 250 mm (high), 250 mm (thick) uninsulated blocks for loadbearing internal walls; and 500 mm (length), 250 mm (high), 300 mm (thick) and 365 mm (thick) insulated blocks for external walls. The widest insulated walling units have a U-value of $0.19 \, W/m^2 \, K$. Due to the lightweight nature of the blocks they can be built to form a wall and do not require the skilled labour of a bricklayer. Both the Styrostone and Durisol units need a render or dry applied cladding to provide an aesthetically pleasing finish. Although walling blocks both systems rely on in situ structural concrete to give loadbearing strength and can be considered as quick assemble permanent formwork systems. It is clear that these modern methods of construction will increase and become important methods of construction.

Photograph 5.26   (a) Durisol recycled timber building blocks. (b) Insulated Durisol blocks. (c) Insulated recycled blocks and reinforcement ready for structural concrete to fill the void. (http://www.durisol.net)

5.11 Cavity wall construction

The idea of forming a vertical cavity in brick walls was first proposed early in the nineteenth century and developed through the twentieth century. The outer leaf and the cavity serve to resist the penetration of rain to the inside face and the inner leaf to support floors, provide a solid internal wall surface and, to some extent, act as insulation against transfer of heat. Various widths of cavity were proposed from the first 6 inch (150 mm) cavity; a later 2 inch (50 mm) cavity followed by proposals for 3, 4 or 5 inch wide cavities. Early cavity walls were constructed with bonding bricks laid across the cavity at intervals to tie the two leaves together. Either whole bricks with end closers or bricks specially made to size and shape for the purpose were used. Later, iron ties were used to tie the two leaves together. From the middle of the twentieth century it became common practice to construct the external walls of houses as a cavity wall with a 50 mm wide cavity and metal wall ties. The need to improve the thermal resistance of cavity walls resulted in the cavity being filled, either partially or fully with insulating materials as the cavity provided a convenient space for adding the thermal insulating material (see Figure 5.68). (Photograph 5.27). With a fully filled cavity the 'cavity wall' has evolved into a sandwich construction. More stringent regulations has led to an increase in the cavity width to accommodate a greater depth of thermal insulating material, although with the development of ultra-thin insulating materials it is likely that the width of the cavity may be reduced in some circumstances. The depth of some cavity walls has built up to over 500 mm thickness of different materials combining to make a highly insulated, fully filled, cavity wall. This composite construction is still referred to as a cavity wall, although there is no longer an air gap (cavity) in the

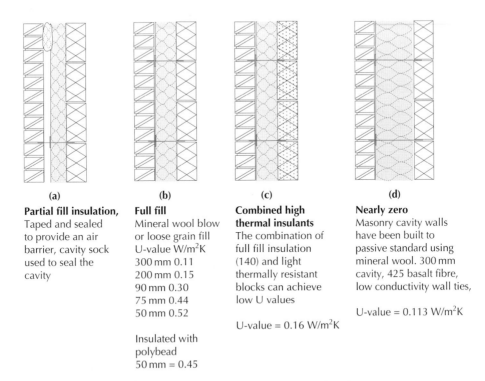

(a)

Partial fill insulation,
Taped and sealed
to provide an air
barrier, cavity sock
used to seal the
cavity

(b)

Full fill
Mineral wool blow
or loose grain fill
U-value W/m²K
300 mm 0.11
200 mm 0.15
90 mm 0.30
75 mm 0.44
50 mm 0.52

Insulated with
polybead
50 mm = 0.45

(c)

**Combined high
thermal insulants**
The combination of
full fill insulation
(140) and light
thermally resistant
blocks can achieve
low U values

U-value = 0.16 W/m²K

(d)

Nearly zero
Masonry cavity walls
have been built to
passive standard using
mineral wool. 300 mm
cavity, 425 basalt fibre,
low conductivity wall ties,

U-value = 0.113 W/m²K

Figure 5.68   Thermally resistant cavity wall construction (courtesy of Felix Thomas).

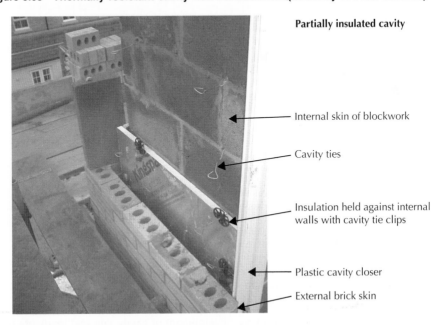

Partially insulated cavity

Internal skin of blockwork

Cavity ties

Insulation held against internal
walls with cavity tie clips

Plastic cavity closer

External brick skin

Photograph 5.27   Cavity wall construction.

construction. Using 300 mm of mineral wool, lightweight blocks with plaster are used to seal and create an airtight fabric. Highly insulated fully filled cavity walls can be constructed to U-values of 0.1 W/m² K, meeting nearly zero fabric standards as shown in Figure 5.68.

Strength and stability

Cavity walls with adequate ties offer strong and stable construction. The external skin, which is not loadbearing, is held stable by the internal skin, which is loadbearing. When tied together the two skins offer increased rigidity and stability (which relies on the two skins remaining connected). The practical guidance in Approved Document A accepts a cavity of 50–100 mm for cavity walls having leaves at least 90 mm thick, built of coursed brickwork or blockwork with wall ties spaced at 450 mm vertically and from 900 to 750 mm horizontally for cavities of 50–100 mm wide, respectively. As the limiting conditions for the thickness of walls related to height and length are the same for a solid bonded wall 190 mm thick as they are for a cavity wall of two leaves each 90 mm thick, it is accepted that the wall ties give the same strength and stability to two separate leaves of brickwork as the bond in solid walls.

Wall ties

Early cavity walls were very simple constructions, with the two leaves connected by bricks that simply crossed the cavity. While these were structurally sound the bricks provided a thermal bridge and passage for water. Iron ties (which rust) were replaced by mild steel ties. However, the early ties were prone to corrosion (Figure 5.69) and many

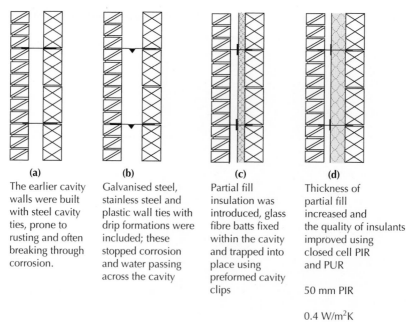

(a)	**(b)**	**(c)**	**(d)**
The earlier cavity walls were built with steel cavity ties, prone to rusting and often breaking through corrosion.	Galvanised steel, stainless steel and plastic wall ties with drip formations were included; these stopped corrosion and water passing across the cavity	Partial fill insulation was introduced, glass fibre batts fixed within the cavity and trapped into place using preformed cavity clips	Thickness of partial fill increased and the quality of insulants improved using closed cell PIR and PUR 50 mm PIR 0.4 W/m²K

Figure 5.69   Cavity wall construction, development of cavity ties and insulants (courtesy of Felix Thomas).

(non-galvanised) steel ties used during the 1930s to 1950s have had to be replaced. Where ties break or corrode the external skin of stone or brick will bow and deform as it expands and contracts with thermal and moisture movement. Galvanised stainless steel and plastic cavity ties were introduced to resolve the problem. The ties had clips and deformations at their centre, known as a drip, designed to encourage any moisture that passed on to the tie to drip off at the deformation. The moisture collected on the drip drops down into the cavity and out through the external leaf of the cavity.

Wall ties must be protected during delivery, storage, handling and use against the inevitable knocks that may perforate the toughest coating to mild steel and the consequent probability of rust occurring. Wall ties made from stainless steel will not suffer corrosion during the useful life of buildings. Sharp edges of steel wall ties can pose a hazard to bricklayers. It is common practice to build up one leaf of the cavity wall to a reasonable height (one lift) with the wall ties built in. The problem that the bricklayers face is that as they build the second leaf of brickwork they have to avoid the protruding sharp metal wall ties as they bend down to lay the second leaf, and a number of eye injuries have resulted. Plastic wall ties, which do not suffer from corrosion, are now much more commonly used because they do not have sharp edges and do not pose the same risk as metal ties.

Standard section wall ties, illustrated in Figure 5.70, are the vertical twist strip, the butterfly and double triangle wire ties. As a check to moisture that may pass across the tie, the butterfly type is laid with the twisted wire ends hanging down into the cavity to act as a drip. The double triangle tie may have a bend in the middle of its length and the strip tie has a twist as a barrier to moisture passing across the tie. Of the three standard types, the butterfly is more likely to collect mortar droppings than the others.

The wall tie illustrated in Figure 5.71 is made from corrosion resistant stainless steel. The ridge at the centre of the length of the tie is designed for strength and to provide as small as possible a surface for the collection of mortar droppings. The perforations are to improve bond to mortar. The length of wall ties varies to accommodate different widths of cavity and the thickness of the leaves of cavity walls. For a 50 mm cavity with brick leaves, a 191 or 200 mm long tie is made. For a 100 mm cavity with brick leaves, a 220 mm long tie is

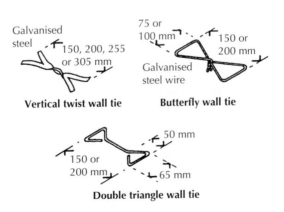

Figure 5.70 Cavity wall ties.

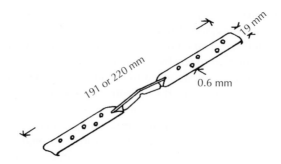

Figure 5.71   Stainless steel wall tie.

used. The increased demands for higher levels of insulation have led to the development of cavity walls with 150 mm cavities; extra long wall ties with increased stiffness are now available for such cavities.

Spacing of ties

The spacing of wall ties built across the cavity is usually 900 mm horizontally and 450 mm vertically, or 2.47 ties per square metre, and staggered, as illustrated in Figure 5.72. The spacing is reduced to 300 mm around the sides of windows, door openings and where movement joints are used. In practice a block with a mortar joint is 225 mm deep; therefore, ties are usually positioned every block course around the openings (Figure 5.73). In Approved Document A to the Building Regulations, the practical guidance for the spacing of ties is given as 900 and 450 mm horizontally and vertically for 50–75 mm cavities, 750 and 450 mm horizontally and vertically for cavities from 76 to 100 mm wide and 300 mm vertically at unbonded jambs of all openings in cavity walls within 150 mm of openings to all widths of cavities.

Openings in walls

Approved Document A states that the number, size and position of openings should not impair the stability of a wall to the extent that the combined width of openings in walls between the centre line of buttressing walls or piers should not exceed two-thirds of the length of that wall together with more detailed requirements limiting the size of opening and recesses. There is a requirement that the bearing end of lintels with a clear span of 1200 mm or less may be 100 mm and above that span, 150 mm. Figure 5.74 is an illustration of a window opening in a brick wall with the terms used to describe the parts noted.

For strength and stability the brickwork in the jambs of openings has to be strengthened with more closely spaced ties and the wall over the head of the opening supported by an arch, lintels or beams. The term jamb is derived from the French word *jambe*, meaning leg. From Figure 5.74, it will be seen that the brickwork on each side of the opening acts like legs, which support brickwork over the head of the opening. The word 'reveal' is used more definitely to describe the thickness of the wall revealed by cutting the opening, and the reveal is a surface of brickwork as long as the height of the opening. The lower part of the

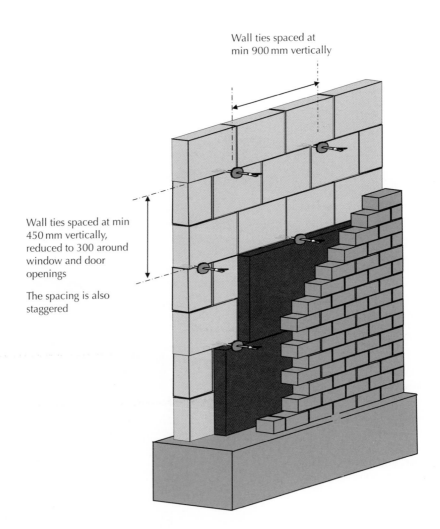

Wall ties spaced at
min 900 mm vertically

Wall ties spaced at min
450 mm vertically,
reduced to 300 around
window and door
openings

The spacing is also
staggered

Figure 5.72   Cavity wall tie spacing.

opening is a cill (alternatively spelt sill) for windows or a threshold for doors. The jambs
of openings may be plain or square into which the door or window frames are built or
fixed or they may be rebated with a recess, behind which the door or window frame is built
or fixed.

It was practice to build in cut bricks or blocks as cavity closers at the jambs of openings
to maintain comparatively still air in the cavity as insulation. To prevent penetration of
water through the solid closing of cavity walls at jambs, a vertical dpc was bedded in the
reveal as illustrated in Figure 5.75. To avoid cold bridging the reveal should be insulated.
dpcs backed with 25-mm insulation are now commonly used to prevent the cold bridge
(see Figure 5.76 and Figure 5.77).

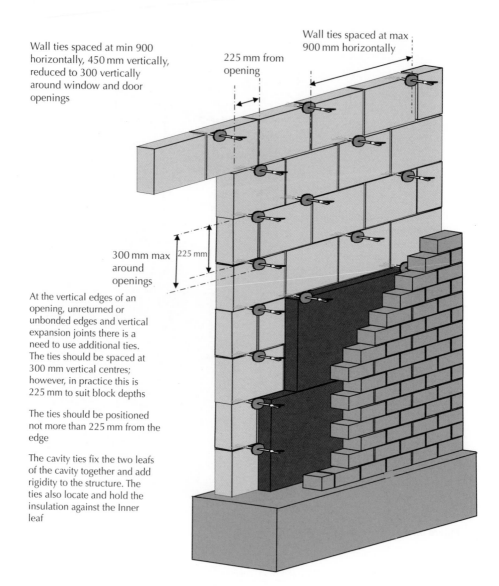

Wall ties spaced at min 900 horizontally, 450 mm vertically, reduced to 300 vertically around window and door openings

225 mm from opening

Wall ties spaced at max 900 mm horizontally

300 mm max around openings

225 mm

At the vertical edges of an opening, unreturned or unbonded edges and vertical expansion joints there is a need to use additional ties. The ties should be spaced at 300 mm vertical centres; however, in practice this is 225 mm to suit block depths

The ties should be positioned not more than 225 mm from the edge

The cavity ties fix the two leafs of the cavity together and add rigidity to the structure. The ties also locate and hold the insulation against the Inner leaf

Figure 5.73 Cavity wall tie spacing around openings.

It has become common to insert insulated plastic cavity closers, which reduce heat flow and prevent the passage of moisture across the cavity (Figure 5.78 and Photograph 5.28).

As an alternative to solidly filling the cavity at jambs with cavity closers, window or door frames were used to cover and seal the cavity. Pressed metal subframes to windows were specifically designed for this purpose, as illustrated in Figure 5.79. With mastic pointing between the metal subframe and the outer reveal, this is a satisfactory way of sealing

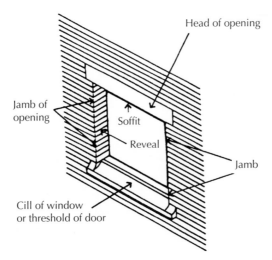

Figure 5.74   Opening in wall.

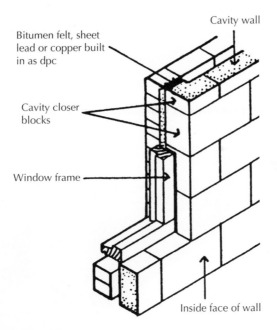

Figure 5.75   Solid closing of cavity at jambs.

(a) Thermal bridge through solid closing

Where there is no thermal break (insulation) at the reveal
heat energy will be conducted out of the building. Solid
bridges across the cavity create a thermal bridge

Where the reveal is made
out of solid masonry with no
thermal break a cold bridge
will form. The internal
surface will become cold,
drawing heat out of the
building

Cold surfaces will lead to
condensation especially if
warm moist air passes over
them

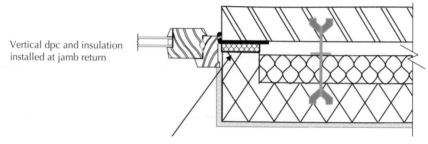

Thermal
bridge

(b) Thermal bridge avoided by insulating the reveal

Vertical dpc and insulation
installed at jamb return

dpc backed with 25 mm of
insulation along the full length of the window reveal
provides a continuous thermal break

Figure 5.76   Solid cavity closing with insulation to avoid thermal bridging.

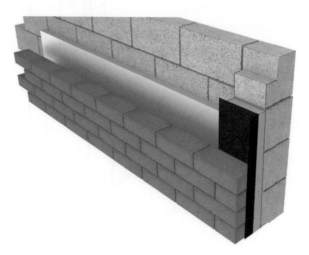

**Figure 5.77   Insulated cavity closer fixed to the vertical dpc (courtesy of http://www
.arcbuildingsolutions.co.uk).**

Typical thermal bridges

Plastic cavity closer
fixed to brickwork and
nailed into mortar joints

Insulated closer
slots into rebate

The plaster will
hide the inside edge
of the closer and the
window or door
frame will hide the
external edge of the
cavity closer

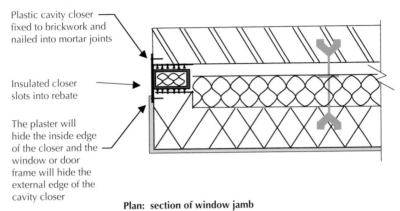

Plan: section of window jamb

Figure 5.78   Cavity closer.

Cavity closer is inserted
between the two masonry
skins
The closer insulates the cavity
and acts as a vertical dpc

Photograph 5.28   Sealing the cavity in a wall opening: cavity closer.

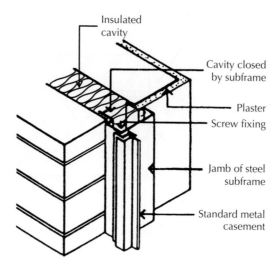

Figure 5.79   Cavity closed with frame.

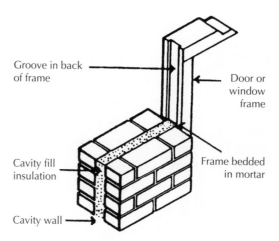

Figure 5.80   Cavity fill insulation.

cavities. With the increasing requirement for insulation it has become practice to use cavity insulation as the most practical position for a layer of lightweight material. If the cavity insulation is to be effective for the whole of the wall it must be continued up to the back of window and door frames, as a solid filling of cavity at jambs would be a less effective insulator and act as a thermal bridge.

It has become common practice to use cavity insulation continued up to the frames of openings, as illustrated in Figure 5.80, and to use cavity closers, as shown in Figure 5.78,

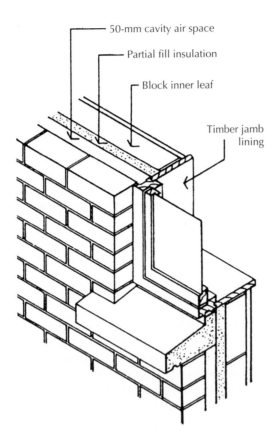

50-mm cavity air space

Partial fill insulation

Block inner leaf

Timber jamb lining

Figure 5.81   Jamb lining to wide cavity.

to avoid the thermal bridge effect caused by solid fill. Door and window frames are set in position to overlap the outer leaf with a resilient mastic pointing as a barrier to rain penetration between the frame and the jamb. With a cavity 100 mm wide and cavity insulation as partial fill, it is necessary to cover the part of the cavity at jambs of openings that is not covered by the frame. This can be affected by covering the cavity with plaster on metal lath or by the use of jamb linings of wood, as illustrated in Figure 5.81. With this form of construction at the jambs of openings there is no purpose in forming a vertical dpc at jambs. The advantages of the wide cavity is that the benefit of the use of the cavity insulation can be combined with the cavity air space as resistance to the penetration of water to the inside face of the wall.

Cills and thresholds of openings
A cill is the horizontal finish to the wall below the lower edge of a window opening. The function of a cill is to protect the wall below a window. Cills are shaped or formed to slope out and project beyond the external face of the wall, so that water runs off. The cill should

project at least 45 mm beyond the face of the wall below and have a drip on the underside of the projection. The cavity insulation shown in Figure 5.81 is carried up behind the stone cill to avoid a thermal bridge and a dpc is fixed behind the cill as a barrier to moisture penetration. A variety of materials may be used as a cill, such as natural stone, cast stone, concrete, tile, brick and non-ferrous metals. The choice of a particular material for a cill depends on cost, availability and, to a large extent, on appearance. As a barrier to the penetration of rain to the inside face of a cavity wall it is good practice to continue the cavity up and behind the cills.

The threshold to door openings serves as a finish to protect a wall or concrete floor slab below the door. Thresholds are commonly formed as part of a step up to external doors as part of the concrete floor slab with the top surface of the threshold sloping out. Alternatively, a natural stone or cast stone threshold may be formed.

Head of openings in cavity walls

The brickwork and blockwork over the head of openings in cavity walls has to be supported by a lintel.

Steel lintels

Most loadbearing brick or blockwork walls over openings, where the cavity insulation is continued down to the head of the window or door frame, are supported by steel section lintels. The advantage of these lintels is that they are comparatively lightweight and easy to handle, they provide adequate support for walling over openings in small buildings and, once they are bedded in place, the work can proceed. The lintels are formed either from a mild steel strip that is pressed to shape, and galvanised with a zinc coating to inhibit rust, or from stainless steel. The lintels for use in cavity walling are formed with either a splay to act as an integral damp-proof tray or as a top hat section over which a damp-proof tray is dressed. Typical sections are illustrated in Figure 5.82, Figure 5.83, Figure 5.84 and Photograph 5.29.

For insulation the splay section and top hat section lintels are filled with EPS. The top hat section steel lintel is built into the jambs of both the inner and outer leaf to provide support for both leaves of the cavity wall, as illustrated in Figure 5.83 and Figure 5.85. The two wings at the bottom of the lintel provide support for the brick outer and block inner leaves over the comparatively narrow openings for windows and doors. Where the cavity

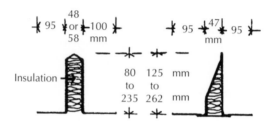

Figure 5.82 Lintels for cavity walls.

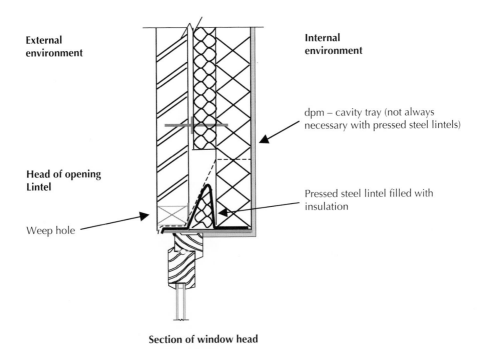

Section of window head

Figure 5.83   **Section of pressed steel lintel.**

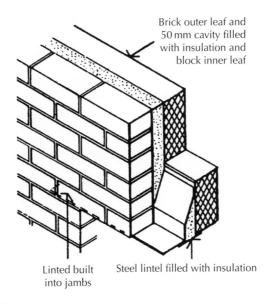

Figure 5.84   **Splay lintel.**

The top of the lintel is shaped so that any water which enters the cavity is shed to the external leaf

Perforations in the steel help improve the bond and slightly reduce the heat flow across the lintel

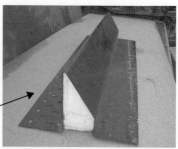

(a) Pressed steel lintel with insulation

(b) Lintel placed on cavity wall ready to receive brickwork

Steel lintel

Insulation

Cavity closer

Plasterboard – ribbon of plaster will be used to seal the gap between the blockwork and plasterboard

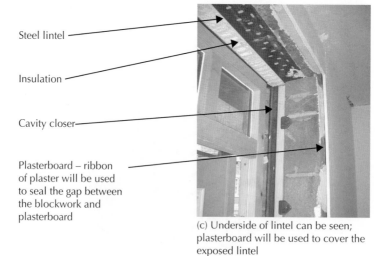

(c) Underside of lintel can be seen; plasterboard will be used to cover the exposed lintel

Photograph 5.29   Pressed steel lintels (cavity walls).

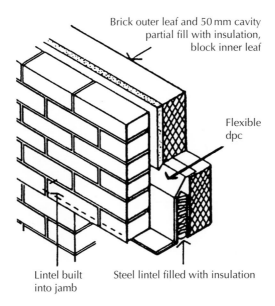

Brick outer leaf and 50 mm cavity
partial fill with insulation,
block inner leaf

Flexible
dpc

Lintel built
into jamb

Steel lintel filled with insulation

Figure 5.85   Top hat lintel.

is partly filled with insulation it is usual to dress a flexible dpc from the block inner leaf down to a lower brick course or down to the underside of the brick outer leaf. The purpose of the dpc or tray is to collect any water that might penetrate the outer leaf and direct it to weep holes in the wall.

The splay section lintel is built into the jambs of openings to provide support for the outer and inner leaf of the cavity wall over the openings, as illustrated in Figure 5.84. Where the cavity is filled with insulation there is no need to build in a dpc or tray. Any water that might penetrate the outer leaf will be directed towards the outside by the splay of the lintel. Unless the window or door frame is built-in or fixed with its external face close to the outside face of the wall, the edge of the wing of the lintel will be exposed on the soffit of the opening. Fairface brickwork supported by steel lintels may be laid as horizontal course brickwork or as a flat brick on edge or end lintel.

Thermal bridging through lintels

Where there is a continuous mass of material across a cavity the heat flow through that material is likely to be considerable. Even with pressed steel lintels, which incorporate insulation (Photograph 5.29), there is likely to be a continuous piece of steel that links the internal and external environment. Figure 5.86 illustrates how heat energy finds its way across the cavity. The use of two separate concrete lintels is more effective at reducing the heat flow and thermal (cold) bridging.

Concrete lintels

As an alternative to the use of steel lintels, reinforced concrete lintels may be used to support the separate leaves over openings (Figure 5.87). A range of precast reinforced

Pressed steel lintel: white area of thermal image
shows transfer of heat (cold bridge) through the lintel

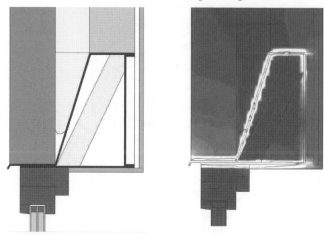

(a) Section through window head with a single pressed steel lintel

With two separate lintels the transfer of heat (white area) does
not flow through the lintel across the cavity. However, the
window and window frame still allow heat to flow

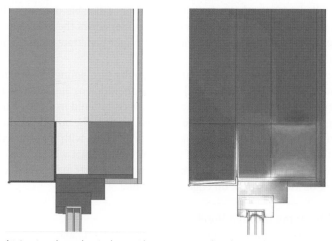

(b) Section through window with two separate lintels

Figure 5.86   Thermal bridging: single lintel and separate lintels (thermal images courtesy of David Roberts).

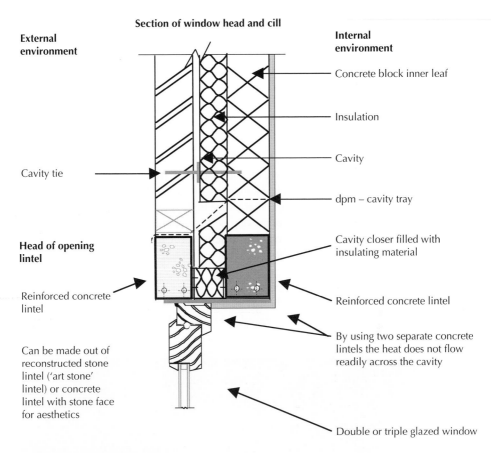

Section of window head and cill

External environment

Internal environment

Concrete block inner leaf

Insulation

Cavity

Cavity tie

dpm – cavity tray

Head of opening lintel

Cavity closer filled with insulating material

Reinforced concrete lintel

Reinforced concrete lintel

By using two separate concrete lintels the heat does not flow readily across the cavity

Can be made out of reconstructed stone lintel ('art stone' lintel) or concrete lintel with stone face for aesthetics

Double or triple glazed window

Figure 5.87   Concrete lintels.

concrete lintels is available to suit the widths of most standard door and window openings with adequate allowance for building in the ends of lintels each side of openings. For use with fairface brickwork the lintel depth should match the depth of brick course heights to avoid cutting of bricks around lintel ends. These comparatively lightweight lintels are bedded on walling as support for outer and inner leaves.

Cavity trays

In positions of severe exposure to wind-driven rain, the outer leaf of a cavity wall may absorb water to the extent that rainwater penetrates to the cavity side of the outer leaf. It is unlikely, however, that water will enter the cavity unless there are faults in construction. Where the mortar joints in the outer leaf of a cavity wall are inadequately flushed up with mortar, or the bricks in the outer leaf are very porous and where the wall is subject to severe or very severe exposure, there is a possibility that wind-driven rain may penetrate the outer leaf to the cavity.

It has become common practice to build in some form of dpc or tray of flexible, impermeable, material to direct any water out to the external face of walls. A strip of polymer-based polythene, bitumen felt or sheet lead is used for the purpose. The dpc and tray shown in Figure 5.87 is built in at the top of the inner lintel and dressed down to the underside of the outer lintel over the head of the window. As an alternative the dpc tray could be built in on top of the second block course and dressed down to the top of the outer lintel, with weep holes in the vertical brick joints.

Cavity wall insulation

Because the resistance to the passage of heat of a cavity wall by itself is poor, it is necessary to introduce a material with high resistance to heat transfer to the wall construction. Most of the materials, thermal insulators, that afford high resistance to heat transfer are fibrous or cellular, lightweight, have comparatively poor mechanical strength and are not suitable by themselves for use as part of the wall structure. The logical position for such material, therefore, is inside the cavity, either as partial or full fill.

Partial fill
Partial fill construction requires the use of insulating material in the form of boards that are sufficiently rigid to be secured against the inner leaf of the cavity (Photograph 5.30). A

Photograph 5.30   Cavity wall construction partial fill.

25 mm wide air space between the outer leaf and the cavity insulation should be adequate to resist the penetration of rain, providing the air space is clear of all mortar droppings and other building debris that might serve as a path for water. In practice, it is difficult to maintain a clear 25 mm wide air gap because of protrusion of mortar from joints in the outer leaf and the difficulty of keeping so narrow a space clear of mortar droppings. Good practice, therefore, is to use a 50 mm wide air space between the outer leaf and the partial fill insulation.

To meet insulation requirements and the use of a 100 mm cavity with partial fill insulation it may be economic to use a lightweight block inner leaf to augment the cavity insulation to bring the wall to the required U-value. The usual practice is to build the inner leaf of the cavity wall first, up to the first horizontal row of wall ties, then to place the insulation boards in position against the inner leaf. As the outer leaf is built, a batten may be suspended in the cavity air space and raised to the level of the first row of wall ties to stop mortar droppings from falling down the cavity. The batten is then withdrawn and cleared of droppings. Insulation retaining wall ties are then bedded across the cavity to tie the leaves and keep the insulation in position. This sequence of operations is repeated at each level of wall ties. The suspension of a batten in the air space and its withdrawal and cleaning at each level of ties do considerably slow the process of brick and block laying, but it is essential that the procedure is followed.

Care should be taken to ensure that mortar does not fall on cavity ties. Where mortar falls on the tie a thermal bridge and potential path for moisture to enter the building are created (Photograph 5.31).

Photograph 5.31 Wall tie covered in mortar – path for moisture and cold bridge.

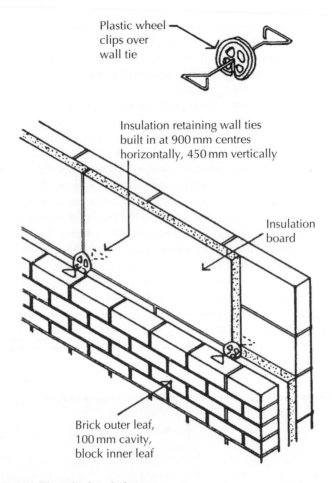

Plastic wheel clips over wall tie

Insulation retaining wall ties built in at 900 mm centres horizontally, 450 mm vertically

Insulation board

Brick outer leaf, 100 mm cavity, block inner leaf

Figure 5.88 Partial fill cavity insulation.

Insulation retaining ties are usually plastic, standard galvanised steel or stainless steel wall ties to which a plastic disc is clipped to retain the edges of the insulation, as illustrated in Figure 5.88. The ties may be set in line one over the other at the edges of boards, so that the retaining clips retain the corners of four insulation boards. The materials used for partial fill insulation should be of boards, slabs or batts that are sufficiently rigid for ease of handling and to be retained in a vertical position against the inner leaf inside the cavity without sagging or losing shape, so that the edges of the boards remain close butted throughout the useful life of the building.

Total fill

Insulation that is built in as the walls are raised, to fill the cavity totally, will to an extent be held in position by the wall ties and the two leaves of the cavity wall. To maintain a continuous, vertical layer of insulation inside the cavity one of the mineral fibre semi-rigid batts or slabs should be used. Fibreglass and rockwool semi-rigid batts or slabs in sizes suited to cavity tie spacing are made specifically for this purpose. As the materials are made in widths to suit vertical wall tie spacing there is no need to push them down into the cavity after the wall is built, as is often the procedure with loose fibre rolls and mats, and so displace freshly laid brick or blockwork.

The most effective way of insulating an existing cavity wall is to fill the cavity with some insulating material that can be blown into the cavity through small holes drilled in the outer leaf of the wall. The injection of the cavity fill is a comparatively simple job. The complication arises in forming sleeves around air vents penetrating the wall and sealing gaps around openings. Glass fibre and granulated rockwool of EPS beads are used for the injection of insulation for existing cavity walls. These materials can also be used for blowing into the cavity of newly built walls.

Insulation materials

The materials used as insulation for the fabric of buildings may be grouped as inorganic and organic insulants. Better thermal performance of buildings has also brought about new thermal insulating materials, such as products made from recycled newspaper and sheep's wool.

Inorganic insulants are made from naturally occurring materials that are formed into fibre, powder or cellular structures that have a high void content as, for example, glass fibre, mineral fibre (rockwool), cellular glass beads, vermiculite, calcium silicate and magnesia or as compressed cork. Inorganic insulants are generally incombustible, do not support spread of flame, are rot and vermin proof and generally have a higher U-value than organic insulants. The inorganic insulants most used in the fabric of buildings are glass fibre and rockwool in the form of loose fibres, mats and rolls of felted fibres and semi-rigid and rigid boards, batts and slabs of compressed fibres, cellular glass beads fused together as rigid boards, compressed cork boards and vermiculite grains.

Organic insulants are based on hydrdocarbon polymers in the form of thermosetting or thermoplastic resins to form structures with a high void content as, for example, polystyrene, polyurethane (PUR), isocyanurate and phenolic. Organic insulants generally have a lower U-value than inorganic insulants, are combustible, support the spread of flame more readily than inorganic insulants and have a comparatively low melting point. Organic insulants most used for the fabric of buildings are EPS in the form of beads or boards, extruded polystyrene (XPS) in the form of boards and PUR, isocyanurate and phenolic foams in the form of preformed boards or spray coatings.

The materials normally used for cavity insulation are glass fibre, rockwool and EPS, in the form of slabs or boards, in sizes to suit cavity tie spacing. With increased requirements for the insulation of walls it may well be advantageous to use one of the more expensive organic insulants such as XPS, PIR or PUR because of their lower U-value, where a 50 mm clear air space can be maintained without greatly increasing the overall width of the cavity.

With growing attention to building in a greener, more sustainable, way some new materials are starting to be used. These include products made from natural materials such as hemp and sheep's wool and products made from recycled products, such as newspaper.

Sheep's wool thermal insulation products are made of 100% wool, or a mixture of wool and other fibres. Wool has a thermal conductivity of around 0.04 W/mK, with 100 and

200 mm depths of insulation giving U-values of 0.40 and 0.20, respectively, and is better suited to roof insulation (see also Chapter 6) or wide cavities afforded by timber-framed construction. Cellulose thermal insulation products are made from recycled newspapers, typically with a thermal conductivity of around 0.036 W/mK.

Calculating the Thickness of Insulation Required in a Wall.

The following formula is useful as reference:

$$R = d/\lambda = \text{Thermal resistance} = (m^2K/W)$$
$$\lambda = \text{thermal conductivity (W/mK)}$$
$$d = \text{thickness of material (in metres)}$$
$$U = \text{U-value, thermal transmittance (units are W/m}^2K)$$

Thermal transmittance (U-value) is expressed as $U = 1/\Sigma R = $ (units are W/m^{2}K).

Thermal resistance of the cavity wall is:

If a wall requires a U-value of 0.35 W/m^{2}K, the resistance required can be calculated by transposing the following formula:

$$U\text{-value} = 1/\Sigma R;$$
$$\text{thus } \Sigma R \text{ (required)} = 1/U\text{-value}$$
$$\textit{Thermal resistance required} = 1/0.35 = 2.86 \text{ m}^2 \text{ K/W}$$

Component/element	Thickness (metres) D	Thermal conductivity (W/mK) λ	Thermal resistance (m² K/W) R = $^d/_\lambda$
External surface resistance RSo			0.04
102 brick outer leaf	0.1025	0.84	0.122
50 mm cavity	0.050	0.27	0.185
100 mm blockwork aircrete	0.100	0.11	0.909
13 mm plasterboard	0.013	0.50	0.026
Internal surface resistance RSi			0.13
ΣR			1.421

Thermal resistance required = 2.86 m² K/W; therefore
2.86 − 1.421 = 1.439 m² K/W
Transposing the formula R = $^d/_\lambda$
thus **R × λ = d**
1.439 × 0.03 = 0.043 m or 0.043 × 1000 = 43 mm

Additional resistance to be provided by insulation $= 2.86 - 1.421 = 1.439 \text{ m}^2 \text{ K/W}$. Assuming that insulation with a thermal conductivity of 0.03 W/mK is used, then the thickness of insulation required will be obtained by rearranging the formula R = $^d/_\lambda$; thus R × λ = d using the simple multiplication $1.439 = 0.03 = 0.043$ m or $0.043 \times 1000 = 43$ mm. So the required thickness of insulation is 43 mm, or the most convenient thickness available from a manufacturer, say, 45 or 50 mm of insulation board.

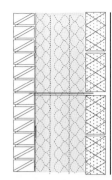

Brick stone or
reconstituted stone

425 mm 7 mm dia Teplo
Basalt ties (0.7 W/m.k)

Nearly zero

Dri-therm ridged slabs
100 mm bats with staggered
joints (0.037 W/m·K)

Wet plaster improves
airtightness

U-value = 0.113 W/m²K

Figure 5.89 Cavity wall built to Passivhaus standards (adapted from http://www.green buildingstore.co.uk).

Achieving low U-values

With different combinations of cavity wall construction it is relatively easy to achieve low U-values (Figure 5.89). The combination of lightweight blocks, insulants with low thermal conductivity and airtight construction, can achieve U-values of less than 0.2 W/m² K. For example, lightweight blocks are now available with thermal conductivities of 0.11 W/m·K (R-value for 100 and 140 mm = 0.1/0.11 = 0.909 m² K/W and 0.14/0.11 = 1.273 m² K/W), used in combination with a PIR of 0.022 W/m·K (R-value for 100 mm = 0.1/0.022 = 4.55 m² K/W), a regular brick thermal conductivity of 0.6 W/m·K (R-value 102.5 mm = 0.102/0.6 = 0.17 m² K/W), internal surface resistance of 0.12 m² K/W and external surface resistance of 0.06 m² K/W can achieve low U-values of 0.16 W/m² K; the U-value for a wall built with 140-mm lightweight block and 100 mm PIR.

Other materials such as aerogel insulants offer much lower conductivities and can achieve low U-values with a thinner cavity/wall section. These are currently relatively expensive compared with other insulants and therefore tend to be used only when space is at a premium.

Some properties are now being built to Passivhaus standards using masonry walls. One example is the Denby Dale passive house, constructed with an outer leaf of stone, an inner leaf of blockwork, 300 mm of low-conductivity Dri-therm ridged slabs of water repellent glasswool (0.037 W/m.K). Extra-long (Teplo basalt fibre) wall ties, 200–425 mm long, which can suit cavities up to 300 mm in width, were used to tie the skins together. Each wall tie has a thermal conductivity (0.7 W/mK) of 7 mm dia, with a sand finish to provide a mortar key and penetrated across the three 100 mm boards of insulation to provide the structural tie between the walls. The low-conductivity wall ties avoid cold bridging and the risk of condensation. With such long protruding cavity ties end caps are a must during construction to avoid risk of injury to the eyes or other skin punctures (Photograph 5.32a–c).

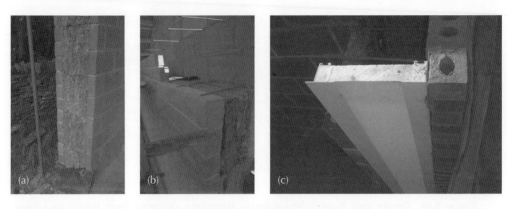

Photograph 5.32   (a) Cavity wall built to passiv standards, 300 mm mineral wool insulation. (b) 300 mm mineral wool cavity insulation, built to passiv standard, basalt Teplo low-conductivity cavity ties and mineral wall used to seal cavity with adjoining property. (c) 300 mm cavity wall built to passiv standard ready to receive blown fibre insulation EPS cavity closer.

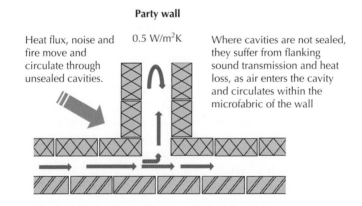

Party wall

Heat flux, noise and fire move and circulate through unsealed cavities.

0.5 W/m^{2}K

Where cavities are not sealed, they suffer from flanking sound transmission and heat loss, as air enters the cavity and circulates within the microfabric of the wall

Figure 5.90   Sound, fire and heat flux flow through unsealed cavities.

Preventing heat loss, flanking sound and ensuring fire integrity across party walls formed with a cavity

Sealing cavities in party walls

Cavity barriers are used to reduce flanking sound and also to reduce air circulation within the cavity (Figure 5.90). By closing the cavity with barriers air movement from one part of the cavity to the other is less likely to occur, and because open spaces within the cavity are reduced, the effect of convection currents, which can manifest as a result of differing temperatures, is also minimised. All parts and edges of the cavity must be sealed, including the top and bottom of the cavity.

Cavity barriers are also required to withhold the passage of smoke and fire. Gaps in the cavity barriers will allow fire to pass around them or flash through the voids as fire follows sources of oxygen. It is important that junctions are designed and detailed to prevent air flow within the cavity and also that the barriers are fitted carefully and then independently checked.

If there are any imperfections in the building envelope that connect to the cavities, cold air will enter the building and circulate around the structure and its connecting cavities. The ingress of cold air (especially when exposed to high wind pressure) and the escape of the warmer air will lead to thermal bypasses, cause cold draughts and result in cold spots within the building (as the cold air cools the fabric). Where cavities are not required to be ventilated they should be sealed to prevent thermal and acoustic exchanges and reduce the potential penetration of smoke and fire.

Effective U values of party walls separated by cavities

The heat flow in relatively stable conditions typically ranges from 0.5 to 0.7 W/m²K, if the cavity is not sealed. Unless edge sealing is used in party walls a default U-value of 0.5 W/m²K must be used (see Part L of the Building Regulations). With effective edge sealing the U-value is assumed to have a value of 0.2 W/m²K. Only a fully filled insulated cavity, effectively sealed at all edges, can be assumed to have a U-value of 0.0 W/m²K (Figure 5.91, Figure 5.92, Figure 5.93, Figure 5.94, Figure 5.95, Figure 5.96, Figure 5.97 and Figure 5.98).

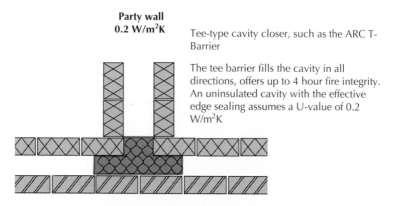

Party wall
0.2 W/m²K

Tee-type cavity closer, such as the ARC T-Barrier

The tee barrier fills the cavity in all directions, offers up to 4 hour fire integrity. An uninsulated cavity with the effective edge sealing assumes a U-value of 0.2 W/m²K

Figure 5.91   Cavity barrier in an uninsulated cavity.

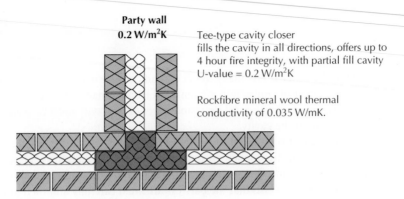

Party wall
0.2 W/m²K

Tee-type cavity closer
fills the cavity in all directions, offers up to
4 hour fire integrity, with partial fill cavity
U-value = 0.2 W/m²K

Rockfibre mineral wool thermal
conductivity of 0.035 W/mK.

Figure 5.92 Cavity barrier in a partial fill cavity.

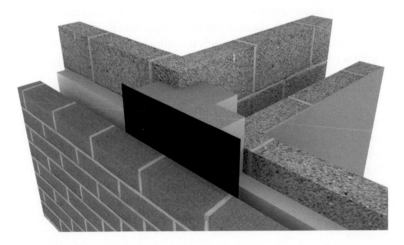

Figure 5.93   Partial fill cavity, party wall junction sealed with a tee cavity closer (courtesy of http://www.arcbuildingsolutions.co.uk).

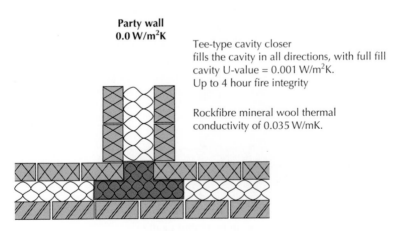

Party wall
0.0 W/m²K

Tee-type cavity closer
fills the cavity in all directions, with full fill
cavity U-value = 0.001 W/m²K.
Up to 4 hour fire integrity

Rockfibre mineral wool thermal
conductivity of 0.035 W/mK.

Figure 5.94 Cavity barrier in fulfil cavity.

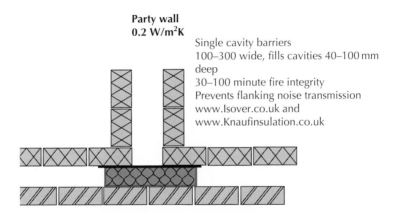

Party wall
0.2 W/m²K

Single cavity barriers
100–300 wide, fills cavities 40–100 mm
deep
30–100 minute fire integrity
Prevents flanking noise transmission
www.Isover.co.uk and
www.Knaufinsulation.co.uk

Figure 5.95   Cavity and party wall sealed with a cavity closer.

Figure 5.96   Cavity sealed with a cavity closer (courtesy of http://www.arcbuilding solutions.co.uk).

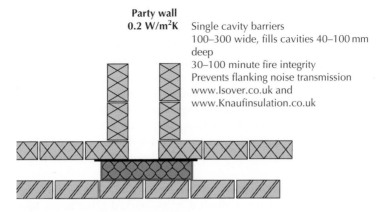

Party wall
0.2 W/m²K

Single cavity barriers
100–300 wide, fills cavities 40–100 mm
deep
30–100 minute fire integrity
Prevents flanking noise transmission
www.Isover.co.uk and
www.Knaufinsulation.co.uk

Figure 5.97   Cavity sealed with a cavity sealer.

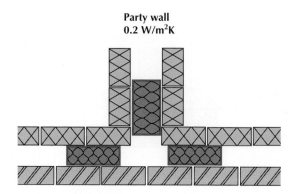

Party wall
0.2 W/m²K

Figure 5.98   Cavity sealed at edges of the external wall and party wall with cavity closers.

5.12 Timber

The word timber describes wood that has been cut for use in building. Timber has many advantages as a building material. It is a lightweight material that is easy to cut, shape and join by relatively cheap and simple hand- or power-operated tools in the production of wall, floor and roof panels, timber joists, and for rafters, walls, floors and roofs, and joinery. As a structural material it has favourable weight to cost, weight to strength and weight to modulus of elasticity ratios and coefficients of thermal expansion, density and thermal resistance. With sensible selection, fabrication and fixing, and adequate weather protection, timber is a reasonably durable material in relation to the life of most buildings and, sourced from sustainable forests, has excellent environmental credentials. Softwood and hardwood are the terms used to classify different timber.

Properties of timber

Up to two-thirds of the weight of growing wood is due to water in the cells of the wood. When the tree is felled and the wood is cut into timber this water begins to evaporate to the air around the timber, and the wood gradually shrinks as water is removed from the cell walls. As the shrinkage in timber is not uniform the timber may lose shape and it is said to warp. It is essential that before timber is used in buildings, either it should be stacked for a sufficient time in the open air for most of the water in it to dry out or it should be artificially dried in a kiln. If unseasoned or wet timber is used in building, it will dry out and shrink, causing twisting of doors and windows and cracking at joints with plastered walls. This means that only seasoned timber should be used and that the timber should be protected from moisture on the building site, when stored and as work proceeds. The process of allowing, or causing, newly cut wood to dry out is called seasoning, and timber that is ready for use in building is said to have been properly seasoned.

Natural dry seasoning
When logs have been cut into timbers they are stacked either in the open or in an open-sided shed. Timbers are stacked with battens between them to allow air to circulate around

them and are left stacked until most of the moisture in the wood has evaporated. Softwoods have to be stacked for a year or two before they are sufficiently dried out or seasoned, and hardwoods for up to 10 years. The lowest moisture content of timber that can be achieved by this method of seasoning is about 18%.

Artificial or kiln seasoning

Because of the great length of time required for natural dry seasoning and because sufficiently low moisture contents of wood cannot be achieved, artificial seasoning is used. After the wood has been converted to timber it is stacked in a kiln with battens between the timber. Air is blown through the kiln, the temperature and humidity of the air being regulated to affect seasoning more rapidly than with natural seasoning, but not so rapidly as to cause damage to the timber. If the timber is seasoned too quickly it shrinks and is liable to crack and lose shape badly. To avoid this it is common to allow timber to season naturally for a time and then to complete the process artificially.

Moisture content of timber

It is necessary to specify the moisture content of timber. Moisture content is stated as a percentage of the dry weight of the timber. The dry weight of any piece of timber is its weight after it has been so dried that further drying causes it to lose no more weight. This dry weight is reasonably constant for a given cubic measure of each type of wood and is used as the constant against which the moisture content can be assessed. Table 5.4 sets out moisture contents for timber. The moisture content of timber should be such that the timber will not appreciably gain or lose moisture in the position in which it is fixed.

Conversion of wood into timber

The method of converting wood to timber affects the timber in two ways: (1) by the change of shape of the timber during seasoning and (2) in the texture and differences in colour on the surface of the wood. Because the spring wood is less dense than the summer wood the shrinkage caused when the wood is seasoned (dried) occurs mainly along the line of the annual rings. Circumferential shrinkage is greater than the radial shrinkage. Because of this the shrinkage of one piece of timber cut from a log may be quite different from that

Table 5.4   Moisture content of timber

Position of timber in building	1 %	2 %
External uses fully exposed	20 or more	–
Covered and generally unheated	18	24
Covered and generally heated	15	20
Internal and continuously heated building	12	20

Source: Column 1 Average moisture content likely to be attained in service conditions Column 2 Moisture content which should not be exceeded in individual pieces at time of erection. Taken from BS 5268: Part 2:1996 (issue 2, May 1997).

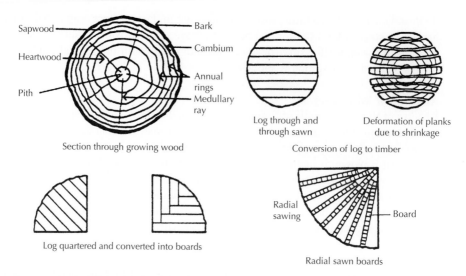

Section through growing wood

Log through and
through sawn

Deformation of planks
due to shrinkage

Conversion of log to timber

Log quartered and converted into boards

Radial
sawing

Board

Radial sawn boards

Figure 5.99   Conversion of wood into timber.

cut from another part of the log. This can be illustrated by showing what happens to the planks of a log converted by the 'through and through' cut method shown in Figure 5.99. When the planks have been thoroughly seasoned their deformation due to shrinkage can be compared by putting them together in the order in which they were cut from the log as in Figure 5.99.

If the face of a timber is cut on a radius of the circle of the log, the cells of the medullary rays may be exposed where the cut is made. With many woods this produces very pleasing texture and colour on the surface of the wood and it is said that the 'figure' of the wood has been exposed (Figure 5.99). Radial cutting of boards as shown is very expensive and is employed only for high-class cabinet making and panelling timbers where the wood will be used decoratively.

Surface finishes for timber

Timber can be left unfinished or a protective decorative finish can be applied to the surface. Unfinished timber is sometimes used on buildings for aesthetic effect and to reduce the environmental impact and maintenance costs of applying finishes to the timber at regular intervals. Care is required in the selection and detailing of suitable timbers for a given location and degree of exposure. There are three types of applied surface finish for wood: paint, varnish and stains (see also *Barry's Advanced Construction of Buildings*, Chapter 10). Paint and varnish are protective and decorative finishes that afford some protection against water and provide a decorative finish that can easily be cleaned. Paints are opaque and hide the surfaces of the wood, whereas varnishes are sufficiently transparent for the texture and grain of the wood to show. There is a wide range of stains available, from those that leave a definite film on the surface to those that penetrate the surface. Stains range from gloss through semi-gloss to matt finish. The purpose of this finish is to give a selected uniform colour to wood without masking the grain and texture of the wood. Most stains contain a preservative to inhibit fungal surface growth. These stains are most effective on rough sawn timbers.

Decay in timber

Decay in timber is caused by fungal decay and insect infestation, both of which can be prevented through careful detailing and specification as well as through careful work both off and on the construction site. Insect attack will depend upon location, moisture content and temperature. Any one of a number of wood-destroying fungi may attack timber that is persistently wet and has a moisture content of over 20%. Fungal decay is classified as being 'dry' or 'wet' rot.

Dry rot and its prevention

This is the most serious form of fungal decay and is caused by *Serpula lacrymans*, which can spread and cause extensive destruction of timber. Airborne spores of this fungus settle on timber, and if its moisture content is greater than 20% they germinate. The spore forms long thread-like cells which pierce the wood cells and use the wood as a food. The thread-like cells multiply, spreading out long white thread-like arms called mycelium, which feed on other wood cells. This fungus can spread many metres from the point where the spore first began to thrive, and is capable of forming thick greyish strands that can find their way through lime mortar and softer bricks. Timber that is affected by this fungus turns dark brown and shrinks and dries into a cracked powdery dry mass, hence the name dry rot. It is generally accepted that there is little risk of fungal decay in softwood if the timber is maintained at a moisture content of 20% or less.

Dry rot can be prevented through careful detailing to allow adequate ventilation and regular maintenance to prevent the moisture content of the timber from rising due to, for example, leaks in the building fabric or service pipes. Unseasoned timber must not be used and effort must be made during the construction work to keep seasoned timber dry. Buildings should be detailed to include horizontal and vertical dpcs and all underfloor areas to suspended timber floors and roof spaces must be adequately ventilated. Manufacturers and suppliers of timber-framed buildings will detail their bespoke systems to minimise the risk of dry and wet rot occurring.

The cause of the persistent dampness that has raised the moisture content of timber above 20%, such as leaking gutters or water pipes, must be corrected. This may involve providing a new dpc, preventing moisture penetration through walls, repairing leaking roofs and plumbing, eliminating condensation, clearing the bridging of dpcs and improving ventilation, including the provision of additional airbricks, and so on. Timber, plaster, dust and debris which have been affected by the fungus or are in close proximity to it should be taken out of the building and be burnt immediately. The purpose of burning the affected timber is to ensure that none of it is used in the repairs and to kill any spore that might cause further rot. All walls and surrounding masonry should be cleaned and treated by surface application of masonry biocide; preservative plugs or pastes should be inserted into holes drilled into localised problem areas and/or irrigation of fungal solution inserted via holes drilled into the wall (PCA, 2008). Although the use of heat has been suggested as a means to sterilising masonry adjacent to the affected area, there is little reported evidence of the use and effectiveness of this. Fungicide fluid for treatment against dry rot in masonry are water based and normally require total saturation; due to the health and safety risks caused by the preservative this should be avoided wherever possible). Preference should be given to the use of preservative plugs, paste or organic solvent-based fungicidal products. Old lime plaster on which, or through which, the rot has spread should be hacked off and renewed. New timber used to replace affected timber should be treated with a wood preservative before it is fixed or built in.

Because dry rot is capable of spreading through other materials and attacking dry timber, eradication involves the replacement of all affected timber with pre-treated timber and the treatment of adjoining timbers, brickwork, plasterwork and adjacent areas well away from the point of decay.

The following operations must be carried out in order to ensure that the fungus is completely eradicated:

❑ Locate and eliminate the dampness responsible for the attack.
❑ Cut out, remove from the site and burn all defective timbers in the vicinity of the attack. All apparently sound timber within 300 mm (HSE, 2001), or to provide a greater safety margin 600 mm (PCA, 2008), of defective timber should also be cut out, removed and burnt.
❑ Note that surface spraying of timbers will not prevent the spread of dry rot from infected timber.
❑ Carry out a thorough check of all other timbers within the building to which the fungus might have spread. This may involve removing flooring, ceilings, skirtings, architraves, wall panels, and so on.
❑ Strip off all wall plaster which may contain fungal strands to 300 mm beyond the observed limit of growth (HSE, 2001; PCA, 2008).
❑ Clean down all exposed masonry by wire brushing. Remove all stripped plaster, debris and dust from the site.
❑ Surrounding masonry should be cleaned and treated by surface application of masonry biocide, preservative plugs or pastes inserted into holes drilled into localised problem areas and/or irrigation of fungal solution inserted via holes drilled into the wall (PCA, 2008).
❑ Replace all timber that has been cut out with new, well-seasoned timber which has been pre-treated with fungicidal preservative.

Wet rot and its prevention

Wet rot is caused principally by *Coniphora puteana*, the cellar fungus, which occurs more frequently, but is less serious, than dry rot. Decay of timber due to wet rot is confined to timbers that are in damp situations such as cellars, ground floors without dpcs and timbers and joinery that is frequently in contact with moisture. The rot causes darkening and longitudinal cracking of timber and there is often little or no visible growth of fungus on the surface of timber.

Timber should not be built into or in contact with any part of the structure that is likely to remain damp. dpcs and dpms above and at ground level and sensibly detailed flashings and gutters to roofs and chimneys will prevent the conditions suited to the growth of wet rot fungus.

Wet rot is not as prolific as dry rot and the outbreaks are usually more localised; thus the treatment and eradication are simpler. It is usually only necessary to replace the affected timber with new pre-treated timber. This combined with the elimination of the wet rot attack and removal of the original cause is normally sufficient.

A wet rot attack is remedied by cutting out the affected wood from the point of decay with a safety clearance, and splicing-in new, preservative-treated timber.

Attacks of wet rot are often caused by the breakdown of the protective paint finish, which allows rainwater to penetrate the timber and create suitable conditions for attack. The cause of wet rot on internal elements is usually water penetration through the roof, walls and leaking plumbing. It is essential that the source of the water or moisture is identified and eliminated.

Insect attack on wood

In the UK the three types of insects which most commonly cause damage to timber are the furniture beetle, the death-watch beetle and the house longhorn beetle. Timber which the larvae of these beetles have affected should be sprayed or painted with a preservative that contains an insecticide during early summer and autumn. These preservatives prevent the larvae changing to beetles at the surface of the wood, which helps to prevent further infestation.

Wood preservatives

Wood may be preserved as a precaution against fungal decay or insect attack. Current practice is based on the premise that prevention is better than cure. The two types of preservative in general use are water borne or organic solvent formulations where water or a volatile solvent serves as a vehicle for the active fungicide or insecticide components.

Where there is timber decay it is advisable to consult one of the many companies which specialise in its diagnosis and treatment. They will carry out a detailed survey and diagnosis of the timber decay and provide a treatment and eradication package backed by an extensive guarantee.

Treatment of insect attack

Where there are only a few insect exit holes and the insect attack is not advanced, it is recommended that the timber be treated with water-based insecticide. The insecticide should be applied so that it minimises the exposure to people, pets and the environment. If the insect attack goes to the heart of the timber, the amount of organic solvent-based treatment must match the degree of penetration. If the attack is widespread a paste or injected solvent-based formulations should be used and applied to all timbers. If the attack is localised treatment need only be applied 300 mm beyond the area of the affected timber. The cure to both insect attack and mould growth is to remove the source of the damp condition and environment. Without the water content the insects cannot survive and the mould cannot grow.

Timbers that have been structurally weakened by insect attack must be treated with insecticide and strengthened. If the timber is beyond use and repair it should be cut out, burnt and completely replaced with new, pre-treated timber.

The primary control method to prevent insect infestation and fungal attack is to keep timber dry. Excessive use of even modern water-based treatment is not recommended.

5.13 Timber-framed walls

The construction of a timber-framed wall is a rapid, clean, dry operation. The timbers can be cut and assembled with simple hand- or power-operated tools, and once the wall is raised into position and fixed it is ready to receive wall finishes. A timber-framed wall has adequate stability and strength to support the floors and roof of small buildings, such as houses. Covered with wall finishes, such as plasterboard, it has sufficient resistance to damage by fire, good thermal insulating properties and reasonable durability providing it is well constructed and protected from decay. Timber-framed houses can be constructed very quickly. Using off-site prefabrication allows the erection of a house within a day, with roofing and external cladding completed soon afterwards. Alternatively, timber-framed houses can be built from timber by carpenters or by self-builders using a variety of proprietary systems (Photograph 5.33). There are a number of web sites that provide information on the design and specification of timber frame housing.

Photograph 5.33   Self-build timber frame house.

- ❏ TRADA: timber research and development association provides details, guidance, reports and updates; http://www.trada.co.uk
- ❏ Virtual site: provides details, 360 images and information on low carbon construction; http://www.leedsmet.ac.uk/teaching/vsite/
- ❏ Woodspec: guidance on designing and specifying timber frame building; http://www.woodspec.ie

Modern methods of timber frame construction were introduced into the UK in the 1960s. Timber frame construction offers flexible planning, energy-efficient construction, economic use of materials and a wide range of finishes. Timber frame construction, especially when light cladding is used, is lighter than a comparable masonry construction, and in some instances foundations can be designed to be smaller and hence less wasteful of materials. The dry construction is fast and there is no need to wait for wet trades to dry out before decorating. The high levels of thermal insulation make timber frame an attractive option given stringent thermal requirements. The structure of timber frame buildings must be designed by a structural engineer to demonstrate structural stability of the structure and compliance with the Building Regulations. It is common to erect buildings to a height of two or three storeys, although it is possible to build higher (six to eight storeys) and still satisfy the Building Regulations.

Strength and stability of timber

The strength of timber varies with species and is generally greater with dense hardwoods than less dense softwoods. Strength is also affected by defects in timber such as knots, shakes, wane and slope of the grain of the wood.

Stress grading of timber

Stress grading of structural timbers, which was first adopted in the Building Regulations 1972, is now generally accepted in selecting building timber. There are two methods of stress grading: visual grading and machine grading.

Visual grading

Trained graders determine the grade of a timber by a visual examination from which they assess the effect on strength of observed defects such as knots, shakes, wane and slope of grain. There are two visual grades, general structural (GS) and special structural (SS), the allowable stress in SS being higher than in GS.

Machine grading

Timbers are subjected to a test for stiffness by measuring deflection under load in a machine that applies a specified load across overlapping metre lengths to determine the stress grade. This mechanical test, which is based on the fact that strength is proportional to stiffness, is a more certain assessment of the true strength of a timber than a visual test. The machine grades, which are comparable to the visual grades, are machine general structural (MGS) and machine special structural (MSS). There are, in addition, two further machine grades: M50 and M75. The stress of M50 lies between MGS and MSS and M75 is the highest stress grade in the series.

Stress-graded timbers are marked GS and SS at least once within the length of each piece for visually graded timber together with a mark to indicate the grader or company. Machine-graded timber is likewise marked MGS, M50, MSS and M75 together with the BS kitemark and the number of the British Standard, 4978. Approved Document A, which gives practical guidance to meeting the requirements of the Building Regulations for small buildings, includes tables of the sizes of timber required for floors and roofs, related to load and span.

Stability

The stability of a timber-framed wall depends on a sound foundation on which a stable structure can be constructed. Figure 5.100 is an illustration of the base of a timber-framed wall set on a brick upstand raised from a strip foundation. The 150 × 50 mm timber sole plate is bedded on a horizontal dpc; shot fired nail with sufficient anchorage is used to locate the sole plate, in exposed positions 13 mm bolts at 2 m centres built into the foundation to anchor the plate against wind uplift. As an alternative the bolts may be shaped so that the bottom flange is built into the wall, run up on the inside face of the wall with a top flange turned over the top of the plate. Angle brackets may also be used to secure the sole plate.

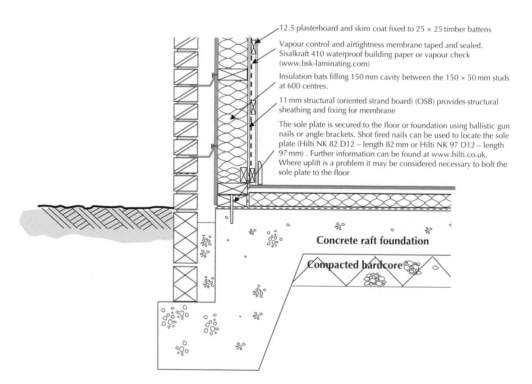

Figure 5.100   Base of timber frame wall: fixing detail.

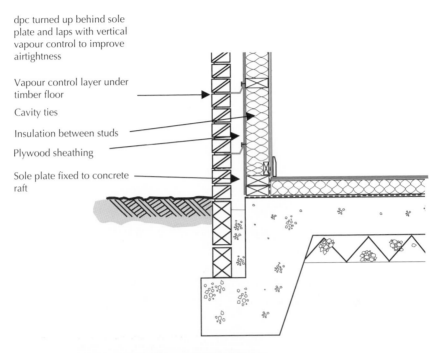

dpc turned up behind sole
plate and laps with vertical
vapour control to improve
airtightness

Vapour control layer under
timber floor

Cavity ties

Insulation between studs

Plywood sheathing

Sole plate fixed to concrete
raft

Figure 5.101   Timber-framed wall on raft foundation.

The upstand kerb of a concrete raft foundation serves as a base for the timber wall with the anchor bolts set into the concrete curbs and turned over the top of the sole plate. The vertical 150 × 50 mm studs are nailed to the sole plate at 400–600 mm centres with double studs at angles to facilitate fixing finishes (illustrated in Figure 5.101). Photograph 5.34 shows the sole plate, a packing piece of timber and the timber stud panel fixed to the floor.

A timber stud wall consists of small section timbers fixed vertically between horizontal timber head and sole plates, as illustrated in Figure 5.102 and Photograph 5.35. The vertical stud members are usually spaced at centres of 400–600 mm to support the anticipated loads and to provide fixing for external and internal linings. The horizontal noggins fixed between studs are used to stiffen the studs against movement that might otherwise cause finishes to crack. A face of plywood sheeting is often applied to both sides of the insulated stud panel to provide considerable lateral stability.

By itself a timber stud wall has poor structural stability along its length because of the non-rigid, nailed connection of the studs to the head and sole plate, which will not strongly resist racking deformation. A timber stud wall must, therefore, be braced (stiffened) against racking. As an internal wall or partition a timber stud frame may be braced by diagonal timbers or by being wedged between solid brick or block walls. As an external wall a timber stud frame may be braced between division walls and braced at angles where one wall butts to another, as illustrated in Figure 5.102. Diagonally fixed boarding or plywood sheathing fixed externally as a background for finishes braces an external stud frame wall.

Photograph 5.34   Timber frame mounted on sole plate.

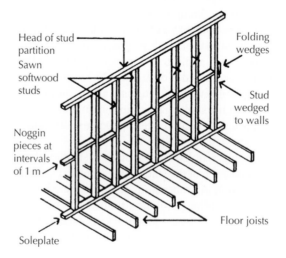

Figure 5.102   Timber stud frame.

Photograph 5.35   Timber stud frame.

Because of its small mass, a timber frame wall has poor lateral stability against forces such as wind that tend to overturn the wall. For stability along the length of the wall, connected external and internal walls or partitions will serve as buttresses. For stability up the height of the wall, timber upper floors and the roof connected to the wall will serve as buttresses. Buttressing to timber walls that run parallel to the span of floor joists and roof frames is provided by steel straps that are fixed across floor joists and roof rafters and fixed to timber walls in the same way that straps are used to buttress solid walls as previously described.

The usual method of supporting and fixing the upper floor joists to the timber wall frame is by using separate room height wall frames. The heads of the ground floor frames provide support for the floor joists on top of which the upper floor wall frame is fixed, as illustrated in Figure 5.103. The roof rafters are notched and fixed to the head of the upper floor wall frame. As an alternative, a system of storey height wall frames may be used with the top of the head of the lower frame in line with the top of the floor joists that are supported by a timber plate nailed to the studs, as illustrated in Figure 5.104. The upper frame is formed on the lower frame. The advantage of this system is that there is continuity of the wall frame and the disadvantage is that there is a less secure connection.

Figure 5.105 is an illustration of a two-storey house with timber walls, floor and roof with a brick outer leaf, with the insulation filling the cavity rather than contained within the timber stud panels. It is rare to see timber frames constructed in this manner, but there is no reason why it cannot be done. Most timber frame panels are prefabricated therefore the site assembly is much quicker if the units come with insulation installed.

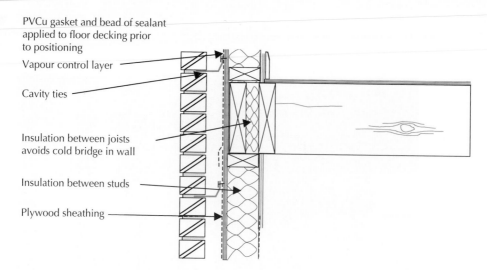

PVCu gasket and bead of sealant applied to floor decking prior to positioning

Vapour control layer

Cavity ties

Insulation between joists avoids cold bridge in wall

Insulation between studs

Plywood sheathing

Figure 5.103   Support for floor joists.

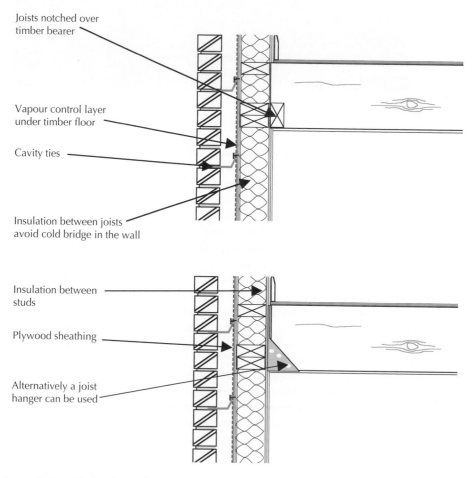

Joists notched over timber bearer

Vapour control layer under timber floor

Cavity ties

Insulation between joists avoid cold bridge in the wall

Insulation between studs

Plywood sheathing

Alternatively a joist hanger can be used

Figure 5.104   Timber-framed wall.

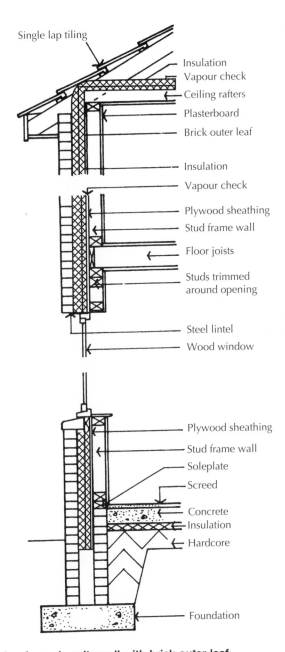

Single lap tiling

Insulation
Vapour check
Ceiling rafters
Plasterboard
Brick outer leaf

Insulation

Vapour check

Plywood sheathing
Stud frame wall

Floor joists

Studs trimmed
around opening

Steel lintel
Wood window

Plywood sheathing
Stud frame wall
Soleplate
Screed
Concrete
Insulation
Hardcore

Foundation

Figure 5.105   Timber-framed cavity wall with brick outer leaf.

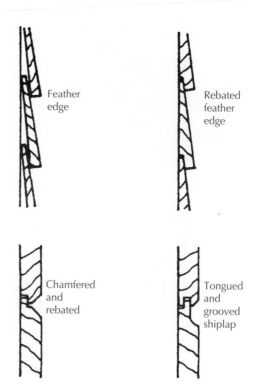

Feather
edge

Rebated
feather
edge

Chamfered
and
rebated

Tongued
and
grooved
shiplap

Figure 5.106   Timber weatherboarding.

Timber or tile clad timber frame

The traditional weather envelope for timber walls is timber weatherboarding nailed horizontally across the stud frame. The weather boards are shaped to overlap to shed water.
Some typical sections of boarding are illustrated in Figure 5.106. The wedge section, feather
edge boarding, is either fixed to a simple overlap or rebated to lie flat against the studs as
illustrated. The shaped chamfered and rebated and tongued and grooved shiplap boarding
is used for appearance, particularly when the boarding is to be painted. To minimise the
possibility of boards twisting it is usual practice to use boards of narrow widths, such as
100 and 150 mm.

As protection against rain and wind penetrating the weatherboarding it is usual to fix
sheets of breather paper behind the weatherboarding. Breather paper serves to act as a
barrier to water and at the same time allows the release of moisture vapour under pressure
to move through the sheet. Instead of nailing weatherboarding directly to the studs of the
wall frame it is usual to fix either diagonally fixed boarding or sheets of plywood across
the external faces of the stud frame. The boarding and ply sheets serve as a brace to the
frame and as a sheath to seal the frame against weather. Figure 5.107 is an illustration of
weatherboarding fixed to plywood sheathing with insulation fixed between studs.

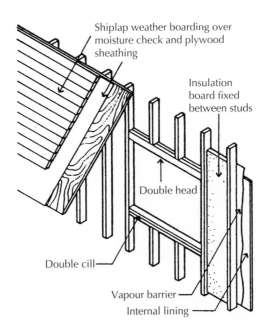

Shiplap weather boarding over moisture check and plywood sheathing

Insulation board fixed between studs

Double head

Double cill

Vapour barrier

Internal lining

Figure 5.107   Weather envelope.

Around openings to windows and doors the weatherboarding and ply sheath may be butted to the back of window and door frames fixed to project beyond the stud frames for the purpose. At the head of the opening the head of the frame may be reduced in depth so that the boarding runs down over the face of the frame. The weatherboarding butts up to the underside of a projecting cill. For extra protection sheet lead may be fixed behind the weatherboarding and nailed and welted to window and door frames. At external angles the weatherboarding may be mitred or finished square edged. An effective weathering is to fix a strip of lead behind the weatherboarding to form a sort of secret gutter.

Tile or slate hanging may be used to provide more durable protection, especially in exposed positions. In the UK it has been common practice in speculative house construction to provide the weather protection with a brick outer leaf, which provides protection to the timber in most situations. A sensible argument for this form of construction could be speed of erection and completion of building work by combining the rapid framing of a timber wall, floor and roof structure that could be completed and covered in a matter of a few days, with a brick outer leaf and speedy installation of electrical, water and heating services and dry linings.

Timber frames do not need to be clad in brickwork and can be covered in timber latts, tiles or render. Photograph 5.36, Photograph 5.37, Photograph 5.38, Photograph 5.39, Photograph 5.40, Photograph 5.41, Photograph 5.42, Photograph 5.43 and Photograph 5.44 show full scale models of timber frame construction. The details (Figure 5.108, Figure 5.109 and Figure 5.110) show the individual elements of these units. Although the timber frame

Photograph 5.36   Timber frame internal face with suspended ground floor.

provides the structural unit, the cladding can be any material that will provide the required protection from the elements and necessary security. As with all construction attention needs to be paid to the control of airtightness. Vapour control and airtightness membranes should be lapped and taped to ensure airtightness at all points of the structure (Figure 5.109).

Internal plasterboard linings to the timber-framed walls, the soffit of the first floor and the ceiling will provide a sufficient period of fire resistance to meet the requirements for a two-floor house. The requirement for barriers in external cavity walls to small houses applies only to the junction of a cavity and a wall separating buildings.

Prefabricated timber frames

There are three types of prefabricated timber frame systems that can be used, stick build, platform frame and balloon frame. Choice is often determined by the desired amount of work to be carried out on the building site, which has to be balanced against the cost of the timber frame. For example, some wall panels are delivered to site complete with sheathing and insulation, whereas other manufacturers provide the frame only, with the sheathing and insulation to be fixed on the site.

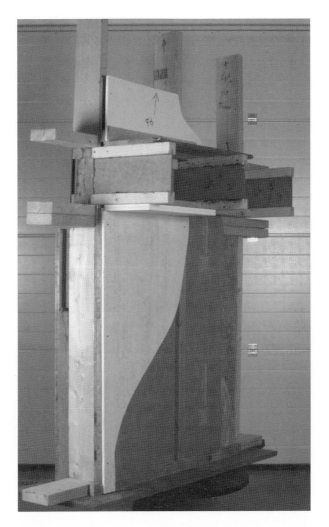

Photograph 5.37   **Timber frame intermediate floor to wall junction.**

Stick build

Using this traditional technique, the timber walls and floors are simply assembled from the individual members and components. Although not pre-assembled, the timber members are often delivered to site pre-cut and identified for ease of assembly. This type of construction is used to construct bespoke timber-framed houses, for example, oak post and beam structures. Apart from small, self-build and/or complicated structures, stick build is rarely used.

Platform frame

The advantage of using frames that are fabricated either on or off site complete with outer and inner finishes is speed of erection. Where a number of houses are to be built it

Photograph 5.38 Timber frame intermediate floor to wall junction.

Photograph 5.39 Timber frame suspended ground floor.

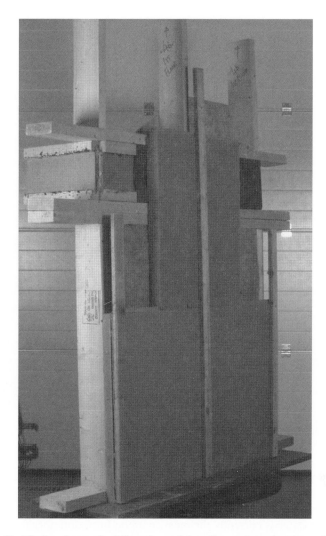

Photograph 5.40 Timber frame first floor to wall junction external perspective.

is possible to complete a building in a matter of days. The systems most used are either platform or storey frames. (The term platform frame equally applies to light steel frame house construction made from cold-formed steel sections.)

The platform frame system of construction employs prefabrication frames that are floor to ceiling level high, with the sole of the lower stud frame bearing on the foundation and the head of the frame supporting first floor joists, as illustrated in Figure 5.111. The first floor can then be used as a working platform from which the upper frames are set on top of the lower. The wall frames or panels may be the full width of the front and rear walls of narrow terrace houses or made in two or more panels. The first floor joists and roof provide

Photograph 5.41   Timber frame external wall and eaves detail.

sufficient bracing up the height and the separating wall will brace across the width of the wall. The wall frames may be prefabricated as stud frames sheathed with plywood or made complete with finishes on both sides.

Platform frame is currently the most common form of timber frame construction. The panel sizes will fit easily on to standard road transport and can be easily lifted into position. Photograph 5.45, Photograph 5.46, Photograph 5.47, Photograph 5.48, Photograph 5.49 and Photograph 5.50 show platform construction used to assemble a multi-storey timber frame building.

Structurally insulated panels are becoming more common and can be designed to achieve passive standards (see Photograph 5.51a–c).

Balloon frames

Storey frames (balloon frames) are made the height of a storey, floor to floor, so that the top of the head of a frame is level with the top of the floor joists. A bearer fixed to the stud frame supports the joists. This arrangement provides continuity of the stud framing up the height of the wall at the expense of some loss of secure anchor of floor joists to wall. A

Photograph 5.42   Timber frame wall interface with roof internal perspective.

Photograph 5.43   Timber frame eaves detail.

Photograph 5.44   Timber frame eaves detail showing tiles.

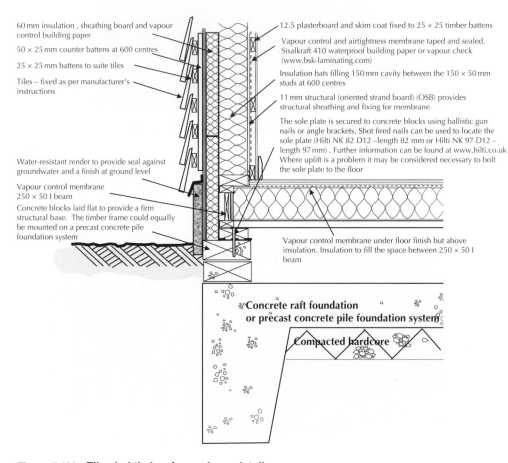

60 mm insulation , sheathing board and vapour control building paper

50 × 25 mm counter battens at 600 centres

25 × 25 mm battens to suite tiles

Tiles – fixed as per manufacturer's instructions

Water-resistant render to provide seal against groundwater and a finish at ground level

Vapour control membrane 250 × 50 I beam

Concrete blocks laid flat to provide a firm structural base. The timber frame could equally be mounted on a precast concrete pile foundation system

12.5 plasterboard and skim coat fixed to 25 × 25 timber battens

Vapour control and airtightness membrane taped and sealed. Sisalkraft 410 waterproof building paper or vapour check (www.bsk-laminating.com)

Insulation bats filling 150 mm cavity between the 150 × 50 mm studs at 600 centres

11 mm structural (oriented strand board) (OSB) provides structural sheathing and fixing for membrane

The sole plate is secured to concrete blocks using ballistic gun nails or angle brackets. Shot fired nails can be used to locate the sole plate (Hilti NK 82 D12 –length 82 mm or Hilti NK 97 D12 – length 97 mm) . Further information can be found at www.hilti.co.uk Where uplift is a problem it may be considered necessary to bolt the sole plate to the floor

Vapour control membrane under floor finish but above insulation. Insulation to fill the space between 250 × 50 I beam

Concrete raft foundation or precast concrete pile foundation system

Compacted hardcore

Figure 5.108   Tile clad timber frame: base detail.

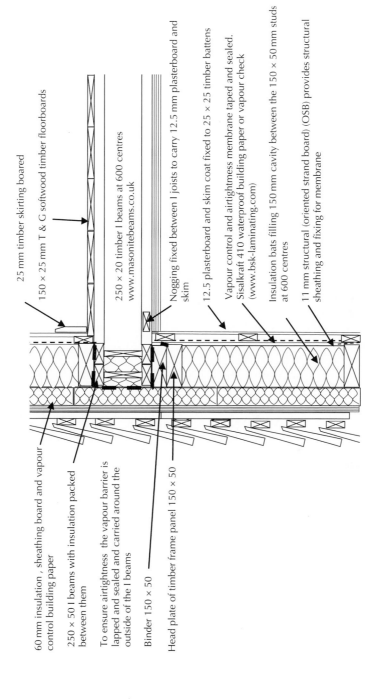

25 mm timber skirting boarded

150 × 25 mm T & G softwood timber floorboards

250 × 20 timber I beams at 600 centres
www.masonitebeams.co.uk

Nogging fixed between I joists to carry 12.5 mm plasterboard and skim

12.5 plasterboard and skim coat fixed to 25 × 25 timber battens

Vapour control and airtightmess membrane taped and sealed. Sisalkraft 410 waterproof building paper or vapour check (www.bsk-laminating.com)

Insulation bats filling 150 mm cavity between the 150 × 50 mm studs at 600 centres

11 mm structural (oriented strand board) (OSB) provides structural sheathing and fixing for membrane

60 mm insulation , sheathing board and vapour control building paper

250 × 50 I beams with insulation packed between them

To ensure airtightness the vapour barrier is lapped and sealed and carried around the outside of the I beams

Binder 150 × 50

Head plate of timber frame panel 150 × 50

Figure 5.109 Tile clad timber frame: wall to upper floor detail.

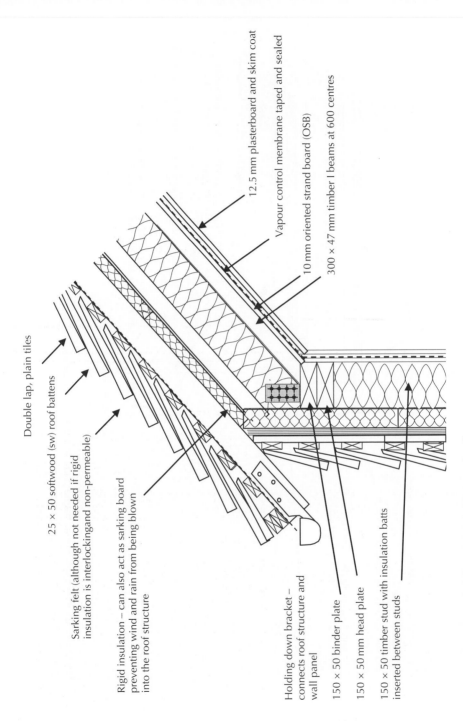

Double lap, plain tiles

25 × 50 softwood (sw) roof battens

Sarking felt (although not needed if rigid insulation is interlockingand non-permeable)

Rigid insulation – can also act as sarking board preventing wind and rain from being blown into the roof structure

Holding down bracket – connects roof structure and wall panel

150 × 50 binder plate

150 × 50 mm head plate

150 × 50 timber stud with insulation batts inserted between studs

12.5 mm plasterboard and skim coat

Vapour control membrane taped and sealed

10 mm oriented strand board (OSB)

300 × 47 mm timber I beams at 600 centres

Figure 5.110 Tile clad timber frame: Eaves detail.

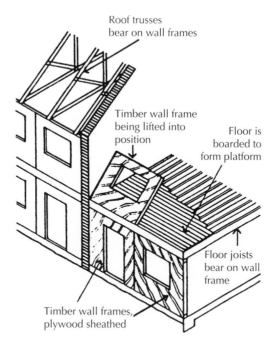

Figure 5.111   Platform frame.

Photograph 5.45   Platform frame construction (courtesy of Shepherd Construction).

Photograph 5.46   Platform frame construction ground floor (courtesy of Shepherd Construction).

Photograph 5.47   Vapour control membrane taped and sealed (courtesy of Shepherd Construction).

Photograph 5.48 **Platform frame third storey with prefabricated bathroom pods in position** (courtesy of Shepherd Construction).

Photograph 5.49 **Brick cladding secured to timber frame** (courtesy of Shepherd Construction).

Photograph 5.50 Internal stud walls with infill panels where fixtures need to be secured (courtesy of Shepherd Construction).

balloon wall frame is fabricated as one continuous panel the height of the two floors of small houses, as illustrated in Figure 5.112. The advantage of the balloon frame is speed of fabrication and erection, and the least number of joints between frames that have to be covered and weathered externally. The term balloon frames also applies to steel frame housing (light steel, cold-formed sections).

Functional requirements specific to timber-framed buildings

Fire safety

Specifying a minimum period of fire resistance for the elements of structure restricts the premature failure of the structural stability of a building. A timber-framed wall covered with plasterboard internally satisfies the requirement for houses of up to two storeys. To prevent the spread of fire between buildings, limits to the size of 'unprotected areas' of walls and finishes to roofs close to boundaries are set out in the Building Regulations. By reference to the boundaries of the site, the control will limit the spread of fire. Unprotected areas are those parts of external walls that may contribute to spread of fire and include glazed windows, doors and those parts of a wall that may have less than a notional fire resistance. Limits are set on the use of roof coverings that will not provide adequate protection against the spread of fire across their surface to adjacent buildings.

The passage of fire through connecting voids in timber frame construction is as significant as it is in masonry (loadbearing) construction. The connecting external and party wall cavity barriers must be sealed to prevent the passage of smoke and fire. The addition of edge seals reduces the passage of heat and improves airtightness.

Photograph 5.51   (a) Timber frame building constructed with structural insulated panel (SIP) to Passivhaus standards. (b) Timber frame building constructed with SIP's panels to Passivhaus standards. (c) Engineered insulated panels under construction.

Figure 5.113 shows a tee barrier used to improve the thermal insulating properties of the cavity, reduce the passage of sound and prevent the passage of fire within a specified period.

Resistance to the passage of heat

Timber is a comparatively good insulator. However, the sections of a timber frame do not by themselves generally afford sufficient insulation and a layer of some insulating material has to be incorporated in the construction. The layer of insulation is fixed either between the vertical studs of the frame or on the outside or inside of the framing. The disadvantage

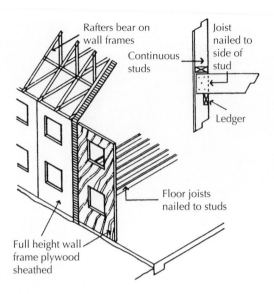

Figure 5.112   Balloon frame.

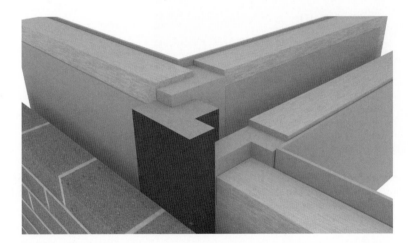

Figure 5.113   Cavity closer in timber frame (courtesy of http://www.arcbuildingsolutions .co.uk).

of fixing the insulation between the studs is that there may be a deal of wasteful cutting of insulation boards to fit them between studs and to the extent that the U-value of the timber stud is less than that of the insulation material, there will be a small degree of thermal bridge across the studs. The advantage of fixing the insulation across the outer face of the timber frame is simplicity in fixing and the least amount of cutting. Also, the void space between the studs will augment insulation and provide space in which to conceal service pipes and cables. The disadvantage of external insulation is that the weathering finish such as weatherboarding has to be fixed to vertical battens screwed or nailed through the insulation to the studs. Unless the insulation is one of the rigid boards it may be difficult to make a fixing for battens sufficiently firm to nail the battens to. Internal insulation is usually in the form of one of the insulation boards that combine insulation with a plasterboard finish.

Inorganic materials are most used for insulation between the studs because there is no advantage in using the more expensive organic materials. Rolls of loosely felted fibres or compressed semi-rigid batts or slabs of glass fibre or rockwool are used. The material in the form of rolls is hung between the studs where it is suspended by top fixing and a loose friction fit between studs, which generally maintains the insulating material in position for the comparatively small floor heights of domestic buildings. The friction fit of semi-rigid slabs or batts between studs is generally sufficient to maintain them, close butted, in position. For insulating lining to the outside face of studs one of the organic insulants such as XPS or PIR provides the advantage of least thickness of insulating material for a given resistance to the transfer of heat. The more expensive organic insulants, in the form of boards, are fixed across the face of studs for ease of fixing and to save wasteful cutting. A vapour check should be fixed onto, or next to, the warm inside face of insulants against penetration of moisture vapour. Organic insulants, such as XPS, which are substantially impervious to moisture vapour, can serve as a vapour check, particularly when rebated edge boards are used and the boards are close butted together.

Vapour check

A high level of insulation required for walls may well encourage moisture vapour held by warm inside air, particularly in bathrooms and kitchens, to find its way due to moisture vapour pressure into a timber-framed wall and condense to water on the cold side of the insulation. Condensate may damage the timber frame so as a barrier to warm moist air, there should be some form of vapour check fixed on the warm side of the insulation. Closed cell insulating materials such as XPS, in the form of rigid boards, are impermeable to moisture vapour and will by themselves act as a vapour check. The boards should either be closely butted together or supplied with rebated or tongued and grooved edges so that they fit tightly and serve as an efficient vapour check.

Where insulation materials that are pervious to moisture vapour, such as mineral fibre, are used for insulation between studs, a vapour check of polythene sheet must be fixed across the warm side of the insulation. The polythene sheet should be lapped at joints and continued up to unite with any vapour check in the roof and should, as far as practical, not be punctured by service pipes.

The vapour check is now also being incorporated as an airtightness control membrane (see Photograph 5.36, Photograph 5.37, Photograph 5.38, Photograph 5.39, Photograph 5.40, Photograph 5.41, Photograph 5.42, Photograph 5.43 and Photograph 5.44). To ensure airtightness and vapour control, timber-framed buildings are now externally wrapped and sealed internally.

To prevent overheating of electrical cables that run through insulation, the cables should be de-rated by a factor of 0.75 by using larger cables than specified, which will generate less heat. So that cables are not run through insulation it is wise to fix the inside dry lining to timber frames that are filled with insulation onto timber battens nailed across the frame so that there is a void space in which cables can be safely run.

Resistance to the passage of sound

The small mass of a timber-framed wall affords little resistance to airborne sound but does not readily conduct impact sound. The insulation necessary for the conservation of heat will give some reduction in airborne sound and the use of a brick or block outer leaf will appreciably reduce the intrusion of airborne sound. Where sound pollution is a problem, acoustic plasterboard is used to reduce sound transmission. Photograph 5.52 shows an internal stud wall which separates a corridor from a bedroom. The additional acoustic board will reduce the sound transmission. To ensure that the acoustic board is effective, joints should be staggered and sealed, holes should not be cut into the board and no services should pass through them.

Photograph 5.52 Acoustic board added to a plasterboard to reduce sound transmission.

5.14 Steel frame wall construction

Although steel is traditionally associated with large commercial and industrial buildings, light gauge steel (cold-formed steel sections) is being used in house construction. The sections of steel are used in a similar way to timber. The rolled steel sections can be assembled using either stick, panel or balloon construction. When using stick construction the individual members are delivered to site pre-cut, pre-punched for holes to be cut or self-tapping screws to be used. Advantages of stick-build construction are:

❏ Slight modifications on sight can be accommodated.
❏ Adjustments can be made so that site tolerances can be achieved.
❏ The structural members can be packed and transported in small tightly packed loads.

Stick build construction is labour intensive and is not therefore widely used.

The panel construction has the advantage that the wall subframes, panels, floors and roof trusses are prefabricated and delivered to site ready assembled. The subframes and panels are connected on site using bolts or self-tapping screws. The main advantages of panel construction are:

❏ Large prefabricated sections – saving labour time
❏ Speed of erection
❏ Good quality control achieved in factory production
❏ Accuracy of the components and panels make them easy to assemble on site.

In balloon construction the panels are much larger, floor to roof, but the components are much the same as panel frame construction. Because the steel frame is a good conductor of heat it is important that the insulation is not placed between the metal frame but is applied to the inside of the cavity (see *Barry's Advanced Construction of Buildings* for further information).

5.15 Internal and party walls

Internal walls may either be loadbearing or non-loadbearing. Non-loadbearing walls are usually referred to as 'partition' walls, although care is required because the term is used very loosely. In loadbearing masonry construction the internal walls were usually constructed from brick or blockwork, although more recently the trend has been to use stud walls made of timber or metal, which are quicker to erect and easier to move at a future date (Figure 5.114).

Party walls

The requirement of Part E of Schedule 1 to the Building Regulations is that walls that separate a dwelling from another building or from another dwelling shall have reasonable resistance to airborne sound. Where solid walls of brick or block are used to separate

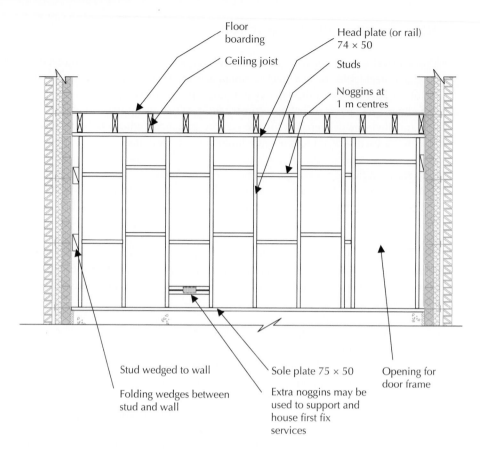

Figure 5.114   Timber stud partition wall.

dwellings, the reduction of airborne sound between dwellings depends mainly on the weight of the wall and its thickness. A cavity wall with two leaves of brick or block does not afford the same sound reduction as a solid wall of the same equivalent thickness because the stiffness of the two separate leaves is less than that of the solid wall and, in consequence, is more readily set into vibration. The joints between bricks or blocks should be solidly filled with mortar and joints between the top of a wall and ceilings should be filled against airborne sound transmission.

Approved Document E provides practical guidance to meeting the requirements of the Building Regulations in relation to walls between dwellings by stipulating the minimum weight of walls to provide adequate airborne sound reduction. For example, a solid brick wall 215 mm thick, plastered on both sides, should weigh at least 300 kg/m^2 including plaster, and a similar cavity wall 255 mm thick, plastered on both sides, should weigh at least 415 kg/m^2 including plaster, and a cavity block wall 250 mm thick, plastered on both sides, should weigh at least 425 kg/m^2, including plaster.

Care should be taken to ensure that the wall and surrounding structure are suitable to achieve the level of sound insulation required. Consideration needs to be given to impact sound, sealing air paths (to prevent airborne sound) and controlling flanking transmission. Thermal insulation also needs to be considered.

In flats and houses separated by party walls it is essential that fire is restricted from passing from one dwelling to another. The Building Regulations outline the requirements of compartment walls (walls that restrict the passage of fire from one area to another).

The Party Wall Act 1996 provides a framework of statutory regulations that must be complied with when undertaking works that affect adjoining property. The three main issues that are addressed are the construction of new walls on boundaries between adjoining owners' land (Section 1: New building on junction); repairs and modification to existing party walls (Section 2: Repair, etc. of party wall: rights of owner); and excavation near to the neighbouring buildings or property (Section 6: Adjacent excavation and construction). In order to comply with the Act, a notice must be served on adjoining owners if works covered by the Act, such as work to party walls or boundaries, are planned. Where there is disagreement between owners, the parties are required to resolve the matter through a dispute resolution procedure.

5.16 Straw wall construction

Straw is a renewable and viable building material, being plentiful and inexpensive. The annual harvest of grain from barley, flax, oats, rye and wheat results in the generation of a considerable quantity of stalks (straw), which is mostly underutilised or even wasted. From an environmental perspective, straw contains carbon, which is trapped in the construction, rather than being released through burning/disposal. The embodied energy in the straw bale construction is also low compared with conventional walls, requiring little processing and little transportation (as use is usually local or regional). Straw is an organic tube made of cellulose that is structurally strong in compression and thus well suited to a number of applications in construction. Straw was first used to reinforce mud and prevent it from cracking, and then as a building block with the invention of the mechanical baler. Straw has reasonably good thermal insulating properties and, because bales are approximately 600 mm thick, a straw bale wall has a high thermal resistance. Straw is currently used as a building material in bale form, or as a pressed panel.

Straw bale construction

Straw bale construction is a relatively old construction method that has become fashionable in recent years, especially in the American Southwest. The drive for a more sustainable approach to the construction of buildings has also resulted in considerable interest and practical application in many countries, including the UK. Straw bale buildings offer a simple and practical method of creating a building with excellent performance characteristics. Super insulated walls (thermal and acoustic), simple construction, low build costs and the conversion of an agricultural by-product into building material are attractive characteristics. With the limited skills required to build a straw bale house, the technique tends to appeal to self-builders and community build groups who can realise low-cost, energy-efficient dwellings, creating a relatively organic addition to the community and at the same time helping to generate income for local farms. Well-detailed, properly constructed and maintained, straw bale houses have a long life, which is comparable with other types of construction. Straw bale construction has also been used in conjunction with other materials in more complex designs for houses and commercial buildings. Here the economies of scale are not available to contractors; thus straw bale walls tend to be more expensive than more traditional approaches.

The bales

Automatic straw balers create tight blocks of straw that provide a relatively easy-to-manoeuvre building block. Sizes of bale vary depending on the baler; however, typical dimensions and weight when dry are:

❑ Two-wire bale: 450 × 350 × 900 mm, approx. 25 kg
❑ Three-wire bale: 600 × 400 × 1050 mm, approx. 35 kg

Although the smaller-size bale is easier to handle, the medium-sized three-wire bale provides better structural, thermal and acoustic performance. Larger cubical and round bales are also available, but these require lifting by mechanical means. There have been moves in the US to establish a 'construction grade' straw bale. Ideally bales should be twice as long as they are wide to simplify and maintain a running bond in the courses.

Advantages of straw bale construction

In addition to simplicity of design and ease of adaptability there are a number of performance advantages a straw bale house has over conventional loadbearing masonry or framed construction.

Straw bale structures are highly fire resistant. The compressed bales contain enough air to provide good insulation values, but because they are compacted firmly they do not hold enough air to permit combustion. Combined with render and plaster surface finishes, a high degree of fire resistance is possible. The type of straw and the moisture content of the bales will mainly determine the thermal characteristics of the wall. With infill construction the frame will also have a determining factor on the overall thermal performance of the wall. Acoustic insulation is considerably better than a conventional wall structure.

Closely packed straw bales, covered with render on the outside face and plaster on the inner face, provide good air leakage control. Coupled with simple geometric design, well-built and maintained straw bale construction provides a building with very little air leakage. The render and plaster finishes can be left unpainted and are inexpensive to maintain. The construction cost of a straw bale wall should be significantly less than that for a comparative wall with the same thermal and acoustic performance.

Disadvantages of straw bale construction

The main disadvantage of straw bale construction relates to concerns over durability, particularly moisture-related damage. Straw bale construction is most suited to dry climates (hot and cold). In the UK the construction of straw bale houses is a relatively recent innovation; thus it is difficult to state with any certainty what the durability of the structure will be over the longer term. As more straw bale buildings are built in the UK and more research is conducted into their durability, designers and builders should have more information from which to make an informed decision.

Fungal rot represents the greatest threat to the life of a straw bale building; thus careful detailing is required to prevent the straw bales from becoming wet. Foundations and roof construction are critical in preventing unwanted rain and moisture penetration, and the construction of the wall must be done in such a way as to avoid any possibility of interstitial condensation. Fungi and mites can live in wet straw; therefore the bales must be bought when they are dry, kept dry until needed and then sealed into the wall construction with plaster and render to eliminate any chance of access for pests. Paint for interior and exterior wall surfaces should be permeable to water vapour so that moisture does not get trapped inside the wall. Services, especially sealing around outlets, need special attention.

Straw bale walls are considerably thicker than more traditional construction methods (at least double), and so where land is at a premium a straw bale house will provide less internal space for the same external dimensions as a traditional construction. This may be of little concern in rural locations. The appearance of straw bale buildings may not always appeal to a more urban environment, although there are a few examples of straw bale construction being used in densely populated urban environments, usually in conjunction with other materials as part of a composite construction.

Selection of straw bales

Straw bales should be purchased immediately after the harvest, when they are abundant, fresh and dry. Dealing directly with a farmer is the cheapest and perhaps most reliable method of selecting the best quality bales since quality is largely judged on visual appearance and touch. Transportation to the building site will need to be addressed, perhaps independently of the arrangement with the farmer if the bales are to be transported out of the local area. Straw bale merchants provide an alternative source of supply and will have an established transport infrastructure and storage facilities. Once again the selection of the bales, especially if sourced from more than one farmer, is an important consideration.

Bales should be tied tightly with polypropylene string or bailing wire and should not twist or sag when lifted. All bales should be uniformly compacted and contain thick, long-stemmed straw; any bales comprising short, thin straw should be rejected. Old and/or damaged bales should be rejected. If practical the construction of a straw bale building should be undertaken immediately after the main grain harvest when the bales are fresh and dry, with bales transported to site, positioned and protected from the weather as quickly as possible. If this is not an option then the bales should be selected and stored under dry, ventilated, conditions. Bales should be tested for moisture content, which must be 14% or below when purchased and when used for building. All bales must be stored under dry conditions until required for building purposes. Protection of the bales from the weather is also necessary during construction.

Foundation construction and site drainage

The base of the straw bale construction must be kept dry at all times; thus the manner in which the straw bale interacts with the foundation will be a crucial factor in determining the durability of the wall. The position of the lowest straw bale in relation to the finished ground level and the detailing of the external wall finish at this junction are particularly important. Similarly, site drainage must be designed to get water away from the base of the walls as quickly as possible (Figure 5.115).

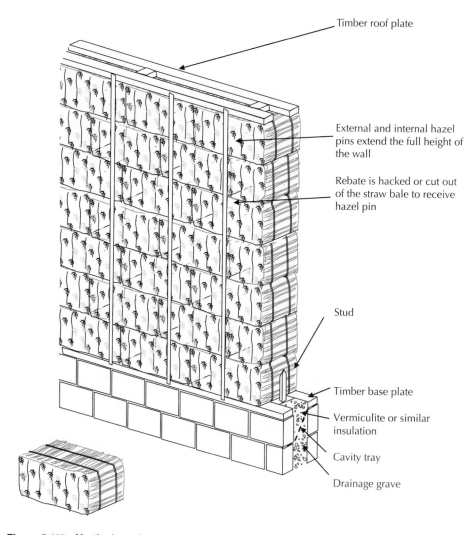

Timber roof plate

External and internal hazel pins extend the full height of the wall

Rebate is hacked or cut out of the straw bale to receive hazel pin

Stud

Timber base plate

Vermiculite or similar insulation

Cavity tray

Drainage grave

Figure 5.115   Vertical section through a straw bale wall.

Guidance on the minimum distance between the finished ground level and the lowest bale position varies from around 225 mm upwards to 300 and 450 mm. Knowledge of local rainfall patterns, the extent of the roof overhang in relation to the height of the wall and the ground finish adjacent to the wall will be determining factors. As a general guide the bales should be positioned at least 450 mm off the finished ground level to avoid splashback of rain, bouncing off the ground and on to the face of the wall. The material under the straw bales should be good quality stone, engineering brick or dense blockwork. The dpc should be positioned between the solid foundation and the timber base or sole plate. The manner in which the external render is finished at the junction of foundation and wall is also important. A drip should be formed at the bottom of the render to help throw any water that runs down the wall off and away from the base of the wall.

Care should be taken so as not to compromise the site drainage through inappropriate surface finishes. The ground around the structure must be well drained and kept dry with adequate surface and below ground drainage. The position of the water table and history of the proposed site with regard to flooding should be investigated and appropriate decisions taken to protect the construction from damp and/or wet conditions. Self-draining foundations are one method of helping to keep the base of the wall free of water. Vermiculite is placed above the cavity tray. Below the cavity tray, the cavity should be filled with drainage gravel and weep holes provided at regular intervals (Figure 5.116).

Roof construction

The design and construction of the roof is another critical area. The roof should be constructed with a large overhang to provide protection to the wall below from rain and snow. Leaks in the roof and/or inappropriate detailing at the wall to roof junction will compromise the long-term durability of the structure. A large roof overhang on single-storey buildings will also provide good protection to the base of the wall. Roof pitch is another consideration, with steep pitches and a good quality roof covering to facilitate the rapid discharge of rainwater from the roof considered important to the durability of the roof. Timber roof construction is the usual method, with the roof loads transferred to the foundations via the loadbearing straw bale wall or via the timber frame construction (described further in more detail). Roof coverings may be of reed thatch, tile or lightweight finishes, with lightweight roof structures and coverings preferred for loadbearing straw bale walls.

Wall construction

The main principle of straw bale design is based on the use of full bales as a building module – simplicity is the key. Half bales will be required for bonding purposes, the same principle as that for brick and block wall construction. Bales smaller than a half size should not be used because of their lack of compressive strength. Careful planning is required before any drawings are finalised, simply because the actual size of bales varies between suppliers. Once a bale supplier and hence dimension has been determined, it is then possible to design and detail the building plan and also the opening sizes (doors and windows) and their position within the wall. The main principles relating to the strength and stability of the wall are similar to those for loadbearing or framed construction. Where possible, door and windows should be selected to suit the bale module size or be manufactured/built to

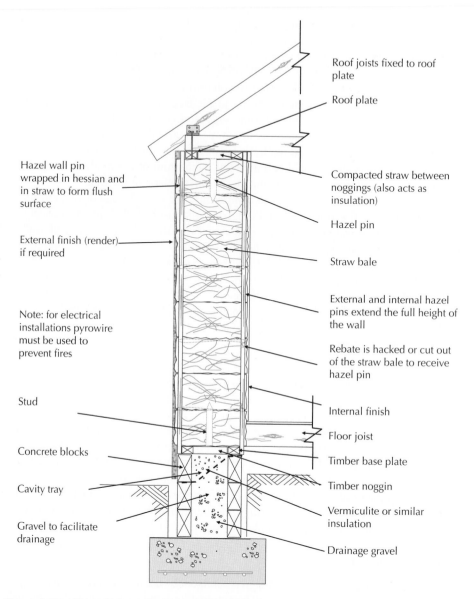

Figure 5.116   Straw bale wall to foundation detail.

suit the bale module. When using loadbearing straw bale construction, there will be some settlement of the wall after the roof has been completed. The straw bales must be sealed against the weather and pests. External surfaces are finished in lime render (stucco). Internal surfaces are plastered. The render and plaster provide an attractive surface finish, which also adds to the structural integrity of the wall.

Loadbearing straw bale construction

In loadbearing straw bale construction the straw bale walls directly support the loads from the roof. The quality of the bales is an important consideration. Hazel pins are used to tie the bales together, two per bale from the fourth course upwards, with staples used to tie the bales at changes of direction, such as corners. The bales are often tied together tightly with threaded metal rods, wire or plastic straps, which helps to compress them and hence minimise settlement of the bales when the roof structure is added. For small buildings, such as small single-storey houses, with relatively short lengths of wall, there may be no need for additional support. For long walls it will be necessary to provide a system of vertical posts at 10 m (or less) intervals to help support the weight of the roof, in which case the infill method may be more suitable. The straw bales will provide lateral restraint at the corners; thus a minimum return of at least two bales is recommended before any openings are formed in the walls for doors or windows. This is a similar principle to that adopted for brick and block walls.

Settlement of the straw bales under load, which may be irregular, must be considered in the design. Settlement in a seven-bale high wall may be up to 50 mm and is difficult to predict because the extent of settlement will depend on the density of the bales (which varies between bales) and the amount of loading applied to them. Thus for practical reasons it is necessary to measure the opening sizes after the main structure is complete and to install doors and windows late in the construction process (after the initial settlement of the bales).

Mortar bale

A structural mortar made of Portland cement and sand is applied between the straw bales to create a lattice structure. When dry, the lattice forms a structural framework between the bales and thus provides a form of backup should a straw bale fail. Bales are finished with stucco on the external face and plastered on the interior to protect the bales and also to provide an attractive finish. The stucco and plaster also add to the structural integrity of the wall. Some designers and builders may feel that resorting to cement will compromise their environmental ideals.

Infill straw bale construction

Non-loadbearing, post and beam, framed and infill are terms used to describe a framed wall construction with the bales bearing their own weight. The wall-framing members carry the weight of the roof. This method is better suited to larger structures. The bales are attached to each other by piercing them with bamboo, hazel or metal reinforcing bars and attached to the timber framework. With the infill technique the settlement of the straw bales may be less (less load on them) and hence the settlement of the bales is less of a concern than with a loadbearing construction. Restricting the straw bales to an infill function only may be a less risky approach to straw bale construction. If there is a problem with any of the straw bales, the structural integrity of the structure will not be compromised. Framed construction provides an opportunity to construct the frame before the grain harvest; thus fresh bales can be taken directly from the field and installed. This helps to ensure dry bales are used and removes the need for storage and unnecessary weather protection.

Straw and clay building

A traditional and durable construction method is to use a 'batter' of clay and water stirred into loose straw to produce a straw-reinforced clay mud. The mixture is packed tightly into a lightweight timber ladder framework to create partition walls or as infill panels for framed wall construction.

Pressed straw panels

Pressed straw panels are made by compressing straw under temperature to produce a panel made of 100% straw. The combination of compressed straw, recycled paper lining and adhesive will form a board to comply generally with BS 4046. The result is a low-cost, versatile product with environmental benefits. The panels can be used as a self-supporting, non-loadbearing partition system and also for roof decking. The technique was first introduced to the UK and developed into a sophisticated straw board product by Stramit, in Suffolk. Pressed straw panels may also be used in roofs and floors. Research and development is currently being conducted by a number of companies and research organisations into the use of pressed straw panels for use in a structural capacity.

5.17 Earth wall construction

While earth or unfired clay brick walls are commonly used in very warm, dry, climates (Photograph 5.53 and Photograph 5.54), earth wall construction is also gaining increasing interest in the UK Rocksand soils are some of the most economic building materials, available locally and with a long record of use in vernacular buildings in the UK, for example, cob construction. There has been renewed interest in the use of earth as a building material in the UK, mainly associated with the drive for a more environment-friendly approach to construction. There remain a number of concerns about durability, but, when detailed to suit the climate and topography of the site, there is no reason why earth-based construction cannot provide an alternative to more traditional materials and methods. The most common method used in the UK is earth-sheltered structures, with growing interest in adobe and rammed earth. To reduce transportation costs, and the associated pollution, earth should be used from the building site, and additional materials sourced locally. It may be an obvious point, but earth construction will be specific to the locale, soil condition, techniques and materials available. There will be a strong reliance on the extent of local knowledge among the workers.

Earth-sheltered construction

Earth-sheltered buildings are an energy-efficient and environmentally conscious approach to construction that may suit certain sites and some clients. The concept of burying the building under the ground, or at least covering a significant proportion of the building with

Photograph 5.53   Non-fired earth – clay walls.

Photograph 5.54   House construction in Egypt using non-fired earth – clay walls.

soil, is not new. Various designs exist, ranging from buildings completely buried beneath the ground to buildings that have one or more walls protected by the earth.

When the entire structure is buried, the term 'underground' structure is used. When earth is banked up (bermed) against an external wall, this is known as a 'bermed' structure. A bermed structure may be built above, or partially above, natural ground level. In the majority of cases the roof will be covered with earth and vegetation to form a green roof. The vegetation, grass or sedums, will help to reduce erosion and provide an attractive finish. Three generic house designs have been developed from the use of underground and/or bermed construction. They are:

❑ Courtyard (or atrium) design. An underground structure planned around a central atrium, from which rooms are accessed, and which forms the focal point of the house. Some designs are based on an open courtyard, which provides natural ventilation but poses some problems with removal of surface water and snow. Atrium designs are built with a glazed roof over the courtyard, which, depending on the design, can also help to provide some natural ventilation. Plan forms tend to be circular, ovoid or hexagonal in nature.

❑ Elevational design. A bermed structure, usually with a glazed south-facing wall. While the glazed facade is exposed, the other walls are covered with earth. The south-facing wall allows penetration of daylight and thermal gain while also providing an opportunity for natural ventilation. Skylights can be used to alleviate the problems of poor light and poor air circulation further into the plan. To reduce the depth of the house, the south (and north) faces are considerably longer than the east and west walls, creating a long, thin house. To improve the space it is common to construct a curved elevation to the south.

❑ Penetrational design. Again a bermed structure that covers the entire house, with the exception of doors and windows. The house is usually built on a flat site with the earth bermed around the entire structure. This allows the opportunity for cross-ventilation and for natural light to all rooms.

In the UK, earth-sheltered construction has been employed on a small number of commercial buildings but tends to be favoured by clients wanting a bespoke house design. The primary concern is with protection from moisture. Careful choice of site (e.g. topography, groundwater position, soil type), along with the provision of adequate drainage at and below ground and waterproofing to walls, is a crucial factor. Earth-sheltered structures need to be well designed and constructed, and designers and contractors with experience of this type of work should be used.

A naturally sloping site is the best location for an earth-sheltered building. The extent of the dig required to excavate a suitable area will be determined by the angle of the slope, thus a moderately steep site is usually preferable to a gently sloping site. Soil type is another determining factor. Granular soils compact well and are permeable, which allows groundwater to drain away. Sites with cohesive soils (such as clays) should be avoided because the clay will expand when wet and has poor permeability. Site investigation is also necessary to establish the height of the water table (the lowest level of the building must be above the water table). Areas in which radon gas naturally occurs should be avoided and special measures may be required where the site is heavily contaminated.

Advantages of earth-sheltered structures

Well-detailed and well-constructed, an earth-sheltered house provides a number of benefits over more traditional structures. The biggest advantage is that the building is protected from the weather and hence is not subject to deterioration by, for example, solar radiation or wind damage. In very exposed locations, protection from adverse weather may be a considerable benefit. Protecting the building with earth will reduce or even eliminate the need for regular painting, cleaning of gutters and repair. Thus maintenance is considerably less onerous for the majority of earth-sheltered designs.

The earth provides natural insulating properties to the structure, providing a relatively stable indoor temperature, which is less affected by external air temperature fluctuations. Thus less energy is required to maintain a stable internal temperature. The earth also provides considerable sound dampening in addition to that provided by the structure of the building.

Earth-sheltered structures will blend into the landscape and may be an ideal solution in areas where the visual appearance of a structure would cause a problem. This does not mean, however, that approval from the local town planning department will be forthcoming since other factors, such as land use and access, will also need to be considered.

Disadvantages of earth-sheltered structures

There are a number of potential disadvantages associated with earth-sheltered structures. Exclusion of moisture is perhaps the most obvious. Waterproofing and tanking is a primary concern and often carries a relatively high initial cost. Surface water and underground drainage is key. Seasonal or regular surface water flows must be channelled away from the structure. Underground drainage should be designed to remove water (and water pressure on the walls) quickly. Regular maintenance of drainage channels to avoid blockages is also essential if problems are to be avoided.

The initial cost of an earth-sheltered structure can be as much as 20% higher than a comparable structure built above the ground. Exact figures will depend on the nature of the ground and topography of the site (e.g. gently or steeply sloping). However, there may be considerable cost savings made during the life of the building, which must be considered in a whole life costing approach.

Earth-sheltered houses tend to be characterised by high levels of humidity; therefore, air exchange is important. Some designs may also result in pockets of stale air unless some form of air flow is introduced. Ensuring adequate air flow can be achieved through the use of passive or mechanical ventilation systems. Passive ventilation can be provided through vented skylights, which are required to get natural light into the internal spaces. Linking the air exchange system to an earth to air heat exchanger may help to keep energy costs to a minimum.

Repairs and remedial work to the structure of the building may be expensive, since it is difficult to get to walls without removing a great deal of earth. This places additional emphasis on the quality of the detailing and the quality of the work undertaken on site. Careful and systematic inspection of the waterproofing work as it proceeds is essential.

Materials

The most commonly used material for the retaining walls is concrete. Reinforced in situ concrete and/or prefabricated concrete can be used. Alternatively masonry (brick, block, stone) can be used, reinforced with steel bars. Attention must be given to joints (i.e. between precast concrete units) to ensure the walls remain watertight. In addition to their strength, walls must provide a good surface for waterproofing and for thermal insulation (where required). Waterproofing to the walls can be achieved with a number of systems. The main materials are:

- ❏ **Rubberised asphalt.** Sheets are applied directly to walls and roofs and have a long life expectancy.
- ❏ **Plastic sheets.** The integrity of these sheet systems relies on the seams between the sheet materials. The seam must be formed and sealed properly; otherwise the membrane will leak. A variety of sheets with different material properties are available.
- ❏ **Liquid PURs.** Usually used in places where it is difficult to apply a membrane. Sometimes applied over the external side of insulation.
- ❏ **Bentonite.** A natural clay which expands on contact with water and forms a barrier to moisture penetration. Formed into panels attached to walls or spray applied in liquid form with a binding agent.

Thermal insulation

Insulation is usually placed on the exterior face of the walls, after the waterproofing system has been completed. In this way the heat generated, captured and absorbed within the earth-sheltered envelope is retained by the building's interior. Rigid insulation sheets help to protect the waterproofing from punctures from sharp stones in the soil. Thin protective boards are often used as a barrier between the insulation and the earth.

Adobe

Adobe is compressed earth, rammed into moulds or pressed into blocks while damp and sun-dried to form a building block or brick. Adobe is a common construction method in many parts of the world (mainly places with a dry climate). The best adobes are made from soil that is high in clay content. Mechanical presses can be used to form adobe blocks directly from the building site's soil. These blocks are then used to build loadbearing (or infill) walls, laid without mortar. Blocks are laid in a walling bond, with the connecting surfaces wetted with water to help provide a bond between the individual units when dry. The walls are then rendered (stuccoed). Compaction of the earth is critical and over-compaction can lead to fracture of the material. Scientific analysis of the soil is therefore crucial.

Adobe and rammed earth walls absorb solar heat during the day and radiate heat back into the air at night, thus providing a comfortable and relatively constant internal temperature. Thus adobe tends to be used in dry and hot locations. Both exterior and internal adobe walls provide good thermal mass and can form part of a passive design. The principles of weather protection and durability of earth are similar to that of straw, namely, protection from moisture. In damp climates and areas of heavy rainfall it might be prudent to use adobe only for internal walls.

Rammed earth

Rammed earth is a term used to describe the mixing of cement and earth, which is then packed into wall forms with a pneumatic tamper. Hence the term 'rammed' earth. The result is a material that resembles a sedimentary rock, with compressive strengths about half that of concrete. The strength of the material will vary, depending on the ratio of cement to soil (and soil type); thus testing of batches is necessary to determine the load-bearing capacity of the resulting wall. By using soil from the building site, rammed earth can provide a relatively cheap material with reasonable environmental credentials. The use of the material is similar to that described for concrete walls. By increasing the thickness of the walls, the loadbearing capacity may be increased.

Photograph 5.55 **(a) Hempcrete providing the insulation of the timber frame building. (b) Shuttering between placed on the outside of the timber frame to allow the hempcrete to be placed and lightly compacted in position. (c) Hempcrete construction within a timber frame drying out. (d) Hempcrete construction within a timber frame. (e) Completed hempcrete building.**

Hempcrete

Hemp is a low-impact, sustainable plant material that is easy to grow, reaching heights of 4 m. The fibres are separated from the woody core, shredded and then graded for use in construction. The fibres are then mixed with quicklime (calcium oxide) to provide a homogeneous material that can be used as insulation and non-structural walling material. When used within a timber frame hempcrete has good insulation properties, and with the application of an air barrier it can achieve the same standard of airtightness as other building materials.

Hempcrete can be sprayed onto a background to provide an insulation layer. Alternatively it can be prepacked into prefabricated panels that can be delivered to site and erected in the same way as other off-site engineered panels, or the panels can be placed in situ between the timber frame. Formwork is used to provide the temporary mould for the hempcrete, which can be removed after 24 hours. The drying times of hemp vary. Typical conductivity of hempcrete is between 0.07 and 0.09 W/mK. For a 500 mm deep wall, in a timber frame it is possible to achieve a U-value of 0.21 W/m^{2}K. Photograph 5.55a–e shows a timber frame bungalow with hempcrete applied in situ between the shuttering.

6 Roofs

The roof is an important element in providing protection from the weather and has a significant role to play in the reduction of heat loss from a building. Roofs are classified as being either pitched or flat. Timber is the most common material used, with some use of concrete for flat roofs. Although it is common practice to construct pitched roofs from prefabricated timber trusses, space has been reserved for the traditional timber roof built on site, primarily as an aid to the understanding of construction and terminology, but also as an aid to those involved in repair and refurbishment projects. However, more traditional methods of roof construction may return as the value of habitable space increases and rooms within the roof space become more common.

6.1 Functional requirements

The functional requirements of a roof are:

- ❑ Strength
- ❑ Stability
- ❑ Resistance to weather
- ❑ Durability and freedom from maintenance
- ❑ Fire safety
- ❑ Resistance to the passage of heat
- ❑ Resistance to the passage of sound
- ❑ Resist air leakage
- ❑ Security
- ❑ Aesthetics

Strength and stability

The strength and stability of a roof depends on the characteristics of the materials from which it is constructed and the way in which the materials are formed as a horizontal (flat) platform or as a triangular (pitched) framework.

The strength and stability of a flat roof depends on adequate support from walls or beams and sufficient depth or thickness of timber joists. Whether it is made of timber, concrete

Barry's Introduction to Construction of Buildings, Third Edition. Stephen Emmitt and Christopher A. Gorse.
© 2014 John Wiley & Sons, Ltd. Published 2014 by John Wiley & Sons, Ltd.

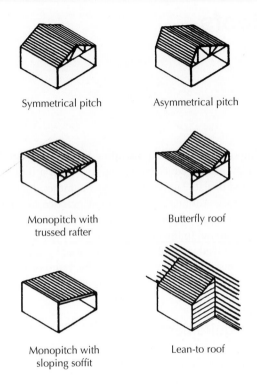

Symmetrical pitch   Asymmetrical pitch

Monopitch with   Butterfly roof
trussed rafter

Monopitch with   Lean-to roof
sloping soffit

Figure 6.1   Sloping roofs.

or steel, the roof should not suffer excessive deflection under the dead load of the roof itself and the load of snow and wind pressure or uplift that it may suffer.

The construction of a pitched roof relies on triangular frames of sloping rafters tied (trussed) together with horizontal ceiling joists, usually with a system of struts. The strength and stability of this form of roof depends on the depth of the triangular frames at mid-span. There is an inherent instability across the slopes of this roof, parallel to the ridge, to the extent that wind pressure may cause the frames to rack or fall over. To resist racking the frames are braced by gable end walls, hipped ends or by cross-bracing (diagonal roof boarding or braces fixed across sloping rafters). Typical forms of pitched roofs are illustrated in Figure 6.1.

Resistance to weather

A roof prevents water entering a building by using a roof covering which prevents rain penetration. The materials that cover the roof range from the continuous impermeable layer of asphalt that can be laid horizontal to exclude rain, to the small units of clay or concrete tiles that are laid overlapping or interlocking so that rain runs off the roof to rainwater gutters.

When using small units to cover roofs the tiles must have considerable overlap and the slope of the roof should be laid at a greater pitch to prevent the rain from penetrating between the joints. Larger units such as profiled metal sheets (see *Barry's Advanced Construction of Buildings*) can be laid at a lower pitch than that required for tiles. Impermeable materials such as asphalt and bitumen are laid flat without joints, and sheet metals such as lead and copper that are overlapped or joined with welts can be laid with a very shallow fall. The small open-jointed units of tile and slate which provide little resistance to the penetration of wind into the roof require a continuous layer of a sheet material beneath them to exclude wind (often called sarking felt).

A roof structure will be subject to movements due to variations in loading by wind pressure or suction, snow loads and movements due to temperature and moisture changes. The advantage of slate and tile is that as the small units are hung, overlapping down the slope of roofs, the joints between the tiles or slates can accommodate movements in the roof structure without breaking slates or tiles or letting in rainwater. Where the roof is covered with large unit size materials or continuous roof coverings they may fail (cracking and allowing rain penetration) if there is inadequate provision of movement joints.

Durability and freedom from maintenance

The durability of a roof depends on the ability of the roof covering to exclude rain, snow and the destructive action of frost and temperature fluctuations. Persistent penetration of water into the roof structure may cause or encourage decay of timber, corrosion of steel or disintegration of concrete. Pitched roof coverings are relatively durable and need little in the way of maintenance. In comparison most flat roof coverings have a shorter service life and will need to be replaced, which can be disruptive to the use of the building.

Fire safety

The requirements for fire safety in the Building Regulations are concerned for the safe escape of occupants to the outside of buildings. The regulations require adequate means of escape and limitation to internal and external fire spread. On the assumption that the structure of a roof will ignite after the occupants have escaped there is no requirement for resistance to fire of most roofs. However, the requirements do limit the external spread of fire across the surface of some roof coverings to adjacent buildings by limits to the proximity of buildings.

Resistance to the passage of heat

The materials of roof structures and roof coverings are generally poor insulators against the transfer of heat so it is necessary to use insulating materials to control excessive loss or gain of heat.

The target requirement of the Building Regulations for the insulation of roofs of dwellings is $0.13\,\text{W/m}^2\,\text{K}$, which allows flexibility in design by providing a maximum limiting U-value of $-0.2\,\text{W/m}^2\,\text{K}$ for all roofs (Figure 6.2).

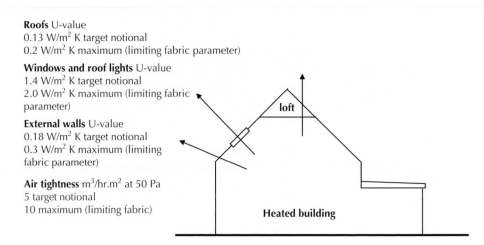

Roofs U-value
0.13 W/m² K target notional
0.2 W/m² K maximum (limiting fabric parameter)

Windows and roof lights U-value
1.4 W/m² K target notional
2.0 W/m² K maximum (limiting fabric parameter)

External walls U-value
0.18 W/m² K target notional
0.3 W/m² K maximum (limiting fabric parameter)

Air tightness m³/hr.m² at 50 Pa
5 target notional
10 maximum (limiting fabric)

loft

Heated building

Figure 6.2   Target and maximum U-values for roofs.

The increased emphasis on low and zero energy buildings requires higher levels of thermal insulation. The nearly zero carbon dwellings and Passivhaus adopt building fabric performance of between 0.8 and 0.15 W/m² K (see Figure 6.3).

The most economical method of insulating a pitched roof is to lay or fix some insulating material between or across the ceiling joists, the area of which is less than that of the roof slope or slopes. This insulating layer will reduce the loss of heat from the building to the roof space, and reduce gain of heat from the roof space to the building. As the roof space is not insulated against loss or gain of heat it is necessary to insulate water storage cisterns and pipes in the roof against possible damage by frost, since the air space will be a similar temperature to that outside the building. This is termed a 'cold roof' construction (Figure 6.4a) and it is necessary to ventilate the roof space to prevent condensation in the cold air space.

Where the space inside a pitched roof is used for storage or as part of the building it is usual to insulate the underside of the roof slope. With the insulation fixed across the roof rafters, under the roof covering, the roof is insulated against changes in outside air temperature and is termed a 'warm roof' construction (Figure 6.4b). The warm and cold roof terminology is also used for flat roof construction, as described later.

Resistance to the passage of sound

The resistance of a roof to the penetration of airborne sound is not generally considered unless the building is close to an airport or busy road or rail network. The mass of the materials of a roof is the main consideration in the reduction of airborne sound. A solid concrete roof will have a greater effect in reducing airborne sound than a similar timber roof. The introduction of mineral fibre slabs, batts or boards to a timber roof will have some effect in reducing intrusive, airborne sound.

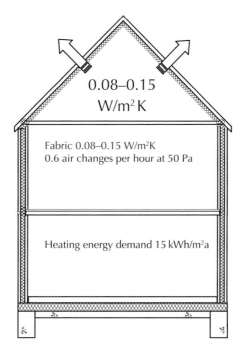

Figure 6.3   Nearly zero carbon and Passivhaus fabric standards.

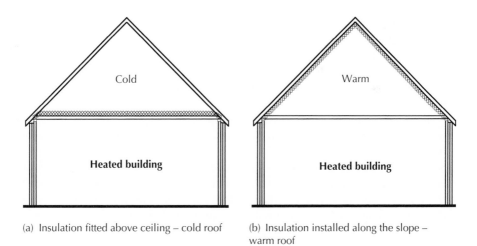

(a) Insulation fitted above ceiling – cold roof

(b) Insulation installed along the slope – warm roof

Figure 6.4   Cold and warm pitched roof construction.

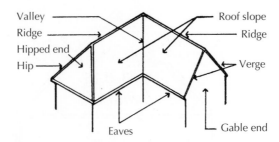

Figure 6.5 Pitched roof.

6.2 Pitched roofs

The majority of pitched roofs are constructed as symmetrical pitch roofs with equal slopes pitched to a central ridge. The least slope of a pitched roof is determined by the minimum slope necessary for the roof covering to exclude rain and snow.

The terms used to describe the parts of a pitched roof are illustrated in Figure 6.5. The ridge is the highest, usually central horizontal part of a pitched roof. Eaves is a general term used to describe the lowest part of a slope from which rainwater drains to a gutter or to the ground. A gable end is the triangular part of a wall that is built up to the underside of roof slopes, and the junction of slopes at right angles to a wall, the verge. A valley is the intersection of two slopes at right angles. A hipped end is formed by the intersection of two, generally similar, slopes at right angles.

Roof types

Couple roof

The simplest form of pitched roof structure consists of timber rafters pitched up from supporting walls to a central ridge. This form of pitched roof is termed a couple roof as each pair of rafters acts like two arms pinned at the top, known as a 'couple' and illustrated in Figure 6.6a. Couple roofs have been used to provide shelter for farm buildings and stores with the ends of such buildings formed with gable ends built in brick or stone or timber framed and rough boarded.

Pairs of rafters are nailed each side of a central ridge board. The lower part or foot of each rafter bears on, and is nailed to, a timber wall plate, which is fixed to the walls. When this form of roof is covered with slates or tiles and subject to wind pressure there is a positive tendency for the foot of the rafters to spread and overturn the walls on which they bear, as illustrated in Figure 6.6b. Spreading of rafters is only weakly resisted by the nailed connection of rafters to the ridge board, which does not act as an effective tie. The maximum span of this roof is generally limited to 3.5 m. The most straightforward way of providing a tie to resist the spread of the foot of rafters is to fix a metal or timber tie rod between the foot of pairs of rafters. By this device working spans of up to 5 m are practical.

Rafters are spaced from 400 to 600 mm apart to provide support for tiling or slating battens. The size of rafter depends in part on the spacing of rafters and mainly on the clear

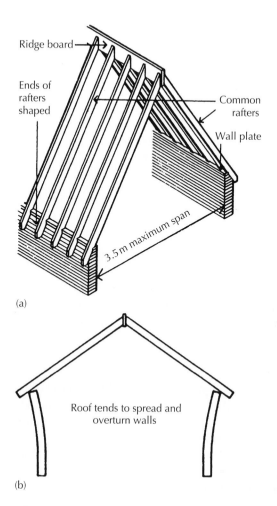

Figure 6.6   Couple roof.

span, measured up the length of the rafter from the support on the wall plate to the ridge. Sawn softwood rafters for a typical couple roof would be 100 mm deep by 38 mm thick to provide reasonable support for the roof, its covering and wind and snow loads.

Ridge board

The purpose of the ridge board is to provide a means of fixing the top of pairs of rafters. A softwood board is fixed with its long axis vertical and its length horizontal. The top of rafters is cut on the splay so that pairs of rafters fit closely to opposite sides of the ridge board to which they are nailed, as illustrated in Figure 6.7.

The ridge board is one continuous length of softwood, usually 32 mm thick. The depth of the ridge board is determined by the depth of the splay cut ends of rafters that must bear fully each side of the board. A steeply pitched roof and deep rafters will need a deeper ridge

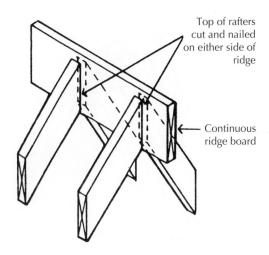

Top of rafters
cut and nailed
on either side of
ridge

Continuous
ridge board

Figure 6.7   Ridge board.

board than a shallow pitched roof. The ridge board is usually some 50 mm deeper than required as bearing for rafter ends, to provide fixing for battens.

Wall plate
A continuous timber wall plate is fixed on walls by steel brackets. The wall plate provides a means of support and fixing for the foot of rafters to the wall. The sawn softwood plate, usually 100 × 75 mm in section, which is laid on its 100 mm face, serves to spread the load from rafters along the length of the wall.

A 'birdsmouth' cut is made in the top of each rafter to fit closely round the wall plate, as illustrated in Figure 6.8 and Photograph 6.1. Rafters are nailed to the wall plate. The birdsmouth cut should not be greater than one-third of the depth of a rafter. This limits the loss of strength in rafters so that the part of the rafter that projects over the face of the wall is still structurally sound; this projection is termed the eaves. The bearing of the rafter ends on the wall plate does not effectively resist the tendency of a couple roof to spread under load, to the extent that the plate may be moved by the spreading action of the roof.

Close couple roof
Pitched roofs to small buildings such as houses and bungalows are framed with rafters pitched to a central ridge board with horizontal ceiling joists nailed to the side of the foot of each pair of rafters, as illustrated in Figure 6.9. The ceiling joists serve the dual purpose of ties to resist the natural tendency of rafters to spread and as support for ceiling finishes. Because the ceiling joists act as ties to the couple of pairs of rafters, this form of roof construction is a close couple or closed couple roof.

Ceiling joists are usually 38 or 50 mm thick and 97–220 mm deep sawn softwood, the size of the joists depending on the spacing and clear span between supports. The maximum span between supporting walls for the close couple roof illustrated is 5.5 m. For this span,

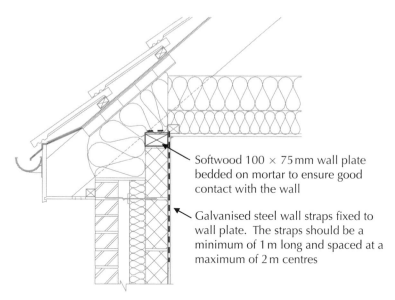

Softwood 100 × 75 mm wall plate bedded on mortar to ensure good contact with the wall

Galvanised steel wall straps fixed to wall plate. The straps should be a minimum of 1 m long and spaced at a maximum of 2 m centres

The wall plate is used to transfer and evenly distribute the loads from the roof to the wall. Once the wall plate is fixed to the wall with holding down straps it also provides a good medium (wood to wood) for the timber roof trusses to be fixed.

Figure 6.8   Wall plate and holding down straps (adapted from http://www.leedsmet.ac.uk/ teaching/vsite).

Loft vent tray

Roof truss with birdsmouth cut over wall plate

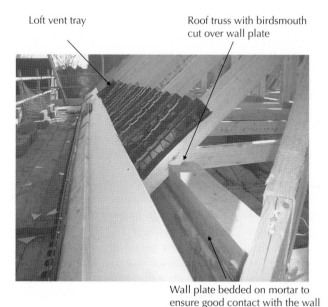

Wall plate bedded on mortar to ensure good contact with the wall

Photograph 6.1   Eaves detail showing trusses fixed to the wall plate and bedded on the block wall.

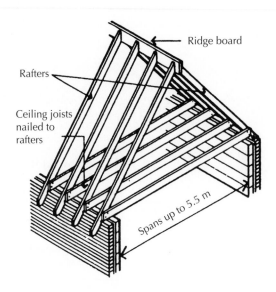

Figure 6.9   Close couple roof.

ceiling joists with no intermediate support from internal walls would be 220 mm deep by 50 mm thick to support ceiling finishes without undue deflection.

The advantage of the triangular space inside the roof above the ceiling joists is that it will to some extent provide insulation, provide a convenient space for water storage cisterns and provide the storage space that is lacking in most modern house designs. The disadvantage of the close couple roof structure by itself is that the considerable clear spans of rafters and ceiling joists require substantial timbers as compared with similar roofs where there is intermediate support to rafters from purlins and to ceiling joists from binders.

Collar roof

Another form of tied couple roof is framed with collars joined across pairs of rafters, at most one-third up the height of the roof, as illustrated in Figure 6.10. The purpose of this arrangement is to extend first-floor rooms into the roof space and so limit the largely unused roof space. A disadvantage of this arrangement is that the head of windows formed in a wall will be some distance below the ceiling and give less penetration of light. To provide normal height windows a form of half dormer window is often used with the window partly built into the wall and partly as a dormer window in the roof.

A collar, fixed at up to a third up the height of a roof, is not as effective in tying the pairs of rafters together as standard ties at the foot of rafters. To provide a secure joint between the ends of collars and rafters, a dovetail half depth joint is formed. The ends of collars are cut to half their width in the shape of a half dovetail to fit into a similar half depth housing in rafters and the two nailed together, as illustrated in Figure 6.10. Because a collar is a less effective tie than a ceiling joist the maximum span of this roof is limited to 4.5 m. To provide solid framing the rafters are usually 125 × 44 mm and the collars

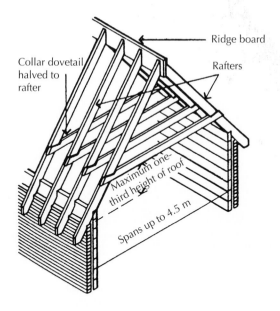

Figure 6.10 Collar roof.

125×44 mm. Photograph 6.2 shows a roof structure constructed with a collar close to the ridge and close couple across the roof span.

Purlin or double roofs

To economise in the section of roof rafters it has been practice to provide intermediate support up the slope of roofs by the use of purlins. Purlins are horizontal timbers supported by end walls or struts to internal loadbearing walls. By the use of a comparatively substantial timber purlin an appreciable saving in timber rafter size can be affected because the clear span of the rafter is halved. Timber purlins are supported on masonry corbels built to project from separating walls or on metal joist hangers fixed into walls.

The roof illustrated in Figure 6.11 is framed with purlins for support for rafters and an internal loadbearing wall to provide intermediate support for ceiling joists to economise in timber sizes. The purlins illustrated are 175×75 mm, the rafters 125×50 mm and the ceiling joists 125×50 mm sawn softwood.

Where there are no separating or gable end walls to provide support for purlins, an internal loadbearing wall can be used to support timber struts fixed between the wall and the rafters, as illustrated in Figure 6.12.

The purlins are fixed with their long axis vertical with rafters notched over and nailed to purlins. Horizontal collars are nailed to the side of every fourth pair of rafters under purlins. Pairs of struts are notched around a wall plate bedded on the internal wall, notched under and around purlins and nailed to the side of collars. Ceiling joists nailed to the foot of rafters and bearing on the internal wall serve as ties to the close couple roof and as ceiling support. Struts 75×75 mm support 150×50 mm purlins with 125×50 mm collars, 150×50 mm rafters and 125×50 mm sawn softwood ceiling joists.

Photograph 6.2 Roof structure with collar at the ridge and closed at the bottom, close couple-type roof with collar (courtesy of J Kangwa).

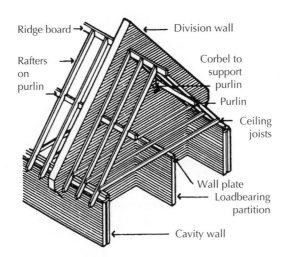

Figure 6.11 Purlin roof.

Hipped ends of roof

To limit the expanse of roof to detached buildings such as houses, for appearance's sake, the ends of a pitched roof are sometimes formed as slopes described as hipped ends. The hipped, sloping ends of pitched roofs are usually framed at the same slope as the main roof. To provide fixing and support for the short length of rafters, which are pitched up to the intersection of roofs, a hip rafter is used as illustrated in Figure 6.13. To provide adequate

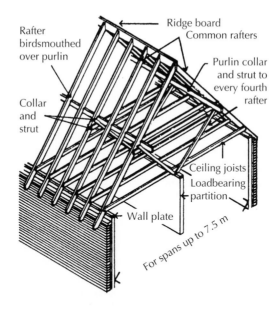

Rafter birdsmouthed over purlin

Collar and strut

Ridge board
Common rafters

Purlin collar and strut to every fourth rafter

Ceiling joists
Loadbearing partition

Wall plate

For spans up to 7.5 m

Figure 6.12 Double roof.

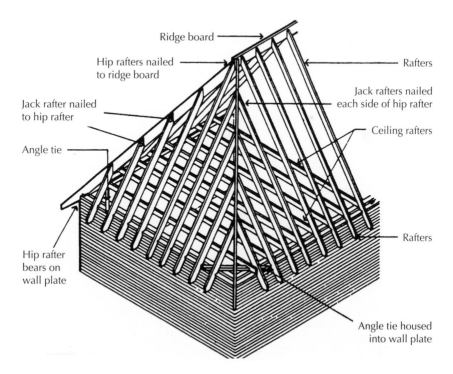

Ridge board

Hip rafters nailed to ridge board

Rafters

Jack rafter nailed to hip rafter

Jack rafters nailed each side of hip rafter

Angle tie

Ceiling rafters

Hip rafter bears on wall plate

Rafters

Angle tie housed into wall plate

Figure 6.13 Hipped end.

bearing for the splay cut ends of rafters pitched up to it, the hip rafter is usually 200–250 mm deep and 38 mm thick. The short lengths of rafter pitched to the hip rafter are the same size as the rafters of the main roof. These shorter lengths of rafter, commonly termed 'jack rafters', are nailed to the hip rafter and finished with shaped ends for gutter and soffit boards.

A hip rafter provides bearing and support for the ends of jack rafters, which support the roof covering and wind and snow loads. Because of the considerable load that a hip rafter carries it will tend to spread and displace the junctions of the wall plates on which it bears and overturn the walls that support it. To maintain the right-angle junction of wall plates and wall against the spread of a hip rafter it is necessary to form a tie across the right-angle junction. Angle ties are cut from 100×75 mm sawn softwood timbers. These ties are either bolted to the wall plate or dovetail housed into the wall plates some 600 mm from the angle across which they are fixed, as illustrated in Figure 6.13.

An advantage of a hip end roof is that the hip end acts to provide stability to the main roof against the tendency of a pitched roof to rack and overturn parallel to its ridge. The disadvantage of a hip end roof is the very considerable extra cost in the wasteful cutting of timber and roof covering at the hips. The hip end illustrated in Figure 6.13 is shown as a close couple roof. More usually this form of roof is framed as a purlin or double roof. Photograph 6.3 shows the internal roof construction of a hipped roof.

Valleys

The valley formed at the internal angle junction of two pitched roofs is framed in much the same way, in reverse, as a hip end (Photograph 6.4, Photograph 6.5, Photograph 6.6 and Photograph 6.7). A valley rafter is pitched up from the junction of the wall plates at the internal angle junction of walls to the intersection of ridge boards, splay end cut to fit to the ridge and notched and housed over the wall plate and nailed in position. The 38 or

Photograph 6.3   Hipped end roof structure (courtesy J Kangwa).

Ridge board

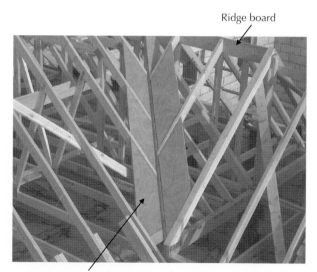

Valley formed at internal junction
between pitched roofs formed out of
roof trusses

Photograph 6.4   Valley formed in trussed roof.

Valley formed with valley lining
overlapped with roof tiles

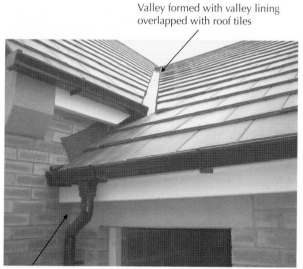

The gutter and downpipe collects rainwater form roof
tiles and the water from the valley

Photograph 6.5   Roof valley and gutter.

Photograph 6.6   Valley construction (courtesy of J Kangwa).

50 mm thick valley rafter is of the depth required to provide a bearing for the full depth of the splay cut ends of the rafters it supports and finish level with the tops of these rafters. The cut jack rafters are nailed to the side of the ridge board and valley rafter, respectively. The valley rafter is fixed in position with its top edge level with adjoining rafters so that it does not obstruct the battens that run down into the valley that will provide the structure for the valley gutter.

Trussed roof construction

A traditional method of constructing pitched roofs that do not need intermediate support from internal partitions was to form timber trusses. The word truss means tied together and a timber roof truss is a triangular frame of timbers securely tied together. The traditional timber roof truss was constructed with large section timbers that were cut and jointed with conventional mortice and tenon joints that were strapped with iron straps screwed or bolted to the truss. The traditional large section timber king post and queen post trusses supporting timber purlins and rafters were designed to be used without intermediate support, over barns, halls and other comparatively wide span buildings. The

Photograph 6.7   Valley formed, breathable sarking felt in place and with double lap tiles (courtesy of J Kangwa).

combination of the shortage of timber that followed the end of the Second World War (1945) and the need for greater freedom in planning internal partitions prompted the development of an economical timber truss designed by the Timber Development Association (http://www.trada.co.uk). The timbers of the truss were bolted together with galvanized iron timber connectors, illustrated in Figure 6.14, bolted between timbers at connections. The strength of the truss was mainly in the rigidity of the connection.

These timber trusses were designed for spans up to 8 m. Each truss was framed with timbers of the same section as the rafters and ceiling joists, which provided intermediate support too, through the purlins and binders they supported. The prefabricated timber trusses were fixed in position at 1.8 m centres bearing on and nailed to wall plates. The 175 × 25 mm ridge board was fixed and nailed to trusses. Purlins, 150 × 50 mm, were placed in position supported by struts and nailed to struts and rafters. Ceiling binders 125 mm deep by 50 mm wide were placed in position on 100 × 38 mm ceiling joists. To complete the roof framing, 100 × 38 mm rafters were cut to bear on the ridge board, notched over purlins and wall plate, and then nailed in position. Ceiling joists were nailed to the foot of rafters and the underside of ceiling binders.

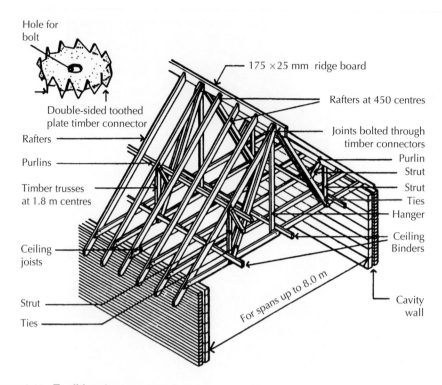

Hole for bolt

Double-sided toothed plate timber connector

175 ×25 mm ridge board

Rafters at 450 centres

Rafters

Purlins

Joints bolted through timber connectors

Purlin
Strut

Timber trusses at 1.8 m centres

Strut
Ties
Hanger

Ceiling joists

Ceiling Binders

Strut

For spans up to 8.0 m

Cavity wall

Ties

Figure 6.14   Traditional trussed roof.

The advantage of this trussed roof construction is that the continuity of the ridge board, purlins and binders along the length of the roof together with roofing battens provided adequate stability against racking. Trussed roof construction has been largely replaced by prefabricated trussed rafter roofs that require somewhat less timber and less skilled labour.

Prefabricated trussed rafters

For the maximum economy in site labour and timber the trussed rafter roof form was first used in the UK in the 1960s and has since been used extensively for housing and smaller commercial developments. Trussed rafters are fabricated from light section, stress graded timbers that are accurately cut to shape, assembled and joined with galvanised steel connector plates. Much of the preparation and fabrication of these trussed rafters is mechanised, resulting in accurately cut and finished rafters that are delivered to site ready to be lifted and fixed as a roof frame with the minimum of site labour.

The members of the truss are joined with steel connector plates with protruding teeth that are pressed into timbers at connections to make a rigid joint. Trussed rafters, which serve as rafters and ceiling joists, are fixed at from 400 to 600 mm centres, as illustrated in Figure 6.15 for spans up to 12 m for roofs pitched at 15°–40°. The trussed rafters bear on and are nailed to wall plates. As the rafters are trussed there is no need for a ridge board to provide a bearing and fixing for rafters.

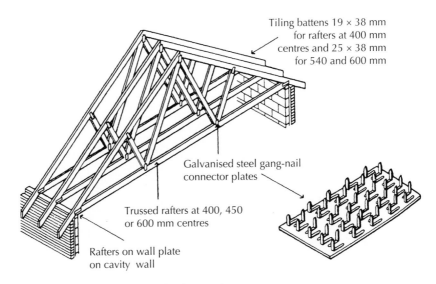

Tiling battens 19 × 38 mm
for rafters at 400 mm
centres and 25 × 38 mm
for 540 and 600 mm

Galvanised steel gang-nail
connector plates

Trussed rafters at 400, 450
or 600 mm centres

Rafters on wall plate
on cavity wall

Figure 6.15   Prefabricated trussed rafter roof.

Approved Document A states that, for stability, trussed rafter roofs should be braced to the recommendations in BS 5268: Part 3. For the roofs of domestic buildings the members of trussed rafters with spans up to 11 m should be not be less than 35 mm thick and those with spans of up to 15 m, 47 m thick. These slender section trusses are liable to damage in storage and handling. They should be stored on site either horizontal on a firm level base or in a vertical position with adequate props to avoid distortion. In handling into position each truss should be supported at eaves rather than mid-span to avoid distortion. The trussed rafters should be fixed vertical on level wall plates at regular centres and maintained in position with temporary longitudinal battens and raking braces. The rafters should be fixed to wall plates with galvanised steel truss clips that are nailed to the sides of trusses and wall plates. A system of bracing should then be permanently nailed to the trussed rafters. Bracing adds stability to the roof and is also designed to maintain the rafters in position and to reduce buckling under load as illustrated in Figure 6.16. The bracing members should be 25 × 100 mm and nailed with two 3.35 × 75 m galvanised, round wire nails at each crossover. The longitudinal brace at the apex of trusses acts in much the same way as a ridge board and those halfway down slopes act like purlins to maintain the vertical stability of trusses. The longitudinal braces, also termed binders, at ceiling level serve to resist buckling of individual trussed rafters. The diagonal under-rafter braces and the diagonal web braces serve to stiffen the whole roof system of trussed rafters by acting as deep timber girders in the roof slopes and in the webs. Manufacturers of roof trusses provide the design and layout of the prefabricated trusses and structural calculations as part of the overall service offered to designers and contractors. Photograph 6.8 shows the main components of a prefabricated truss roof. The gang nail plate which ties the members together can be seen in Photograph 6.8a.

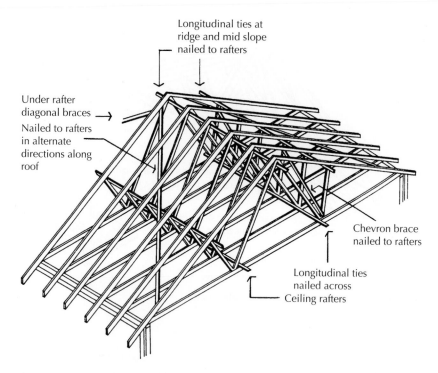

Figure 6.16   Stability bracing to trussed rafters.

To further increase the speed of construction and reduce the amount of work that is carried out at a height, many trussed rafters are erected and braced together on the ground, then the whole roof structure is lifted on to the walls in one go (Photograph 6.9a and b). To do this extra bracing is required to ensure the roof can be lifted safely by a crane.

Eaves
Eaves is a general term used to describe the lowest courses of slates or tiles and the timber supporting them. The eaves of most pitched roofs are made to project some 150–300 mm beyond the external face of walls. In this way the roof provides the top of the wall with some protection from rain (Figure 6.17a). 'Closed' or 'flush' eaves provide little protection to the wall but are used to economise in timber and roof covering (Figure 6.17b). The ends of the rafters and ceiling joists are cut flush with the outside face of the wall below. A fascia or gutter board 25 mm thick is nailed to rafters and ceiling joist ends.

The necessary depth of the gutter board is such that it covers the cut ends of rafters and projects some 25–30 mm above the top of rafters to act as a bearing for the lowest courses of tiles or slates which project over it some 25 mm to discharge rainwater into a gutter fixed to the gutter board (Figure 6.17c). Insulation is not shown in the previous figure allowing various components to be seen. Figure 6.18 shows the general arrangement of insulation and ventilation components at the eaves.

(a) Gang nail plate holds members of the truss in position

(b) Cavity sock seals the top of the cavity at gable end

(c) Diagonal bracing across roof trusses

(d) Forming valley at intersection of roofs

Photograph 6.8   Truss roof construction.

To prevent the build-up of stagnant air and condensation forming, ventilation is required. This is normally provided by ventilation gaps at the eaves and the ridge of the roof structure (see Photograph 6.10, which shows the roof construction at the eaves). The Building Regulations require that the ventilation gap varies with the pitch of the roof. If there is a ventilation gap of minimum 25 mm along the length of the eaves any pitch can be used. If the roof is greater than 35° pitch, 10 mm ventilation is sufficient (Figure 6.18). Where the attic is converted into a room, then 25 mm ventilation at the eaves is required.

Dormer windows

Dormer windows are framed as a vertical window, whereas roof lights are formed as a glazed opening in the slope of the roof. Dormer windows may be framed as a projection from the roof slope, recessed behind the slope or partly projecting from the roof and partly in the face of the wall below, as illustrated in Figure 6.19.

A recessed dormer window is particularly suited to a Mansard form of roof. To provide the maximum number of floors it is often practice to form one or several floors inside a

Photograph 6.9   (a) Trussed roof being assembled on the ground. (b) Roof assembled at ground level, braced ready to be craned into position.

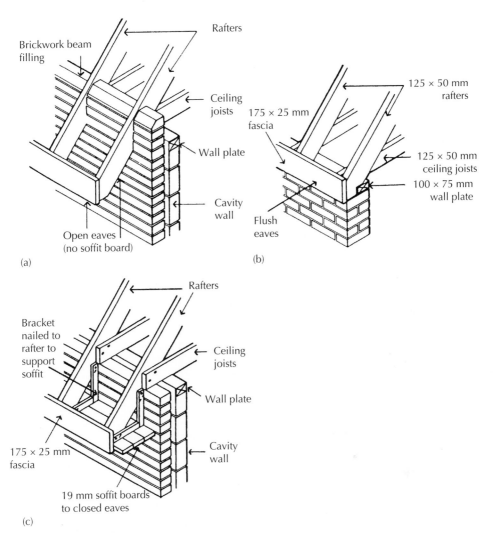

Figure 6.17   (a) Open eaves. (b) Flush eaves. (c) Closed eaves.

roof sloping at 60° with recessed dormer windows to provide daylight. The advantage of the Mansard slope form of pitched roof, illustrated in Figure 6.20, is that it provides maximum useful floor area. Where there are no restricting angles of light a Mansard roof space may be formed with projecting dormer windows.

The opening in roof rafters for dormer windows is framed between trimming rafters and the head and cill trimmers, which provide support for the trimmed rafters, as illustrated in Figure 6.21. The roof and cheeks of the dormer are boarded for lead, copper sheet or vertically hung tiles. The junction of the main roof covering with the sheet metal covering of the roof and cheeks of the dormer is weathered with lead or copper flashings and gutter.

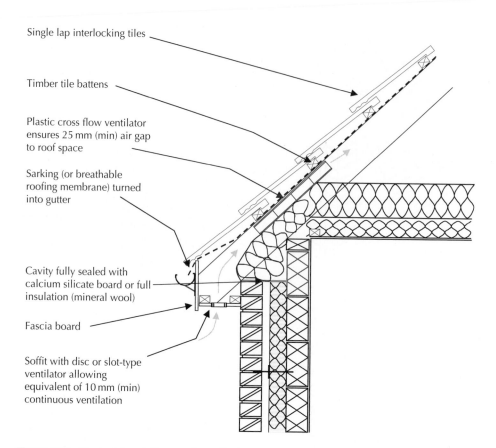

Single lap interlocking tiles

Timber tile battens

Plastic cross flow ventilator
ensures 25 mm (min) air gap
to roof space

Sarking (or breathable
roofing membrane) turned
into gutter

Cavity fully sealed with
calcium silicate board or full
insulation (mineral wool)

Fascia board

Soffit with disc or slot-type
ventilator allowing
equivalent of 10 mm (min)
continuous ventilation

Figure 6.18   Typical insulation and ventilation requirements at eaves.

Roof light

Windows that are housed within the roof rafters and follow the line of the roof are termed roof lights. Where the roof space is large enough to accommodate an extra room, roof lights offer a cost-effective method of incorporating a window within the attic space. The most common form of roof light is the centre-pivot window, which can be rotated through 160°, allowing cleaning from the inside. A wide range of roof windows such as fixed, top-hung, tilt and turn and bottom-hung windows are available (Figure 6.22). Modern roof lights are supplied as prefabricated units with flashings to suit; simply fitted between the roof joists, these units help to facilitate rapid construction of the roof. Figure 6.23a and b and Photograph 6.11 show a typical roof light.

Fire safety

The space inside a pitched roof is a void space that should be separated from other void spaces or cavities by cavity barriers, sealing the junction of the cavity and preventing spread of smoke and flames. The cavity in a wall is generally separated from the cavity in a pitched

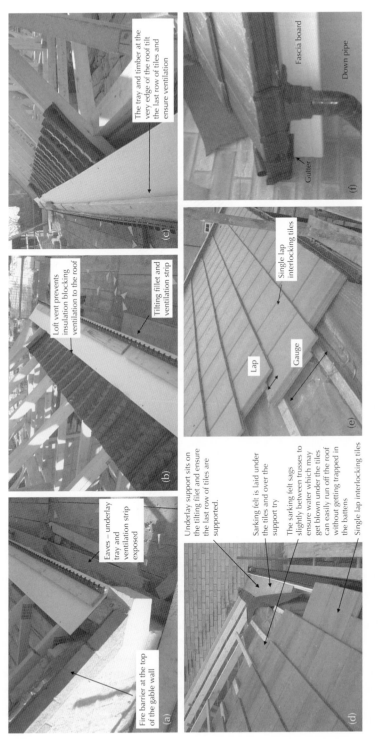

Photograph 6.10 (a) View of eaves at the gable end – fire protection seals the cavity wall. (b) Tray to ensure the insulation does not block the ventilation is in place. (c) Eaves detail – tilting fillet. (d) Tiles and sarking felt are laid on the roof. (e) Single lap interlocking tiles fixed to battens with required overlap. (f) Fascia boards, gutters and downpipes are fixed in position.

The following labels appear on the photographs:

(a) Fire barrier at the top of the gable wall
Eaves – underlay tray and ventilation strip exposed

(b) Loft vent prevents insulation blocking ventilation to the roof
Tilting fillet and ventilation strip

(c) The tray and timber at the very edge of the roof tilt the last row of tiles and ensure ventilation

(d) Underlay support sits on the tilting filet and ensure the last row of tiles are supported.
Sarking felt is laid under the tiles and over the support try
The sarking felt sags slightly between trusses to ensure water which may get blown under the tiles can easily run off the roof without getting trapped in the battens
Single lap interlocking tiles

(e) Single lap interlocking tiles
Lap
Gauge

(f) Fascia board
Down pipe
Gutter

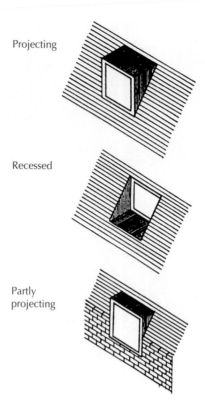

Projecting

Recessed

Partly
projecting

Figure 6.19   Dormer windows.

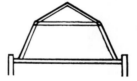

Figure 6.20   Mansard roof.

roof by the cavity barrier at the top of the cavity in the wall. Cavities in walls should also be sealed to improve airtightness and to increase thermal properties (still air is a good insulator).

Separating walls, party walls, between semi-detached and terraced houses should resist the spread of fire from one house to another by being raised above the level of the roof covering or by being built up to the underside of the roof covering so that there is a continuous fire break. Other patented systems of fire barriers are available for industrial and commercial buildings. Slate, tile and non-ferrous sheet metal do not encourage spread of flame across their surface and are, therefore, not limited in use in relation to spread of fire between adjacent buildings.

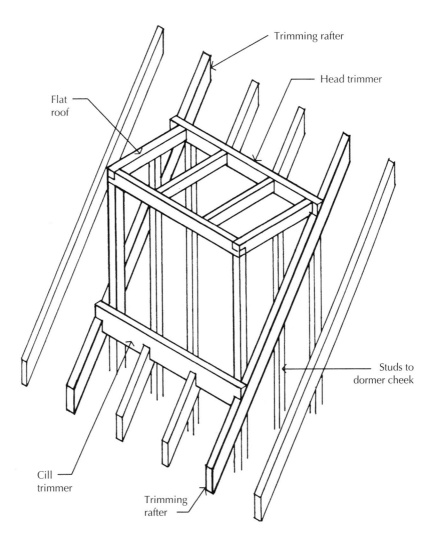

Trimming rafter

Head trimmer

Flat roof

Studs to dormer cheek

Cill trimmer

Trimming rafter

Figure 6.21 Dormer windows framing.

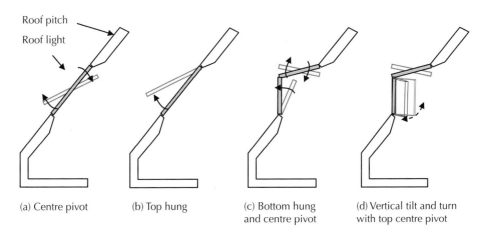

Roof pitch

Roof light

(a) Centre pivot

(b) Top hung

(c) Bottom hung and centre pivot

(d) Vertical tilt and turn with top centre pivot

Figure 6.22 Types of roof windows.

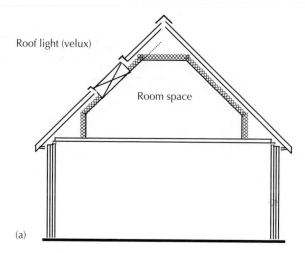

Roof light (velux)

Room space

(a)

Figure 6.23   (a) Roof light in attic conversion (warm roof construction). (b) Detail of roof light in warm roof.

6.3 Pitched roof coverings

The traditional covering for pitched roofs in England and in northern European countries has been clay tile and natural slate. These are still extensively used for new houses and other small buildings for their appearance, durability and freedom from maintenance.

Thatch

Thatch is a traditional building material that is rarely used in modern developments, although this sustainable material has seen something of a revival in recent years. Long straight stalks of reeds are cut, dried, bound together and laid up the slopes of pitched (sloping) roofs as thatch. Thatch efficiently drains rainwater, excludes wind and acts as a very effective insulator against transfer of heat, a combination of advantages that no other roof covering offers. The disadvantage of thatch is that the dry material readily ignites and burns vigorously and the thick layer of thatch is an ideal home for small birds, rodents and insects. Fire protection is required to the underside of the thatched roof to comply with regulations and building insurance policies. The durability of thatch will depend on the exposure of the site, the material used and the skills of the thatcher; replacement should be expected every 20–30 years.

Plain tiles

A plain tile is a rectangular roofing unit of burnt clay. Plain tiles are made with a small upward camber so that when laid overlapping, the tail of a tile bears directly on the back of a tile below. Plain tiles are now made with two nibs for hanging to battens, as illustrated in Figure 6.24a. The two holes near the head of the tile are for nailing tiles to timber battens.

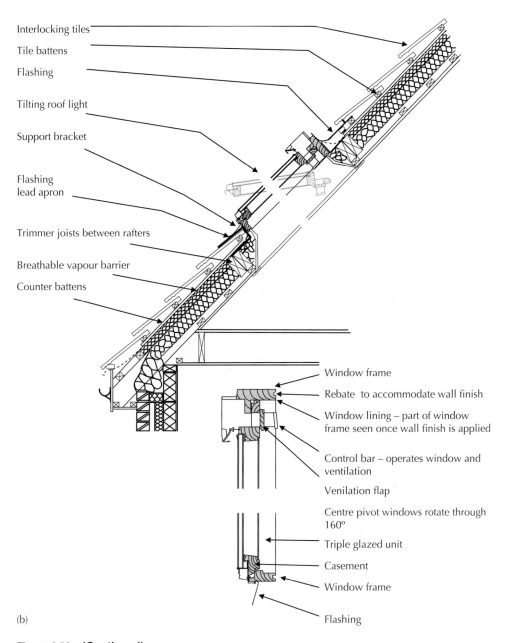

Interlocking tiles

Tile battens

Flashing

Tilting roof light

Support bracket

Flashing
lead apron

Trimmer joists between rafters

Breathable vapour barrier

Counter battens

Window frame

Rebate to accommodate wall finish

Window lining – part of window
frame seen once wall finish is applied

Control bar – operates window and
ventilation

Venilation flap

Centre pivot windows rotate through
160°

Triple glazed unit

Casement

Window frame

Flashing

(b)

Figure 6.23   (Continued)

Photograph 6.11   Rooflight (velux).

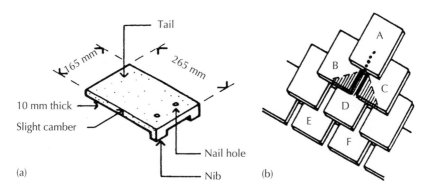

Figure 6.24   (a) Plain tile. (b) Rain flow over plain tiles.

Plain tiles are hung on timber battens in overlapping horizontal courses with the side joints between tiles bonded up the roof so that rain runs down the slope from tile to tile to the eaves. There are at least two thicknesses of tile at any point on the roof and also a 65 mm overlap of the tail of each tile over the head of tiles two courses below. The reason for this extra overlap is that, were the tiles laid in a straightforward overlap, rain might penetrate the roof. If plain roof tiles were only lapped once, rain running off tile A in Figure 6.24b will run into the gap between tiles B and C, and spread over the back of tile D, as indicated by the hatching, and probably drip over the top of tiles E and F into the roof. To avoid this all tiles are double lapped.

Lap, gauge and margin
With a 65 mm double end lap, plain tiles are laid to overlap 100 mm up the roof slope. The softwood battens on which the tiles are hung must therefore be fixed at a gauge (measurement) of 100 mm centres. The tail end of each tile shows 100 mm of the length, which is described as the margin, as illustrated in Figure 6.25.

Minimum slope (pitch) of roof
The minimum roof slope (or pitch) for plain tiles depends on the density of the burnt tile and is supplied by manufacturers for each tile they produce. Handmade tiles, which fairly

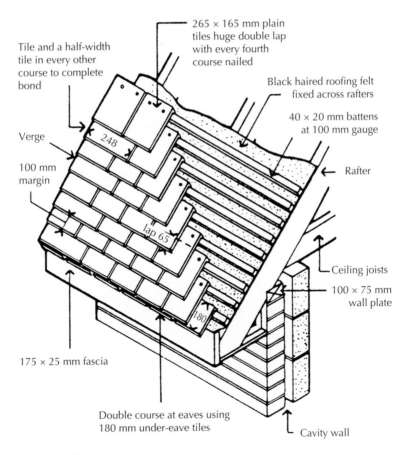

Figure 6.25   Plain tiling.

readily absorb water, should not be laid on a roof slope pitched at less than 45° and the denser machine pressed tiles at not less than 35°. Because of the thickness and double lap of these tiles, the actual slope of a tile on a roof will be a few degrees less than that of the roof slope.

Roofing felt (sarking and sarking felt)

The open butt, side joints of tiles and inexact fit of tile over tile would allow wind to blow into the roof space. To exclude wind rolls of sheet sarking are laid across roof rafters from the eaves upwards with widths of felt lapped 75–150 mm up to the ridge. Timber battens secure the sarking covering, as illustrated in Figure 6.25. Various materials are available, including breathable membranes, reinforced plastic and flexible felt material. Breathable membranes allow unwanted water vapour to escape from the roof space while still excluding wind. Ventilation through the roof and breathable membranes reduce the potential for condensation to form on the cold surfaces within the roof.

When it is planned to use a roof space for dry storage it is usual to cover roof slopes with either plain edge or tongued and grooved boarding laid across and nailed to rafters to prevent wind and dust entering. Counter battens are then fixed up the slopes, over roofing felt as illustrated in Figure 6.26. The roofing felt will then conduct any water that penetrates the tiles down to the eaves. Roof boarding, sometimes called sarking, acts to brace a pitched roof against its inherent instability across the slope of the roof.

Tiling battens

Plain tiles are hung on 38 × 19 mm or 50 × 25 mm sawn softwood battens, which are nailed across rafters at 100 mm centres (gauge). Battens should be impregnated with a preservative to prevent rot and attack by insects. So that nails do not rust and perish they should be galvanised or made of a non-ferrous alloy. The tiles in every fourth course are nailed to battens as a precaution against strong gusts of wind lifting and dislodging tiles. In exposed positions every tile should be nailed.

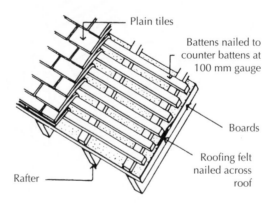

Figure 6.26 Roof boarding and counter battens.

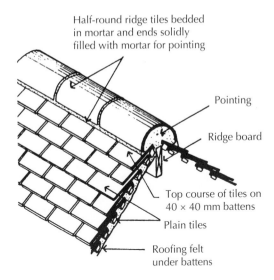

Half-round ridge tiles bedded
in mortar and ends solidly
filled with mortar for pointing

Pointing

Ridge board

Top course of tiles on
40 × 40 mm battens

Plain tiles

Roofing felt
under battens

Figure 6.27   Ridge.

Eaves and ridge tiles

So that there are two thicknesses of tile at the eaves a course of eaves tile is used. These special length tiles are 190 mm long so that when hung to battens their tails lie directly under the tail of the course of full tiles above, as illustrated in Figure 6.25. The tails of both the eaves tiles and full tiles above bear on the gutter board and project beyond the face of the board to discharge run-off water into the gutter.

At the ridge a top course of tiles is used to overlap the course of full length tiles below to maintain the necessary two thicknesses of tile. These top course tiles are 190 mm long and hung on 40 × 40 mm battens, which are thicker than normal battens, so that top course tiles ride over the tiles below. The ridge tiles are bedded on the back of the top course tiles so that the usual margin of tile is showing. The ends of the ridge tiles are solidly filled with mortar as a backing for the mortar pointing, as illustrated in Figure 6.27.

Figure 6.28 illustrates four different ridge tiles. Which one is used depends in part on the slope of the roof and also on the desired appearance. Specials can also be produced for decorative purposes.

Verges

The bonding of tiles at verges of plain tile roofing at the junction of a roof with a gable end and the junction of square abutments of slopes to parapet walls is completed with tile and half width tiles 248 mm wide in every other course. The use of tile and half width tiles at the verge of a tiled roof and a gable end is illustrated in Figure 6.29. The verge tiling is hung to overhang the gable end wall by some 25 mm and the tiles are tilted slightly towards the roof slope to encourage rain to run down the roof slope rather than down the gable end wall.

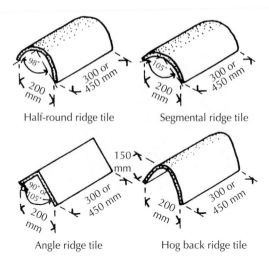

Half-round ridge tile

Segmental ridge tile

Angle ridge tile

Hog back ridge tile

Figure 6.28   Ridge tiles.

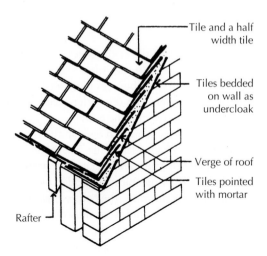

Tile and a half width tile

Tiles bedded on wall as undercloak

Verge of roof

Tiles pointed with mortar

Rafter

Figure 6.29   Verge.

Hip ends

Hip tiles are manufactured to bond in with the courses of tile in the adjacent slopes to provide a more pleasing appearance to the roof. Two sections of hip tile are produced. The angular and the bonnet hip tile, illustrated in Figure 6.30, have holes for nailing the tiles to the hip rafter. The hip tiles are nailed to the hip rafter up the slope of the hip and overlap so that the tail of each tile courses in with a course of plain tiles. The hip tiles are bedded in mortar on the back of plain tiles which are cut to fit close to the hip, and the end of hip

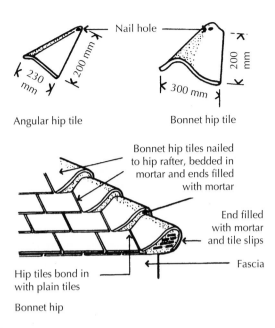

Angular hip tile   Bonnet hip tile

Bonnet hip tiles nailed to hip rafter, bedded in mortar and ends filled with mortar

End filled with mortar and tile slips

Fascia

Hip tiles bond in with plain tiles

Bonnet hip

Figure 6.30   Hip tiles.

tiles is filled with cement mortar as illustrated in Figure 6.30. For the sake of appearance the end hip tile is filled with mortar with slips of cut tile bedded in the mortar.

Valleys

At the junction of plain tile covered roof slopes in a valley, formed by an internal angle of walling, either a lead gutter is formed or special valley tiles are used. The valley is formed as a lead lined gutter dressed into a gutter formed by timber valley boards. The valley gutter should be wide enough to allow such debris as dust and leaves to be washed down to eaves without being obstructed and so blocking the gutter. A clear width of at least 125 mm is usual.

Gutter boards 25 or 38 mm thick and some 200 mm wide are nailed to the top of rafters, each side of the valley, with a triangular fillet of wood in the bed of the gutter as illustrated in Figure 6.31. Tiling battens are continued down over the edges of the gutter boards. Roofing felt is laid under battens and over a double depth batten nailed at the cut ends of roofing battens.

As an alternative to a valley gutter, special valley tiles may be used to course in with the tile slopes each side of the valley. Some of the manufacturers of tiles provide valley tiles of the same material as their plain tiles. These valley tiles are shaped so that they bond in with tiles on main slopes pitched at specific angles. The length of the tile provides for the normal end lap. The dished tile tapers towards its tail, as illustrated in Figure 6.32. The valley tiles are nailed to tiling battens continued down to the valley with plain tiling hung to side butt to valley tiles using tile and half width tiles as necessary.

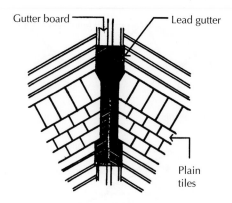

Figure 6.31   Lead valley gutter.

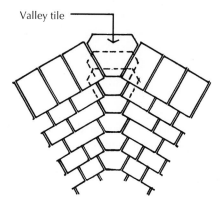

Figure 6.32   Valley tiles.

Concrete plain tiles

Concrete plain tiles are made from a mixture of carefully graded sand, cement and water, which is compressed in a mould. A thin top dressing of sand, cement and colouring matter is pressed into the top surface of the tile. The moulded tiles are then left under cover for some days to allow them to harden. Concrete tiles are uniform in shape, texture and colour; however, concrete tiles are more normally made as interlocking units as described further.

Single lap tiles

Single lap tiling is so called because each tile is laid to end lap the tile below down the slope of a roof and to side lap over and under the tiles next to it, in every course of tiles. The two most common types of single lap tile are the rounded unders (channel) and overs (cover) and the flat unders (channel) and round overs (cover) tiles illustrated in Figure 6.33. These

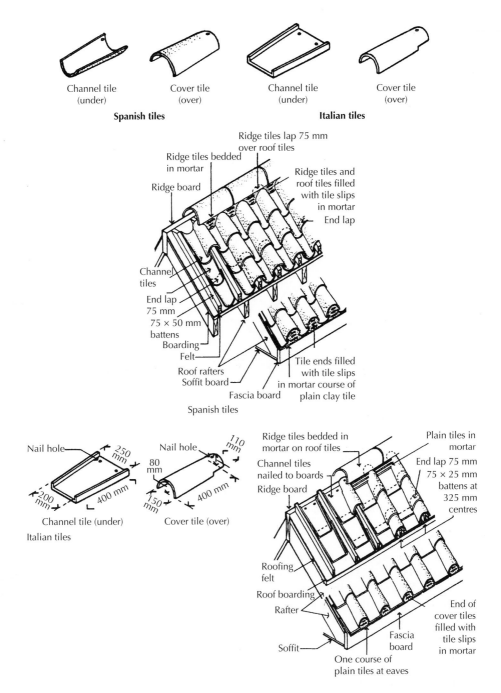

Figure 6.33 Single lap tiles.

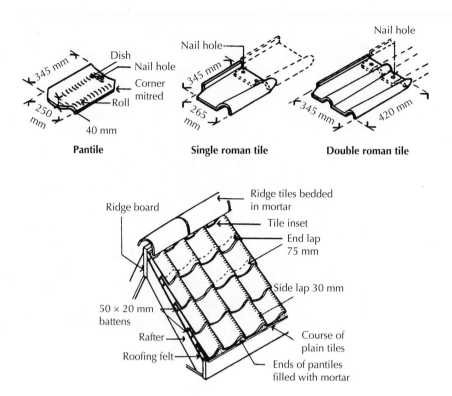

Figure 6.34 Pantiles.

traditional tiles are called Spanish and Italian tiles, respectively, even though they are not specific to those countries. In the UK these tiles have been replaced by pantiles in which the rounded unders and overs are combined as one pantile and the flat unders and rounded overs of Italian tiles are combined as single and double Roman tiles.

Pantiles

A pantile incorporating the rounded unders and overs of Spanish tiles in one tile is illustrated in Figure 6.34 and Photograph 6.12. By the use of selected clays or cement and sand and machine pressing this larger tile can be produced in standard sizes of uniform shape. The advantage of this tile is that it can be hung to be comparatively close fitting to exclude wind and rain. Both clay and concrete pantiles were fairly extensively used before the advent of interlocking tiles. Clay pantiles have been made with a fired on glazed finish to all exposed faces and edges, in a range of colours.

Pantiles are made with a nib for hanging on to battens and a nail hole for securing tiles by nailing to battens against wind uplift. The tiles are hung on to 50 × 20 mm sawn softwood battens, nailed to rafters over roofing felt. The tiles are hung with an end lap of 75 mm at a gauge of 270 mm. The purpose of the mitred corners of pantiles is to facilitate fixing and proper bedding of the tiles.

Photograph 6.12 Pantile roof.

Ridge tiles are bedded over the backs of pantiles and pointed at butt joints in cement and sand. As a bed for mortar filling, a course of plain tiles is hung at eaves on which mortar and tile slips are used to end fill pantiles. The pitch of roof for pantiles may be as low as 20°–35°, depending on exposure. An illustration of a roof covered with pantiles can be seen in Figure 6.34.

Interlocking single lap tiles

Interlocking single lap tiles are illustrated in Photograph 6.13. Grooves in the vertical long edges of tiles are designed to interlock under and over the edges of adjacent tiles to exclude wind and rain. A range of profiles is made from the interlocking double pantile to the flat pantile illustrated in Figure 6.35. There is a wider range of concrete than clay pantiles because of the difficulty of controlling shrinkage of clay during firing to produce a satisfactory interlock at edges.

There is no accepted standard size of interlocking tile, although they are of similar size of about 420 mm long by 330 mm wide for concrete and about 320 mm long by 210 mm wide for clay. These comparatively thick tiles are made with nibs and hung on 38×25 mm sawn softwood battens on a roofing felt underlay on rafters at a pitch from 15° to 22½°. A single end lap of 75 mm is usual with the side butt lap dictated by the system of interlock, as illustrated in Figure 6.36.

These tiles are not usually made with nail holes; instead a system of aluminium clips is used. The aluminium clips hook over an upstand side edge at the head of a tile, under the

(a) Truss rafters ready to receive sarkin felt and roof battens

(b) Ventilator installed over rafters at eaves

(c) Interlocking single tiles fixed to battens

(d) Cavity tray protruding from wall. Lead flashing will be inserted under the trays and lapped over the tiles

Photograph 6.13 Roof construction with single lap tiles.

end lap, and are nailed or screwed to the back edge of battens. Clips are used in every course in severe exposure or every other course or more for less exposed conditions. At the ridge (Figure 6.37) a half-round ridge tile is bedded in mortar with the end joints pointed. At the eaves a course of plain tiles provides a bed for mortar filling to the open ends of profiled tiles. Alternatively a number of proprietary 'dry' systems are available that allow the eaves and sometimes the ridge to be constructed without the use of cement mortar.

At hips and valleys profiled tiles have to be cut to fit either under hip tiles or to a lead lined gutter to valleys. Because of the thickness of the tiles they have to be cut mechanically to provide a neat edge.

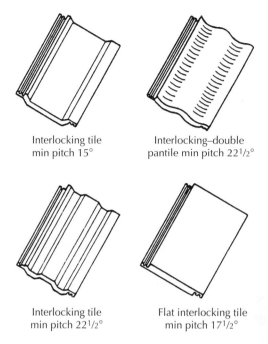

Interlocking tile
min pitch 15°

Interlocking–double
pantile min pitch 22¹/₂°

Interlocking tile
min pitch 22¹/₂°

Flat interlocking tile
min pitch 17¹/₂°

Figure 6.35   Interlocking tile profiles.

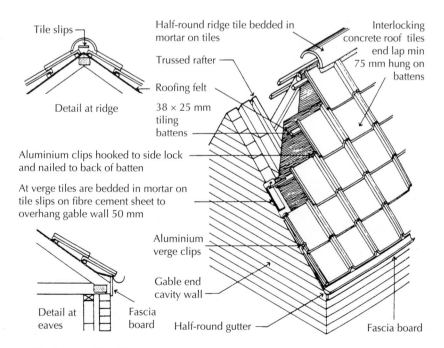

Tile slips

Half-round ridge tile bedded in
mortar on tiles

Interlocking
concrete roof tiles
end lap min
75 mm hung on
battens

Detail at ridge

Trussed rafter

Roofing felt

38 × 25 mm
tiling
battens

Aluminium clips hooked to side lock
and nailed to back of batten

At verge tiles are bedded in mortar on
tile slips on fibre cement sheet to
overhang gable wall 50 mm

Aluminium
verge clips

Gable end
cavity wall

Detail at
eaves

Fascia
board

Half-round gutter

Fascia board

Figure 6.36   Interlocking tiles.

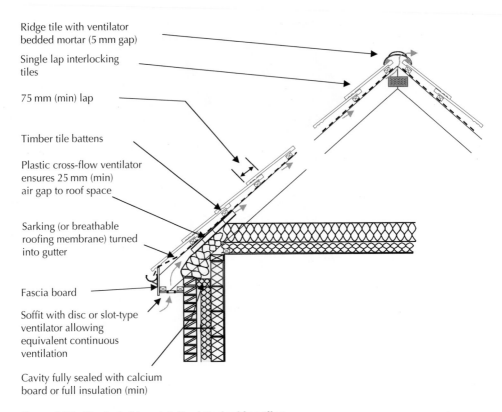

Ridge tile with ventilator bedded mortar (5 mm gap)

Single lap interlocking tiles

75 mm (min) lap

Timber tile battens

Plastic cross-flow ventilator ensures 25 mm (min) air gap to roof space

Sarking (or breathable roofing membrane) turned into gutter

Fascia board

Soffit with disc or slot-type ventilator allowing equivalent continuous ventilation

Cavity fully sealed with calcium board or full insulation (min)

Figure 6.37   Typical ridge detail – interlocking tiles.

Slates

Thin slabs of natural stone have for many centuries been a traditional roof covering material in areas where natural stone can be split and shaped as a slate. These have traditionally come from Wales and Cumbria (Westmorland), although more recently it has become common to import slates from countries where production costs are less expensive. The thickness, size and durability of slate depend on the nature of the stone and its ease of splitting into a usable size. Owing to the nature of the stone, the slates have irregular thickness, which can range from 4 to 10 mm for thin dense slates and 30–40 mm for thick stone tiles ('stonewolds'). Slate sizes are 600 × 300, 500 × 250, 450 × 225 and 400 × 200 mm.

Reclaimed slates

Because of the cost of new slates it has been a practice for some years to use slates recovered from demolition work. The disadvantage of reclaimed slates is that they may vary in quality, be damaged by being stripped from old roofs (nail holes are particularly susceptible to damage) and there is little likelihood of recovering the slate and half width slates necessary to complete the bond at verges and abutments. Therefore salvaged roof slates are usually sorted and reconditioned before they can be reused. Although this increases the cost of reclaimed slates, the environmental impact is less compared with the use of new slates, and the weathered slates will provide a better match for repair and conservation work.

Slate substitutes

From the middle of the twentieth century, slates made from asbestos fibres, cement and water were used as cheap substitutes for natural slates. They are now made with natural and synthetic fibres, often incorporating slate dust for an authentic and more durable finish. These slates are made from pigmented cement and water, reinforced with natural and synthetic fibres. The wet mix is compressed to slates, which are cured to control the set and hardening of the material. The standard slate is 4 mm thick, with a matt coating of acrylic finish and a protective seal on the underside against efflorescence and algal growth. The standard slate is rectangular and holed for nail and rivet fixing. Standard fibre cement slates are made in sizes of 600 × 300, 500 × 250 and 400 × 200 mm, as illustrated in Figure 6.38, with a comprehensive range of fittings for ridge, eaves, verges and valleys. The standard fibre cement slate is uniform in shape, colour and texture. Fibre cement slates have a life of about 30 years in normal circumstances.

Fixing natural slates

Slates are nailed to 50 × 25 mm sawn softwood battens with copper composition or aluminium nails driven through holes, which are punched in the head of each slate. Galvanised steel nails should not be used as they will in time rust and allow the slates to slip out of position. Two holes are punched in each slate some 25 mm from the head of the slate and about 40 mm in from the side of the slate, as illustrated in Figure 6.39.

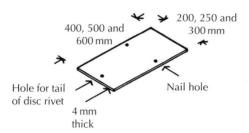

Figure 6.38   Fibre cement slate.

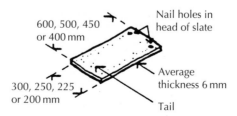

Figure 6.39   Natural slate.

The battens are nailed across the roof rafters over roofing felt and the slates nailed to them so that at every point on the roof there are at least two thicknesses of slate, and so that the tail of each slate laps 75 mm over the head of the slate two courses below. This is similar to the arrangement of plain tiles and is done for the same reason. Because the length of slates varies and the end lap is usually constant it is necessary to calculate the spacing or gauge of the battens. The formula for this calculation is:

$$\text{Gauge} = \frac{\text{Length of slate} - (\text{lap} + 25)}{2} \text{mm}$$

For example, the gauge of the spacing of the battens for 500 mm long slates is:

$$\text{Gauge} = \frac{500 - (75 + 25)}{2} = 200 \text{ mm}$$

The 25 that is added to the lap represents the 25 mm that the nail holes are punched below the head of the slate so that the 75 mm end lap is measured from the nail hole.

Figure 6.40 is an illustration of natural slates head nailed to preservative treated softwood battens with an end lap of 75 mm and the side butt joints between slates breaking joints up the slope of the roof. The roof is pitched at 25° to the horizontal, which is the least slope generally recommended for slates. At verges a slate, which is one and a half times greater than the standard width of slates, is used in every other course (Figure 6.40). Oversized slates avoid using half-width slates to complete the bond.

So that there shall be two thicknesses of slates at the eaves, a course of under-eave slate is used. These slates are cut to a length equal to the gauge plus lap plus 25 of the slating, as illustrated in Figure 6.40.

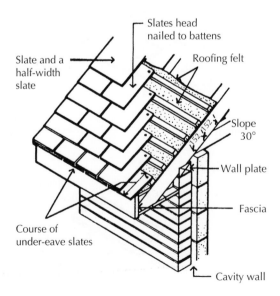

Figure 6.40   Head nailed slates.

Slates are usually fixed by means of nails driven through holes in the head of slates. This is the best method of fixing slates as the nail holes are covered with two thicknesses of slate so that, even if one slate cracks, water will not get in. But if long slates such as 600 mm are head nailed on a shallow slope of, say, 30° or less, it is possible that in a high wind the slates may be lifted so much that they snap off at the nail holes.

In exposed positions on low pitch roofs it is common to fix the slates by centre nailing them to battens. The nails are not driven through holes exactly in the centre of the length of the slate but at a distance equal to the gauge down from the head of the slate, so that the slate can double lap at tails as illustrated in Figure 6.41. With this method of fixing there is only one thickness of slate over each nail hole so that if that slate cracks water may get into the roof.

Ridge

Common practice is to cover the ridge at the intersection of two slated slopes with clay ridge tiles bedded in cement mortar. So that there is a double thickness of slate a top course of slates is used at ridges. These shorter slates are usually the gauge plus lap plus 50 or 75 mm long and head nailed to a double thickness batten. Their ends are solidly filled with and joints pointed in cement mortar, as illustrated in Figure 6.42.

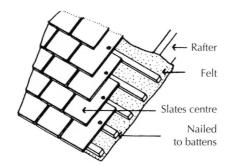

Figure 6.41   Centre nailed slates.

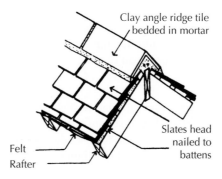

Figure 6.42   Tile ridge.

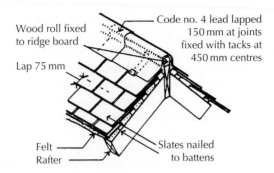

Figure 6.43 Lead ridge.

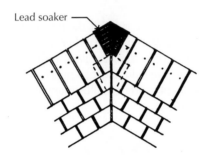

Figure 6.44 Slate valley.

A somewhat more expensive, and durable, finish is to use a sheet lead ridge capping, as illustrated in Figure 6.43. A wood roll, cut from a 50 mm square softwood section, is nailed to the ridge board, which is deeper than usual. The lead (about 450 mm wide) is dressed around the wood roll and down the slates both sides. At joints between sheets of lead capping, there is a 150 mm overlap. To prevent the wings of the sheet lead being blown up in high wind, a system of 50 mm wide strips of sheet lead is nailed to the roll under the lead at 450 mm centres. These lead tacks are turned up and around the edges of the wings of the lead capping to keep them in place.

Hips
At hipped ends of a roof, slates and one and a half width slates in every other course are cut to the splay angle up to the hip. The hip is then weathered with clay ridge tiles bedded in cement mortar or with a sheet lead capping similar to the ridge capping.

Valleys
At the internal angle intersection of slated roof slopes the full and one and a half width slates are splay cut up to the valley. A system of shaped lead soakers is used to weather the mitred valley. A shaped lead soaker cut from Code No. 4 sheet lead is hung over the head of each course of slates. A similar soaker is hung over the head of slate courses above to overlap the soaker below, as illustrated in Figure 6.44. Careful cutting of slates and the careful arrangement of bonding of slates will create a neat, mitred, watertight valley.

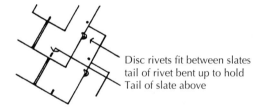

Disc rivets fit between slates
tail of rivet bent up to hold
Tail of slate above

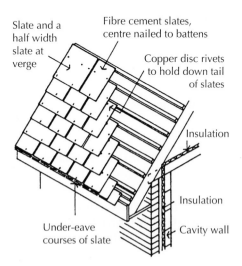

Slate and a
half width
slate at
verge

Fibre cement slates,
centre nailed to battens

Copper disc rivets
to hold down tail
of slates

Insulation

Insulation

Under-eave
courses of slate

Cavity wall

Figure 6.45   Fibre cement slating.

Fixing fibre cement slates

Fibre cement slates are fixed double lap up the slope of roofs pitched at a minimum slope of 20° with a usual slope of 25° to the horizontal. The slates are fixed with the double end lap used with natural slates. The end lap may be the usual 75 mm, which is usually increased to 100 mm to provide cover for the centre nail fixing and the rivet hole. These comparatively thin, brittle slates are centre nailed to avoid cracking the slate. Each slate is nailed with copper composition or aluminium nails to 38 × 25 mm or 50 × 25 mm, impregnated, sawn softwood battens nailed over roofing felt to rafters, as illustrated in Figure 6.45. The gauge, spacing, of battens is calculated by the same formula used for natural slates.

To secure these lightweight slates against wind, uplift copper disc rivets are used. The disc of these rivets fits between slates in the under course and the tail of the rivet fits through a hole in the tail of the slate above. The tail of the rivet is bent up to hold the slate in position, as illustrated in Figure 6.45. At verges and square abutments, one and a half width slates are used in every other course to complete the bond of slates up roof slopes. At eaves, two under-eave courses of slate are used. These cut or specially made short lengths of slate are head nailed to battens, as illustrated in Figure 6.45. At the ridge, special flanged

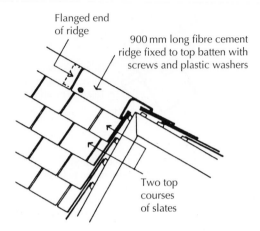

Figure 6.46   Fibre cement slate ridge.

end, fibre cement angle ridges are used. These 900-mm-long ridge fittings are weathered by the overlap of the flanged ends, as illustrated in Figure 6.46 and are secured as illustrated with screws and plastic washers. The wings of the ridge fittings bear on the back of an under-ridge course of slates without the need for cement and sand pointing. At hips and valleys the slates are mitre cut to fit up to the hip or valley and the mitre cut slates weathered with lead soakers hung over the head of slates.

6.4 Sheet metal covering to low pitch roofs

A low-pitched roof has slopes of from 10° to 30° to the horizontal and may serve as a good compromise between flat and more steeply pitched roofs. The principal coverings for low-pitched roofs are copper and aluminium strips 450 or 600 mm wide, both of which are comparatively light in weight and therefore do not require heavy timbers to support them. Complicated and labour intensive joints (such as drips or double lock cross welts or other horizontal joints) are avoided by using strips with lengths up to 8.0 m. Because of the great length of each strip the fixing cleats used to hold the metal strips in position have to be designed to allow the metal to contract and expand freely.

Standing seams

Both copper and aluminium sheets are sufficiently malleable (workable) to be bent and folded without damage to the material. The strips of metal are jointed by means of a standing seam joint, which is a form of double welt and is left standing up from the roof as shown in Figure 6.47. The completed standing seam is constructed so that there is a gap of some 13 mm at its base, which allows the metal to expand without restraint, as illustrated in Figure 6.47. The lightweight metal strips have to be secured to the roof surface

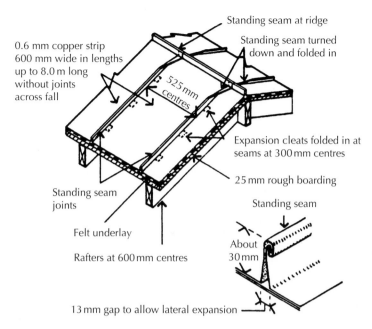

0.6 mm copper strip 600 mm wide in lengths up to 8.0 m long without joints across fall

525 mm centres

Standing seam at ridge

Standing seam turned down and folded in

Expansion cleats folded in at seams at 300 mm centres

25 mm rough boarding

Standing seam

Standing seam joints

Felt underlay

Rafters at 600 mm centres

About 30 mm

13 mm gap to allow lateral expansion

Figure 6.47   Copper standing seam roofing.

at intervals of 300 mm along the length of the standing seams. This close spacing of the fixing cleats is necessary to prevent the metal drumming due to uplift in windy weather.

Two types of cleats are used: fixed cleats and expansion cleats. Five fixed cleats are fixed in the centre of the length of each strip and the rest of the cleats are expansion cleats. Figure 6.48 illustrates the arrangement of these cleats. The fixed cleats are nailed to the roof boarding through the felt underlay; Figure 6.48 illustrates the formation of a standing seam and shows how the fixed cleat is folded in. The expansion cleats are made of two pieces of copper strip folded together so that one part can be nailed to the roof and the second piece, which is folded in at the standing seam, can move inside the fixed piece.

The ridge is usually finished with a standing seam joint, as illustrated in Figure 6.49a, but as an alternative a batten roll or conical roll may be used. Whichever joint is used at the ridge, the standing seams on the slopes of the roof have to be turned down so that they can be folded in at the ridge. This is illustrated in Figure 6.49b. Because copper and aluminium are generally considered to be attractive coverings to roofs, the roof is not hidden behind a parapet wall, and the roof slopes discharge to an eaves gutter, as illustrated in Figure 6.49a.

Where the slope of the roof finishes at the parapet or wall, the strips of metal are turned up as an upstand and finished with an apron flashing. This is illustrated in Figure 6.50, which also illustrates the cutting and turning down of the standing seam. The verge of low-pitched roofs can be finished with batten roll or conical roll, as illustrated for flat roof coverings.

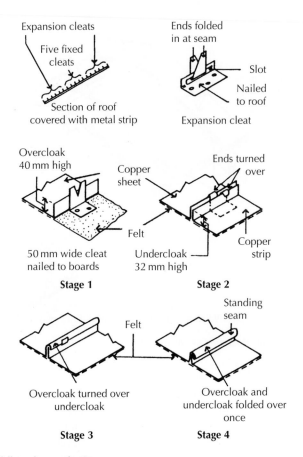

Figure 6.48   Holding down cleats.

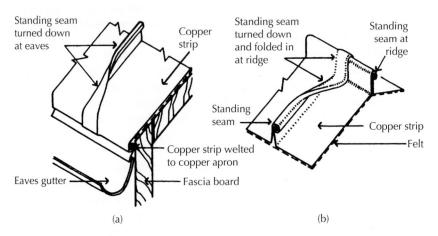

Figure 6.49   (a) Eaves. (b) Ridge.

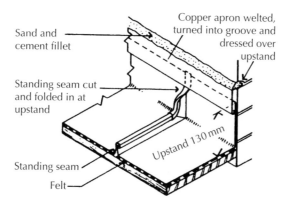

Copper apron welted, turned into groove and dressed over upstand

Sand and cement fillet

Standing seam cut and folded in at upstand

Upstand 130 mm

Standing seam

Felt

Figure 6.50   Upstand.

Felt underlay

Strip metal coverings must be laid on an underlay (bitumen felt or similar) laid across the roof boards and nailed to the boards with butt side joints. The felt allows the metal strips to expand and contract freely as they respond to changes in temperature.

6.5 Thermal insulation to pitched roofs

There are two approaches to insulating pitched roofs: constructing either a 'cold roof' or a 'warm roof', as described later.

Cold roof

The most convenient, economical and usual place for insulation in a pitched roof is either across the top of, or between, the ceiling joists to create a cold roof construction. The disadvantage of spreading or laying mineral fibre insulation or insulation boards between ceiling joists is that there will be a deal of wasteful cutting in fitting the material closely between joists. Therefore rolls of insulation (such as fibreglass, rockwool or sheep's wool) are laid between the joists or loose fill (such as cellulose fibre products) is spread between the joists. Where insulation is fixed between joists an allowance for the different thermal conductivity of wood should be made.

The most straightforward way of providing a layer of insulation is to spread rolls of mineral fibre across the top of ceiling joists right across ceilings in both directions and extended up to and overlapping insulation in walls, as illustrated in Figure 6.51. So that the layer of insulation is continuous over the whole of the ceiling area it is recommended that loft hatches (illustrated in Figure 6.52), giving access to roof spaces, be insulated and draught sealed, and that where service pipes penetrate ceiling finishes some effective form of draught seal be formed. All water carrying service pipes, water storage cisterns or tanks

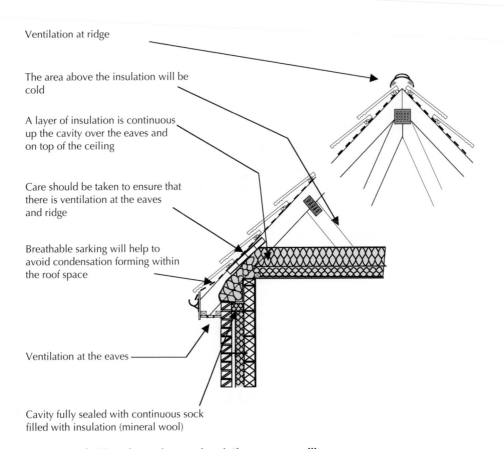

Ventilation at ridge

The area above the insulation will be cold

A layer of insulation is continuous up the cavity over the eaves and on top of the ceiling

Care should be taken to ensure that there is ventilation at the eaves and ridge

Breathable sarking will help to avoid condensation forming within the roof space

Ventilation at the eaves

Cavity fully sealed with continuous sock filled with insulation (mineral wool)

Figure 6.51   Cold roof: continuous insulation across ceiling.

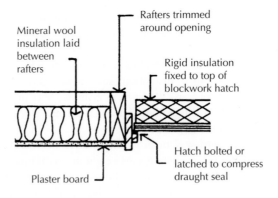

Rafters trimmed around opening

Mineral wool insulation laid between rafters

Rigid insulation fixed to top of blockwork hatch

Hatch bolted or latched to compress draught seal

Plaster board

Figure 6.52   Insulation to loft hatch.

inside cold roof spaces must be effectively protected with insulating material (lagging) against the possibility of damage from the water freezing.

With mats or rolls of mineral fibre spread across ceiling joists there is a possibility of the loose material being compressed and losing efficiency as an insulator, especially under walkways in roof spaces. Boarded access inside roofs should, therefore, be raised on battens above the level of the insulation.

Warm roof

Thermal insulation is fixed above, between or below the rafters of a pitched roof (Figure 6.53) to create a warm roof construction. Because the area of a pitched roof surface is greater than that of a horizontal ceiling there is a greater area to cover with insulation. The advantage of this arrangement is that the roof space will be warmed by heat rising from the heated building below and will, in consequence, be comparatively warm and dry for use.

The most practical and economic place to fix insulation for a warm roof is on top of rafters where there will be the least cutting of the insulation boards. However, the latest robust details show insulation on top of, and between, roof rafters (similar to that shown in Figure 6.53), which further improves the thermal resistance of the roof. The insulation on top of the rafters should cover the whole area of the roof preventing any cold bridges. The rigid insulation boards, which are placed on top of the rafters, also act as a form of

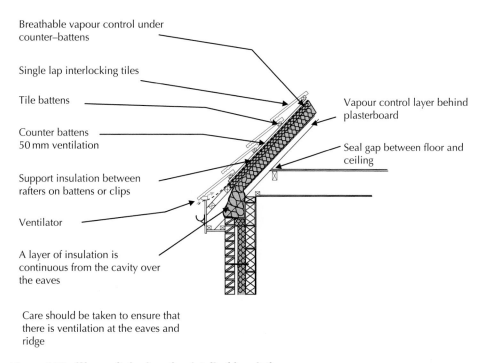

Breathable vapour control under counter–battens

Single lap interlocking tiles

Tile battens

Counter battens 50 mm ventilation

Support insulation between rafters on battens or clips

Ventilator

A layer of insulation is continuous from the cavity over the eaves

Vapour control layer behind plasterboard

Seal gap between floor and ceiling

Care should be taken to ensure that there is ventilation at the eaves and ridge

Figure 6.53   Warm pitched roof – detail of insulation.

sarking excluding the wind that would otherwise blow through tiled and slated roofs. To limit thickness, a material with a high thermal resistance, such as extruded polystyrene (XPS), polyisocyanurate (PIR) or polyurethane (PUR), is used. These insulation boards are nailed directly to the roof rafters with their edges close butted or with tongued and grooved edges for a tight fit. Roofing felt is spread over the insulation boards and sawn softwood counter battens are nailed through the insulation boards to rafters to provide a fixing for tiling or slating battens.

Vapour check

To control the movement of warm moist air from inside a building to the cold side of insulation, it is practice to fix a layer of some impermeable material, such as polythene sheeting, to the underside of insulation. Water vapour on the warm side of insulation is normally not a problem as there are no cold surfaces for the condensation to form. If water vapour is allowed to penetrate cold areas of the building or structure the air will not be able to hold the moisture and condensation will form on the cold surfaces. By definition a vapour check is any material that is sufficiently impermeable to prevent the movement of moisture vapour without being an impenetrable barrier (may allow the passage of air but not moisture laden air). The usual form of vapour check is sheets of 250 gauge polythene sheet with the edges overlapped and spread under insulation over ceiling joists with edges taped around pipes and cables that penetrate ceiling finishes.

Some insulating materials, such as the organic, closed cell boards, e.g. extruded polystyrene, are impermeable to moisture vapour. When these boards can be close butted together or provided with rebated or tongued and grooved edges and cut and close fitted to junctions with walls and seals around pipes, they will serve as a vapour check. However, site practice is rarely so precise and it is advisable to have a vapour check on the warm side of insulation. Breathable sarking that lets water vapour from inside the roof escape, but which prevents water and wind from penetrating from outside the dwelling, can also be used.

Ventilation

A requirement from Part F to the Building Regulations is that adequate provision be made to prevent excessive condensation in a roof (there is no definition of what constitutes 'excessive'). Cold roof spaces should be adequately ventilated to outside air to reduce the possibility of condensation. A more rational approach to reducing the possibility of warm, moist air penetrating a cold roof space is the recommendation for provision of both background and mechanical extract ventilation to kitchens and bathrooms as recommended in Approved Document F.

Cold roof spaces should have ventilation openings to promote cross-ventilation, as illustrated in Figure 6.54, Figure 6.55 and Figure 6.56. The suppliers of tiles and slates offer a range of special fittings to provide roof ventilation through eaves, ridges and in roof slopes. A typical under-eave ventilator for slate is illustrated in Figure 6.55. The plastic ventilator is fixed over roof rafters before the roofing felt is laid. It is fitted with insect screens over the openings. The roofing felt and slate are laid over the ventilators. A sufficient number of ventilators are used to provide the requisite ventilation area to comply with the regulations.

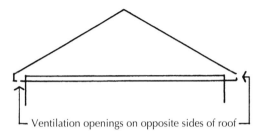

Ventilation openings on opposite sides of roof

Figure 6.54 Roof ventilation.

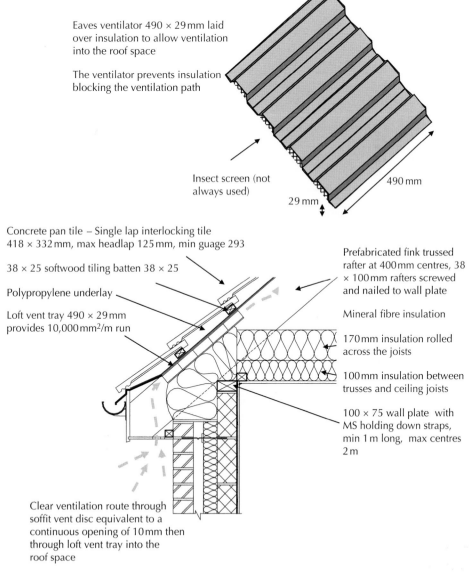

Eaves ventilator 490 × 29 mm laid over insulation to allow ventilation into the roof space

The ventilator prevents insulation blocking the ventilation path

Insect screen (not always used)

490 mm

29 mm

Concrete pan tile – Single lap interlocking tile 418 × 332 mm, max headlap 125 mm, min guage 293

38 × 25 softwood tiling batten 38 × 25

Polypropylene underlay

Loft vent tray 490 × 29 mm provides 10,000 mm²/m run

Prefabricated fink trussed rafter at 400 mm centres, 38 × 100 mm rafters screwed and nailed to wall plate

Mineral fibre insulation

170 mm insulation rolled across the joists

100 mm insulation between trusses and ceiling joists

100 × 75 wall plate with MS holding down straps, min 1 m long, max centres 2 m

Clear ventilation route through soffit vent disc equivalent to a continuous opening of 10 mm then through loft vent tray into the roof space

Figure 6.55 Eaves ventilator and cold roof detail.

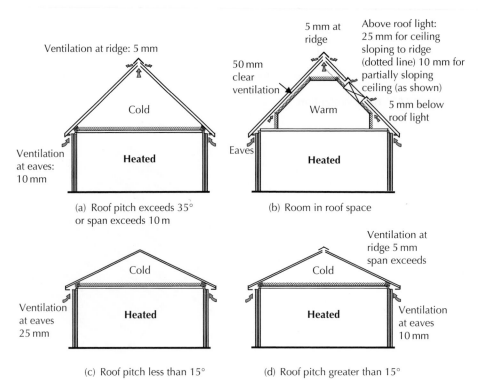

Figure 6.56 Pitched roof ventilation.

The National House Building Council (NHBC) Standards (2003) states that all roof voids should be ventilated to prevent condensation. Figure 6.56 shows illustrations of the guidance provided in the NHBC Standards.

Uncontrolled ventilation

Care should be taken not to over-ventilate a property. There is no guidance on the maximum level of ventilation; however, it would be wrong to over-ventilate a building when attempts are being made to seal all other parts of the building structure. The increased emphasis on airtight buildings is to seal the building and only allow air to escape into and out of the building by prescribed ventilation duct. This approach provides controlled ventilation, ventilation controlled by design and not defects that allow air to leak between the structures.

Thermal bypass at the eaves in warm roofs

Where a room is constructed within the roof attention should be given to the connection between the air barrier and the thermal barrier. If the air barrier does not correctly interface with the thermal barrier the heat energy will escape. The air and thermal barrier should be in continuous contact with the insulation.

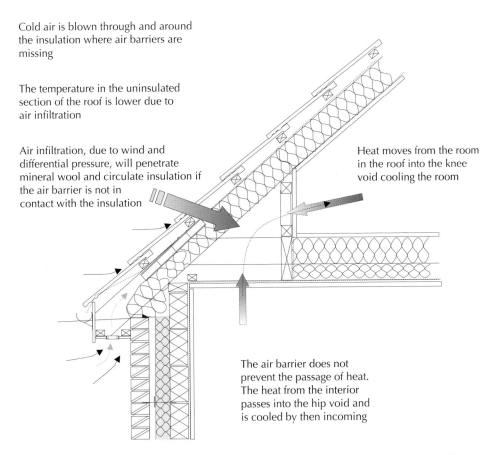

Cold air is blown through and around the insulation where air barriers are missing

The temperature in the uninsulated section of the roof is lower due to air infiltration

Air infiltration, due to wind and differential pressure, will penetrate mineral wool and circulate insulation if the air barrier is not in contact with the insulation

Heat moves from the room in the roof into the knee void cooling the room

The air barrier does not prevent the passage of heat. The heat from the interior passes into the hip void and is cooled by then incoming

Figure 6.57   Thermal bypass at the hip of a roof, failure to maintain thermal and air barrier interface.

A potential weakness of designs that feature a room in the roof occurs at the eaves, where the narrow angle between the floor and rafter junction at the eaves is boxed-out (Figure 6.57). The air barrier must remain in contact with the insulation, ensuring that heat does not simply jump the air barrier. Figure 6.57 illustrates the thermal bypass that may occur. During high winds or due to pressure differences air will be forced through open cell insulation or around closed cell insulation. Figure 6.58 shows a typical problem when insulation is poorly positioned at the eaves.

6.6 Flat roofs

A roof is defined as flat when its weather plane is finished at a slope of 1°–5° to the horizontal. The shallow slope to flat roofs is necessary to encourage rainwater to flow towards rainwater gutters or outlets and to avoid the effect known as ponding. Where there is no

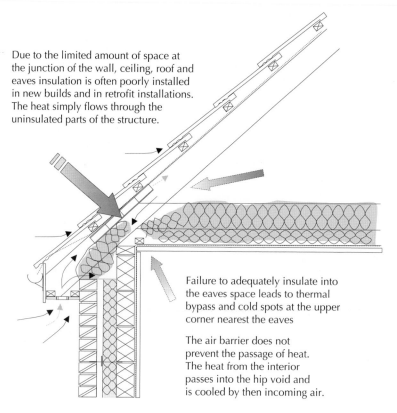

Due to the limited amount of space at the junction of the wall, ceiling, roof and eaves insulation is often poorly installed in new builds and in retrofit installations. The heat simply flows through the uninsulated parts of the structure.

Failure to adequately insulate into the eaves space leads to thermal bypass and cold spots at the upper corner nearest the eaves

The air barrier does not prevent the passage of heat. The heat from the interior passes into the hip void and is cooled by then incoming air.

Figure 6.58 Thermal bypass and bridge due to gaps in insulation barrier and incorrectly positioned insulation.

slope or fall or a very shallow slope it is possible that rainwater may not run off the roof. The increase in load generated by rain pooling may cause the roof to deflect leaving the rainwater lying in a shallow pond. This water pressure may in time cause deterioration of the roof membrane.

The minimum falls (slope) for flat roof coverings are 1:60 for copper, aluminium and zinc sheet, and 1:80 for sheet lead and for mastic asphalt and built-up bitumen felt membranes. To allow for deflection under load and for inaccuracies in construction it is recommended that the actual fall or slope should be twice the minimum and allowance be made where slopes intersect so that the fall at the mitre of intersection is maintained.

Lead, copper and zinc sheet are laid as comparatively small sheets dressed over rolls at junctions of sheet (down the slope of the roof) and with small steps (drips) at junctions of sheet across the slope. The size of the sheet is limited to prevent excessive expansion and contraction that might otherwise cause the sheet to tear. The upstand rolls down the slope and the steps or drips across the slope provide a means of securing the sheets against wind uplift and as a weathering to shed water away from the laps between sheets. More recently aluminium sheet has also been used.

Because of the cost of the material and labour involved in covering a roof with these metals, the continuous membrane materials, asphalt and bitumen felt, are normally used. Asphalt is a dense material that is spread while hot over the surface of a roof to form a continuous membrane, which is impermeable to water. It is generally preferred as a roof covering to concrete flat roofs. Bitumen impregnated felt is laid in layers bonded in hot bitumen to form a continuous membrane that is impermeable to water. Bitumen felt is comparatively lightweight and commonly preferred for timber flat roofs for that reason. Both asphalt and bitumen felt will oxidise and gradually become brittle and have a useful life of at most 20 years. Bitumen felt becomes brittle in time and may tear due to expansion and contraction caused by temperature fluctuations. Fibre-based bitumen felts have lost favour due to premature failures and polyester-based felts are now more used for their greater durability.

6.7 Timber flat roof construction

The construction of a timber flat roof is similar to the construction of a timber upper floor. Sawn softwood timber joists 38–75 mm thick and from 97 to 220 mm deep are placed on edge, spaced at 400–600 mm centres, with the ends of the joists built into or on to or against brick walls and partitions. Tables in Approved Document A give sizes of joists for flat roofs related to span and loads for roofs with access only for maintenance and repair and also for roofs not limited to access for repair and maintenance. Strutting should be fixed between the roof joists to provide lateral restraint as described for upper timber floors.

Flat roof construction

Roof deck
Boards that are left rough surfaced from the saw are the traditional material used to board timber flat roofs. This is called rough boarding and is usually 19 mm thick and cut with square, that is, plain, edges. Plain edged rough boarding was the cheapest obtainable and used for that reason. Because square edged boards often shrink and twist out of level as they dry, chipboard or plywood is mostly used today to provide a level roof deck. For best quality work, tongued and grooved boards were often used.

End support of joists
If there is a parapet wall around the roof, the ends of the roof joists may be built into the inner skin of cavity walls or supported in metal hangers. The joists can bear on a timber or metal wall plate or be packed up on slate or tile slips as described for upper floors. The ends of the roof joists are sometimes carried on brick corbel courses, timber plate and corbel brackets or on hangers in precisely the same way that upper floor joists are supported. The end of roof joists built into solid brick walls should be given some protection from dampness by treating them with a preservative.

Timber firring
Flat roofs should be constructed so that the surface has a slight slope or fall towards rainwater outlets. The sloping surface is created with the use of firring pieces. These consist of

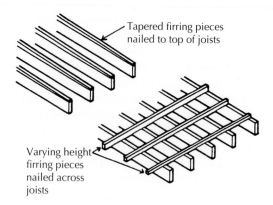

Tapered firring pieces
nailed to top of joists

Varying height
firring pieces
nailed across
joists

Figure 6.59   Firring to timber flat roof.

either tapered lengths of fir (softwood) nailed to the top of each joist or varying depth lengths of softwood nailed across the joists, as shown in Figure 6.59.

Varying depth firring is used where rough boarding is fixed as the deck for flat roof sheet metal coverings. By this arrangement the boards are fixed down the slope so that any variation in the surface of the boards, due to shrinkage or twisting, does not impede the flow of rainwater down the shallow slope. Tapered firring is used where the roof deck is formed by chipboard or plywood sheets nailed to firring with close butted edges to form a level surface. As an alternative to timber firring, insulation boards that are made or cut to a shallow wedge section can be used to provide the necessary shallow fall.

Sheet metal covering to timber flat roofs

Sheet metal is used as a covering because it gives excellent protection against wind and rain; it is durable and lighter in weight than asphalt, tiles or slates. Four metals in sheet form are used: lead, copper, zinc and aluminium.

Sheet lead
Lead is a heavy metal that is comparatively soft, has poor resistance to tearing and crushing and has to be used in comparatively thick sheets as a roof covering. It is malleable and can easily be bent and beaten into quite complicated shapes without damage to the sheets. Lead is resistant to all weathering agents including mild acids in rainwater in industrially polluted atmospheres. On exposure to the atmosphere, a film of basic carbonate of lead oxide forms on the surface of the sheets. These films adhere strongly to the lead and, as they are non-absorbent, they prevent further corrosion of the lead below them. The useful life of sheet lead as a roof covering is upwards of a hundred years.

Rolled sheet lead for roof work is used in thicknesses of 1.8, 2.24 and 2.5 mm. These thicknesses are described as Code Numbers 4, 5 and 6, respectively, the code corresponding to the imperial weight of a given area of sheet. No sheet of lead should be larger than 1.6 m^2 so that the joints between the sheets are sufficiently closely spaced to allow the metal to contract without tearing away from its fixing. Another reason for limiting the size of sheet, which is peculiar to lead, is to prevent the sheet from creeping down the roof under gravity.

The joints across the fall of the roof are made in the form of a 50 mm drip or step down to encourage flow of water. To reduce excessive increases in the thickness of the roof due to these drips, they are spaced up to 2.3 m apart and the rolls (joint longitudinal to fall) 600–800 mm apart. Figure 6.60 illustrates part of a lead covered flat roof showing the general layout of the sheets and a parapet wall around two sides of the roof.

Wood rolls
The edges of sheets longitudinal to the fall are lapped over a timber, which is cut from lengths of timber 50 mm square to form a wood roll. Two edges of the batten are rounded so that the soft metal can be dressed over it without damage from sharp edges. Two sides of the batten are slightly splayed and the waist so formed allows the sheet to be clenched over the roll. Figure 6.61 is an illustration of a wood roll. An underlay of bitumen

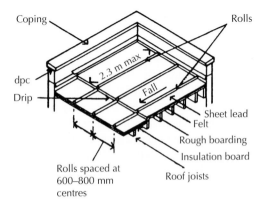

Figure 6.60   Lead flat roof.

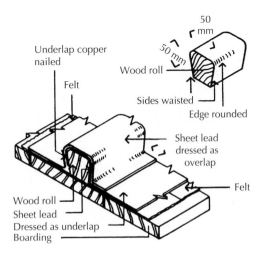

Figure 6.61   Wood rolls.

impregnated felt or stout waterproof building paper is first laid across the whole of the roof boarding and the wood rolls are then nailed to the roof at from 600 to 800 mm centres. The purpose of the underlay of felt or building paper is to provide a smooth surface on which the sheet lead can contract and expand freely.

The edges of adjacent sheets are dressed over the wood roll in turn. In sheet metalwork the word dressed is used to describe the shaping of the sheet. The edge of the sheet is first dressed over as underlap or undercloak and is nailed with copper nails to the side of the roll. The edge of the next sheet is then dressed as overlap or overcloak. A section through one roll is shown in Figure 6.61. In this way no sheet is secured with nails on both sides, so that if it contracts it does not tear away from the nails.

Drips

Drips 50 mm deep are formed in the boarded roof by nailing a 50 × 25 mm fir batten between the roof boards of the higher and lower bays. The drips are spaced at not more than 2.3 m apart down the fall of the roof. The edges of adjacent sheets are overlapped at the drip (forming the underlap and overlap). The underlap edge is copper nailed to the boarding in a cross-grained rebate, as shown in Figure 6.62. An anti-capillary groove formed in the 50 × 25 mm batten is shown into which the underlap is dressed. This groove is formed to ensure that no water rises between the sheets by capillary action.

Figure 6.62 also shows the junction of wood rolls with a drip and illustrates the way in which the edges of the four sheets overlap. This arrangement is peculiar to sheet lead covering, which is a soft, very ductile material that can be dressed as shown without damage. The end of the wood roll on the higher level is cut back on the splay (called a bossed end) to facilitate dressing the lead over it without damage. This seemingly complicated junction of four sheets of lead, which provides a watertight overlap and makes allowance for thermal movement of lead, can be quickly made by a skilled plumber.

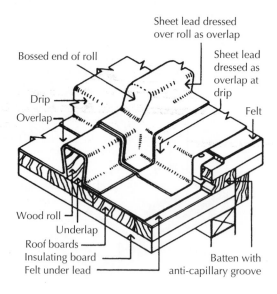

Figure 6.62   Drip.

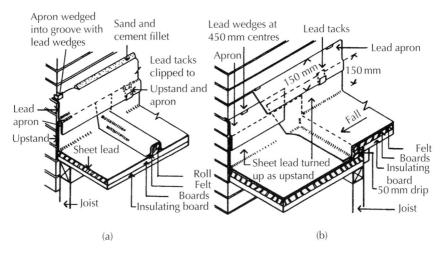

Figure 6.63 **(a) Junction of roll and upstand. (b) Junction of drip and upstand.**

Where there is a parapet wall around the roof or where the roof is built up against a wall the sheets of lead are turned up against the wall about 150 mm as an upstand (no fixing is needed); 170 mm strips of sheet lead are then placed into the rebated (raked) brickwork. The horizontal joint of the brickwork is cut or raked out (25 mm) deep enabling the lead to be tucked into the brickwork (water which runs down the face of the brickwork will not get behind or under the leadwork). To hold the strips in place wedges made of lead are used and the lead strip is folded down (dressed down) over the lower leadwork (upturned apron flashing). To prevent the apron from being blown up by the wind, lead clips are fixed as shown in Figure 6.63a and b.

Lead gutter
Where the lead flat roof is surrounded on all sides by parapet walls it is necessary to collect the rainwater falling off at the lowest point of the roof. A shallow timber-framed gutter is constructed and this gutter is lined with sheets of lead jointed at drips and with upstand and flashings similar to those on the roof itself. The gutter is constructed to slope or fall towards one or more rainwater outlets. The gutter is usually made 300 mm wide and is formed between one roof joist, spaced 300 mm from a wall, and the wall itself (Figure 6.64).

The gutter bed is supported by 50 × 50 mm gutter bearers fixed at 450 mm centres supported by 50 × 25 mm battens which are nailed to the wall and the joist to provide the necessary fall in the gutter. Gutter boards, 19 mm thick, are fixed along the length of the gutter as a gutter bed.

Where the roof is surrounded by a parapet wall and a rainwater pipe can be fixed to the outside of the wall, an opening is formed in the wall as an outlet usually 225 × 225 mm square. A rainwater head is fixed to the wall below the outlet. The sheet lead gutter lining is continued through the outlet and dressed to discharge into the rainwater head. Both upstands to the gutter lining are dressed into the sides of the rainwater outlet in the parapet

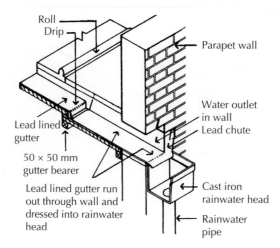

Figure 6.64   Rainwater outlet.

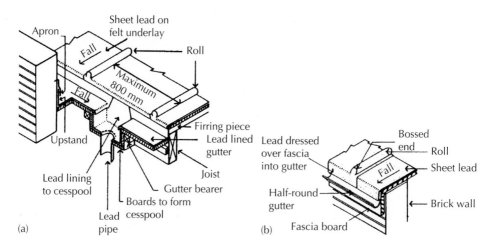

Figure 6.65   (a) Cesspool. (b) Gutter.

wall and dressed against the wall face to form a chute to direct water into the rainwater head, as illustrated in Figure 6.64.

Where it is not possible or desirable to fix a rainwater pipe and head externally the gutter has to be formed to discharge to a cesspool (Figure 6.65). The purpose of this lead lined box, cesspool or catchpit is as a reservoir so that, during a heavy storm, when the rainwater pipe may not be able to carry the water away quickly enough, the cesspool prevents flooding of the roof. A lead downpipe is connected to the cesspool and run down as a downpipe or connected to a downpipe. Where there is no parapet wall on one side of the roof it can discharge rainwater directly to a gutter fixed to a fascia board on the external face of the wall. The lead flat roof covering is dressed to discharge into the gutter, as illustrated in Figure 6.65.

Copper sheet

Copper is a metal with good mechanical strength and is sufficiently malleable in sheet form that it can be bent and folded without damage. Copper is resistant to all normal weathering agents and its useful life as a roof covering is as long as that of lead.

The usual thickness of copper sheet for roofing is 0.6 mm. An oxide of copper forms on the surface of copper sheet and, in the course of some few years, the sheets become entirely covered with a light green compound of copper. This light green 'patina' is impervious to all normal weathering agents and protects the copper below it. The standard sizes of sheet supplied for roofing are 1.2 m and 1.8 m × 900 mm. The minimum fall for a copper covered roof is 1 in 60 and the fall is provided by means of firring pieces just as it is for lead covered roofs.

The traditional method of fixing copper sheet to a flat roof has been by the use of wood rolls down the shallow slope, over which adjacent sheets are shaped with welted joints and drips across the slope. Copper sheets cannot be shaped over rolls and drips easily, so they are joined by the use of double lock welts, which are a double fold at the edges of adjacent sheets. Double lock welts are used for joints of sheet over the wood rolls and at joints across the fall, as illustrated in Figure 6.66. Because of the difficulty of forming a double lock welt it is necessary to stagger these joints to avoid the difficulty of forming this joint at the junction of four sheets of copper.

Figure 6.67 is an illustration of a copper sheet covered flat roof. The wood rolls along the fall of the roof are fixed at centres to suit the standard width of sheet, e.g. 750 mm. At drips the rolls are staggered to avoid welts at the junction of four sheets. Drips are formed across the slope at from about 2 to 3 m to suit the size of standard sheet with staggered double lock cross welts between drips and drip and parapet wall. Drips are formed with a 40 × 50 mm batten, which is nailed between the upper and lower roof boarding on the firring pieces on joists, as illustrated in Figure 6.67.

At the drip, the sheets of copper are welted with the upstand of the lower sheet shaped to the end of the roll and welted to the overcloak as illustrated in Figure 6.67. The conical rolls formed down the fall of the roof are formed around 63 × 50 mm sections of softwood,

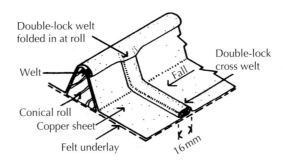

Figure 6.66   Double lock welt.

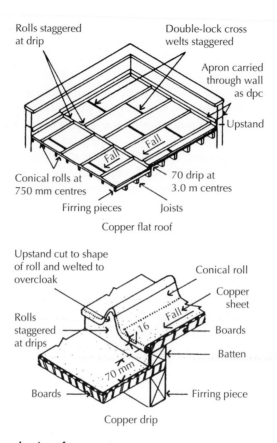

Figure 6.67 Copper sheet roof.

which are shaped in the section of a cone with a round top. The shape is selected to facilitate shaping the metal. The wood rolls are nailed through a felt underlay to the roof boarding and also 50 mm wide strips of copper sheet at 450 mm centres as cleats. The felt underlay serves to allow the sheets to expand and contract without restraint.

The copper strips are used to restrain the sheets against wind uplift by being folded into the welted joint formed over the roll, as illustrated in Figure 6.68. At upstands to parapet walls the sheets are turned up at the upstand and covered with an apron flashing which is wedged into a brick joint. At the junction of the conical roll and the upstand it is necessary to use a separate saddle piece of copper sheet that is welded to the capping and dressed up to cover the joint between adjacent upstands, as illustrated in Figure 6.68.

The copper covered roof may discharge to a verge gutter or to a parapet wall gutter formed in the roof and lined with copper. As an alternative to the use of a conical roll down the slope or fall of copper roofs a batten roll may be used with drips and double lock cross welts across the fall or slope of the roof. A wood roll with slightly shaped sides is nailed through a felt underlay and also 50 mm wide strips of copper at 450 mm centres as cleats.

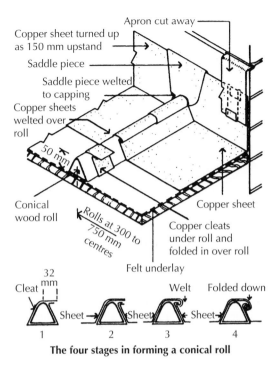

The four stages in forming a conical roll

Figure 6.68   Conical roll.

The sides of adjacent sheets of copper are shaped as upstand sides to the roll and the roll is covered with a separate strip of copper as a capping. The cleats are folded into the double lock welts formed between the upstand of sheets and the edges of the capping. This is the type of joint longitudinal to falls that is also used with zinc sheet roof covering.

Zinc sheet

Zinc, a dull, light grey metal, is used in sheet form as a covering for timber flat roofs. It is the cheapest of the metals used for roofs and has been extensively used in northern European countries, but less so in the UK. Zinc sheet is appreciably more difficult to bend and fold than copper. Being a brittle material it is liable to crack if bent or folded too closely. Standard 2.4 m long by 900 mm wide sheets, either 1 or 0.8 mm thick, are bent up to the sides of softwood batten rolls. The batten rolls are nailed at 850 mm centres through felt underlay and clips at 750 mm centres. The clips are welted over the upstand edges of the sheets.

Because of the difficulty of making a double lock welt of zinc sheet the capping is shaped to fit over the batten rolls. The lower edges of the capping are bent in, feinted, to grip the sheets. To secure the capping, zinc holding down clips are nailed to the batten over the end of a lower capping and the end of the upper capping tucked into the fold, as illustrated in Figure 6.69a. Drips are formed across the slope at 2.3 m intervals, with beaded drips and

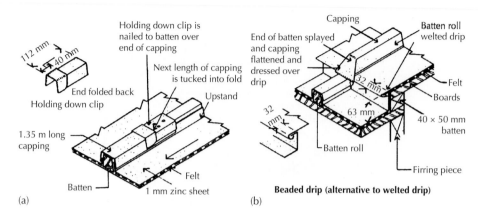

Figure 6.69   (a) Zinc batten rolls. (b) Zinc drip.

splayed cappings to roll ends, as illustrated in Figure 6.69b. At parapets the sheets are turned up as upstand and covered with a zinc apron that is wedged into a brick joint. The end of a batten capping is folded and flattened to cover the edges of adjacent sheets.

Waterproof membranes for timber flat roofs

The two materials that have been used as a waterproof membrane for timber-framed roofs are asphalt as a continuous membrane and bitumen felt laid in layers to form a membrane or skin.

Mastic asphalt

Asphalt is described in the relevant British Standard as mastic asphalt. The material, which is soft and has a low softening (melting) point, is an effective barrier to the penetration of water. Asphalt is manufactured either by crushing natural rock asphalt and mixing it with natural lake asphalt, or by crushing natural limestone and mixing it with bitumen while the two materials are sufficiently hot to run together. The heated asphalt mixture is run into moulds in which it solidifies as it cools. The solid blocks of asphalt are heated on the building site and the hot plastic material is spread over the surface of the roof in two layers breaking joints to a finished thickness of 20 mm. As it cools it hardens and forms a continuous, hard, waterproof surface.

If there is no parapet wall around the roof it is usually designed to overhang the external walls, to give them some protection, and the asphalt drains to a gutter over the strip of sheet lead, which is dressed down into a gutter, as illustrated in Figure 6.70.

If the roof has parapet walls around it, or adjoins the wall of a higher building, an asphalt skirting or upstand, 150 mm high, is formed and this skirting is turned into horizontal brick joints purposely cut about 25 mm deep to take the turn-in of the asphalt skirting, as illustrated in Figure 6.71. For strengthening, an internal angle fillet is run at the junction of the flat and the asphalt skirting. An alternative detail is to form the skirting as an upstand only, with the upstand weathered by a sheet lead apron that is wedged into a raked out

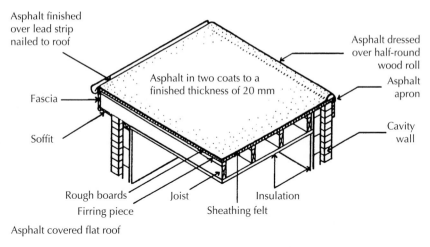

Asphalt finished over lead strip nailed to roof

Asphalt dressed over half-round wood roll

Asphalt in two coats to a finished thickness of 20 mm

Asphalt apron

Fascia

Soffit

Cavity wall

Rough boards   Joist   Insulation

Firring piece   Sheathing felt

Asphalt covered flat roof

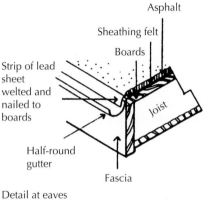

Asphalt

Sheathing felt

Boards

Strip of lead sheet welted and nailed to boards

Joist

Half-round gutter

Fascia

Detail at eaves

Figure 6.70   Asphalt covered flat roof.

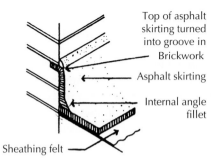

Top of asphalt skirting turned into groove in Brickwork

Asphalt skirting

Internal angle fillet

Sheathing felt

Figure 6.71   Asphalt skirting.

brick joint and dressed down over the upstand. The advantage of this finish is that structural, thermal or moisture movements of the roof are less likely to cause cracking of the asphalt either at the internal angle fillet or at the turn in the wall.

It is essential that asphalt be laid on an isolating membrane underlay so that slight movements in the structure are not reflected in the asphalt membrane. A properly laid asphalt covering to a roof will not absorb water at all and any rainwater that does lie on it due to ponding will eventually evaporate. The asphalt is laid to a slight fall, of at least 1:80, so that the rainwater drains away to a rainwater outlet or gutter. Asphalt is a comparatively cheap roof covering and if the asphalt is of good quality and is properly laid, it should have a useful life of around 20 years, after which it will need to be removed and replaced to remain watertight.

Built-up bitumen felt roofing
The word felt is used for fibres that are spread at random around a large, slowly rotating drum on which a mat of loosely entwined fibres is built up. The mat is cut, rolled off the drum and compressed to form a sheet of felted (matted) fibres. For use as a roof surface material the felt is impregnated or saturated with bitumen. A variety of felts is made from animal, vegetable, mineral or polyester fibre or filament for use as flat roof covering.

Sheathing and hair felts
These felts are made from long staple fibres, loosely felted and impregnated with bitumen (black felt) or brown wood tars or wood pitches (brown felt). They are used as underlays for mastic asphalt roofing and flooring to isolate the asphalt from the structure.

Fibre-based bitumen felts
These felts are made as a base of animal or vegetable fibres that are felted, lightly compressed and saturated with bitumen. Fine granule surfaced felts may be used as a lower layer in built-up roofing and as a top layer on flat roofs that are subsequently surfaced with bitumen and mineral aggregate finish as a comparatively cheap roof finish.

Mineral surfaced fibre-based felt is finished on one side with mineral granules for appearance and protection for use as a top layer on sloping roofs. Fibre-based bitumen felts have been used as a covering for small area pitched roofs. The material is either bonded to a boarded finish with bitumen in two layers, breaking joint or nailed and bonded at joints and edges for the first coat and bonded with bitumen as a top coat.

Glass fibre-based bitumen felts
Glass fibre-based felts have for many years largely replaced natural fibre-based felts as the material used for built-up felt roofing for flat roofs. The material is made from felted glass fibres that are saturated and coated with bitumen.

The felted glass fibre base forms a tenacious mat that is largely unaffected by structural, thermal and moisture movement of the roof deck and does not deteriorate as a result of moisture penetration. The fine granule coated, glass fibre-based felt is used as underlay and as a top layer on flat roofs that are subsequently surfaced with bitumen and mineral aggregate. Mineral surfaced glass fibre-based felts may be used as a top layer on flat roofs as a low-cost finish and on sloping roofs as a finish layer. A perforated, glass fibre-based felt is produced for use as a venting first layer on roofs where partial bonding is used. The durability of glass fibre-based felt depends principally on the bitumen with which it is saturated. In time the bitumen coating will oxidise on exposure to the radiant energy from the sun, harden, and ultimately crack and let in water. A layer of insulation under the felt will appreciably increase the rise in temperature and expansion of the surface and so accelerate the cracking of the hardened top surface. A layer of a light colour mineral aggregate dressed over the surface will provide some protection from the sun. The serviceable life of this weathering is of the order of 20–30 years.

Polyester-based bitumen felts

A polyester base of staple fibre or filament, formed by needling or spin bonding, is impregnated with bitumen. The fibre or filament base of polyester has higher tensile strength than the true felt bases. Because of this greater strength this 'felt' is better able to withstand the strains due to structural, thermal and moisture movement without the rupture that a flat roof covering will suffer.

The fine granule surfaced felt is used as a base, intermediate or top layer of built-up roofing, which is to be subsequently covered with bitumen and mineral aggregate finish. The mineral surface felt is for use as a top layer where there is no additional surface treatment. Polyester-based felt, which is generally used for the three layers of built-up roofing, is sometimes used as a top coat to underlays of glass fibre-based felt.

High-performance roofing

High-performance bitumen coated bases are made with a polyester fabric that is coated with polymer modified bitumen. The bitumen is modified with styrene butadiene styrene (SBS) or atactic polypropylene (APP) to provide improved low-temperature flexibility, improved creep resistance at high temperatures and greatly improved fatigue endurance to the bitumen. These high-performance bitumen bases, which are generally used in two layers, are more expensive than other built-up roofings and are used for their appreciably improved durability.

Single-ply roofing

For details of single-ply, polymeric roof membranes see *Barry's Advanced Construction of Buildings*.

Laying built-up bitumen felt roofing

Bitumen felt is applied to flat roofs in three layers: the first, intermediate and top layers. Practice is to bond the first layer of felt to the roof deck with bitumen spread over the whole surface of the roof, or more usually in a system of partial bonding with bitumen to allow some freedom of movement relative to the roof. The rolls of felt are spread across the roof with a side lap of 75 mm minimum between the long edges of the rolls and with a head, or end, lap of at least 75 mm for felts and 150 mm for polyester-based felts. To avoid an excessive build-up of thickness of felt at laps, the side lap of rolls of felt is staggered by one-third of the width of each roll between layers so that the side lap of each layer does not lie below or above that of other layers, as illustrated in Figure 6.72.

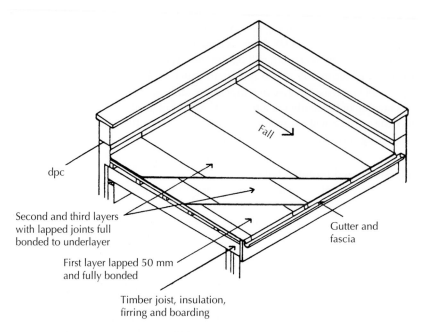

dpc

Second and third layers
with lapped joints full
bonded to underlayer

First layer lapped 50 mm
and fully bonded

Timber joist, insulation,
firring and boarding

Fall

Gutter and
fascia

Figure 6.72   Built-up felt covering to timber flat roof.

Full bond

The traditional method of fixing felt roofing is by fully bonding the felt to the roof deck
and the layers of felt to one another by the 'pour and roll' technique, in which hot bitumen
is poured on the roof deck and the rolls of felt are continuously rolled out as the bitumen
is poured. The pour and roll method of bonding is used for all three layers in the full
bonding method and for the two top layers in the partial bond method. The purpose of
the bitumen is initially to bond the first layer to the roof deck against wind uplift and then
to bond the succeeding layers to each other and form a watertight seal at overlaps of rolls
of felt. The full bond method is used where appreciable wind uplift is anticipated in exposed
positions and the perforated bottom layer where wind uplift is low in sheltered positions
that will also allow some freedom of movement of the felt relative to the roof.

Partial bond

Where it is anticipated that wind uplift will be moderate it is usual to use a perforated, glass
fibre-based felt as a first layer to allow some movement of the deck independent of the felt
covering. The perforated first layer is laid loose over the deck and the intermediate layer is
fully bonded to it. The hot bitumen bed will penetrate the perforated bottom layer to
achieve a partial bond to the deck. The top layer is then fully bonded to the intermediate
layer.

On a roof deck covered with insulation boards the method of bonding the first layer
depends on the composition of the insulation. The majority of insulation boards either have
a surface to which hot bitumen can be applied or are coated to assist the bond of bitumen.

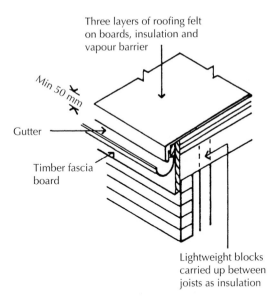

Three layers of roofing felt
on boards, insulation and
vapour barrier

Min 50 mm

Gutter

Timber fascia
board

Lightweight blocks
carried up between
joists as insulation

Figure 6.73   Eaves gutter.

On polyurethane and polyisocyanurate boards, provision should be made for the escape of gases generated by the use of hot bitumen, by using a perforated venting first layer of felt that is laid loose over the boards.

Built-up bitumen felt roof coverings should be laid to a shallow fall so that rainwater will run off to a gutter. The most straightforward way of draining these flat roofs is by a single fall to one side to discharge directly to a gutter fixed right across one side of the roof, as illustrated in Figure 6.73. Here the roof is enclosed inside a parapet wall on three sides. The roof joist ends are carried over the wall to provide a fixing for a softwood timber fascia board, which supports a gutter (Figure 6.73). To direct the run-off of water into the gutter, a strip of felt is nailed to the fascia board, welted and then turned up on the roof and bonded between the first and intermediate layers of felt. Because felt is difficult to welt without damage, a strip of sheet lead may be used as an alternative to the felt strip. For continuity of insulation a lightweight insulation inner leaf of wall or cavity insulation is carried up to the level of the roof insulation.

At the junction of a felt covered flat roof and a parapet or a wall the top two layers of three-ply bitumen felt are dressed up the wall over an angle fillet as an upstand 150 mm high. The top of the felt upstand is covered with a lead flashing, which is tucked into a raked out brick joint and dressed over the upstand (Figure 6.74).

Mineral surface dressing
The finished top surface of built-up felt roof covering is usually finished with a dressing of mineral aggregate spread over a bitumen coating. The purpose of this mineral dressing is to act as a reflective, protective layer of light coloured particles to reduce the oxidising and

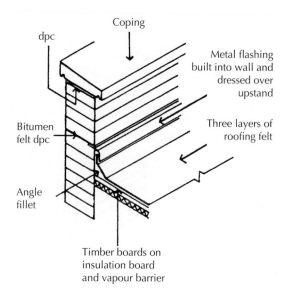

Figure 6.74   Upstand and flashing.

hardening effect of direct sunlight on the bitumen bonding compound and also to improve the appearance of the roof. Light coloured mineral aggregate, graded in size from 16 to 32 mm, is spread to a thickness of at least 12.5 mm on a layer of bitumen. This also provides the necessary cover to prevent the surface spread of fire.

Inverted or upside down roof

Thermal insulation laid or fixed below a flat roof coverings will appreciably increase the temperature fluctuations that a roof covering will suffer between hot days and cold nights. Bitumen felt coverings are particularly affected by direct sunlight, which causes oxidisation of the bitumen, which hardens and becomes brittle and liable to crack. The most rational place for the layer of insulation on flat roofs, particularly those covered with bitumen felt and, to a lesser extent, asphalt, is on top of the covering to reduce temperature fluctuations. This type of construction is termed an inverted or upside down roof (Photograph 6.14a–c).

In this form of construction one of the organic closed cell insulation boards, which do not absorb water and are sufficiently dense to support the overlay of stones or paving slabs, are used. The roof covering may be laid to slight falls to rainwater outlets or laid flat to outlets. The insulation boards are laid on top of the roof covering. To protect the insulation against wind uplift the roof is covered with a layer of mineral aggregate, stones, of sufficient depth to hold the insulation in place. Where the flat roof serves as a terrace, paving stones are used as a loading coat (Figure 6.75). The paving stones are laid, open jointed, on a filter layer of dry sand or mineral fibre mat as a bed and filter for rainwater. To avoid heat flowing through the concrete roof structure and passing up the parapet wall (forming a thermal bridge) lightweight thermal concrete blocks should be used.

Photograph 6.14   (a) Inverted roof also used as a terrace. (b) Inverted roof 20-mm asphalt waterproof membrane, built up in two layers sits below the insulation. (c) Inverted roof paving slabs positioned on top of rigid insulation boards.

Rain falling on the roof will in part be retained by the stone or paving slab covering and the filter layer and will, in part, drain through the insulation to the covering. During dry periods much of the rain retained by the stones or paving slabs will evaporate. For access to clear any blockages that might occur in rainwater outlets it is wise to provide a margin, around the edge of the roof, of loose aggregate over a narrow strip of insulation that can be lifted to clear blockages. The disadvantage of the inverted roof is that where leaks occur

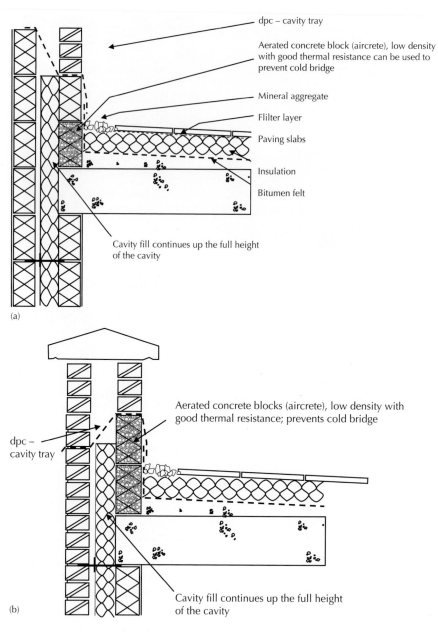

Figure 6.75 labels (a):
- dpc – cavity tray
- Aerated concrete block (aircrete), low density with good thermal resistance can be used to prevent cold bridge
- Mineral aggregate
- Flilter layer
- Paving slabs
- Insulation
- Bitumen felt
- Cavity fill continues up the full height of the cavity

(a)

Figure 6.75 labels (b):
- dpc – cavity tray
- Aerated concrete blocks (aircrete), low density with good thermal resistance; prevents cold bridge
- Cavity fill continues up the full height of the cavity

(b)

Figure 6.75   (a) Inverted roof abutting an internal wall. (b) Inverted roof abutting parapet wall.

in the covering the whole top layer may have to be removed. Because of the protection afforded by the two top layers, faults in the covering are less likely than with coverings laid on the insulation.

6.8 Concrete flat roofs

Reinforced concrete flat roofs for small buildings are constructed in the same way as reinforced concrete floors with hollow beam, beam and filler blocks or in situ cast concrete slab (see Chapter 4). Roofs are designed to support the self-weight of the roof together with loading from rain and snow, resist wind pressure and provide support for access for maintenance or for use as a roof terrace. The concrete roof is supported on external walls with intermediate support from internal loadbearing walls or beams as necessary.

Roof coverings

To provide a smooth level surface ready for the roof covering the concrete or concrete topping may be power floated. More usually a cement and sand screed is used, which can be spread and finished to the level or levels necessary to provide a fall to drain rainwater from the roof. A screed can be finished to one-, two- or four-way falls to rainwater outlets with the necessary currents at the intersections of falls.

Asphalt roof covering
Hot asphalt is spread over a layer of loose laid sheathing felt on a dry screed finish in two layers to a finished thickness of 20 mm. The asphalt is then covered with a dusting of dry, fine, sharp sand to absorb the 'fat' of the neat asphaltic bitumen that is worked to the surface by hand spreading. On a screed laid to falls the asphalt drains to outlets formed in parapet walls (Figure 6.76). The asphalt skirting is dressed into the outlet in the parapet over a lead chute dressed down over a rainwater head.

A reinforcing, internal angle, fillet of asphalt is formed at the junction of the flat roof and the parapet wall and the 150 mm high asphalt skirting. The top of the skirting may be turned into a groove cut in a horizontal brick joint (Figure 6.77). As an alternative the asphalt skirting may be run up the face of the wall by itself or over plastic vents placed at intervals to provide ventilation for moisture vapour. The top of the skirting is then covered with a sheet lead flashing, tucked into and wedged in a raked out brick joint and dressed down over the upstand asphalt skirting.

Built-up bitumen felt covering
As a precaution against the possibility of water drying out from screeded concrete roofs and rising to the surface under bitumen felt coverings, expanding and causing blisters, it is usual to use the partial bonding method.

With the partial bonding method the first layer of felt is bonded to the surface of the screed with perimeter and intermediate strips of bitumen with 180 mm wide vents between the strips of bitumen to allow moisture vapour to vent to perimeter and central vents. The surface of the screed is first coated with a bitumen primer to improve the bond of bitumen

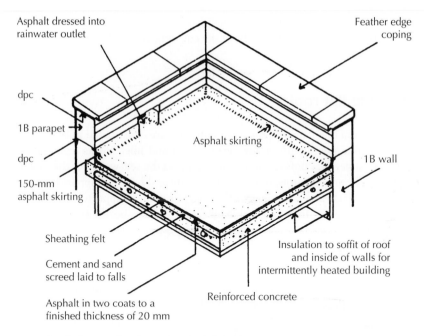

Asphalt dressed into
rainwater outlet

Feather edge
coping

dpc

1B parapet

dpc

150-mm
asphalt skirting

Asphalt skirting

1B wall

Sheathing felt

Cement and sand
screed laid to falls

Insulation to soffit of roof
and inside of walls for
intermittently heated building

Reinforced concrete

Asphalt in two coats to a
finished thickness of 20 mm

Figure 6.76   Asphalt covering to concrete flat roof.

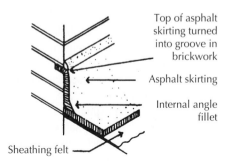

Top of asphalt
skirting turned
into groove in
brickwork

Asphalt skirting

Internal angle
fillet

Sheathing felt

Figure 6.77   Asphalt skirting.

to screed. Perimeter strips of bitumen 450 mm wide are spread around the roof with 150 mm wide vents at intervals for moisture vapour to vent to the perimeter. Strips of bitumen are spread over the body of the roof, as illustrated in Figure 6.78. The principal adhesion of the felt covering, against wind uplift, is affected by the perimeter bonding. The size and spacing of the strips of bitumen is chosen as a matter of judgement between the need for adhesion to keep the felt covering flat and the assumed need to provide ventilation paths for moisture vapour pressure.

At verges the intermediate and final layers of felt can be shaped over a splayed wood block and then covered with a strip of felt that is welted to a timber batten and turned over

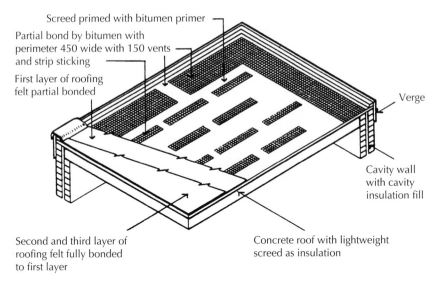

Screed primed with bitumen primer

Partial bond by bitumen with
perimeter 450 wide with 150 vents
and strip sticking

First layer of roofing
felt partial bonded

Verge

Cavity wall
with cavity
insulation fill

Second and third layer of
roofing felt fully bonded
to first layer

Concrete roof with lightweight
screed as insulation

Figure 6.78   Partial bond of felt roofing to concrete.

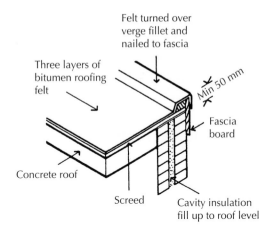

Felt turned over
verge fillet and
nailed to fascia

Three layers of
bitumen roofing
felt

Min 50 mm

Fascia
board

Concrete roof

Screed

Cavity insulation
fill up to roof level

Figure 6.79   Verge.

on to the roof (Figure 6.79). Where there is a parapet wall around the roof the intermediate and top layers of felt are turned up against the wall some 150 mm and covered with a lead sheet flashing which is turned into and wedged in a raked out brick joint.

Because the end laps of rolls of felt are made to overlap down the slope or fall of a roof it is difficult to form cross falls and currents to slopes of roof to a rainwater outlet; therefore, it is usual to drain felt roofing to one continuous verge gutter. Where a built-up bitumen felt covering is laid to falls to parapet wall outlets, it is impossible to avoid an untidy build-up of overlaps at oblique cuts of felt.

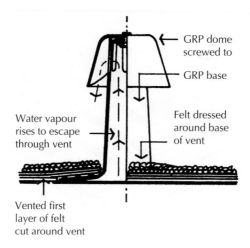

Figure 6.80   Ventilator. GRP, glass reinforced plastic.

The usual method of providing ventilation for moisture vapour pressure is plastic ventilators that are fixed behind the felt upstand to parapet walls and to the felt overlap at verges. The parapet ventilators are covered with the apron flashing. On larger roofs, where it is deemed necessary to provide additional ventilators in the surface of the roof, glass fibre reinforced plastic vents are fixed in the roof at about 6 m centres. These ventilators are fixed to the roof and the intermediate and top layers of felt are cut and bitumen bonded around the vents (Figure 6.80).

6.9 Thermal insulation to flat roofs

Most of the materials used in the construction of flat roofs, separately or together, provide insufficient resistance to the transfer of heat to meet the requirement of the Building Regulations. It is necessary, therefore, to build in or fix some material with high resistance to heat transfer to act as a thermal insulation.

Position of the insulation

The most practical position for a layer of insulation for a flat roof is on top of the roof structure either under or over the weathering cover. In this position the insulation boards can be fixed or laid without undue wasteful cutting to provide insulation for the roof structure and utilise the heat store capacity of a concrete roof to provide some heat during periods when the heating is turned off.

As an alternative to fixing insulation on top of a flat roof it may be fixed between the ceiling joists of a timber roof so that the thermal resistance of the timber joists combines with the greater resistance of the insulation material (Figure 6.81). Tables in Approved Document L provide details of the thickness of insulation required for common insulating

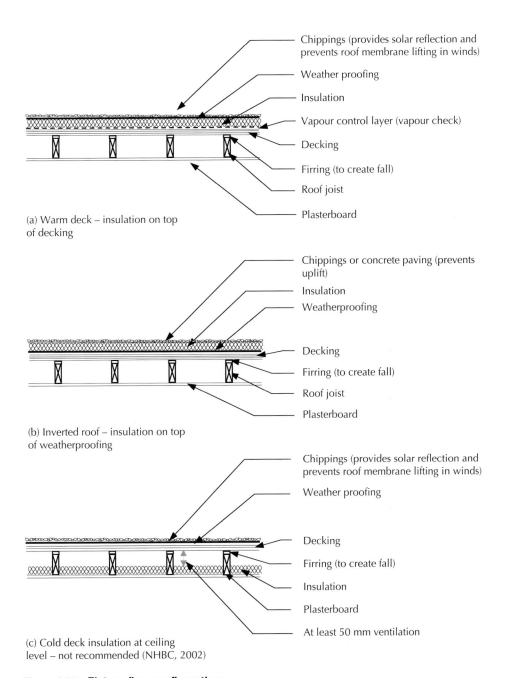

(a) Warm deck – insulation on top of decking

Chippings (provides solar reflection and prevents roof membrane lifting in winds)

Weather proofing

Insulation

Vapour control layer (vapour check)

Decking

Firring (to create fall)

Roof joist

Plasterboard

(b) Inverted roof – insulation on top of weatherproofing

Chippings or concrete paving (prevents uplift)

Insulation

Weatherproofing

Decking

Firring (to create fall)

Roof joist

Plasterboard

(c) Cold deck insulation at ceiling level – not recommended (NHBC, 2002)

Chippings (provides solar reflection and prevents roof membrane lifting in winds)

Weather proofing

Decking

Firring (to create fall)

Insulation

Plasterboard

At least 50 mm ventilation

Figure 6.81   Flat roofing configurations.

materials for a range of U-values for insulation between joists. The disadvantage of fixing insulation between the joists of timber flat roofs is the wasteful cutting necessary to fit the material between the joists, the labour necessary to support or wedge the material in position and the need to oversize electrical cables (necessary to prevent overheating) that run within the insulation.

The disadvantage of the warm roof is that as the insulation is directly under the roof covering, the material of the covering will suffer very considerable temperature fluctuations between hot sunny days and cold nights, as described earlier. An inverted or upside down warm roof, with the insulation on top of the roof covering, will protect the roof covering from severe temperature fluctuations.

Where the insulation is below the roof structure the roof is referred to as a cold roof construction. The disadvantage of cold roof is that the moisture vapour pressure of warm internal air may cause vapour to penetrate the insulation and condense to water on the cold side of the insulation, where it may adversely affect the performance of the insulant and on the cold surfaces of the roof structure. Cold roofs are often protected by a vapour check (vapour barrier) on the warm side of the insulation and roofs are ventilated to prevent any build-up of moisture.

Avoiding thermal bridges

For efficient insulation against heat loss through the fabric of a building it is necessary to unite the insulation used in walls with that used in roofs and minimise or eliminate those parts of the construction that provide a low resistance path to transfer of heat across thermal bridges (Figure 6.82). The extent to which a detail of construction will act as a thermal bridge will depend on the difference in thermal resistance of the thermal bridge, and that of the adjoining construction. Where the difference is large an appreciable concentration of condensation may appear on the colder surface of the bridge and may cause unsightly stains and mould growth and adversely affect materials such as iron and steel.

Where the connection between concrete and masonry components cannot be avoided and a thermal bridge occurs, it may be possible to use aerated blocks, which have a high thermal resistance. The blocks are available in a range of 75–355 mm thicknesses. Where the strength of the block is a problem it may be possible to distribute the load over a larger area using a thicker block, so reducing the load per unit area. Figure 6.83 shows how aerated blocks can be used to prevent thermal bridging above a concrete roof.

In buildings such as offices and places of assembly where the building is intermittently heated with high-temperature radiant heaters it is often the practice to fix the insulation to the fabric on the inside face of walls and ceilings in the form of insulation backed plasterboard, as illustrated in Figure 6.84. In this way the radiant heat, when first turned on, immediately heats inside instead of expending some of its energy on heating the fabric.

Insulation materials

To provide the required thermal resistance for a flat roof, one of the semi-rigid insulation boards is used for roof level insulation. The cheapest material is mineral wool slabs, either of uniform thickness or cut to provide falls for roof drainage, particularly for asphalt finishes. Expanded and extruded polystyrene boards are the cheapest of the inorganic materi-

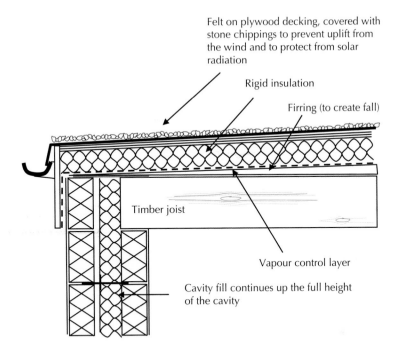

Felt on plywood decking, covered with stone chippings to prevent uplift from the wind and to protect from solar radiation

Rigid insulation

Firring (to create fall)

Timber joist

Vapour control layer

Cavity fill continues up the full height of the cavity

Figure 6.82   Wall insulation joined to roof insulation.

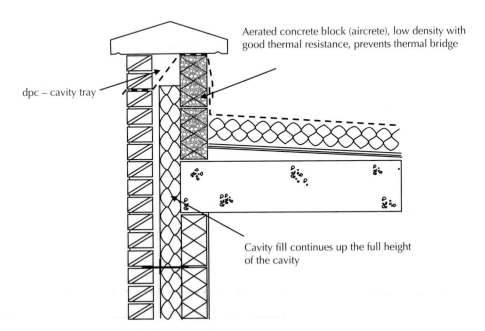

Aerated concrete block (aircrete), low density with good thermal resistance, prevents thermal bridge

dpc – cavity tray

Cavity fill continues up the full height of the cavity

Figure 6.83   Junction with flat roof and parapet wall.

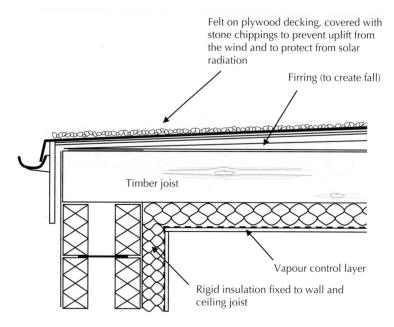

Felt on plywood decking, covered with stone chippings to prevent uplift from the wind and to protect from solar radiation

Firring (to create fall)

Timber joist

Vapour control layer

Rigid insulation fixed to wall and ceiling joist

Figure 6.84   Internal insulation.

als made in boards of uniform thickness or tapered for falls to roofs with expanded polystyrene. The inorganic material boards PIR and PUR have the lowest U-value. They are the more expensive of the materials and are faced with glass fibre tissue for protection and as a finish impermeable to moisture vapour so that they may be used without the need for a moisture vapour check.

Ventilation

Where there is a likelihood of excessive condensation in roof voids above insulated ceilings, the 'cold roof' space should be ventilated to outside air with a clear air space of at least 50 mm above the insulation to meet the requirements of Approved Document F. This space should be ventilated by continuous strips at least equal to continuous strips 25 mm wide running the full length of eaves on opposite sides of the roof. The ventilating openings are formed in the soffit of the overhang by plastic ventilators fitted with insect screens.

Where the insulation is laid or fixed between the joists there should be a clear space above the top of the insulation and underside of the roof of at least 50 mm for air to circulate across the roof from opposite sides, as illustrated in Figure 6.85. So that electric cables are not run in the insulation, the ceiling is fixed to battens to provide a space in which to run cables.

Where rolls of mineral fibre are laid between joists to the extent that there is not a clear 50 mm space between joists above the insulation, it is necessary to fix timber counter battens across the ceiling joists to provide the recommended 50 mm minimum ventilation space as illustrated in Figure 6.86.

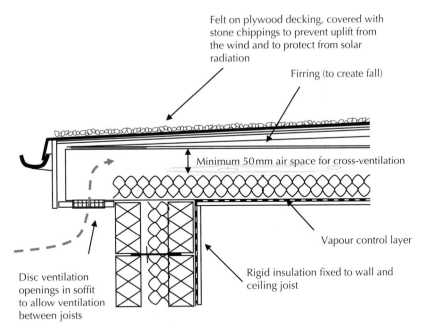

Felt on plywood decking, covered with stone chippings to prevent uplift from the wind and to protect from solar radiation

Firring (to create fall)

Minimum 50 mm air space for cross-ventilation

Vapour control layer

Disc ventilation openings in soffit to allow ventilation between joists

Rigid insulation fixed to wall and ceiling joist

Figure 6.85   Ventilation of flat roof.

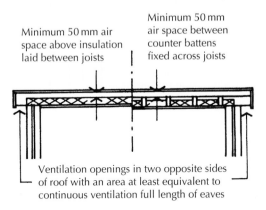

Minimum 50 mm air space above insulation laid between joists

Minimum 50 mm air space between counter battens fixed across joists

Ventilation openings in two opposite sides of roof with an area at least equivalent to continuous ventilation full length of eaves

Figure 6.86   Ventilation.

6.10 Parapet walls

External walls of buildings are sometimes raised above the level of the roof as parapet walls for the sake of the appearance. Parapet walls are exposed on all faces to driving rain, wind and frost and are more liable to damage than external walls below eaves level. Because parapet walls are free-standing, their height is limited in relation to their thickness to retain

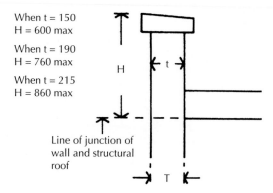

When t = 150
H = 600 max

When t = 190
H = 760 max

When t = 215
H = 860 max

H

t

Line of junction of
wall and structural
roof

T

Figure 6.87   Solid parapet wall.

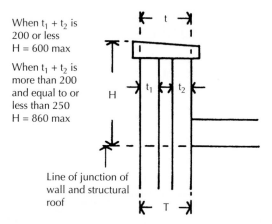

When $t_1 + t_2$ is
200 or less
H = 600 max

When $t_1 + t_2$ is
more than 200
and equal to or
less than 250
H = 860 max

H

t

t_1

t_2

Line of junction of
wall and structural
roof

T

Figure 6.88   Cavity parapet wall.

stability. Approved Document A sets limits to the thickness and height of solid parapet walls, as illustrated in Figure 6.87. Where the height (H) of the parapet walls is not more than 600 mm it should be not less than 150 mm thick; where H is 760 mm, the thickness should be not less than 190 mm thick; and where it is not more than 860 mm the thickness should not be less than 215 mm.

Where an external cavity wall is carried up as a parapet wall, as illustrated in Figure 6.88, the limits of height (H) are not more than 600 mm where the combined thickness of the two leaves ($t_2 + t_2$) is equal to or less than 200 and 860 mm where the combined thickness of the two leaves ($t_2 + t_2$) is greater than 200 mm and equal to or less than 250 mm.

Weather protection

To protect the top surface of a parapet wall, which is exposed directly to rain, it is essential that it should be covered or capped with some dense material to prevent rain saturating the wall. Natural stone was commonly used; termed coping stones, they usually project some 50 mm or more each side of the parapet wall so that rainwater running from them drips clear of the face of the wall. It is practice to cut semi-circular grooves in the underside of the over-hang edges of the stones so that water runs off the extreme drip edges of the stones. Feather edge copings are laid so that the weathered top surface slopes towards the roof to minimise staining fair face external brick faces. The stones are bedded in cement mortar on the parapet wall and butt end joints between stones are filled and pointed in cement mortar. For economy, cast stone copings are used instead of natural stone copings. Three common sections employed for coping stones are shown in Figure 6.89.

Coping stones are usually in lengths of 600 mm with the joints between them filled with cement mortar. In time the mortar between the joints may crack and rainwater may penetrate and saturate the parapet wall below. To prevent the possibility of rainwater saturating the parapet through the cracks in coping stones it is common practice to build in a continuous damp-proof course (dpc) of bituminous felt, copper or lead below the stones (Figure 6.89).

Another method of capping parapet walls is to form a brick on edge and tile creasing capping. This consists of a top course of bricks laid on edge, and two courses of clay creasing tiles laid breaking joint in cement mortar, as illustrated in Figure 6.90. The bricks of the capping are laid on edge, rather than on bed, because many facing bricks have sand faced stretcher and header faces. By laying the bricks on edge only the sanded faces show, whereas if the bricks were laid on bed, the bed face, which is not sanded, would show. Also a brick on edge capping looks better than one laid on bed. Creasing tiles are usually 265 mm

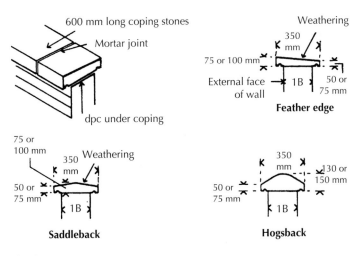

Figure 6.89 Coping stones.

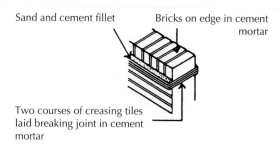

Sand and cement fillet   Bricks on edge in cement mortar

Two courses of creasing tiles laid breaking joint in cement mortar

Figure 6.90   Brick capping.

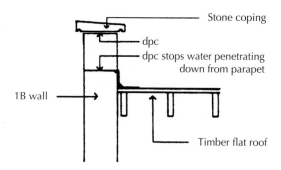

Stone coping

dpc

dpc stops water penetrating down from parapet

1B wall

Timber flat roof

Figure 6.91   Solid parapet.

long by 165 mm wide and 10 mm thick and are laid in two courses breaking joint in cement mortar.

The tiles overhang the wall by 25 mm to throw water away from the parapet below. A weathered fillet of cement and sand is formed on top of the projecting tile edges to assist in throwing water away from the wall. Two courses of good creasing tiles are generally sufficient to prevent water soaking down into the wall and no dpc is usually necessary under them. Parapet walls should be built with engineering bricks or sound facing bricks that are not liable to frost damage and laid in a cement mortar mix 1 part cement to 3 of sand.

Metal flashings can also be used, although these can look unsightly unless well detailed.

Parapet wall dpc

The purpose of the horizontal dpc is to prevent moisture in the exposed parapet wall penetrating to the roof (see Figure 6.91).

To provide protection to the roof structure it is usual to extend a cavity external wall up to the level of the top of the upstand of the roof covering to flat roofs as illustrated in Figure 6.92. In this construction, cavity insulation can be continued up to the level of insulation under roof coverings. A horizontal dpc is usually built in at the level that the solid brick parapet wall is raised on the cavity external wall, as illustrated in Figure 6.92.

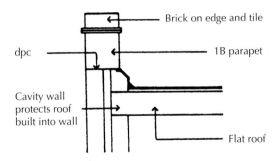

Figure 6.92   Cavity parapet wall showing dpc position.

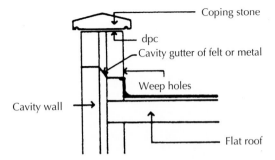

Figure 6.93   Cavity carried above roof.

A cavity external wall is sometimes continued up as a parapet wall, as illustrated in Figure 6.93. The only advantage of this arrangement is to continue the normal stretcher bond of external brickwork up to the parapet instead of having to change from stretcher to English or Flemish bond in a solid parapet, for the sake of appearance. As a precaution against water penetrating to the cavity in a parapet wall it is practice to form a cavity dpc and tray which is continued across the cavity and built in one course lower in the roof side, as illustrated in Figure 6.93. To collect and drain any water that may enter the cavity, weep holes are formed with raked out vertical joints in brickwork so that water runs down on to the roof.

6.11 Green roofs

The concept of the 'green' or 'living' roof is not new, but the technologies associated with the concept have evolved and matured (Photograph 6.15a–c). Largely associated with vernacular architecture until relatively recently, plant and soil layers (sod) have been used in cold climates to retain heat and in warm climates to keep buildings cool. The combined layers of soil and plants (usually grass or sedums) provide excellent insulation to roofs,

Photograph 6.15   (a) Green roof. (b) Green roof construction on timber glue lam roof. (c) Green roof with access stair for maintenance.

both pitched and flat. By combining new technologies (hi tech) with old principles (low tech) it is possible to create a highly durable roof structure with good environmental credentials. Green roofs provide a habitat for insects, birds and other small animals. This is particularly useful in brownfield and urban environments to help improve biodiversity. The terms turf roof and green roof are used, sometimes interchangeably. A turf roof usually refers to a simple roof, usually pitched, with grass and wild flowers growing on it. A green roof usually refers to a roof, pitched or flat, with sedums and other larger plants growing on it. There are two basic types of green roof systems, the extensive system and the intensive system.

Extensive green roofs

Extensive green roofs are characterised by low weight and capital cost. They have limited plant diversity and usually require minimal maintenance once established. The growing medium is usually from 50 to 150 mm deep, comprising a mixture of sand, gravel, peat, organic matter and some soil. Typically these roofs are pitched (up to 30° from the horizontal) and are covered with turf or sedums, which grow well on a thin medium. This cheap and natural looking construction relies on rainfall for irrigation and can become dry and unattractive in long, dry summers. Similarly, the grass and sedums can look unattractive in the winter. These roofs may be constructed on timber, steel or concrete roof structures. Textured waterproofing membranes help to stop the turf or plant layer from slipping down the slope for gentle pitches. On slopes greater than 20° to the horizontal the sod or plant layer must be strapped horizontally to the roof structure to prevent the material from slipping under its own weight when saturated. Some manufacturers provide support grid systems that have been specially designed for this purpose.

Intensive green roofs

Intensive green roofs tend to be characterised by deeper soil depth and greater weight. These roofs are built off a flat roof deck and are accessible to building users, thus increasing the amenity value of the building. Plant diversity tends to be greater than with extensive roofs, costs are higher and maintenance requirements more extensive. The growing medium tends to be from 200 to 600 mm deep, and is often soil (loam) based. The increased depth of the growing medium allows bushes and trees to be grown, creating a more extensive ecosystem. Regular watering is usually provided through automated irrigation systems. Some systems are designed to include storm water retention facilities. Typically, these roofs are constructed off reinforced concrete decks.

Typical construction

Green roofs are constructed in a series of layers (Figure 6.94 and Photograph 6.16), which typically include, from the top-down:

❏ Vegetation layer (e.g., grass, sedums and specially selected plants)
❏ Growing medium (usually 'engineered' lightweight material)
❏ Filter cloth (fleece) allows water in but helps to contain roots and the growing medium
❏ Drainage/water retention layer (may contain built-in reservoirs)
❏ Waterproof roofing membrane (with integral root repellent and/or metal foil between membrane layers and joints to prevent root damage)
❏ Rigid insulation
❏ Vapour control layer
❏ Structural deck
❏ Internal finish to underside of deck

The structural integrity of a green roof is dependent upon the interaction between the different layers and the quality of the waterproof membrane. Single-ply (heat-seamed reinforced plastic) roofing systems are regarded as the most efficient and cost-effective. The material also provides additional protection against root penetration, which tends to offset concerns about the material's environmental credentials. Other suitable materials include rubber membranes [ethylene propylene diene monomer (EPDM)] and thermoplastic polyolifins (TPOs). Bitumen-based roofing systems must be isolated from the roots because the organic product is a potential food source for organisms.

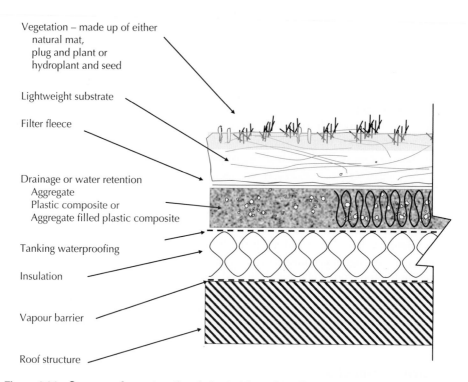

Vegetation – made up of either
  natural mat,
  plug and plant or
  hydroplant and seed

Lightweight substrate

Filter fleece

Drainage or water retention
  Aggregate
  Plastic composite or
  Aggregate filled plastic composite

Tanking waterproofing

Insulation

Vapour barrier

Roof structure

Figure 6.94   Green roof construction (adapted from http://www.greenroof.co.uk).

Photograph 6.16   Mock-up of green roof with single-ply membrane using Knauf extra strong insulation (http://www.knaufinsulation.co.uk).

Rainfall is largely maintained within the roof; however, proper drainage of the roof area is still important. Penetrations through the roof deck, for skylights, vents and chimneys must be well protected with flashings and a gravel skirt around the penetration. Similarly parapets need careful detailing.

Advantages

Green roofs provide a number of environmental benefits, which include the following:

❑ Increased energy efficiency. In the summer the vegetation layer shades the building from solar radiation, helping to keep the interior cool. In the winter the additional insulation provided by the growing medium helps to limit the amount of energy required to heat the building interior.

❑ Improved sound insulation. Primarily from the growing medium (low frequencies) but some reduction of higher frequency sound waves from vegetation.

❑ Longer life of the waterproof roof membrane. Green roof design helps to protect the waterproof roof membrane from temperature fluctuations, ultraviolet radiation and accidental damage from pedestrian traffic. Well-constructed, green roofs will last at least twice as long as a conventional roof, thus reducing the need for replacement and reducing the associated waste of materials. Manufacturers provide guidance on life expectancy and replacement cycle of the waterproof membrane.

❑ Storm water run-off. Green roofs will retain some storm water and help to reduce the amount of water run-off from roofs, thus helping to reduce the risk of flooding. Intensive systems can be constructed to include storm water reservoirs.

❑ Environmental benefits. The vegetation layer will filter particulates from the air, helping to improve air quality. The vegetation will also help to reduce the 'urban heat island effect', the effect of solar radiation on hard surfaces in urban areas.

❑ Habitat creation and opportunities for biodiversity. The loss of natural habitat through building activity can be addressed through the use of a green roof. The vegetation layer provides habitat for wildlife and helps to encourage biodiversity.

❑ Amenity space for building users. Intensive systems provide an environment for building users to interact with nature, usually in an urban environment such as a city centre.

❑ Aesthetic benefits. Well-designed and maintained green roofs are visually attractive.

Disadvantages

There are a few disadvantages associated with green roofs. They are:

❑ Fire resistance. While a saturated green roof may well help to limit the surface spread of flame, a dry roof can present a fire hazard, especially in a built-up area. On large roof structures a series of fire breaks 600 mm wide, at 40 m intervals, made from non-combustible material, for example, concrete pavers, should be used. The use of sedums can help (which have a high water content), although a sprinkler irrigation system, linked to a fire alarm, is a sensible precaution. Advice should be sought from the appropriate fire authority and building control office.

❑ Initial cost. Although extensive roof systems are a relatively cheap method of construction, the increased load of the roof requires a more substantial roof structure, for which there is an initial cost premium. Intensive roof systems are more expensive than a traditional roof construction due to their weight and extensive planting. However, life cycle costing shows that green roofs are as cost-effective as a traditional roof over the life of a building.

❑ Maintenance. More maintenance is required than that for a traditional construction.

Access to the roof construction under the vegetation for repair and replacement of membranes can be difficult to do without removing the upper layers of the roof and hence tends to be expensive. Membranes will need to be completely replaced after approximately 30–50 years, depending on the quality of the membrane and the design, construction and maintenance of the roof.

Structural considerations

The additional loading placed on the structure by the green roof construction is a primary consideration in assessing the viability of the roof. Wet soil weighs around $1600 \, kg/m^3$, which is quite a considerable loading to place on a structure. This has led to the development of many lightweight growing mediums, some of which weigh approximately $300 \, kg/m^2$ when saturated. For new build projects the loading can be considered at the design stage and the structure designed to accommodate any additional loading. When installing a green roof on an existing building, the structure and foundations must be checked to see what the maximum design load is and how this affects the design of the green roof. Some structural upgrading should be anticipated.

Different systems use slightly different methods, but it is not uncommon for the green roof system to be built up from the finished waterproofing membrane by a specialist subcontractor. There are a number of well-known proprietary systems on the market, so description here is limited to the basic principles. Manufacturers of proprietary systems provide an extensive technical design service and warranties for waterproof membranes and/or complete roofing systems.

7 Windows

A window is an opening formed in a wall or roof primarily to admit daylight through some transparent or translucent material. Windows also serve an important function in providing controlled natural ventilation to buildings and make a major contribution to the visual appearance of buildings. As the window is part of the wall or roof envelope, it should serve to exclude wind and rain, and act as a barrier to excessive transfer of heat, sound and spread of fire in much the same way that the surrounding wall or roof does.

7.1 Functional requirements

The primary function of a window is to:

❑   Admit daylight and provide a view

  Additional performance requirements include:

❑   Safety – comply with relevant health and safety legislation including the Construction Design and Management (CDM) Regulations
❑   Strength, stability and airtightness
❑   Fire safety
❑   Provision of ventilation
❑   Resistance to the passage of heat
❑   Resistance to the passage of sound
❑   Safety
❑   Security
❑   Aesthetics
❑   Durability and freedom from maintenance

The key elements of a window within a typical wall detail are shown in Figure 7.1.

Daylight

The prime function of a window is to admit adequate daylight for the efficient performance of daytime activities. The quantity of light admitted depends in general terms on the size

Barry's Introduction to Construction of Buildings, Third Edition. Stephen Emmitt and Christopher A. Gorse.
© 2014 John Wiley & Sons, Ltd. Published 2014 by John Wiley & Sons, Ltd.

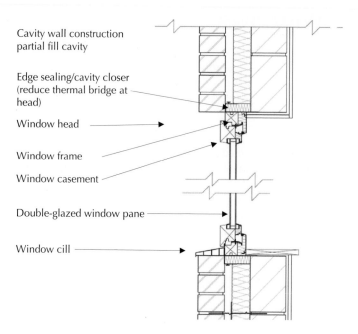

Cavity wall construction
partial fill cavity

Edge sealing/cavity closer
(reduce thermal bridge at
head)

Window head

Window frame

Window casement

Double-glazed window pane

Window cill

Figure 7.1   Typical double-glazing window detail (adapted from Gorse and Thomas, 2013).

of the window or windows in relation to the area of the room lit, and the depth inside the room to which useful light will penetrate depends on the height of the head of windows above floor level. The quantity of daylight in rooms is proportional to the area of glass in windows relative to floor area. The intensity of daylight at a given point diminishes progressively into the depth of the room; the deeper the room, the greater the requirement for supplementary artificial light.

Daylight factor

Daylight varies considerably in intensity, both hourly and daily, due to the rotation of the earth and the consequent relative position of the sun, and also due to climatic variations from clear to overcast skies. In Britain and north-west Europe it is common to calculate daylight in terms of a 'daylight factor', which is the ratio of internal illumination to the illumination occurring simultaneously out of doors from an unobstructed sky. In the calculation of the daylight factor it is assumed that the illumination from an unobstructed sky, in the latitude of Britain, is 5000 lux and that a daylight factor of 2% means that 2% of the 5000 lux outdoors is available as daylight illumination at a specified point inside.

The assumption of a standard overcast sky (poor outdoor illumination) is taken as a minimum standard on which to make assumptions. The term 'unobstructed sky' defines the illumination available from a hemisphere of sky free of obstructions such as other buildings, trees and variations in ground level, a condition that rarely occurs in practice.

Table 7.1   Recommended average daylight factors

Building type	Location	Daylight factor
Dwellings	Living rooms	1.5
	Bedrooms	1
	Kitchens	2
Work places	Offices	
	Libraries	
	Schools	5
	Hospitals	
	Factories	
All buildings	Residential	2
All buildings	Entrances	
	Public areas	2
	Stairs	

The International Commission on Illumination (CIE) defines daylight factor as 'the ratio of the daylight illumination at a given point on a given plane due to the light received directly or indirectly from a sky of assumed or known luminance distribution, to the illumination on a horizontal plane due to an unobstructed hemisphere of this sky'. Direct sunlight is excluded for both values of illumination. The intensity of illumination or luminance of the standard sky is assumed to be uniform to facilitate calculation of levels of daylight. In practice sky luminance varies, with luminance at the horizon being about one-third of that at the sun's zenith (the sun at its highest point). Average daylight factors for various activities are given in Table 7.1.

Where artificial illumination is used to supplement daylighting it is often practice to determine a working level of illumination in values of lux and convert this value to an equivalent daylight factor by dividing the lux value by 50 to give the daylight factor. For example, a lux value of 100 is equivalent to a daylight factor of 2.

In a room with a window on one long side, as illustrated in Figure 7.1, with no external obstructions and a room surface reflectance of 40%, where the glass area is one-fifth or 20% of floor area, the average daylight factor will be 4 and the minimum about half that figure. Conversely, to obtain an average daylight factor of, say 6, in a room with a floor area of $12\,m^2$, a glass area of about $6 \times 12 \times 5/100 = 3.6\,m^2$ will be required. This method of calculation is generally sufficient when the room is used for general activity purposes such as in living rooms, and it is an adequate base for preliminary assumptions of window to floor area, which can be adjusted later by a more accurate calculation of the light required for activities in which the lighting is critical.

Daylight penetration

A broad measure of the penetration of useful daylight into rooms is where the depth of penetration in line with the centre of the window is taken as being equal to the height of the window above floor level, as illustrated in Figure 7.2 and Figure 7.3. The quantity and quality of daylight illumination in side-lit rooms is affected to an extent by the light

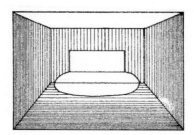

Figure 7.2   Long low window.

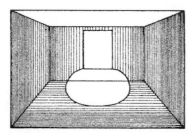

Figure 7.3   Tall narrow window.

reflected from floors, walls and ceilings, which will augment light coming directly through windows. The effect of this reflected light would be affected by the colour and texture of the reflective surfaces. Similarly some daylight, reflected from pavings and nearby external obstructions such as buildings and trees, will add to both the direct penetration of light and internally reflected light.

Reflected light

In the assumption of a daylight factor, account is taken of the contribution of what is termed 'the internally reflected component' and the 'externally reflected component' of indoor daylight illumination. The extent to which both the internal and external reflected light adds to the indoor lighting will be least with low levels of overall daylight and dark, rough textured reflective surfaces, and will be most with higher levels of overall daylight and light coloured, smooth textured reflective surfaces.

The shape, size and position of windows affect the distribution of daylight in rooms and the view out. Tall windows give a better penetration of light than low windows, as illustrated in Figure 7.4. The tall, narrow windows illustrated in Figure 7.4 provide good penetration of daylight into rooms that may be enhanced by the reflection from white painted, internal reveals to the windows. Some distribution of daylight between the windows is provided by the overlap of penetration between the two windows.

Windows in adjacent walls give good penetration and reduce glare by lighting the area of wall surrounding the adjacent window, as illustrated in Figure 7.5. Windows in opposite walls of narrow rooms give good penetration and reduce glare by lighting opposite walls

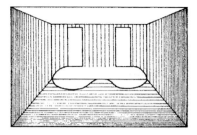

Figure 7.4   Tall narrow windows.

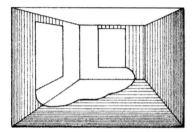

Figure 7.5   Windows in adjacent walls.

around windows. In the calculation of daylight factors it is usual to determine the quantity of daylight falling on a horizontal working plane 850 mm above floor level to correspond with the height of working surfaces such as tables, desks or benches.

Area of glass
The averaged or average daylight factor represents the overall visual impression of the daylighting in a room or space taking into account the distribution of light in the space and the effect of reflected light. The penetration and distribution of daylight in rooms will increase by internal reflection of light from ceilings, walls and floors. For example, in a room 3 m square with a 3 m ceiling height, if the reflectance from light coloured smooth surfaces is good, the net area of glass required to provide a daylight factor of 2, will be 1.28 m as compared with 1.60 m where reflectance is low from dark rough surfaces.

Calculation of daylight factor
Where daylighting by itself or in combination with artificial lighting is critical for the performance of activities, such as drawing at a fixed point or points in a room, it is necessary to estimate the minimum daylighting available at a point. For this purpose there are a number of aids, such as the artificial sky and the overlays for scale drawings. These are described later, although it should be noted that computer software is available to help calculate the daylight factor.

An artificial sky provides luminance comparable to the standard overcast sky, through an artificially lit dome which is laid over a scale model of the building in which photometers are used to measure the light available. The graphical aids in the form of overlays include Waldram diagrams, BRS protractors and the dot or pepper pot diagrams of which the dot diagram is the most straightforward to use. The dots represent a small proportion of the daylight illumination available at that point. The greater the density of dots, the greater the illumination.

The dot or pepper pot diagram

The pepper pot diagram is a transparent overlay on which dots are printed above a horizontal line representing the horizon. The diagram is drawn to a scale of 1 : 100 as an overlay to drawings to the same scale. Each dot represents 0.1% of the sky component. The overlay shown in Figure 7.6 is for daylight through side-lit windows with the CIE standard overcast sky. To use the overlay, draw the outline of a window to a scale of 1 : 100 so that the outline represents the glass area to scale. The diagram is designed to determine the sky component of daylight on a line 3 m back from the window. Place the overlay on the scale elevation of the window with the horizontal line of the overlay on the line of the working plane (850 mm above the floor), drawn to scale on the window elevation. To determine the sky component at a point 3 m back from the centre of the window, place the vertical line of the overlay on

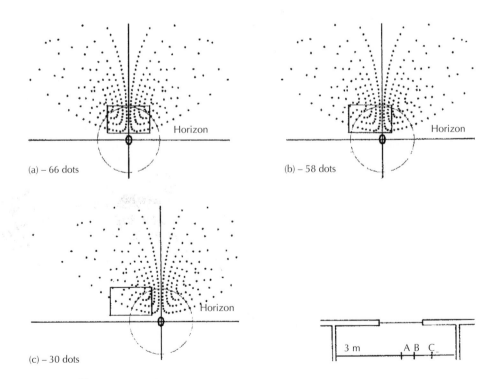

Figure 7.6   Sky component.

the centre of the window as illustrated in Figure 7.6a, then count the dots inside the window outline. The 66 dots inside the window outline represent a sky component of 66/10, that is, 6.6% at a point 3 m back from the centre of the window on the working plane.

To find the sky component on the line 3 m back from the window at other points, slide the overlay horizontally across the window outline until the vertical line of the overlay coincides with the chosen point inside the room, either inside or outside the window outline, as illustrated in Figure 7.6b and c. Count the dots inside the window outline to determine the sky component at the chosen points.

Where there are obstructions outside windows, such as adjacent buildings, which obscure some of the daylight, the overlay can be used to determine both the loss of light due to the obstruction and the externally reflected component of light due to reflection of light off the obstruction and into the room through the window. A simple example of this is where a long low building will obstruct daylight at a point 3 m inside the room on the centre of the window at the working plane. The outline of the long obstruction is shown in Figure 7.7 by the shaded area. The height of the obstruction above the horizon is represented by the height to distance ratio of the obstruction relative to the point on the working plane inside the window. This ratio is 0.1 for each 3 mm above the horizon on the scale drawing of the window. The number of dots inside the window outline above the shaded obstruction gives the sky component as 40% and the number of dots 12 inside the shaded area, the externally reflected component. These dots represent 0.01% of the externally reflected component.

To find the sky component at points on a line other than the line 3 m back from the window drawn to a scale of 1 : 100, it is necessary to adjust the scale of the window outline. If the scale of the window is doubled, it will represent the sky component at points 1.5 m back from the window, and if the scale is halved, 6 m back from the window, as illustrated in Figure 7.8 for points on the centre of the window. In adjusting the scale of the window outline it is also necessary to adjust the scale height of the working plane above the floor by doubling or halving the scale as shown in Figure 7.8.

The particular use of this diagram is to test the sky component of daylight inside rooms at an early stage in the design of buildings. By the use of window outlines drawn freehand

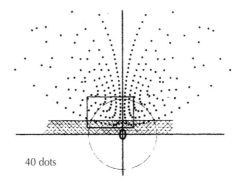

40 dots

Figure 7.7 Sky component.

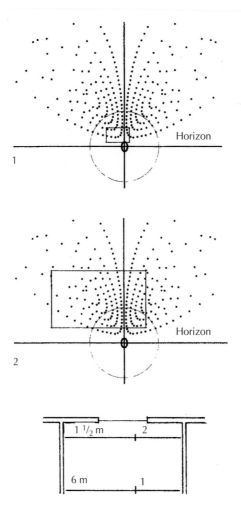

Figure 7.8   Sky component.

to scale on graph paper, with the overlay, a comparative assessment of the effect of window size and position on the sky component of daylight inside rooms can quickly be made. This will provide a reasonably accurate assessment of comparative daylight levels in rooms to be used for many activities where the exact level of daylight is not critical.

Quality of daylight

Consideration must also be given to the quality of the light to avoid glare. Glare is defined as 'a condition of vision in which there is discomfort or a reduction in the ability to see significant objects or both, due to an unsuitable distribution or range of luminance, or to extreme contrasts in space or time'. The two distinct aspects of glare are defined as disability glare and discomfort glare.

Disability glare

Disability glare is defined as 'glare which impairs the vision of objects without necessarily causing discomfort', and is caused when a view of bright sky obscures objects close to the source of glare. An example of this is where a lecturer is standing with his or her back to a window so that he or she is obscured by the bright sky behind. Disability glare can be avoided by a sensible arrangement of the position of windows and people, whose vision of objects might otherwise be obscured.

Discomfort glare

Discomfort glare, defined as 'glare which causes discomfort without necessarily impairing the vision of objects', is created by large areas of very bright sky viewed from inside a building which causes distraction, dazzle and even pain. With vertical windows discomfort glare is caused, in the main, by the contrast between visible sky and the room lighting. This can be reduced by splaying window reveals and painting them a light colour to provide a graded contrast between the bright sky and the darker interior. This 'contrast grading' effect can be used with many window shapes and sizes. With very large windows such as the continuous horizontal strip windows which face southwards, discomfort glare is difficult to avoid owing to the large unbroken area of glazing. Here some form of shading device will be required. The degree of glare can be determined numerically and stated in the form of a 'glare index'.

View out and privacy

As well as admitting daylight windows perform the useful function of providing a view out and hence a visual link with the outside. Windows also allow a view in, and this may not always be desirable. Depending on the siting of the building and its use, the balance between views out and maintaining privacy will need to be met. Different types of glazing and/or shutters can be used to allow a view out but limit the view in, as can the careful positioning of windows in the building design.

Sunlight

Sunlight causes most coloured materials to fade. It is the ultraviolet radiation in sunlight that has the most pronounced effect on coloured materials by causing the chemical breakdown of the colour in such materials as textiles, paints and plastics by oxidative bleaching. The bleaching effect is more rapid and more noticeable with bright colours. The lining of colour-sensitive curtains on the window side with a neutral coloured material and the use of window blinds are necessary precautions to prolong the life of colour-sensitive materials.

Solar heat gain

The term 'radiation' describes the transfer of heat from one body through space to another. When radiant energy from the sun passing through a window reaches, for example, a floor, part of the radiant energy is reflected and part is absorbed and converted to heat. The radiant energy reflected from the floor will in part be absorbed by a wall and converted into heat and partly reflected. The heat absorbed by the floor and wall will in turn radiate energy that will be absorbed and converted to heat. In the calculation of energy use in maintaining equable indoor temperature and necessary insulation to limit heat loss, described later, allowance is made for solar heat gain. A calculation is made of the probable solar heat gain as part of the necessary energy input to maintain indoor temperature.

The degree of solar heat gain is affected by the size and orientation of windows. Large windows facing south in the northern hemisphere will be more affected than those facing east or west. The time of year will also have some effect between the more intense summer radiation, which will not penetrate deeply into rooms at midday to the less intense but more deeply penetrating radiation of spring and autumn. Discomfort from solar heat gain has mainly been a consequence of the fashion to use large areas of glass as a sealed walling material for offices and other non-domestic buildings, where the build-up of heat can make working conditions uncomfortable. The transmission of solar radiation can be effectively reduced by the use of body tinted, surface modified or surface coated glass.

There are geometric sunpath diagrams that may be used to check whether the face of a building will receive sunlight and when, the depth of penetration and the resultant patch of sunlight on room surfaces and the shading by obstructions at various times of the day throughout the year. An example of the use of 'gnomonic' projections to deduce sunlight patterns on room surfaces throughout the day is shown in Figure 7.9. The diagram shows the floor of a single room with a southeast facing window and the walls on which the sun will shine at half-hourly intervals on 15 January. These sunpath diagrams may also, with suitable overlays, be used to predict the intensities of direct and diffuse solar radiation and the consequent solar heat gain. Computer programs are available that will predict energy consumption for heat loss and heat input calculations and will make allowances for the variable of solar heat gain through windows so that modifications in both window sizes and the heat input from heating plant can be adjusted at the design stage.

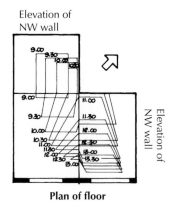

Plan of floor

Figure 7.9   Gnomonic projection.

Sun controls and shading devices

The traditional means of controlling the penetration of sunlight to rooms are slatted wooden louvre shutters common to the French window, and awnings and blinds that can be opened or closed to vary the amount of shade provided. These controls are adjustable manually or electronically. These shading controls are adjustable between winter and summer conditions, graduated from no shade and the maximum penetration of daylight in winter through some shade and some daylight to full shading in high summer. Fixed projections above windows, such as canopies and balconies, are also used to provide shade from summer sun, while allowing penetration of sun at other times of the year. Such fixtures help to control sunlight, glare and solar heat gain.

Strength, stability and airtightness

Windows should be securely fixed in the wall opening for security and weathertightness. A window should be strong enough when closed to resist the likely pressures and suctions due to wind, and when open be strong and stiff enough to resist the effect of gale force winds on opening lights. A window should also have sufficient strength and stiffness against pressures and knocks due to normal use and appear to be safe, particularly to occupants in high buildings.

Wind loading

To determine probable wind loads on buildings, the method given in BS 6262 can be used for buildings that are of simple rectangular shape and up to 10 m high from eaves to ground level. The basic wind speed is determined from the map of the UK (Figure 7.10). The basic wind speed is then multiplied by a correction factor that takes account of the shelter

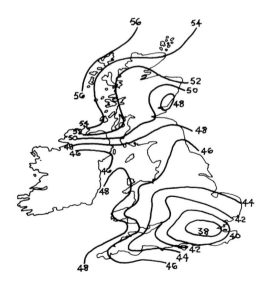

Figure 7.10   Basic wind speeds.

Table 7.2 Correction factors for ground roughness and height above ground

Height above ground (m)	Category 1	Category 2	Category 3	Category 4
3 or less	0.83	0.72	0.64	0.56
5	0.88	0.79	0.70	0.60
10	1.00	0.93	0.78	0.67

Table 7.3 Probable maximum wind loading

Design wind speed (m/s)	Wind loading (N/m²)	Design wind speed (m/s)	Wind loading (N/m²)
28	670	42	1510
30	770	44	1660
32	880	46	1820
34	990	48	1980
36	1110	50	2150
38	1240	52	2320
40	1370		

afforded by obstructions and ground roughness as set out in Table 7.2 to arrive at a design wind speed. The left-hand column in Table 7.2, 'Height above ground', relates to the height of the window above ground as plainly the higher above ground, the less will ground roughness and obstructions provide shelter. The four categories of protection by obstructions and ground roughness run from 1 with effectively no protection in open country to 4 with maximum protection from surrounding buildings in city centres. A degree of judgement is necessary in selecting the correction category suited to the site of a particular building as the purpose is to select a window construction suited to the most adverse conditions that will occur on average once in 50 years. The probable maximum wind loading is then obtained from Table 7.3 by reference to the design wind speed. The wind loading is used to select the test pressure class of window construction necessary and graphs are used to select the required thickness of glass.

Windows are tested in a laboratory to determine test pressure classes; a sample of manufactured windows complete with opening lights and glass is mounted in a frame to represent the surrounding walls. The criterion of success in the pressure test is that, after the test, the window should show no permanent deformation or other damage and there should be no failure of fastenings.

Air permeability (airtightness)

Air leakage (infiltration) can occur around openings such as windows and roof lights, accounting for up to around 15% of a building's total air leakage. Air leakage around window frames and through glazing joints can be avoided by careful detailing, good construction (workmanship and site supervision) and regular maintenance. The necessary clearance gaps around opening lights can be made reasonably airtight by care in design and the use of weather-stripping, and if necessary injecting expanding polyurethane foam into the joint around the window frame.

Close attention should be paid to the solid filling or sealing of all potential construction gaps and cracks as well as controlling leakage between the window frame and the wall (Photograph 7.1). The flow of air through windows is caused by changes in pressure and suction. To control this air movement systems of check rebates and weather-stripping are used in windows, as illustrated in Figure 7.11. The performance of windows with regard to airtightness is based on predicted internal and external pressure coefficients, which depend on the height and plan of the building. These are related to the design wind pressure, which is determined from the exposure of the window and basic wind speed from the map in Figure 7.10. From these, test pressure classes are established for use in the tests for air permeability and watertightness to set performance grades.

Photograph 7.2 shows the effect of air entering at the top of a sash window due to poor sealing around the casement. Photograph 7.3 shows a sash window that has been renovated with double-glazed units; one unit remains single glazed and the effect of the thermal bridging can clearly be seen.

Watertightness

Penetration of rain through cracks around opening lights, frames or glass occurs when rain is driven on to vertical windows by wind, so that the more the window is exposed to driving rain, the greater the likelihood of rain penetration. Because of the smooth, impermeable surface of glass, rain will be driven down, across and up the surface of glass, thus making seals around glass and clearance gaps around opening lights vulnerable to rain penetration.

Tests for watertightness of windows are based on predictions similar to those used for air infiltration in determining design wind speed, exposure grades and test pressure classes to set performance standards. To minimise the penetration of driven rain through windows, it is advantageous to:

❑ Set the face of the window back from the wall face so that the projecting head and jamb will to some extent give protection by dispersing rain.
❑ Ensure that external horizontal surfaces below openings are as few and as narrow as practicable to avoid water being driven into the gaps.
❑ Ensure that there are no open gaps around opening lights by the use of lapped and rebated joints, and where there are narrow joints that may act as capillary paths it will be necessary to add capillary grooves.
❑ Restrict air penetration by means of weather-stripping on the room side of the window. If weather-stripping is not used pressure differences may drive water into the joint.
❑ Ensure that any water entering the joints is drained to the outside of the window by open drainage channels that run to the outside.

In modern window design weather-stripping is used on the room side of the gaps around opening lights to exclude wind and reduce air filtration, and rebates and drain channels are used on the outside to exclude rain as illustrated in Figure 7.12.

Any gaps around the perimeter of the window are filled with expanding polyurethane foam. Once set, the foam will be trimmed away and a silicon sealant will provide a neat watertight finish

Polyurethane foam fills gap between window frame and wall

dpc tray over stone work; once the brickwork is finished the excess dpc will be trimmed off

Horizontal dpc tray under window

Trickle ventilation gap

Polyurethane foam fills the gap all the way through the window frame

Photograph 7.1   Sealing the gap between window frame and wall.

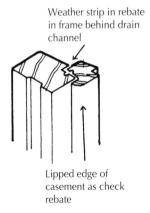

Weather strip in rebate in frame behind drain channel

Lipped edge of casement as check rebate

Figure 7.11   Weather strip and check rebates.

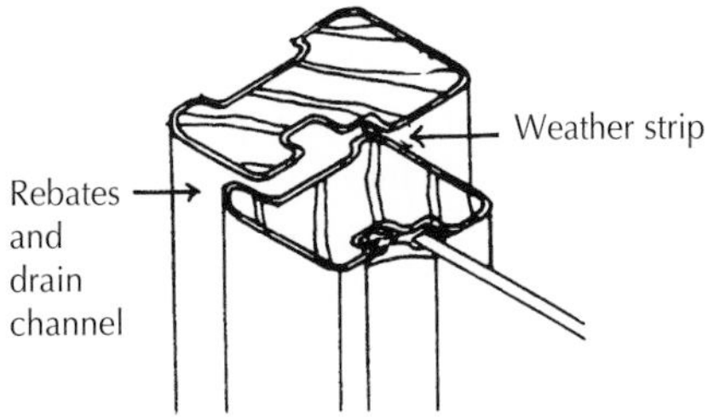

Weather strip

Rebates and drain channel

Figure 7.12   Drainage channel.

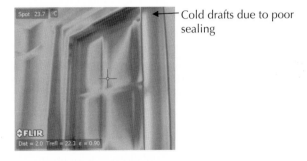

Cold drafts due to poor sealing

Photograph 7.2   Thermal image: sash window poor sealing allows air in at the top of the frame.

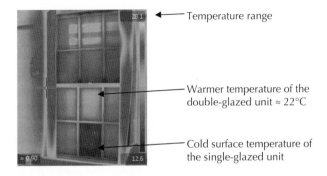

Temperature range

Warmer temperature of the double-glazed unit ≈ 22°C

Cold surface temperature of the single-glazed unit

Photograph 7.3   Thermal image: sash window fitted with double- and single-glazed units.

Fire safety

The requirement in the Regulations that concerns windows is external fire spread. To limit the spread of fire between buildings, limits to the area of 'unprotected areas' in walls and finishes to roofs, close to boundaries, are imposed by the Building Regulations. The term 'unprotected area' is used to include those parts of external walls that may contribute to the spread of fire between buildings. Windows are unprotected areas, as glass offers negligible resistance to the spread of fire. In Approved Document B rules are set out that give practical guidance to meeting the requirements of the Building Regulations in regard to minimum distances of walls from boundaries and maximum unprotected areas.

Ventilation

For comfort in living and working conditions in buildings some regular change of air is necessary. The necessary ventilation should be provided through controlled ventilators, through opening lights or by mechanical ventilation. It is not satisfactory to rely on leakage of air through windows for ventilation as this leakage cannot be controlled, and it may be excessive for ventilation and conservation of heat or too little for ventilation.

Windows are usually designed to provide ventilation to rooms through one or more parts that open, encouraging an exchange of air between inside and outside. Ventilation can also be provided through openings in walls and roofs that are either separate from windows or linked to them to perform the separate function of ventilation. The advantage of separating the functions of daylighting and ventilation is that windows may be made more effectively wind- and weathertight and ventilation can be more accurately controlled. This may be an important consideration with regard to safety and security.

The move towards conservation of energy has led to the installation of double and triple glazing to windows in both new and existing buildings and the fitting of effective weatherstripping around the opening parts of windows and doors to reduce draughts of cold air entering the building. The Building Regulations require means of ventilation to habitable rooms, kitchens, bathrooms and sanitary accommodation to provide air change by natural or mechanical ventilation and also to reduce condensation in rooms.

Air changes

The number of air changes will depend on the activities and number of people in the room. The rate of change of air may be given as air changes per hour, for example, one per hour for living and up to four for work places, or as litres per second as a more exact requirement where mechanical ventilation is used, since it gives a clear indication of the size of inlets, extracts, ducts and pressures required.

The rate of exchange of air will depend on variations between inside and outside pressure and heat, and the size and position of other openings in the room such as doors and open fireplaces that may play a part in air exchange. An open window, by itself, may not thoroughly ventilate a room. For thorough ventilation (complete air change) circulation of air is necessary between the window and one or more openings distant from the window. The probable ventilating action of the various types of window in comparatively still air conditions due to the exchange of warmed inside and cooler outside air is illustrated in Figure 7.13.

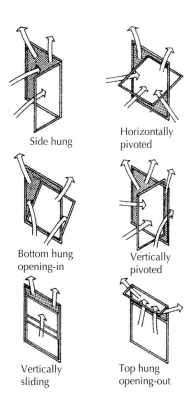

Side hung

Horizontally pivoted

Bottom hung opening-in

Vertically pivoted

Vertically sliding

Top hung opening-out

Figure 7.13   Ventilation.

Approved Document F gives practical guidance to meeting the requirements of the Building Regulations for the provision of means of ventilation for dwellings. The requirements are satisfied for habitable rooms, such as living rooms and bedrooms when there are:

❏ For rapid ventilation. One or more ventilation openings, such as windows, with a total area of at least $^1/_{20}$th of the floor area of the room, with some part of the ventilating opening at least 1.75 m above the floor.

❏ For background ventilation. A ventilation opening or openings having a total area of not less than 4000 mm^2, which is controllable, secure and located so as to avoid undue draughts, such as the trickle ventilator, illustrated in Figure 7.14. Trickle ventilation above windows helps to allow air movement across the face of the glass, thus helping to prevent condensation and allowing water vapour to pass through the ventilator to the outside air.

Trickle ventilators can be included in windows, usually as part of the window head or cill construction, or they may be fixed separate from the windows. For ventilation alone these ventilators need only be small apertures that can be opened and closed by means of simple manual or automatic controls.

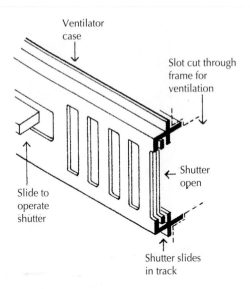

Figure 7.14 Trickle ventilator.

For kitchens the requirements are satisfied when there is both:

❑ Mechanical extract ventilation for rapid ventilation, rated as capable of extracting at a rate of not less than 60 L/s (or incorporated within a cooker hood and capable of extracting at 30 L/s), which may be operated intermittently, for instance, during cooking, and

❑ Background ventilation, either by a controllable and secure ventilation opening or openings having a total area of not less than 4000 mm², located so as to avoid draughts, such as a trickle ventilator or by the mechanical ventilation being in addition capable of operating continuously at nominally one air change per hour.

For bathrooms the requirements are satisfied by the provision of mechanical extract ventilation capable of extracting at a rate of not less than 15 L/s, which may be operated intermittently.

For sanitary accommodation one of the following satisfies the requirements:

❑ Provision for rapid ventilation by one or more ventilation openings with a total area of at least $1/_{20}$th of the floor area of the room and with some part of the ventilation opening at least 1.75 m above the floor level, or

❑ Mechanical extract ventilation, capable of extracting air at a rate of not less than three air changes per hour, which may be operated intermittently with 15 minutes overrun.

Resistance to the passage of heat

A window will affect thermal comfort in two ways: first by transmission (passage) of heat and secondly through the penetration of radiant heat from the sun, which causes solar heat gain. Glass, which forms the major part of a window, offers poor resistance to the passage of heat and readily allows penetration of solar radiation. Transfer of heat also occurs through the window frame, and this varies depending on the size of the frame and the materials used in its manufacture. The transfer of heat through a window is through conduction, convection and radiation. Conduction is the direct transmission of heat through a material, convection the transmission of heat in gases by circulation of the gases and radiation the transfer of heat from one body of radiant energy through space to another.

Window U-values

The U-value of a window depends on the type of glazing and the materials used for the window framing. Glass has low insulation and high transmittance values. For example, the U-value of a single-glazed window of 6 mm thick glass is 5.4 W/m² K and that of a double-glazed unit with two 6 mm thick sheets of glass spaced 12 mm apart is 2.8 W/m² K. The overall U-value of a window varies to some extent on the materials used in window framing, as wood and unplasticated polyvinyl chloride (uPVC) frames tend to provide better insulation against heat transfer than metal. However, advances in window frame technologies now rely on combination of materials to improve thermal performance.

Single glazing does not provide a sufficiently low U-value to satisfy the requirements of Approved Document Part L; thus double or triple glazing is necessary. The notional target regulations will be met where the U-values do not exceed 1.4 W/m² K, with a g-value = 0.63. The maximum allowable U-value for windows is 2.00, but this can only be used where other fabric improvements are made. A number of low-energy houses have been constructed with glazing systems that have a U-value of 1.20 W/m² K; also prototype windows have been constructed with U-values as low as 0.08 W/m² K.

Modification of the basic allowance

The notional targets are set as an acceptable guide (windows having a U-value of 1.4 W/m² K), however with a triple glazed window system with a U-value 0.09 W/m² K (g = 0.57) the amount of window area can be increased or the fabric performance of other elements reduced. The adjustments allow design flexibility when calculating using SAP (standard assessment procedure). The highest U-value (lowest thermal performance) that is allowed in any circumstance is 2.00 W/m² K.

Thermal bridging

Thermal bridges lower the overall thermal insulation of the structure and create cold spots where condensation may form. At window openings in walls thermal bridging may occur through:

❑   Window frames – it is likely to be worse in metal windows than in uPVC or wooden frames, although wooden frames will still provide a thermal bridge

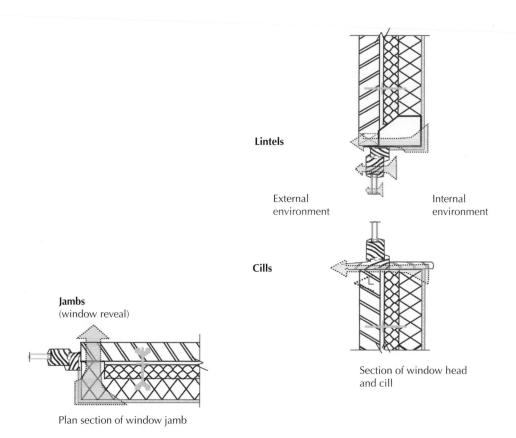

Lintels

External
environment

Internal
environment

Cills

Jambs
(window reveal)

Section of window head
and cill

Plan section of window jamb

Figure 7.15   Risk of thermal bridging around window and door openings.

❏   Window and door jambs, lintels and cills (Figure 7.15, Figure 7.16 and Figure 7.17)
❏   The edge seal of double- or triple-glazing units

Figure 7.18 shows the window fitted in line with the insulation within the cavity, which helps to maintain the continuation of the thermal barrier. At the edge of the window opening an insulated cavity barrier is used rather than returning the blockwork, reducing thermal bridging at the window/wall interface (Figure 7.19).

Positioning of windows – exposure
Depending on level of exposure, the window frame can be installed in different positions to provide additional protection. Exposure to wind and rain can be reduced by recessing the window from the external face of the wall, as illustrated in Figure 7.20. Alternatively, careful detailing can provide additional interest to the window surround while also helping to protect the window frame.

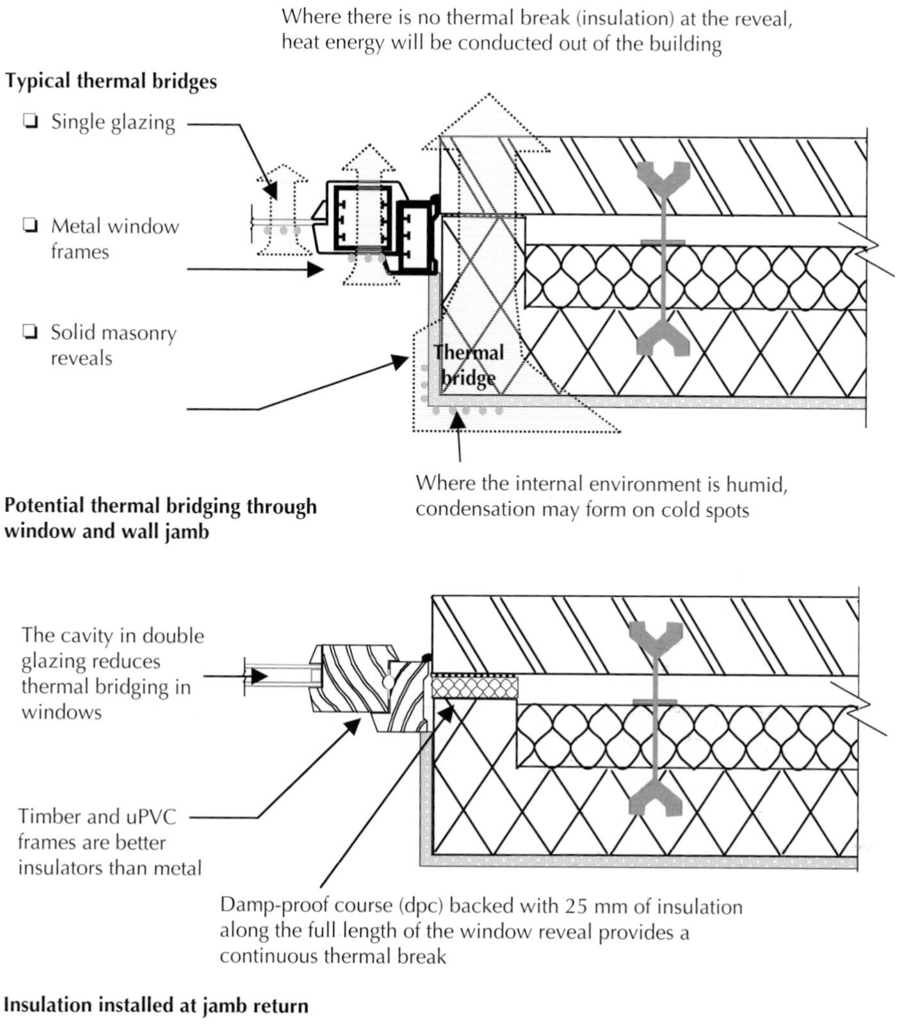

Where there is no thermal break (insulation) at the reveal,
heat energy will be conducted out of the building

Typical thermal bridges

❏ Single glazing

❏ Metal window
   frames

❏ Solid masonry
   reveals

Thermal
bridge

**Potential thermal bridging through
window and wall jamb**

Where the internal environment is humid,
condensation may form on cold spots

The cavity in double
glazing reduces
thermal bridging in
windows

Timber and uPVC
frames are better
insulators than metal

Damp-proof course (dpc) backed with 25 mm of insulation
along the full length of the window reveal provides a
continuous thermal break

Insulation installed at jamb return

Figure 7.16   Prevention of thermal bridging: window jamb detail.

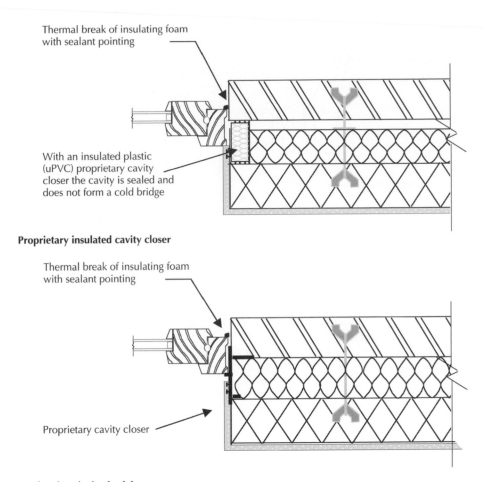

Thermal break of insulating foam with sealant pointing

With an insulated plastic (uPVC) proprietary cavity closer the cavity is sealed and does not form a cold bridge

Proprietary insulated cavity closer

Thermal break of insulating foam with sealant pointing

Proprietary cavity closer

Insulated to the back of the window or door frame

Figure 7.17   Prevention of thermal bridging: window jamb detail.

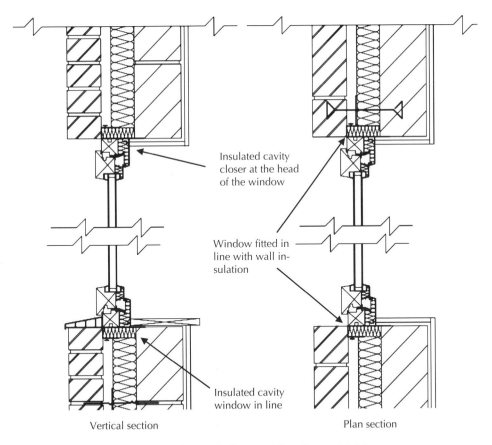

Figure 7.18 Cavity closer and window in line avoiding thermal bridge.

Insulated cavity closer at the head of the window

Window fitted in line with wall in-sulation

Insulated cavity window in line

Vertical section

Plan section

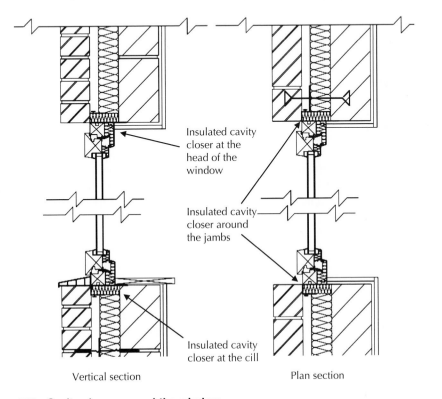

Insulated cavity closer at the head of the window

Insulated cavity closer around the jambs

Insulated cavity closer at the cill

Vertical section

Plan section

Figure 7.19 Cavity closer around the window.

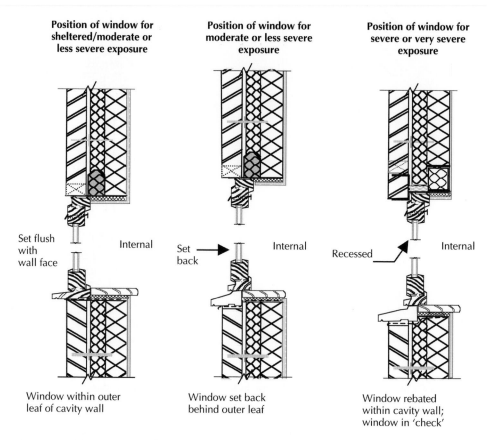

Figure 7.20 Exposure to external environment and location of window in the external wall (adapted from BRECSU, 1995).

Resistance to the passage of sound

Sound is the sensation produced through the ear by vibrations caused by air pressure changes superimposed on the comparatively steady atmospheric pressure. The rate or frequency of the air pressure changes determines the pitch as high-pitch to low-pitch sounds. The audible frequencies of sound are from about 20 Hz (hertz) to 15,000 or 20,000 Hz; 1 Hz is numerically equal to one cycle per second. The sound pressure required for audibility is generally greater at very low frequencies than at high frequencies.

Because of the variation in the measured sound pressure and that perceived by the ear over the range of audible frequencies, a simple linear scale will not suffice for the measurement of sound. The measurement that is used is based on a logarithmic scale that is adjusted to correspond to the ear's response to sound pressure.

The unit of measurement used for ascribing values to sound levels is the decibel (dB). Table 7.4 gives sound pressure levels in decibels for some typical sounds. Because the sensation of sounds at different frequencies, although having the same pressure or energy,

Table 7.4   Sound pressure levels for some typical sounds

Sound	Sound pressure level (dB)
Threshold of hearing	0
Leaves rustling in the wind	10
Whisper or ticking of a watch	30
Inside average house, quiet street	50
A large shop or busy street	70
An underground train	90
A pop group at 1.25 m	110
Threshold of pain	120
A jet engine at 30 m	130

Table 7.5   Tolerance noise levels

Location	dB(A)
Large rooms for speech such as lecture theatre, conference rooms, etc.	30
Bedrooms in urban areas	35
Living rooms in country areas	40
Living rooms in suburban areas	45
Living rooms in busy urban areas	50
School classrooms	45
Private offices	45–50
General offices	55–60

generally appears to have different loudness, a sound of 100 dB is not twice as loud as one of 50 dB; it is very much louder. The scale of measurement used to correlate to the subjective judgement of loudness, which is particularly suitable for traffic noise, is the A weighting with levels of sound stated in dB(A) units. To provide a measure of generally accepted tolerable levels of audible sound that will not distract attention or be grossly intrusive, tolerance noise levels are set out in Table 7.5.

Airborne sound
Sound is produced when a body vibrates, causes pressure changes in the air around it and these pressure changes are translated through the ear into the sensation of sound. Sound is transmitted to the ear directly by vibrations in air pressure – airborne sound, or partly by vibrations through a solid body that in turn causes vibrations of air that are heard as sound – impact sound. The distinction between airborne and impact sound is made in order to differentiate the paths along which sound travels, so that construction may be designed to interrupt the sound path and so reduce sound levels. Airborne sound is, for example, noise transmitted by air from traffic through an open window into a room and impact sound from a door slamming shut that causes vibrations in a rigid structure that may be heard some distance from the source. The sensation of sound is affected by the general background level of noise.

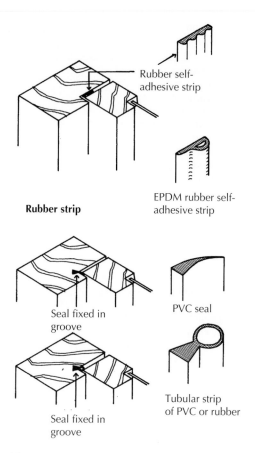

Rubber self-adhesive strip

Rubber strip

EPDM rubber self-adhesive strip

Seal fixed in groove

PVC seal

Seal fixed in groove

Tubular strip of PVC or rubber

Figure 7.21   Weather strips.

For the majority of people who live and work in built-up areas, the principal sources of noise are external, from road traffic and airborne sound, and internal, radios, televisions, the impact of doors and footsteps on hard surfaces. Windows (and doors) are a prime source for the entry of airborne sound both through glass, which affords little insulation against sound, and by clearance gaps around opening parts of windows and doors. Reduction of airborne sound can be achieved by weather-stripping around the opening parts of windows and doors (Figure 7.21).

The transmission of sound through materials depends mainly on their mass; the denser and heavier the material, the more effective it is in reducing sound. The thin material of a single sheet of glass provides poor insulation against airborne sound. A small increase in insulation or sound reduction of glass can be achieved by the use of thicker glass, where an average reduction of 5 dB is obtained by doubling the thickness of the glass. There is no appreciable sound reduction by using the sealed double-glazed units that are effective in heat insulation, as the small cavity is of no advantage, so that sealed double glazing is no more effective than the combined thickness of the two sheets of glass. For appreciable reduction in sound transmission, double windows are used where two separate sheets of glass are spaced from 100 to 300 mm apart. An average reduction of 39 dB with 100 mm

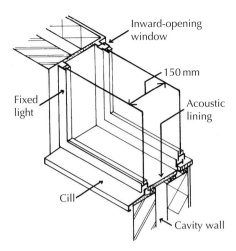

Figure 7.22   Double window for sound insulation.

space and 43 dB with 200 mm space can be obtained with 4 mm glass. This width of air space is more than the usual window section can accommodate and it is necessary to use some form of double window. The double window illustrated in Figure 7.22 comprises two windows, a fixed outer and an inward-opening inner window with the glass spaced 50 mm apart. Acoustic lining to the cill, jambs and head between the windows absorbs sound. The hinged inner sash facilitates cleaning glass.

Safety requirements

Two safety requirements from Approved Documents K concern the opening parts of windows in buildings other than dwellings. The requirement is that measures be taken to prevent people, moving in or about the building, from colliding with open windows. This requirement is met where the projection of a window, either internally or externally, is not more than 100 mm horizontally and the lowest part of the projection is more than 2 m above the floor or ground. The other requirements are that windows, skylights and ventilators can be opened, closed or adjusted safely and that there is safe access for cleaning windows. The requirement for access for operating applies to controls that are more than 1.9 m above floor. The requirement for access of cleaning windows, inside and out, where there is a danger of falling more than 2 m, will be met if provision is made for safe means of access. Safety issues relating to glass are discussed later.

Security

Windows and doors are the principal route for illegal entry to buildings. Of the recorded cases of illegal entry, burglary, about 30% involves entry through unlocked doors and windows. Of the remaining 70%, some 20% involves breaking glass to gain entry by opening catches, and the remaining 80% by forcing frames or locks. Window manufacturers and ironmongery suppliers have responded to the increased need for security with product improvements that make unauthorised entry more difficult. Advice is also available from the police through their Secure by Design initiative.

Aesthetics

Windows, along with doors, are an essential element in the appearance of buildings, and therefore their size and shape will also be determined by the designer's aesthetic requirements, which have to be balanced against performance requirements such as thermal insulation, security and maintenance.

Materials and durability

The main materials used for the construction of window frames are timber, metal and plastics. These materials are sometimes used in isolation, e.g. timber window frames and cills, and sometimes in conjunction with other materials, e.g. polyvinyl chloride (PVC) windows are usually manufactured with a metal support. The durability of these materials will be determined by the quality of the material, the quality of the applied finish (e.g. paint finish to timber), the detailing of the window and window reveal to protect it from exposure to weather and the amount of maintenance required and manner in which maintenance is implemented. The translucent material most commonly fixed within the frame is glass (see Section 7.4).

7.2 Window types

There are two terms used to describe windows, fixed lights and opening lights. The term fixed light or dead light is used to describe the whole or part of a window in which glass is fixed so that no part of the glazing can be opened.

An opening light is the whole or part of a window that can be opened by being hinged or pivoted to the frame or that can slide open inside the frame. Windows with opening lights may be classified according to the manner in which the opening lights are arranged to open inside the frame, as illustrated in Figure 7.23.

Hinged	Side hung
	Top hung
	Bottom hung
Pivoted	Horizontally pivoted
	Vertically pivoted
	Louvre
Sliding	Vertically sliding
	Horizontally sliding
Composite action	Side-hung projected
	Top-hung projected
	Bottom-hung projected
	Sliding folding

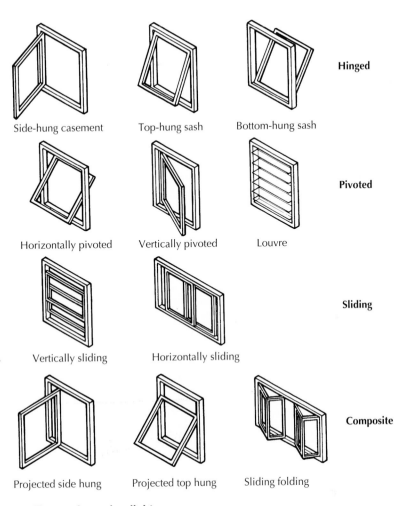

Hinged

Side-hung casement   Top-hung sash   Bottom-hung sash

Pivoted

Horizontally pivoted   Vertically pivoted   Louvre

Sliding

Vertically sliding   Horizontally sliding

Composite

Projected side hung   Projected top hung   Sliding folding

Figure 7.23   Types of opening light.

Hinged opening lights

Side-hung casement

A casement consists of a square or rectangular window frame of wood with the opening light or casement hinged at one side of the frame to open in or out. The side-hung opening part of the window is termed the casement and it consists of glass surrounded and supported by a wooden frame as shown in Figure 7.24, which is an illustration of a simple one-light casement opening out.

The traditional English casement is hinged to open out, primarily because an outward-opening casement can be made to exclude wind and rain more easily than one opening inwards. With an outward-opening window the casement is forced into the outward facing rebate of the window frame by wind pressure, whereas with an inward-opening casement

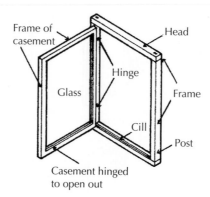

Figure 7.24   Side-hung casement window.

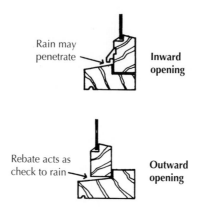

Figure 7.25   Inward- and outward-opening casements.

the casement is forced away from the inward facing rebate of the window frame, as illustrated in Figure 7.25, and so acts as a less effective seal against wind and rain. Another advantage of the outward facing casement is that it will not obstruct curtains or blinds when they are drawn together. French casements, illustrated in Figure 7.26, have been adopted for warmer climates where the casements may be opened inward and louvres, which are fixed externally, can be closed over the opening to exclude sun and allow some ventilation, and also provide some security against unwanted entry.

A casement window may be framed with a pair of casements hinged to close together inside the frame as illustrated in Figure 7.27. In the closed position, the rebated vertical stiles of the two casements, which meet at the centre of the window frame, overlap excluding the wind and rain. The casements will need top and bottom bolts, in addition to a central catch, to close them firmly into the window frame to exclude wind and rain. Any distortion or slight loss of shape of either one or both of the casements may cause them to

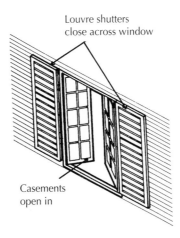

Louvre shutters
close across window

Casements
open in

Figure 7.26   French casement.

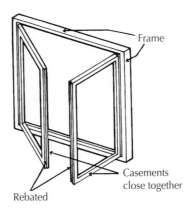

Frame

Casements
close together

Rebated

Figure 7.27   Pair of casements.

bind (stick) inside the frame, making them difficult to open or close. Poorly fitting casements will be ineffective in excluding wind and rain.

A method of framing a two casement window, which is prone to fewer problems, is to use a central member (mullion) so that each casement is hinged to open and close into a separate frame, as illustrated in Figure 7.28. The frame member that separates the casement is termed a mullion. The advantage of this arrangement is that distortion of one casement will not affect the closing of the other and that each casement can be adequately secured with a latch to exclude wind and rain.

It has been common to provide small opening lights, called ventlights, which are usually hinged at the top of the widow frame to open out. So that the ventlights can be opened independently of the casements, the window frame is made with a horizontal member,

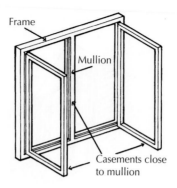

Figure 7.28 Casements and mullion.

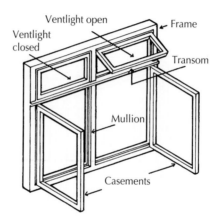

Figure 7.29 Casements and ventlights.

called a transom, into which casements and ventlights close (Figure 7.29). Casements with ventlights are usually designed so that the transom is above the average eye level of people using the room, for obvious reasons. Ventlights that are left open may provide a means of unwanted entry and attention should be given to the size of casement. Small casements may restrict entry but may still allow someone to reach in and open the catches of larger casements. It is common practice to provide locks on all casements.

The disadvantages of a casement window are that the casements, ventlights, mullions and transom reduce the possible unobstructed area of glass and therefore daylight, and that the many clearance gaps around opening casements and ventlights increase the problem of making the window weathertight.

An outward-opening casement may be difficult to clean from inside and is not suited to tall buildings where there is no outside access. The many corners of glass to the comparatively small casements and ventlights make window cleaning laborious.

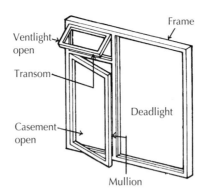

Figure 7.30   Casement window with deadlight.

Dead lights

The manufacturers of standard casement windows now make a range of windows which provide a large dead (fixed) light by itself, or a dead light with a casement alongside it and a ventlight above, as illustrated in Figure 7.30. This type of window combines the advantage of a large area of glazing for maximum daylight with the facility for ventilation from a casement or ventlight.

Top- and bottom-hung windows

These opening lights are principally used for ventilation, the ventilation being controlled by the degree to which the light is opened. Top-hung lights open outwards and bottom-hung open inwards, so that the slope of the sash and its glass directs rain outside the building. The usual practice is to position top-hung lights at a high level, as in the casement window, to encourage warmed air from inside to escape at the sides of the open sash and cold replacement air to enter below the sash (Figure 7.13). Top-hung outward-opening lights are also fixed at high level so that their projection outside is at a high level. Bottom-hung opening-in lights are generally fixed at a low level, so that cold air can enter above the open light and some warmed air from inside can escape at the sides of the sash. Bottom-hung opening-in lights are sometimes described as hoppers.

Top- and bottom-hung lights are often used in schools, places of assembly and factories, either opened by hand or by winding gear to control circulation of air between inside and outside. Because they are top or bottom hung these lights must have a positive opening and stay mechanism; otherwise they bang shut or fully open and would be subject to wind pressure. Top- or bottom-hung opening lights are not so subject to distortion due to their weight as is the side-hung casement. While the bottom-hung lights may be cleaned on both sides from the inside, the top-hung lights cannot.

Pivoted opening lights

The width of a casement is limited by the widow frame's ability to support the weight of the casement. The advantage of a pivoted opening light is that the weight of the frame and glass is balanced over the pivots that are fixed centrally. The sashes may be either

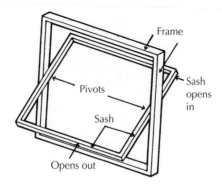

Figure 7.31   Horizontally pivoted sash.

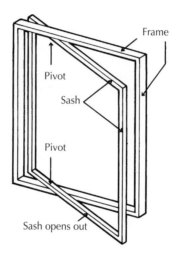

Figure 7.32   Vertically pivoted sash.

horizontally or vertically pivoted to open. Horizontally pivoted sashes are usually pivoted at the centre of the height of the window, as illustrated in Figure 7.31, to balance the weight of the sash over the pivots, and vertically pivoted to open in by one-third of their width to provide least obstruction inside as illustrated in Figure 7.32. Because the weight of the sash is balanced over the pivots, a large sash with small section framing is possible and cleaning the glass on both sides of the window is possible from inside the building. As part of a pivoted sash opens inwards, it may obstruct the movement of curtains. Close control of ventilation with these windows is not possible as they have to open both top and bottom or both sides, and they may act like a sail and catch and direct gusts of wind into the building.

An advantage of pivoted windows is that the glass both inside and outside can be cleaned from inside the building. It is, however, necessary to ensure that windows can be cleaned safely without the risk of people falling out of the building.

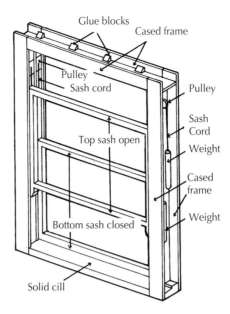

Figure 7.33   **Vertically sliding sash window.**

Sliding windows

Vertically sliding sash window

The word casement is properly used to describe the framing material and glass of a side-hung window. The frame material for other opening lights is termed a sash. A sash window is shown in Figure 7.33.

The advantage of the vertically sliding sash is that as the weight of the sashes is hung vertically on ropes or chains, the sashes do not tend to distort; in consequence large sashes can be framed from small sections and large unobstructed areas of glass are possible. By setting the bulky box frame of these windows behind a rebate in the surrounding wall, the external appearance of the window is of a large area of glass framed in slim members.

Because of the sliding action, the sashes neither project into nor out of the building and close control of ventilation is possible. The sliding action facilitates the use of draught seals between sashes and frame. The disadvantage of this window is that it is not easy to clean the glass on both sides from inside the building. This difficulty has been overcome in recent window design in which it is possible to swing the sashes inwards for cleaning. In time the sash cords will fray and break and it is comparatively laborious to fit new ones. Sashes suspended in spring balances avoid this.

Horizontally sliding sash window

The horizontally sliding wood window is illustrated in Figure 7.34. The window comprises two timber-framed sashes that slide horizontally on runners inside a solid timber frame. As there had to be clearance for moving the sashes it is difficult to make this window weathertight and because of the tendency of the sashes to rack, i.e. move out of the vertical,

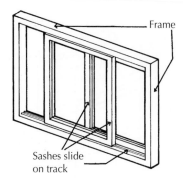

Figure 7.34   Horizontally sliding sash window.

they are liable to jam and can be difficult to open and close. The advantage of this type of window is that there are no internal or external projections from opening sashes and they can be opened to give reasonable control of ventilation. It is difficult to clean the glass both sides from inside.

An adaptation of the horizontally sliding window is the 'patio window', which is in effect a combined fully glazed door and window. The large area of glass provides daylight and a full view. Patio windows or doors are made as two full height sliding sashes or frames, one or both of which slide horizontally on an overhead track from which the sash hangs and slide on guide runners at the bottom. Because of the large area of glass, double- or triple-glazing units are used to reduce heat loss and weather-stripping is fitted around sashes to exclude wind.

Composite action windows

Composite action windows are designed to act like side-, top- or bottom-hung windows for normal ventilation purposes, by opening on pivots which can be unlocked so that the pivots then slide in grooves in the frame and open on hinged side stays to facilitate clean-ing, as illustrated in Figure 7.35. Of the three methods of opening, the top-hung projected window has been the most popular.

Tilt and turn window
This type of window is made specifically for ease of cleaning window glass both sides in safety, from inside the room. For normal operation the sash is bottom-hinged (hung) to open in for ventilation, as illustrated in Figure 7.36. A stay limits the extent to which the head of the sash will open for safety reasons. For window cleaning the window can be converted to a side-hung sash when closed. A lever operates to release bolts, which disen-gage one bottom hinge and simultaneously shoots a side bolt in to engage a top hinge. The sash may then be opened inwards for cleaning the glass on both sides from within.

Sliding folding windows
The sashes in this type of opening window are hinged to each other and fold horizontally in concertina fashion to one or both sides of the window to provide a clear unobstructed

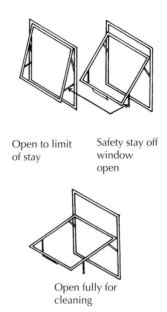

Open to limit   Safety stay off
of stay   window
                     open

Open fully for
cleaning

Figure 7.35   Projected top-hung window.

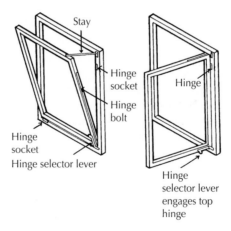

Figure 7.36   Tilt and turn window.

opening as illustrated in Figure 7.37. This opening light system is used as either a horizontal window or fully glazed doors where indoor and outdoor areas can be combined. Each glazed sash is hung on a pivoted wheel that runs in an overhead track fixed to the top of the window frame. The lower edge of each sash is fixed to a pivoted wheel that runs in a track to guide the movement of the sash and maintain it in the vertical position.

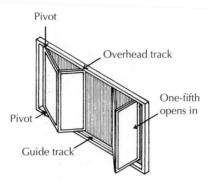

Figure 7.37   Sliding folding sash window.

7.3 Window frames

Their type, e.g. timber, and their construction, e.g. casement, distinguishes window frames. An overview of the main window types commonly used in housing and small-scale developments is provided as follows.

Timber casement windows

The material traditionally used for windows is wood, which is easy to work by hand or machine and can readily be shaped for rebates, drips, grooves and mouldings. It has a favourable strength to weight ratio, and relatively good thermal properties. The disadvantages of wood are the considerable moisture movement that occurs across the grain with moderate moisture changes, and its liability to rot. The dimensional changes can cause joints to open to admit water, which increases the moisture content that can lead to wet rot. The moisture content of timber at the time of assembly should be 17% or less; the timber should be treated with a preservative, and the assembled window should be protected with paint or stain and maintained on a regular basis. Well manufactured, installed and maintained timber windows are very durable.

Figure 7.38 illustrates the arrangement of the parts of a wood casement window, the members of the frames, casements and ventlights being joined with mortice and tenon joints. The casements and ventlights fit into rebates cut in the members of the frame. These rebates serve as a check to wind and rain in normal positions of exposure.

The traditional joint used in timber windows is the mortice and tenon joint illustrated in Figure 7.39. While traditionally the mortice and tenon was produced by the skilled joiner, woodworking machines are now used to prepare, cut and assemble windows and doors with mortice and tenon or dowel joints.

The casements of mass-produced wood windows are often joined with the combed joint illustrated in Figure 7.40, which consists of interlocking tongues cut on the ends of members which are put together, glued and pinned. With the use of modern glue techniques this joint is as strong as a mortice and tenon joint.

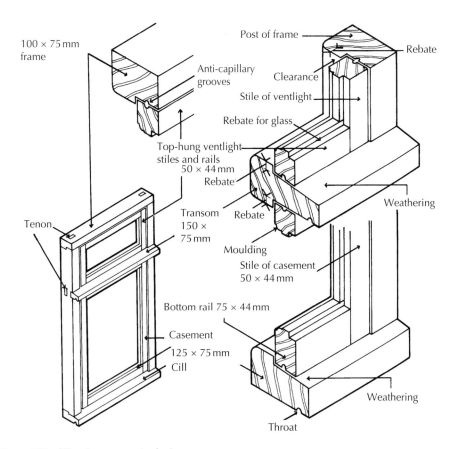

Figure 7.38   Wood casement window.

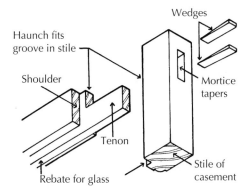

Figure 7.39   Mortice and tenon joint.

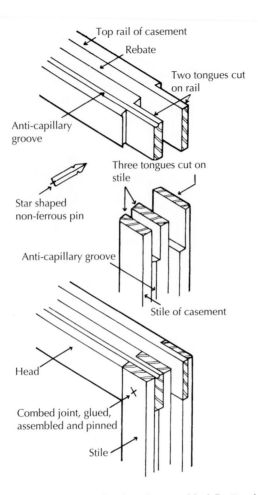

Figure 7.40   Combed joint. Taken apart (top) and assembled (bottom).

Specifying sizes for handmade and machined windows

It is usual to specify the sizes of timber for joinery for windows, doors and frames as being ex 100 × 75 mm, for example. The description 'ex' denotes that the member is to be cut from a rough sawn timber size 100 × 75 mm, which after being planed on all four faces would be about 95 × 70 mm finished size. This system of specifying the sawn sizes of members is used when joinery is to be prepared by hand-operated tools so that the member may be wrought or planed down to a good surface finish without limitation of a precise finished size, yet maintaining the specified size of window. Where joinery is wrought or planed by machine it is practice to specify the precise finished size of each member as this is the dimension the operator needs to know when setting up the machine and it is up to them to select the size of sawn timber to be used to produce the finished size.

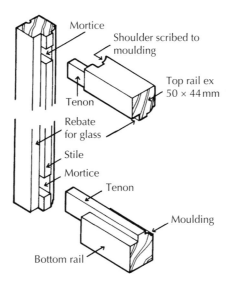

Figure 7.41   Wood casement taken apart.

Casement
A casement is framed from four members, the two vertical stiles and top and bottom rails. Figure 7.41 is an illustration of the framing members of a casement taken apart.

Ventlight
The four members of the ventlight are cut from timbers the same size as the stiles of the casement and are rebated, moulded and joined in the same way as the casement.

Window frame
A casement window frame consists of a head, two posts (or jambs) and a cill joined with mortice and tenon joints, together with one or more mullions and a transom, depending on the number of casements and ventlights. The members of the frame are joined with wedged mortice and tenon joints as illustrated in Figure 7.42. The posts (jambs) of the frame are tenoned to the head and cill with the ends of the cill and head projecting some 40 mm or more each side of the frame as horns. These projecting horns can be built into the wall in the jambs of openings or they may be cut-off on site if the frame is built in flush with the outside of the wall. The reason for using a haunched tenon joint between posts and head is so that when the horn is cut off there will still be a complete mortice and tenon left.

Standard wood casement
The manufacturers of wood windows produce a wide range of standard, mass produced, windows in a variety of sizes and designs. The casements and ventlights are cut so that their edges lip over the outside faces of the frame by means of a rebate in their edges, as illustrated

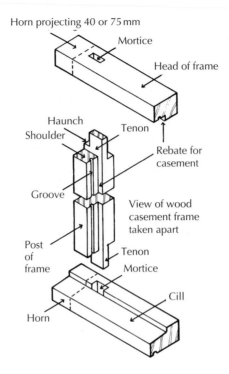

Figure 7.42   Joints of casement window frame.

in Figure 7.43. These lipped edges are in addition to the rebate in the frame so that there are two checks to the entry of wind and rain between opening lights and the frame. The members of the frame and of the opening lights may be joined with mortice and tenon or combed joints.

Weather-stripping

The majority of modern windows include systems of weather-stripping around all opening parts of windows to exclude wind, such as those illustrated in Figure 7.21 and Figure 7.44. Weather-stripping serves as an effective seal against the uncontrolled exchange of cold outside and warmed internal air. Controlled permanent ventilation is provided with specially designed trickle ventilation. In addition to acting as an effective barrier to the entry of draughts of cold air, weather-stripping also serves as an effective barrier to airborne sound.

The two forms of weather-stripping that are commonly used are a flexible bulb or strip of rubber, synthetic rubber or plastic that is compressed between the frame and opening light (Figure 7.21 and Figure 7.44), or a strip of nylon filament pile between the frame and opening light (Figure 7.44). For a maximum effect these seals should be fitted or fixed on the back face of the rebate or the inner face of the frame so that the rebate acts as the first defence against wind and driven rain.

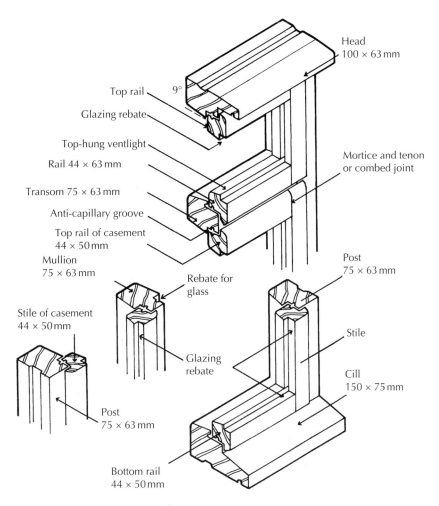

Figure 7.43   Standard wood casement window.

The synthetic rubber strips illustrated in Figure 7.44 are tacked inside the rebate of a wood window frame up to the outward facing rebate or may be self-adhesive for fixing to metal or plastic windows. The advantage of these tacked in place or stuck on strips is that they can easily be replaced when they have lost elasticity in use.

The weather strips illustrated in Figure 7.21 are designed specifically to fit into shallow grooves in wood, metal or plastic windows. The strip is fitted to the dovetail groove with a machine that forces the end of the strip into the groove to make a tight fit. Because of the tight fit, these strips are difficult to replace when they have lost elasticity. The weather-stripping system illustrated in Figure 7.44 consists of an aluminium section into which a strip of nylon filament is fitted. The aluminium section is tacked or screwed to the wood frame so that the flexible bulb bears on the sash when closed, as illustrated in Figure 7.44.

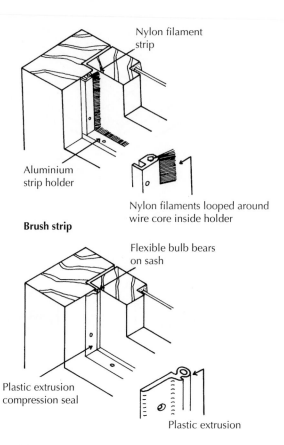

Nylon filament strip

Aluminium strip holder

Nylon filaments looped around wire core inside holder

Brush strip

Flexible bulb bears on sash

Plastic extrusion compression seal

Plastic extrusion

Figure 7.44   Weather strips.

Both weather-stripping systems illustrated in Figure 7.44 are supplied mainly for fixing to existing windows.

Fitting windows

The traditional method of fixing windows in position in a wall is to build solid walling around them as construction proceeds, thus the window is said to be 'built-in'. The advantages of this are that there is a good fit of the wall to the window and that secure fixings may be solidly bedded in horizontal courses as the wall is raised (built) around the window. The alternative method is to 'fix-in' (fit) the window after the wall has been built. The wall is constructed leaving an opening with extra clearance for fitting the window in position. To ensure that the opening is the correct size (window plus adequate tolerance for fitting) a timber window profile is used as a temporary guide (Photograph 7.4). The bricklayer constructs the wall around the profile, then the temporary profile is removed and the window is fixed towards the end of building operations to avoid damage to the frame and/ or glazing.

(a) Temporary window and door template, used to ensure that the wall opening is built to the correct size to receive the window and doorframe

(b) Alternatively, cavity closer can be assembled providing a permanent opening former. The opening formers are held in place with temporary props

Photograph 7.4   Fitting windows: temporary and permanent opening former.

Softwood window frames can be secured in position in solid walls by means of 'L'-shaped galvanised steel cramps or lugs that are screwed to the back of the frame and built into horizontal brick or block courses as the wall is raised (illustrated in Photograph 7.5a and b). Figure 7.45 is an illustration of a fishtail-ended lug 50 × 75 mm in size built into a horizontal course of a brick cavity wall. Where the cavity of a wall is continued up to the jambs of a window opening, a system of plastic cavity closers and ties may be used. A preformed uPVC cavity closer (shown in Photograph 7.6) can be screwed to the back of the window frame as illustrated in Figure 7.46. Nylon wall ties are slotted into the sides of the cavity closer and built into horizontal courses to secure the frame in place. One cramp, lug or tie is used for each 300 or 450 mm of height of window each side of the frame.

Where frames are fixed-in after the walling is completed, one method of fixing frames is to leave pockets in the jambs of the wall into which lugs can be fitted and the walling then made up. The term 'pocket' is used to describe the operation of bedding a few bricks in dry sand so that they may be removed after the wall is built for the building in of lugs at a later stage. As an alternative, the window frames may be secured by galvanised iron straps screwed to the back of the frame and screwed to plugs in the inner reveal of the opening where they will be hidden by subsequent plastering.

Photograph 7.5   (a) uPVC Window fitted using fixing lugs. (b) Window fitted using fixing lugs.

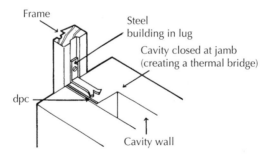

Figure 7.45   Fixing wood window.

Perimeter sealing to wood windows

The gap between the back of the frame and the surrounding walling is sealed against weather with an elastic sealant, described later.

Steel casement windows

Steel casement windows are made either of the standard Z section hot-rolled steel or the universal section illustrated in Figure 7.47. The casement and frame sections fit together

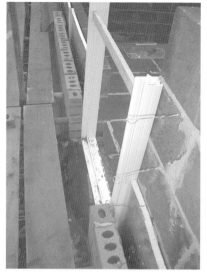

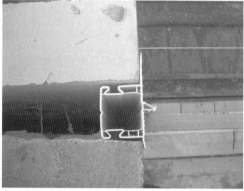

(a) Cavity closer and window opening former

(b) Plan view of un-insulated cavity closer (also available filled with insulation)

Photograph 7.6   Cavity closer.

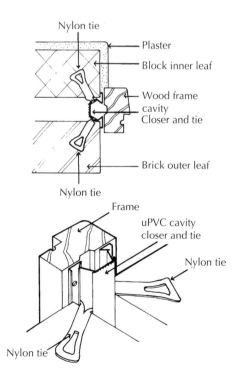

Figure 7.46   uPVC cavity closer and ties.

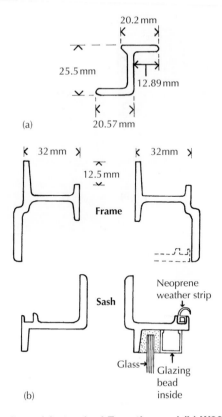

Figure 7.47   Window sections: (a) standard Z section and (b) W20 steel window sections.

as illustrated in Figure 7.48, making a reasonably close fit to exclude rain in all but exposed positions. As an alternative to the standard Z section the universal or W20 steel section may be used, Figure 7.47b.

There is no straightforward method of overcoming the disadvantage of the conductivity of steel windows to the transfer of heat and the possibility of some condensation forming on inside faces, particularly in humid atmospheres such as kitchens and bathrooms. Where there is adequate ventilation in kitchens and bathrooms, condensation can be minimised and rust inhibited by sound protective coating.

Steel casement windows are fitted with steel butt hinges or projecting hinges welded to the frame and opening lights. The projecting hinge illustrated in Figure 7.49 is projected outside the face of the window by steel plate brackets and an angle, which are welded to the frame. The pin around which the opening light hinges is offset so that when the casement is open there is a sufficient gap to make it possible to clean the outside of the glass from inside the room. Lever fasteners and peg stays similar to those used for wood windows are welded to the frame and opening lights as illustrated in Figure 7.49.

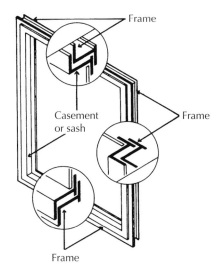

Figure 7.48   Standard metal casement.

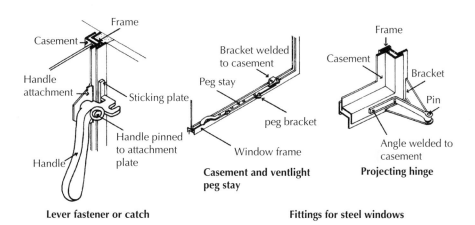

Lever fastener or catch

Casement and ventlight peg stay

Fittings for steel windows

Figure 7.49   Hinges and fasteners.

Fixing steel windows

Standard steel casement windows are usually built in to openings in solid walls and secured in position with 'L'-shaped lugs that are bolted to the frame as illustrated in Figure 7.50. The lugs are adjustable to suit brickwork courses. Where these steel windows are fixed in, after the walls have been built, a galvanised steel lug is bolted to the back of the frame and its projecting arm is then screwed to a plug in the inner reveal of the wall. These lugs will

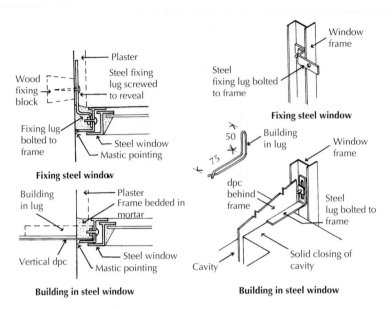

Figure 7.50   Fixing steel windows.

later be obscured by plaster. Although not common, timber sub-frames have also been used for steel windows (Figure 7.51).

Aluminium casement windows

The majority of aluminium windows that are made today are of sections extruded from aluminium alloy in a wide range of channel and box sections with grooves for lips for weather-stripping and double-glazing. The sections are mitre cut and mechanically cleated or screwed at joints, which are sealed against entry of water as illustrated in Figure 7.52. These thin-walled channel and box sections give the material adequate strength and stiffness for use as window sections. The material can be readily welded and has good resistance to corrosion. Aluminium window sections are usually finished with anodised, polyester powder or liquid organic coatings as protection against oxidisation and as a decorative coating that can be easily cleaned.

A disadvantage of aluminium as a window material is that it is a good conductor of heat and a potential thermal bridge. Some manufacturers have introduced a 'thermal break' illustrated in Figure 7.53 and Photograph 7.7. The separate aluminium window sections are mechanically linked to the main window sections through plastic thermal break sections; however, the effectiveness of the small thermal break in reducing heat flow is minimal. Double- and triple-glazed units [known as insulating glass (IG) units] are secured with aluminium beads and the window is weather-stripped with preformed synthetic rubber seals. The aluminium frame is secured to the surrounding wall by aluminium lugs that clip to the back of the frame at centres of up to 600 mm and also adjacent to hinges and fasteners, with the lugs screwed to plugs in the wall.

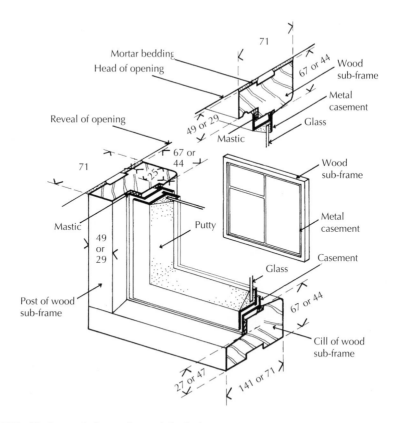

Figure 7.51   Timber sub-frame for metal window.

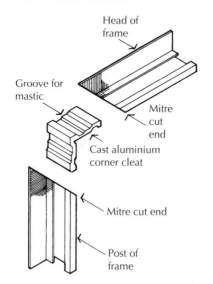

Figure 7.52   Corner cleat for aluminium window.

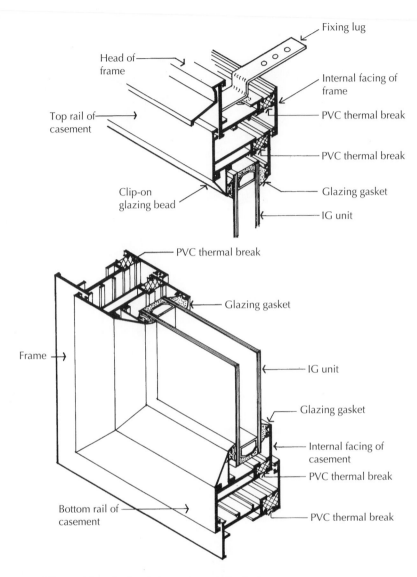

Figure 7.53   Thermal break aluminium casement window.

Sealants

A sealant is a material that is initially sufficiently liquid or plastic for application and which cures or changes to a material that will adhere to surrounding surfaces, retain its shape and accommodate some small movement without loss of seal against wind and rain. Sealants used for sealing perimeter joints around window frames are classed as plastoelastic, elastoplastic or elastic.

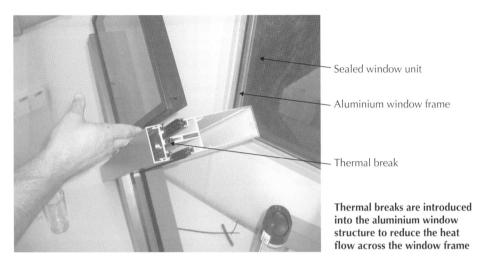

Sealed window unit

Aluminium window frame

Thermal break

Thermal breaks are introduced into the aluminium window structure to reduce the heat flow across the window frame

Photograph 7.7   Thermal break in metal window.

❑ Plastoelastic sealants, which have some elastic property, remain predominantly plastic and can be moulded.
❑ Elastoplastic sealants, which develop predominantly elastic properties as they cure, will return to their former shape when stress is removed and also retain some plastic property when stressed over long periods.
❑ Elastic sealants will, after curing, have predominantly elastic properties in that they will continue to resume their former shape once stress is removed, during the anticipated useful life of the material.

Acrylic, polysulphide, polyurethane and silicone

The materials that may be used for perimeter sealing around window and door frames are acrylic, polysulphide, polyurethane and silicone. Of these, acrylic is classed as plastoelastic, polysulphide as elastoplastic, and polyurethane and silicone as elastic. In general the plastoelastic material is easier to use because of its predominantly plastic nature, but it will not form as tough and elastic a surface as elastoplastic materials that have some plastic property. The elastic materials need some experience and skill in use for successful application.

One- and two-part sealants

Polysulphide and polyurethane sealants are produced as either one-part sealants ready to use, or as two-part sealants that have to be mixed before use. The one-part sealants are more straightforward to use as there is no mixing and the material cures or loses plasticity fairly slowly, allowing adequate time for running into joints and compacting by tooling.

The two-part sealants require careful, thorough mixing and, as they cure fairly rapidly, require skill in application. The advantage of the two-part sealants is that as they cure fairly rapidly, they are less likely to slump and lose shape and adhesion than the more slow curing one-part silicone sealants which cure fairly rapidly to form a tough, elastic material, require rapid application and tooling for compaction.

Window sealant functions

As the prime function of a sealant to perimeter gaps around window frames in traditional walling is as a filler to exclude wind and rain, it should adhere strongly to enclosing surfaces, be resistant to the scouring action of weather and sufficiently elastic to accommodate small thermal movements for the anticipated life of the material. The expected useful life of sealants, after which they should be renewed, is up to 15 years for acrylic and up to 20 years for polysulphide, polyurethane and silicone.

To ensure maximum adhesion, the surfaces onto which a sealant is run should be clean, dry and free from dust, dirt and grease. Sealants are usually run into joints from a gun operated by hand pressure or air pump and, for appearance, the sealant should not be too obvious.

The form of sealant joint used depends on the width of the perimeter gap between the window frame and the surrounding wall and whether the frame is set in a rebate. The types of joint used are butt joint, lap joint and fillet seal (Figure 7.54, Figure 7.55 and Figure 7.56). The butt joint (Figure 7.54) is formed between the back of the frame and the reveal of the opening. Foamed polyethylene is first run into the gap as a backing for the sealant. The sealant is then run into the joint and tooled with a spatula to compact the material

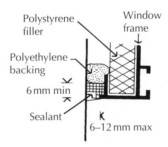

Figure 7.54   Butt joint.

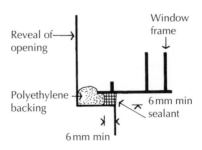

Figure 7.55   Sealed lap joint.

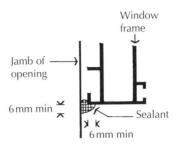

Figure 7.56   Fillet seal.

and make good adhesion to the two surfaces. It is finished with a slight concave finish up to the edge of the window frame.

The best gap width for the joint is from 6 to 12 mm; this is wide enough for application of the sealant, yet small enough to contain the sealant and not too obvious. Butt joints up to 25 mm are practical with the depth of the sealant being half that of the gap. These wider joints tend to look somewhat unsightly. To prevent the sealant from adhering to the outside face of the window frame it is good practice to use masking tape up to the edge of the outside face of the frame. Once the sealant is sufficiently cured the masking tape is stripped towards the sealant. A lap point is formed where the window frame is set behind an accurately formed rebate in wood, metal, masonry or concrete surrounds (Figure 7.55). The sealant is run into the gap over a polyethylene backing and tooled to a slight concave finish to masking tape. It is more difficult to form or renew this joint, which is less obvious than a butt joint.

A gap of less than 6 mm between the window frame and the opening in the wall is too narrow for gunned-in sealant. Here a fillet seal is used, which is formed to adhere to the outside face of the frame and the wall opening (Figure 7.56). The fillet seal is run as a convex fillet to provide sufficient depth of sealant, which is finished as it comes from the gun.

uPVC casement windows

These windows are fabricated from extruded, high-impact strength, white uPVC. Modifiers, such as acrylic, are added to the PVC material to improve impact strength. Pigment may be added to produce body coloured uPVC. The heated, plastic material is forced through dies from which it extrudes as thin-walled hollow box sections, complete with rebates, grooves and nibs for beads, weather seals, glazing seals and for fixing hardware. The word plastics is used in a general sense to embrace a wide range of semi-synthetic and synthetic materials that soften and become plastic at comparatively low temperatures so that they can be shaped by extrusion and pressure moulding. uPVC windows are used extensively as 'replacement windows' in existing domestic buildings and have also become standard in much new development.

uPVC is relatively maintenance free and will maintain its smooth textured surface for the useful life of the material with occasional washing to remove grime. As the material is formed by extrusion it is practical to form a variety of rebates and grooves to accommodate draught seals. The basic colour of the material is off-white, which is colourfast on exposure to ultraviolet light for the useful life of the material. A range of coloured plastics can be produced either with the colour integral to the whole of the material or as a surface finish. Dark colours are more susceptible to bleaching and loss of colour in ultraviolet light than light colours.

Because uPVC has less strength and rigidity than metal sections, it is formed in comparatively bulky, hollow box sections that are not well suited for use in small windows. The comparatively large coefficient of expansion and contraction of the material with the change of temperature and its poor rigidity require the use of reinforcing metal sections fitted into the hollow core of the sections to strengthen it and, to an extent, restrain expansion and contraction. The uPVC sections are screwed to the galvanised steel or aluminium reinforcement to fix the reinforcement in position, restrain deformation due to temperature movement and serve as secure fixing for hardware such as hinges, stays and bolts. Some manufacturers use reinforcement only for frame sections over 1500 mm in length and casement or sash sections over 900 mm in length. For the advantage of a secure fixing for hardware and fixing bolts it is wise to use reinforcement for all uPVC sections.

Metal reinforcement
The extruded sections are mitre cut to length, metal reinforcement is fitted and secured inside the main central cell, and the corner joins are welded together by an electrically heated plate that melts the end material, with the ends then brought together to fuse weld. The process of cutting and welding is fully automated, which makes it a comparatively simple operation to set the machine to make one-off sizes of windows for the replacement window market.

Mitred, welded corners
Reinforcement is fixed inside the hollows of the uPVC cells to provide rigidity to the sections that might otherwise distort due to thermal movement, handling, fixing and in use as opening lights. Reinforcement should be fitted to all frames more than 1500 mm long and all opening lights more than 900 mm long. In fire, uPVC, which does not readily ignite, will only burn when the source of heat is close to the material and will not appreciably contribute to the spread of flame. The rate of generation of smoke and fumes produced when uPVC is subject to fire is no greater than that of other combustible materials used in building.

The uPVC casement illustrated in Figure 7.57 is glazed with an IG double-glazed unit set in synthetic rubber seals and fitted with weather-stripping and reinforcement of galvanised steel or aluminium sections.

Fixing uPVC frames
To avoid damage to the frames during building operations these windows are usually fixed in position after the surrounding walls have been built. Fixing is usually by driving strong screws through holes in the frame and reinforcement into surrounding walls or by means of lugs bolted to the back of frames, which are screwed to plugs in walls. Fixings are at 250 and 600 mm centres and from 150 to 250 mm from corners.

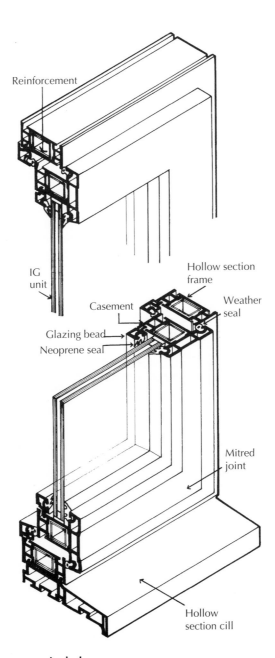

Figure 7.57   uPVC casement window.

Perimeter seals to uPVC windows

The gap between the window and the surrounding wall is sealed with silicone or poly-urethane sealant with backing of foamed, compressible, preformed strips or gunned-in expanded insulation foam, adhesive foam for joints more than 6 mm wide.

Pivoted windows

Horizontally pivoted wood window

Figure 7.58 is an illustration of a double-glazed horizontally pivoted wood window. The frame is solid and rebated. Stop beads are fixed to the sash above and the frame below the pivots. The sash is made as separate inner and outer sashes, each of which is glazed and the sashes are normally locked together.

The purpose of forming the sash as two separate parts is so that the two parts of the sash may be opened for cleaning glass inside the space between the two parts. Reversing the

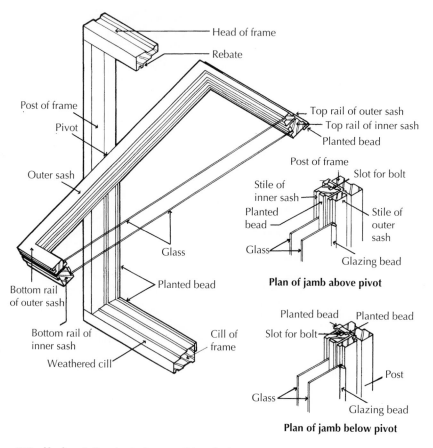

Figure 7.58 Horizontally pivoted reversible window.

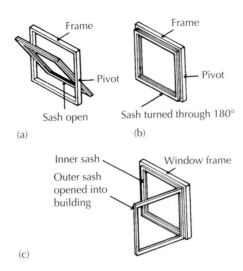

Figure 7.59 Sash reversed for cleaning glass.

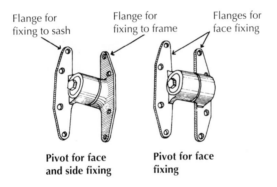

Figure 7.60 Pivots for horizontally pivoted windows.

sash through 180° so that the outer part of the sash may be unlocked and hinge to open into the room as illustrated in Figure 7.59 achieves this. In this position both sides of the glass in the outer part of the sash and the side of the glass of the inner part, facing the air space, may be cleaned in safety from inside the building. The window is opened against the action of the friction pivots illustrated in Figure 7.60 and locked shut with lever operated espanolite bolts that secure the sash at four points, top and bottom, against weather-stripping.

Vertically pivoted steel windows

Figure 7.61 is an illustration of a steel section, hot-dip galvanised pivot window, with single glazing.

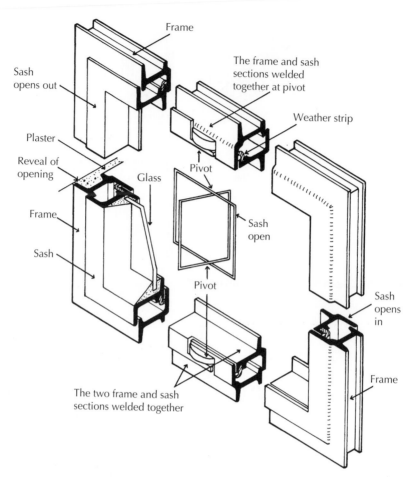

Figure 7.61   Vertically pivoted steel window.

Sliding windows

Vertically sliding wood sash window (double-hung sash)

This traditional window is framed from thin section timbers to form a box or cased frame inside which the counterbalance weights are suspended to support the sliding sashes, which are suspended on cords that run over pulleys fixed to the frame. The construction of the frame is shown in Figure 7.62.

Vertically sliding wood sash window with solid frame

As an alternative to the traditional system of cords, pulleys and weights to hang vertically sliding sashes, spiral sash balances have been used for the past 50 years. The spiral balance consists of a metal tube inside which a spiral spring is fixed at one end. Fixed to the other

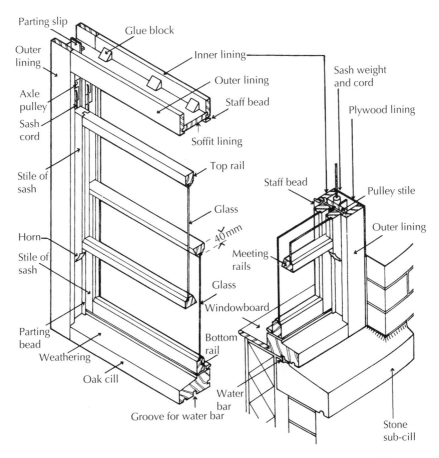

Figure 7.62 Wood vertically sliding (double-hung) sash window with cased frame.

end of the spring is a metal cap through which a twisted metal bar runs. The tube is fixed to the window frame and the twisted bar to the bottom of the sashes. As the sash is raised or lowered the twisted bar tensions the spring which supports the weight of the sashes, enabling the sashes to be raised or lowered with little effort. Figure 7.63 shows one of these sash balances.

Because of the sash balance there is no need for hollow cased frames to take counterbalances and the frame members can be made of solid sections as illustrated in Figure 7.64. The window frame is constructed from four solid rectangular sections of timber, two posts (jambs), head and cill. The posts are joined to the head and cill with combed joints glued and pinned, similar to those described for standard casements. The sashes are similar to those for windows with cased frames, the members being joined with mortice and tenon or combed joints.

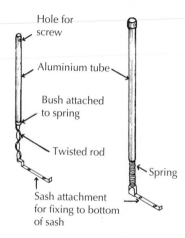

Figure 7.63　Sash balances for vertically sliding windows.

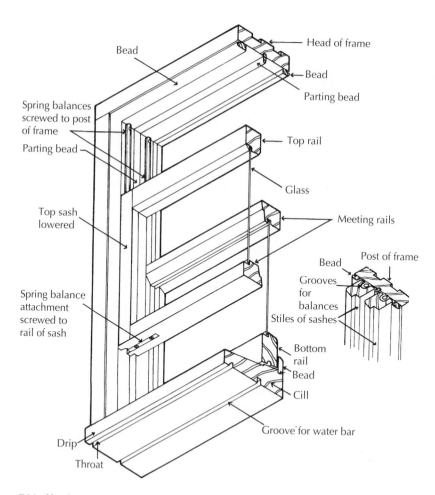

Figure 7.64　Vertically sliding sash window with solid frame and balances.

Weather-stripping

Vertically sliding sash windows are comparatively simple to weather-strip to exclude wind and rain. Because of the vertically sliding movement of the sashes, a system of wiping, sliding seals is effective for the stiles or sides of the window, with compression seals to the head and cill of the window (Figure 7.65).

Aluminium vertically sliding sash window

With the slender sections of extruded aluminium alloy practical for use in the frame and sashes of this type of window, and because the material needs no painting at frequent

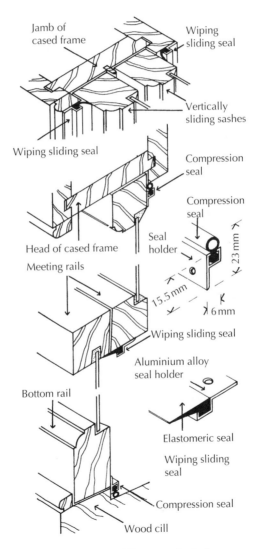

Figure 7.65 Weather seals for vertically sliding wood windows.

intervals, aluminium vertically sliding windows have become increasingly popular. The extruded aluminium sections are joined with screw nailed butt joints, mechanical mortice and tenons, or mitred and cleated joints with stainless steel screws. Figure 7.66 is an illustration of a typical vertically sliding aluminium window.

uPVC vertically sliding sash window

uPVC vertically sliding sash windows have been used as replacement windows for old wood-framed windows for the advantage of reduced maintenance and the facility for fitting weather-stripping to grooves formed in the window sections. The vertically sliding, uPVC sash window illustrated in Figure 7.67 is framed with comparatively bulky hollow sections that are reinforced with metal box sections, which are designed to take double-glazing units.

Horizontal sliding windows

The aluminium section horizontally sliding sash window illustrated in Figure 7.68 slides on a bottom track with nylon filament, pile weather-stripping acting as weather seal and as guide to both top and bottom rails and with pile weather-stripping to stiles. As grit may in time collect around the track, it is often somewhat difficult to open these windows, which, if forced open or closed, may tend to jam on the track and so be more difficult to open. The most common use of this type of window is as a fully glazed, horizontally sliding, 'patio door'.

7.4 Glass and glazing

Glass is made by heating soda, lime and silica (sand) to a temperature at which they melt and fuse. Molten glass is drawn, cast, rolled or run on to a bed of molten tin to form flat glass. Glass may be classified into three groups; annealed flat glasses, processed flat glasses or miscellaneous glasses. Annealed and processed flat glasses are described in more detail as follows. Miscellaneous glasses include double and triple glazing, roof and pavement lens lights, copper lights, leaded lights and hollow glass blocks. Structural glazing is explained in *Barry's Advanced Construction of Buildings*.

Annealed flat glasses

There are two types of annealed flat glass: float and sheet glass.

Float glass

Float glass is made by running molten glass continuously on to a bed of molten tin, on which the glass floats and flows until the surfaces are flat and parallel. The continuous ribbon of molten glass is then run into an annealing lehr or chamber in which the temperature is gradually reduced to avoid distortion of the molten glass as it gradually solidifies and is then cut. The natural thickness of the sheet of glass is 6.5 mm. To produce thinner glass, the molten ribbon of glass is cooled and stretched between rollers. To make thicker glass the spread of the molten ribbon is restricted to produce the required thickness. Float glass is made in thicknesses of 3, 4, 5, 6, 10, 12, 15, 19 and 25 mm.

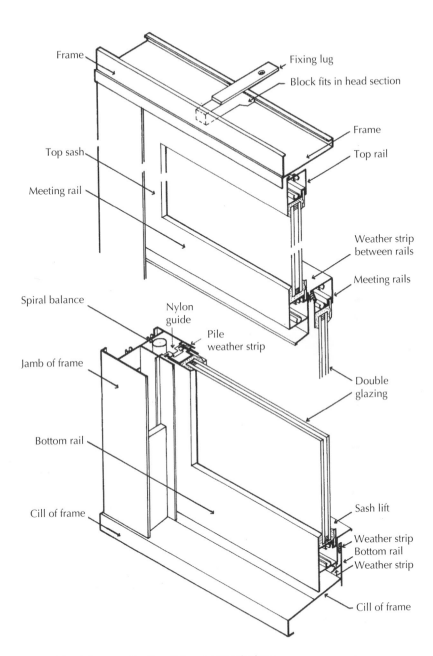

Figure 7.66 Aluminium vertically sliding sash window.

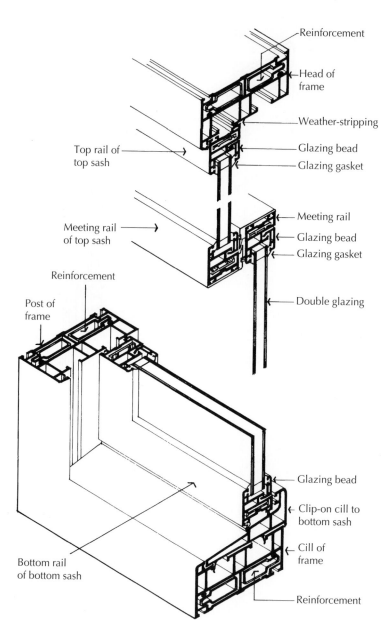

Figure 7.67   uPVC vertically sliding sash window.

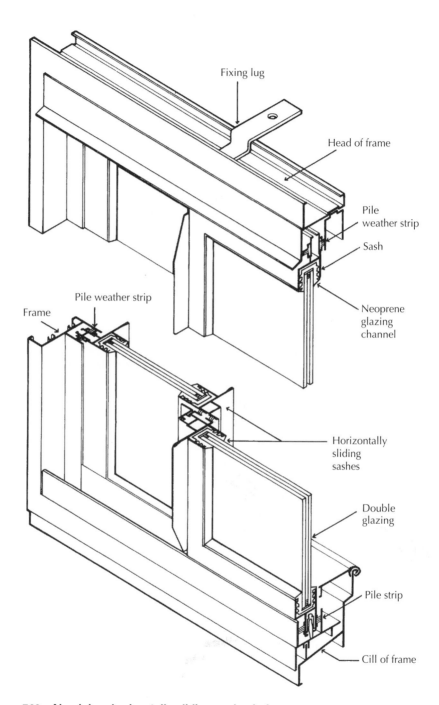

Figure 7.68 Aluminium horizontally sliding sash window.

Solar control float glass

Body tinted glass
Body tinted float glass is transparent glass in which the whole body of the glass is tinted. This type of glass reduces solar radiation transmission by increased absorption. This material is commonly termed 'solar control glass'. Tints are usually green, grey, blue or bronze and thicknesses 4, 6, 10 and 12 mm.

Surface modified glass
Surface modified tinted float glass is transparent glass that, during manufacture, has a coloured layer of metal ions injected on to the glass. Solar control properties are provided by an increase in reflection and absorption. Thicknesses are 6, 10 and 12 mm.

Surface coated glass
Surface coated float glass (reflective float glass) is transparent glass that has a reflective surface layer applied either during or after manufacture. The reflective layer may be on a clear or a body tinted glass. Transmission of solar radiation is reduced by increase in reflection and absorption, and the glass has a coloured metallic appearance. Colours are silver, blue and bronze by reflection.

Low emissivity glass
Surface modified and surface coated glasses are solar control glasses that are also referred to as low-emissivity glasses (low E glass). Manufacturers have various names for this glass, which is marketed as energy-efficient glazing. The effect of the surface coating is to reflect back into the building the long wave energy generated by heating, lighting and occupants, (helping to save energy) while permitting the transmission of short wave solar energy from outside (see Figure 7.69 and Figure 7.70). Low emissivity glass can also be used to reduce the amount of heat entering a building, thus helping to minimise overheating. Low emis-

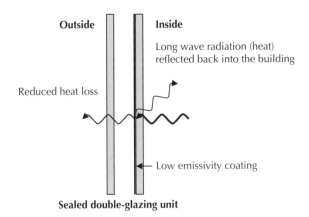

Figure 7.69   Inside pane coated with low emissivity coating (adapted from Gorse et al., 2013).

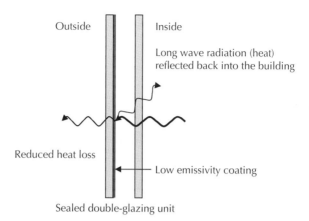

Figure 7.70 Inside wall of the outside pane coated with low emissivity coating (adapted from Gorse et al., 2013).

sivity glass has a slight tint which varies depending on the manufacturer. Two types of emissivity glass exist: one is hard coat, which is applied during the manufacturing process (pyrolitic); the other is soft coat, which is applied after the glass has been manufactured.

Sheet glass

Clear sheet glass
Clear sheet (drawn sheet) glass is transparent glass manufactured by the flat drawn process in which a continuous sheet is drawn from a bath of molten glass in thicknesses of 3, 4, 5 and 6 mm. The continuous sheet is gradually cooled to minimise distortion and then cut into sheets as it solidifies. The drawn sheet is not exactly flat or uniform in thickness and will cause some distortion of vision.

Body tinted sheet glass
Body tinted sheet glass is transparent glass in which the whole body of the glass is tinted to give solar control properties. Tints are usually green, grey or bronze and thicknesses are 3, 4, 5 and 6 mm.

Cast glass (also known as patterned glass)
Clear cast glass is translucent glass made by the rolling process: the deeper the pattern, the greater the obscuration and diffusion.

Body tinted glass
Body tinted glass is similar, with the whole of the glass tinted for solar control and decorative purposes.

Wired glass
Wired glass is cast or rolled with wire completely embedded in it. This type of wired glass is referred to as Georgian glass. The mesh inside the glass is 13 mm square. Cast wired glass

is translucent with a cast or patterned surface. Polished wired glass is transparent, through grinding and polishing.

Processed flat glasses

Toughened glass
Toughened (tempered) glass is made by heating annealed glass and then rapidly cooling it to cause high compression in the surfaces and compensating tension in the centre of the thickness of the glass. This is a safety glazing material that is less liable to break on impact and, when broken, it fragments into comparatively harmless small pieces. Clear float, sheet, polished plate and solar control glass may be toughened.

Laminated glass
Laminated glass is made of two or more sheets (panes) of glass with an interlayer of reinforcing material between the sheets. The interlayers are permanently bonded to the enclosing sheets of glass. This glass is resistant to impact shock and when broken the reinforcing layer prevents extensive spalling of fragments. The reinforcing interlayer is usually in sheet form and made of polyvinyl butyral. This type of glass is specified as three ply, i.e. two sheets of glass and one of reinforcement, and similarly five ply, with three sheets of glass and two of reinforcement. This type of glass is often described as safety or security glass.

Safety glass
Toughened glass and laminated glass are described as flat safety glass which, on breaking, result in a small clear opening by disintegration into small detached fragments that are neither sharp nor pointed and are unlikely to cause cutting or piercing injuries. Approved Document N to the Building Regulations defines critical locations where safety glass should be used. These critical locations are in glazed panels in internal walls and partitions between floor and 800 mm above that level and in glazed doors and door-side panels, between floor and 1500 mm above that level.

Polycarbonate sheet
Polycarbonate sheet is an alternative to glass. Flat plastic sheets made of polycarbonate are manufactured as transparent, translucent and colour tinted sheets for use as safety glazing. The sheets are 2, 3, 4, 5, 6, 8, 9.5 and 12 mm thick. The principal characteristic of this material is its high-impact resistance to breakage. These sheets do not have the lustrous, fire glazed finish of glass nor are they as resistant to abrasion scratching and defacing. A special abrasion resistant grade is produced. Polycarbonate sheet, which is about half the weight of a comparable glass sheet, has a high coefficient of thermal expansion. To allow for this, deeper rebates and greater edge clearance are recommended than for glass. To allow for the flexibility of the material and thermal expansion, one of the silicone compounds is recommended for use with solid bedding.

Glazing

The operation of fixing glass in windows, doors and other openings is termed glazing. The purpose of glazing is to secure glass in position in window frames and sashes and to make a weathertight seal against penetration of rain around the edges of the glass. The choice of a method of glazing depends on:

❑ The anticipated structural and thermal movement of the window
❑ The degree of exposure of the window to wind and rain

The traditional method of glazing for single glazing to softwood and metal frames is putty glazing, and to hardwood, aluminium and uPVC frames is bead glazing with non-setting compounds and tapes for both single and double glazing and gasket glazing for extruded, hollow section aluminium and uPVC frames. Modern windows are, mostly, factory made with sealed double or triple glazed units installed in the factory. The traditional approach is included here for those interested in conservation and upgrading of existing buildings.

Single glazing – putty glazing

Putty is a material that is initially sufficiently plastic to be moulded by hand, spread in the glazing rebate as a bed for glass and finished outside as a weathered front or face putty. The putty sets or hardens over the course of a few days to secure the glass in position and serve as an effective seal against rainwater penetration.

Glazing with putty is used to secure glass in wood or steel frames where there is little structural and thermal movement. Putty is not used on aluminium or plastic frames where the larger structural and thermal movement of these materials might cause the putty to crack, lose adhesion and allow rain to penetrate.

Linseed oil putty

Linseed oil putty is used for glazing to softwood and absorbent hardwood frames. It adheres to both glass and wood and hardens by absorption of some of the oil into the wood and by oxidisation. To prevent too great an absorption, softwood windows should be primed before glazing. Putty is spread by hand in the glazing rebate.

Setting and location blocks

To provide an edge clearance of 2 mm between the edges of the glass and the rebate, to allow for variations in the sash and in the glass and to facilitate setting the glass in place, setting and location blocks are used. Setting and location blocks, which are of PVC, hammered lead, hard nylon or hardwood, are 2 mm thick and up to 150 mm long. Setting blocks are pushed into the soft rebate putty in the bottom edge rebate to support the glass and the location blocks into both side rebates to centre the glass (Figure 7.71). The glass is placed on the setting blocks and pushed firmly into the putty to squeeze surplus putty between the glass and the upstand of the rebate as a back putty bed 1.5 mm thick. Glass is then secured in position, until the putty has hardened, with metal sprigs (cut headless nails) that are tapped into the rebate at not more than 300 mm spacing.

Additional putty is then spread by hand in the glazing rebate around the edges of the glass and finished with a putty knife at an angle from the edge of the glazing rebate up to about 2 mm below the sight line as a seal against rain penetration. Surplus back putty is

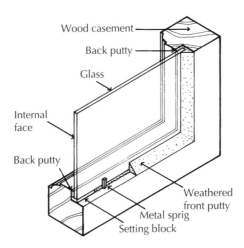

Figure 7.71   Putty glazing to wood sash.

stripped and the back putty is finished at an angle up to the glass to shed any condensation water from the inside face of the glass.

The finished putty should be left to harden for at least 7 days and then painted with the usual undercoat and finish coats of paint to prevent further hardening of the putty. The painted surface should be finished on to the glass as a seal against rain penetration behind the putty.

Metal casement putty

Metal casement putty is designed specifically for use on non-porous surfaces such as galvanised steel frames, sealed timber and sealed concrete with or without glazing beads. It is made from a blend of vegetable oils selected to adhere to and set on non-porous surfaces. It is not suitable for glazing to aluminium, stainless steel, bronze or plastic finishes. When fixing glass in metal frames special metal glazing clips are required (Figure 7.72). This putty hardens and sets and will accommodate the relatively small amount of movement that occurs in steel frames due to temperature change.

Glazing with beads

As an alternative to putty glazing for single glazing, which may not always provide a neat finish, bead glazing may be used, where a bead secures the glass in place. The choice of internal or external fixing of glazing beads depends on the material of the bead, the ease of access for reglazing, appearance and security.

For durability inside glazing is preferable, particularly when softwood beads are used because the joints between the bead, the glass and the frame are vulnerable to the penetration of rain. For access for reglazing and security reasons inside beads are best for ground floor windows.

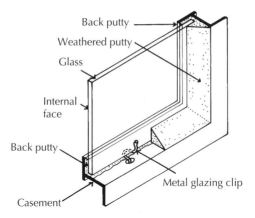

Figure 7.72   Putty glazing to metal sash.

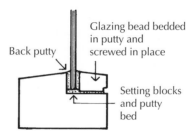

Figure 7.73   Glazing with putty and beads.

Glazing with putty and beads

Glazing with putty and beads may be used for softwood, absorbent hardwood and galvanised steel frames. In sheltered positions the beads may be external; for other exposures they should be internal.

Sufficient putty is spread all round the rebate to provide adequate back putty. Setting blocks are pressed into the putty in the platform rebate and location blocks as necessary. The glass is placed on the setting blocks and pressed firmly into the rebate so that putty 1.5 mm thick is squeezed between the glass and the back of the rebate as back putty bedding as illustrated in Figure 7.73.

For fixing externally, beads should be bedded with putty against the glass and to the bed of the rebate. The beads are pressed firmly against the glass and then secured with pins or screws. Where beads are fixed internally it is not usual to bed them in putty. Back and front putty beds are trimmed and finished with a slope up to the glass. The exposed putty should be painted some 7–14 days after glazing. For putty glazing to non-absorbent hardwood and galvanised steel frames metal glazing putty is used.

Glazing with tape and beads

As an alternative to putty glazing for large squares of glass for both wood and galvanised steel frames, loadbearing mastic tape and a sealant may be used.

Loadbearing mastic tape is made from compressed fabric and butyl in various widths and thicknesses. The tape used as bedding for glass will also serve as a sealant to exclude rain when it is adequately compressed. It is comparatively easy to use and often preferred to putty. A strip of the mastic tape is pressed into position all round the rebate upstand so that it finishes some 3–6 mm below the sight line and setting blocks are placed in the rebate base. Glass is placed on the setting blocks and firmly pressed into the rebate to compress the tape to the rebate upstand. Location blocks are fixed in the side edge clearances to centre the glass.

A second strip of mastic tape is fixed around the outside edges of the glass and the internal glazing beads are fitted in place, pressed firmly against the mastic tape and screwed into position. A sealant is run, by gun, into the edges of the outside of the glass and finished with a smooth chamfer or slope to shed water as illustrated in Figure 7.74.

Double and triple glazing

The term double glazing describes the use of two sheets, or panes, of glass in a window or door with an air gap between them. Triple glazing refers to the use of three sheets of glass with air gaps between the panes of glass. The air gap serves to improve the thermal resistance of the glazed unit. For additional thermal resistance the cavity is filled with inert gas, such as argon, krypton or xenon.

The U-value (thermal transmittance) of a single sheet of 6-mm-thick glass is $5.4\,\text{W/m}^2\,\text{K}$ and that of an IG unit with two sheets of 6-mm-thick glass spaced 12 mm apart is

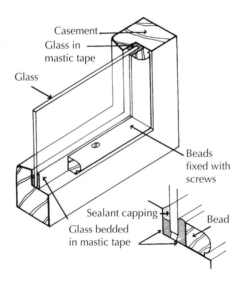

Casement

Glass in mastic tape

Glass

Beads fixed with screws

Sealant capping

Bead

Glass bedded in mastic tape

Figure 7.74 Glazing to wood with beads.

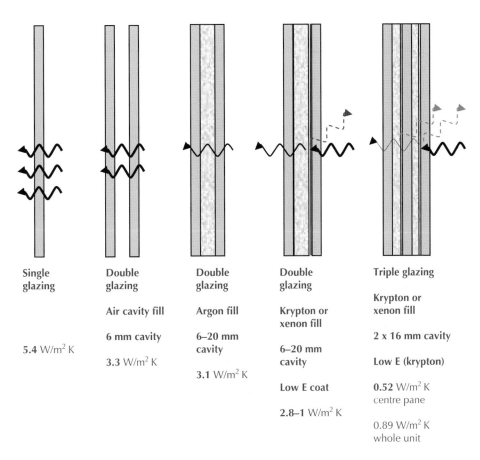

Figure 7.75   Single-/double- and triple-glazing U-values.

$3.0 \, \text{W/m}^2 \, \text{K}$. Using low emissivity coats, inert gas within the cavity and triple glazing, centre pane U-values of $0.52 \, \text{W/m}^2 \, \text{K}$ and whole window casement values of $0.89 \, \text{W/m}^2 \, \text{K}$ can be achieved (Figure 7.75, Table 7.6 and Table 7.7).

The thermal image in Photograph 7.3 shows a sash window retrofitted with slender fit double glazing (3.8 mm cavity, U-value $1.9 \, \text{W/m}^2 \, \text{K}$). The centre middle pane at the bottom of the sash window was left single glazed ($5.4 \, \text{W/m}^2 \, \text{K}$), and the darker parts of the thermal image clearly indicate the cold parts of the window. The difference in surface temperature between the double-glazed unit and the single glazing was recorded at nearly 10°C.

IG units, sealed double-glazing units

The terms double glazing, sealed double glazing and IG are generally interchangeable. The term double glazing embraces all systems of double glazing whether the glazing is unsealed as in double windows or sealed as in IG units. IG units are made up from two sheets (panes or squares) of glass that are hermetically sealed to a continuous spacer around

Table 7.6 Single/double glazing comparison

Single/double and triple glazing approximate comparison						
Panes of glass	**Single**	**Double**	**Double**	**Double**	**Double**	**Double**
Cavity (mm)	No cavity	6	6	6	3.8	5.0
Emissivity coat			Hard coat E	Hard coat E	Low E	Low E
Cavity fill	N/A	Air	Argon	Krypton	Krypton or xenon	Krypton or xenon
Centre pane U-value (W/m^{2}K)	5.4	3.3	3.1	2.8	1.9	1.8
					Slender fit	Slender fit

Table 7.7 Double/triple glazing comparison

Single/double and triple glazing approximate comparison						
Panes of glass	**Double**	**Double**	**Triple**	**Triple**	**Triple**	**Triple**
Cavity (mm)	10 mm	20 mm	2 × 8 mm (three 4 mm panes of glass)	2 × 16 mm	2 × 16 mm (three 4 mm panes of glass, total 44 mm)	2 × 16 mm (three 4 mm panes of glass, total 44 mm)
Emissivity	Hard coat E	Soft coat E	Low E	Low E (not specified)	Soft coat	Low E (not specified)
Cavity fill	Krypton	Krypton	Argon	Krypton or xenon	Argon	Krypton
Centre pane U-value (W/m^{2}K)	1.6	1.0	1.0	0.7	>0.8	0.52
Whole unit U-value				1.1		0.89
Further information	Weather-proof windows	Weather-proof windows	Sureglaze	NorDan	Sureglaze	Envior therm

the perimeter of the unit. The usual space between the two sheets of glass is 6, 10, 12, 16 or 20 mm. Condensation will form inside the IG unit if the units are not properly sealed. Seals may break down due to thermal movement and this means that the entire glazing unit will need to be replaced.

Space tube or bar

The spacer which serves to seal and support the adhesive that holds the sheets of glass together is either a hollow aluminium section or a butyl-based bar with an integral aluminium strip, as illustrated in Figure 7.76 and Figure 7.77.

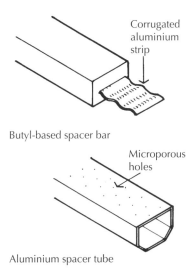

Figure 7.76   **Spacer bar and tube.**

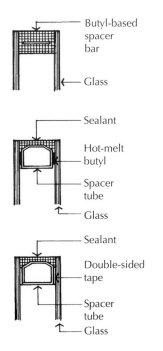

Figure 7.77   **Edge seals.**

The butyl-based bar, with integral aluminium strip, consists of a preformed section with a corrugated, integral aluminium strip as illustrated in Figure 7.76. The aluminium strip serves as reinforcement and in part as a barrier to moisture. A desiccant is embedded in the surface of the bar: this absorbs any moisture vapour that might otherwise condense to water on the inside faces of the glass of the unit. The advantage of the bar, commonly known as a 'swiggle strip', is that it is continuous and being self-adhesive makes fitting more rapid than using a spacer tube and sealant.

Desiccant, dehydrated air and argon gas

The visible face of the spacer tubes is perforated with microporous holes to facilitate absorption of moisture vapour by the desiccant. The butyl-based spacer bar has desiccant in the face of the bar. The sealed space between the two sheets of glass is usually filled with dehydrated air. To provide better thermal insulation the space may be filled with argon gas.

Durability

The useful life of a double- or triple-glazed unit depends mainly on the integrity of the seal as a barrier to moisture vapour penetrating the space between the sheets of glass. It is generally accepted that the useful life of the units is at best up to 20 years. Some assemblies give conditional guarantees of up to 10 years, conditional on workmanship in handling and fixing the units and on the glazing materials used. Replacement of the glazing allows the opportunity to upgrade the thermal performance of the window.

Setting, location blocks and distance pieces

IG units are hermetically sealed and subject to continuous flexing due to changes in atmosphere and temperature. The glazing materials that are used must allow for thermal and structural movement of both the IG unit and the window framing. The glazing method used should allow for movement and prevent water penetrating to the edges of the IG unit.

To accommodate different thermal and mechanical movements, a minimum clearance must be allowed all round IG units of from 3 mm for glass up to 2 m wide to 5 mm for glass over 2 m wide at sides and top of unit, and 6 mm at cill level.

Setting blocks and location blocks should be of some resilient, non-absorbent material such as sealed teak or mahogany, hammered lead, extruded uPVC, plasticised PVC or neoprene. The width of setting blocks should be equal to the thickness of the IG unit plus the backface clearance and at least 25 mm long. Location blocks should be 3 mm wider than the IG unit and at least 25 mm long.

Distance pieces are used to prevent displacement of glazing compounds or sealants by wind pressure on the glass, by retaining the IG unit firmly in the window. Distance pieces should be used except where loadbearing tapes or putty are used for stepped units. Distance pieces should be the same thickness as face clearance and made of a resilient, non-absorbent material similar to that for setting blocks (Figure 7.78).

Factory glazing, site glazing

The advantage of factory glazing is that the operation of glazing can be carried out under cover in conditions most suited to making a good job, with the glazed window delivered to site ready for fixing. The disadvantage of factory glazing is the possibility of damage to

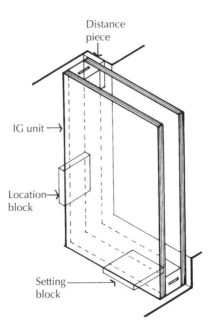

Figure 7.78   Blocks for IG units.

the glazed window in transit and through handling on site. The best conditions for glazing are clean, dry surfaces to which bedding and sealant materials can adhere. Because such conditions are rare on most building sites, factory glazing should have advantage over the other option, site glazing.

Glazing methods

The two systems of glazing used for IG units are solid bedding and drained methods.

(1) *Solid bedding.* The solid bedding method depends on the use of mastic bedding materials around the edges of the IG unit, inside the glazing rebate, with sealants to prevent the penetration of water to the edges of the unit. This method of bedding is most suited to window frames with plain, square sections and rebates such as wood or steel windows into which the unit may be bedded and secured with beads.

(2) *Drained glazing.* The drained method of glazing is designed to encourage water, which may have penetrated the glazing rebate around the edge of the IG unit, to drain to the outside. This method of glazing, which is particularly suited for use with the systems of gasket glazing commonly used with extruded, hollow sections of uPVC and aluminium windows, may be employed with plain section wood, steel, aluminium and uPVC windows.

Whichever method is used, the first defence against penetration of water to the edge seal of the IG units is the mastic or rubberised edge seals to glass, which should at once accommodate movement and act as a seal against water. Drainage is a back up to drain any water that has penetrated. To be effective, the drainage system should at once be adequate to drain water, be protected against wind-blown rain and remain clear of obstruction during the useful life of the window.

Beads with non-setting compound

This is the most straightforward method of bedding IG units in window frames of wood or metal, where one material is used both as bedding and sealant.

The bedding–sealant materials that may be used are non-setting compounds of synthetic rubber, oils and filters or low permeability one- or two-part curing sealants such as polysulphide, silicone or urethane base.

The clean rebates and wood beads are first primed or sealed. A generous fillet of bedding material is spread in the rebate into which setting blocks and distance pieces are pressed. The distance pieces are necessary to maintain the correct thickness of bedding, behind the glass, against wind pressure on the glass. The glass is then placed on the setting blocks and pressed firmly into the rebate against the distance pieces to provide a 3 mm thick back bedding. Location blocks are fixed in the side edge clearance. Bedding material is spread around the edges of the glass to fill the edge clearance gaps around the IG unit. A substantial fillet of bedding material is spread in the glazing rebate. The wood beads are bedded in position and pressed firmly into the bedding material so that there is 3 mm of bedding between the glass and the bead, and a thin bedding below the bead. The wood beads are secured in place with screws at maximum 200 mm centres and no more than 75 mm from corners. The bedding inside and out is trimmed and finished with a smooth chamfer or slope to shed water as illustrated in Figure 7.79.

Beads with loadbearing tape and sealant

An alternative method of solid bedding for IG units glazed to wood and metal frames uses loadbearing mastic tape or cellular, adhesive sections as face bedding and non-setting sealant as capping and bed for units.

Preformed loadbearing mastic tape is made from a fabric base, saturated in butyl (synthetic rubber) or polyisobutylene polymers in various widths and thicknesses. The tape is compressed during manufacture so that it has adequate loadbearing capacity as bedding

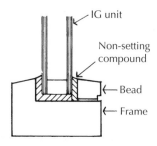

Figure 7.79 Solid bedding glazing.

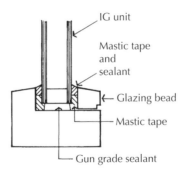

IG unit

Mastic tape
and
sealant

Glazing bead

Mastic tape

Gun grade sealant

Figure 7.80   Solid bedding glazing.

for glass, to resist wind pressure on the unit. A strip of mastic tape is fixed to the rebate upstand with its top edge about 6 mm below the sight line and setting blocks are placed in the rebate platform to support the IG unit. The IG unit is placed on the setting blocks and pressed firmly into the rebate up to the mastic tape in the upstand of the rebate.

One- or two-part sealant of polysulphide, silicone or urethane is run into the clearance gap between the lower edge of the IG unit and the rebate as bedding and as a thin bed for the glazing bead. A strip of loadbearing tape is run around the outside edge of the unit or the back face of the bead so that its top edge finishes some 6 mm below the sight line. The beads are fixed in place and pressed in firmly to compress the tape and bed the wood beads, which are screwed in place at a maximum of 200 mm and no more than 75 mm from corners.

Sealant capping is run around both sides of the glass and finished with a slope or chamfer as illustrated in Figure 7.80. Various combinations of loadbearing tapes and sealants, non-setting compounds and sealants and sealants by themselves may be used as bedding and sealants for solid bedding glazing of IG units, depending on the nature of the materials of the window framing and convenience in the operation of glazing.

Drained glazing

The drained method of glazing for IG units is designed to remove any water that has penetrated to the bottom rebate by drainage and to some extent by ventilation. Any water that lies for some time in the bottom rebate will adversely affect the adhesive edge seal to IG units.

Insert gasket glazing

The majority of extruded, hollow section uPVC and aluminium window frames employ gasket glazing systems. Gaskets are preformed sections of synthetic rubber that are shaped for insertion between the nibs around a groove in the glazing bead or the frame of the window and so positioned that the blades or bearing edges of the gasket make firm contact with glass faces as illustrated in Figure 7.81, to exclude rain and allow for some structural and thermal movement between frame and IG unit.

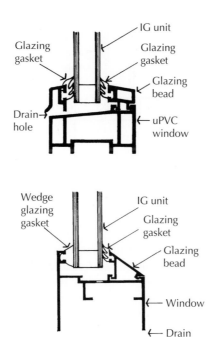

Figure 7.81   Drained glazing (uPVC and metal frames).

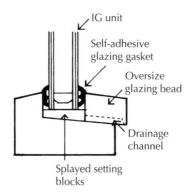

Figure 7.82   Drained glazing (timber frame).

Drainage to square section frames

As an alternative to solid bedding for square section wood and metal frames, systems of face bedding or gasket glazing with drainage to the bottom rebate may be used. For drainage the bottom rebate platform should be cut to slope out at an angle 10° to the horizontal. Drainage of water is through holes in the underside of the oversize bottom bead (Figure 7.82).

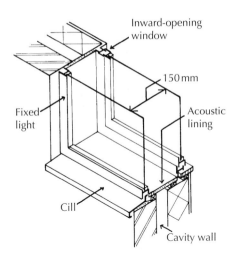

Figure 7.83   Double window for sound insulation.

Double and triple glazing for sound insulation

A major part of the penetration of sound through windows is transmitted as airborne sound through gaps between opening sashes and frames of windows. Appreciable reduction of sound transmission through windows can be achieved by fitting weather seals around all opening parts. Double glazing with a comparatively wide air space between sheets of glass may be used for additional reduction of sound transmission. In effect two windows are built into the opening, separated and not connected, by some 150 mm between the two sheets of glass as illustrated in Figure 7.83.

The outer window consists of a fixed light and the inner window is hinged to open inwards for the purpose of cleaning glass facing the cavity. This opening light is fitted with weatherstripping. The inside of the cavity is lined with an acoustic lining made of a material that will absorb some of the energy causing the air movement that causes sound. By fixing the two windows separately, vibrations of the outer sheet of glass, caused by airborne sound, will be reduced by the still air in the wide cavity and so have less effect on the inner sheet of glass. Triple glazing with large cavities will provide considerable sound insulation.

7.5 Hardware

Hardware is the general term used to describe the window furniture, such as hinges, locks and stays. Ironmongery and window furniture are also terms in common usage.

Hardware for timber windows

Wood casements, ventlights and sashes are hung on a pair of pressed steel butt hinges similar to those used for doors. To inhibit rusting the hinges are galvanised and finished with a lacquer coating. As an alternative, metal offset hinges may be used for casements,

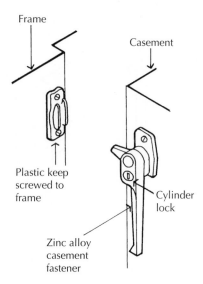

Frame

Casement

Plastic keep
screwed to
frame

Cylinder
lock

Zinc alloy
casement
fastener

Figure 7.84   Lockable casement fastener.

in which the pin is offset outside the casement so that when the casement is open, there is a gap between the hinged edge of the casement and the frame sufficient to allow for cleaning the outside of the glass from inside the building.

To secure casements, sashes and ventlights in the closed position a casement or window fastener or latch is fitted halfway up the height of casements and in the centre of ventlights. These fasteners operate through a latch, which is fixed to the opening casement, and engages a keep fixed to the frame as illustrated in Figure 7.84. The handle of the latch is raised to release the latch to open the window. For security the majority of window latches are lockable by a loose key that operates a lock in the latch.

Some care should be taken in fitting casement fasteners and keeps so that when the fastener is closed it firmly closes the casement on to flexible weather-stripping fixed in the rebate of the window frame around the opening light. Casement fasteners are made of cast zinc, aluminium or steel, usually finished with a protective coating of anodising powder or liquid organic coating or plastic.

To maintain opening lights in a window in a chosen open position, casement stays are fitted to the bottom rail of opening lights, as illustrated in Figure 7.85. The conventional form of these stays is a casement stay fixed to the bottom rail of the sash, which engages a casement peg fixed to the cill of the window frame. Holes in the stay provide a selection of possible openings. The stay, which pivots in its fixing, can be secured in a catch, fixed to the bottom rail of the sash when the window is in the closed position. Stays are made of cast zinc, aluminium or steel which is usually finished to match the protective coating of fasteners.

Hardware for aluminium frames

The hardware of hinges, lockable casement fasteners and stays are made of anodised finish, cast aluminium or die-cast zinc alloy, chromium plated.

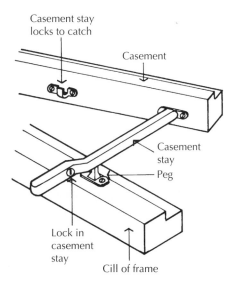

Figure 7.85   Lockable casement peg stay.

Hardware for uPVC frames

Hardware is made from cast aluminium alloy and die-cast zinc alloy with anodised, powder or organic liquid coatings for lockable fasteners and stays that are screwed through the outer wall of the uPVC sections into the reinforcement.

7.6 Window cills

It is good practice to set the outside face of windows back from the outside face of the wall in which they are set, so that the reveals of the opening give some protection against driving rain. The function of an external window cill (alternatively spelt sill) is to conduct the water that runs down from the windows away from the window, and to help prevent the wall immediately below the window from getting wet. The material from which the cill is made should be sufficiently impermeable and durable to perform this function during the life of the building. The internal cill of a window serves the purpose of a finish to cover the wall below the window inside the building, and as a stop for wall plaster. The material used for internal cills should be easy to clean and materials commonly used are painted softwood [or medium-density fibreboard (MDF)], plastic or ceramic tiles.

External cills

External window cills are formed either as an integral part of the window frame, as an attachment to the underside of the window, or as a sub-cill, which is in effect a part of the wall designed to serve as a cill. Most materials used for external cills are preformed so that the dimensions of the cill determine the position of the window in relation to the face of the wall. As a component part of a wall, external cills should serve to exclude wind and rain and provide adequate thermal insulation to the extent that the cill does not act as a thermal bridge. The materials used are natural stone, cast stone, concrete, tile and brick. Natural sedimentary igneous and metamorphic stone cills are less used than cast stone, concrete, tile and brick cills due to the cost of the material. Slate cills, which are readily available, but comparatively expensive, are used to some extent.

Natural stone cills

Natural stone cills are specially cut to section to provide a weathered surface, a groove for a water bar and an overhanging drip edge as illustrated in Figure 7.86. The top surface of the cill is finished flat, as a bed for the window cill. A groove is cut in the top of the cill to take a metal water bar that is set in mastic in the stone cill and the underside of the wood window cill. The water bar acts as a check against wind-driven rainwater that might otherwise penetrate between the stone sub-cill and the timber window cill. A shallow sinking is cut in the top of the cill down to a weathered face that slopes out to shed water. The shallow sinking acts as a check against wind-driven rain being blown up the weathered cill face. A shallow groove is cut near the outside edge of the underside of the cill to form a drip edge to encourage rainwater to run off.

Stooled ends to stone cills provide a solid bearing to walling at jambs and a weather seal against water penetrating between cill and jamb. Cills with stooled ends are commonly used with stone walling so that the cill may bond in with stonework. Stone sub-cills are

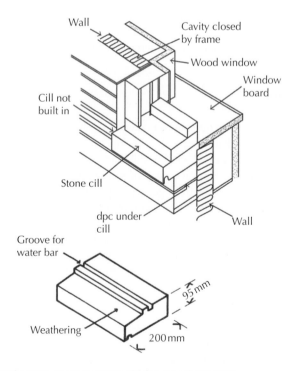

Figure 7.86   Natural stone or cast stone cill for wood window.

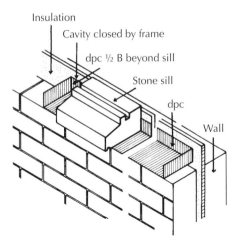

Figure 7.87   Under-cill dpc.

bedded in mortar on the under-cill damp-proof course (dpc), bearing on the outer brick leaf of the cavity wall. As the cill does not extend across the total width of the cavity wall it should not act as a thermal bridge. Reconstituted and precast concrete cills represent a cheaper alternative to natural stone cills. Cills of stone, reconstructed stone and precast concrete with square ends and cills of small units such as brick and tile may not be entirely effective in excluding rain that may penetrate joints in and at the ends of cills to the wall below, particularly in exposed positions. It is practice, therefore, to build in a sub-cill dpc as illustrated in Figure 7.87.

Slate cills
Slate cills are available in a range of standard and purpose-made sections to suit either timber or metal windows. This dense, durable material is impermeable to water and requires no maintenance; however, it is a poor thermal insulator and, being brittle, may crack due to movement of the building fabric. Slate cills should be cut and finished in one length to avoid the difficulty of making a weathertight joint in the material, a limitation that restricts their use to comparatively narrow windows unless a generous thickness of slate is used. The two standard sections illustrated in Figure 7.88 are for use with standard steel windows, which are screwed to the fillet fixed to the cill. Cills for use with wood windows are finished with a groove for a water bar. These cills are bedded in mortar on the wall below the window with the drip edge projected some 38 mm beyond the face of the wall.

Brick cills
A method of forming sub-cills to window openings in walls of fairface brickwork is with a course of bricks laid on edge or a course of brick specials. Whichever method is used the bricks should be sufficiently dense and weather resistant to stand the appreciable volume of rainwater that will run off the impermeable surface of the glass above. Standard size

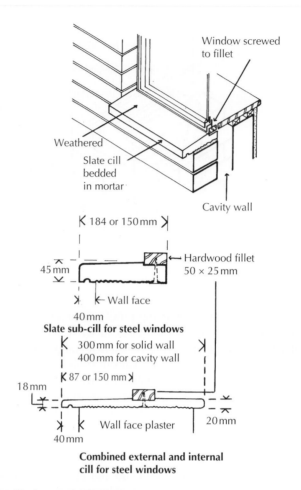

Figure 7.88 Slate cills for steel windows.

bricks are laid on edge as cills with one stretcher face and one header face exposed. The stretcher face top surface of the cill may be laid to a slight fall to shed water. The bricks are laid and pointed in cement mortar with the back edge of the bricks at least 30 mm behind the window cill edge and finished either flush with or projecting from the wall face as a drip edge. The special plinth bricks, illustrated in Figure 7.89, are made from dense clay and formed with a sloping weather face to shed rainwater. The bricks are laid on bed in cement mortar with the back face of the bricks set well back under the window cill so that there is a generous overhang of the drip edge of the wood cill above. A dpc is bedded below the cill bricks, turned up behind them and continued ½ B beyond the jambs of the window opening.

Tile cills

As a comparatively cheap form of cill, two courses of clay or concrete plain tiles are laid, breaking joint, in cement mortar. Machine pressed or concrete plain tiles, which may resist

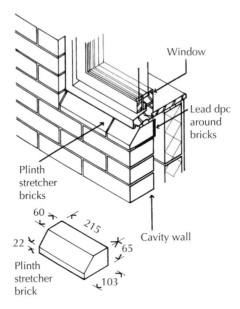

Figure 7.89   Brick cill.

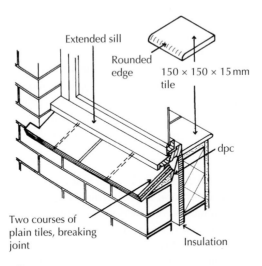

Figure 7.90   Plain tile cill.

the likelihood of frost damage at a slope of 30° are used. The tiles may be laid with their long axis parallel to the wall face to avoid cutting tiles or with their long axis at right angles to the wall and cut as necessary to suit the position of the window. The tiles may be finished square to the jambs of the window opening or notched around the angle of the jambs for appearance's sake, as illustrated in Figure 7.90. The tiles should be laid with their back edge well back under the window cill, so that there is a generous overhang of the window cill, and project out beyond the wall face about 38 mm as a drip edge. A dpc should be bedded below the tile cill and bedding.

Metal cills

Most metal window manufacturers provide standard section metal cills for fixing to the frame of their windows to give cover and protection to the wall below the window. The projection of the cill beyond the face of the wall is determined by the 25-mm width of the welded-on stop ends, which in turn determines the position of the window in the thickness of the wall. The joint between the ends of the cill and the jambs should be pointed with mastic. The steel cill itself will exclude rainwater but the end joints may be vulnerable to water, particularly in positions of severe exposure. A dpc in the course below the cill, extending each side of the opening, might be a wise precaution in conditions of severe exposure. Similarly, extruded aluminium section cills are made to suit aluminium windows as an integral part of the window.

Plastic cills

Most plastic windows have an integral cill as part of the window, which fits over some form of sub-cill. Some manufacturers provide a separate hollow section plastic cill, which is weathered to slope out and is designed to cover and protect the wall below the window. These separate cill sections are clipped or screwed to the frame, as illustrated in Figure 7.57.

Timber cills

Most standard section timber windows can be supplied with a timber cill section that is tongued to a groove in the cill of the frame so that it projects beyond the window either to cover and protect the wall below or to overlap a sub-cill. The cill is designed to project some 25–38 mm from the face of the wall as a drip edge and should be protected with paint or stain and regularly maintained. It is sensible to use hardwood cills in positions of moderate and severe exposure. The cill fits between the jambs of the opening, and the end butt joints should be bedded in mortar and pointed with mastic.

Internal cills

The surface of the internal cill should be such that it can easily be kept clean. A common form of internal cill is a softwood board, termed a windowboard, cut from 19 or 25 mm boards and wrought smooth on one face and square or rounded on one edge. The board may be tongued to fit to a groove in wood window frames. Alternatively, a MDF may be used. The board is nailed to plugs or bearers nailed to the wall so that it projects some 25 mm or more from the finished face of plaster, as illustrated in Figure 7.86, and then painted with gloss paint to provide a durable surface finish.

Clay or concrete tiles may be used as an internal cill. The tiles are bedded in mortar on the wall and pointed in cement, as illustrated in Figure 7.91. Rounded edge tiles are used and laid to project beyond the plaster face.

Various sections of plastic windowboard are made for use with uPVC and other windows and as replacement windowboards. The thin sections are of co-extruded uPVC with a closed cell, cellular core and an integral, impact modified uPVC skin. The thicker sections

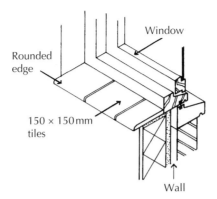

Figure 7.91   Tile internal cill.

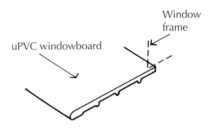

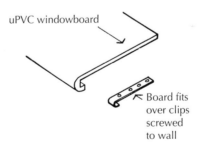

Figure 7.92   Plastic windowboards.

are of chipboard to which a uPVC finish is applied on exposed faces. The advantage of these windowboards is that they do not require painting and are easily cleaned. The disadvantage is that sharp objects fairly readily deface them and the damage cannot be made good. Plastic windowboards are cut to length and width with woodworking tools and fixed with concealed fixing clips, mortar bedding or silicone sealant adhesive, or are nailed or screwed to prepared timber grounds with the nail or screw heads covered with plastic caps (Figure 7.92).

8 Doors

A door is a solid barrier that is fixed to a frame in an opening in a wall to hinge, pivot or slide open (or to close). The door opening (the doorway) allows access and egress from buildings and between rooms, compartments and corridors. As a component part of a wall, the door and its frame are integral to the overall performance of the wall.

8.1 Functional requirements

The functional requirements of doors are specified, relating to both the component parts and the whole of the door sets or assemblies. The primary function of a door when open is to allow:

❑ Safe means of access and egress
❑ The functional requirements of a door, when closed, are:
❑ Security, prevention of unauthorised entry
❑ Privacy
❑ Strength and stability
❑ Resistance to weather
❑ Durability and freedom from maintenance
❑ Fire safety
❑ Resistance to the passage of heat
❑ Resistance to the passage of sound
❑ Airtightness
❑ Aesthetics

Safe means of access and egress

Approved Document M requires that there should be reasonable provision for disabled people to gain access and to use the building and its facilities in all new dwellings, as well as other types of new buildings used by the general public.

A door opening should be sufficiently wide and high for reasonably comfortable access of people, regardless of their ease of mobility. The standard width and height of 762 and 1981 mm (the metric equivalent of the former imperial sizes of 2'6″ × 6'6″) for single-leaf

---

Barry's Introduction to Construction of Buildings, Third Edition. Stephen Emmitt and Christopher A. Gorse.
© 2014 John Wiley & Sons, Ltd. Published 2014 by John Wiley & Sons, Ltd.

doors will be found in a majority of existing houses. To allow comfortable access for people with disabilities, the minimum clear opening for a single door should be 800 mm. Clear opening widths can be achieved by selecting a 1000 mm single-leaf door set (850 mm clear opening width of leaf) or a 1.8 m double-leaf door set (810 mm clear opening width of each).

Doors and design considerations (Approved Document M)

Approved Document M requires that the external entrance door to dwellings has a minimum clear opening width of 775 mm. Table 8.1, also from Approved Document M, gives minimum effective clear widths of entrance doors used by the general public as 1000 mm in new dwellings and 775 mm in existing buildings.

For door widths of 800 mm or less, Approved Document M gives guidance on the widths of corridors. Table 8.2 shows the minimum requirements for access routes within buildings for disabled use in relation to a range of door widths. Before selecting a door size, consideration should be given to the width of the corridor in which the door is to be positioned.

Double-leaf doors are commonly used for access to large spaces or rooms and for convenience in busy corridors. Where the door is an entrance door allowing disabled access, glazed panels as shown in Figure 8.1 should be used.

Table 8.1   Effective clear widths of door openings

Minimum effective width of doors		
Approved Document M		
Direction and approach width	New buildings (mm)	Existing buildings (mm)
When approaching head-on, without a turn	800	750
At right angles to an access route at least 1500 mm wide	800	750
At right angles to an access route at least 1200 mm wide	825	775
External doors to buildings used by the general public	1000	775

Table 8.2   Minimum widths of passageways for a range of door sizes

Minimum width of passageways in relation to door openings	
Approved Document M	
Doorway – clear opening width (mm)	Passageway width (mm)
750 or wider	900 (when approaching head-on)
750	1200 (when approaching at an angle)
775	1050 (when approaching at an angle)
800	900 (when approaching at an angle)

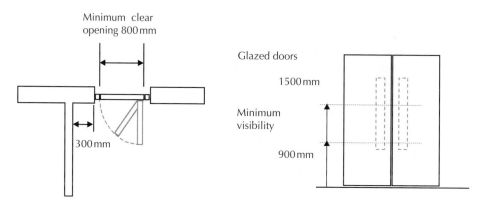

(a) Clear opening (single door) – plan   (b) Glazed panels in entrance doorways – elevation

Figure 8.1   Doors: disabled access.

Doors can be hazardous and consideration should be given to the ability of people to distinguish a door from a clear opening (which may be a problem with glass doors). The presence of a door should be apparent when it is both open and closed; measures should be taken to avoid people walking into doors which are retained in the open position. Consideration should be given to those who are visually impaired who will use the doors. Further considerations are required under Approved Document M when glass doors are used; this is discussed later.

Door closers in public buildings should be avoided, as they can be difficult for disabled people to operate. Where door closers are required to comply with, for example, fire regulations, the door opening should be power assisted to reduce the physical effort required to operate the door.

Security

External doors should be designed so as to provide a secure barrier against unauthorised entry. An external door, particularly at the rear or sides of buildings (out of sight), is obviously a prime target for forced entry. Solid hinges, locks and key bolts to a solidly framed door in a soundly fixed solid frame are the best security against forced entry. Insurance companies offer advice and may set minimum performance requirements for door types and especially the locks. Advice is also available from the police via their Secure by Design initiative.

In addition to physical barriers to unauthorised entry closed circuit TV may be used to help control ingress to a building. The units are positioned close to points of entry to enhance the level of scrutiny and provide records of those that enter and exit the building or different areas within the building. Face, eye, fingerprint and hand scanning are also becoming more popular, as are locking devices that are unlocked with keypad-enabled codes, providing data to security managers as to who is in the building at any one time.

Privacy

Doors should serve to maintain visual privacy inside rooms to the same extent that the enclosing walls or partitions do. For acoustic privacy, doors should offer the same reduction in sound as the surrounding walls or partitions, should be close fitting to the door frame or lining and be fitted with flexible air seals all round. These seals should fit sufficiently to serve as an airborne sound barrier and to improve the airtightness of the door. For internal rooms, where airtightness is not a consideration, seals can make doors difficult to open and care is required in deciding when and where to use the appropriate type of seal.

Strength, stability and durability

A door assembly should be strong enough to sustain the conditions of use without undue damage. The suggested tests are for resistance to damage by slamming shut or open, heavy body impact, hard body impact, torsion due to the leaf being stuck in the frame, resistance to jarring vibrations and misuse of door handles. Categories of duty related to use are suggested, from light duty (LD) through medium duty (MD) and heavy duty (HD) to severe duty (SD). A door should be easy to open, close, fasten or unfasten and should stay closed when shut.

Dimensional stability

A door should not bow, twist or deform in normal use to the extent that its appearance is unacceptable or it is difficult to open or close. The dimensional stability of wood, metal and plastic doors is affected by temperature and humidity differences. Wood doors are affected mainly by temperature and humidity, metal doors by temperature, and plastic by thermal and hygrothermal movements.

Bow in doors is caused by differences in temperature and humidity on opposite faces, which may cause the door to bow with a curvature that is mainly in the height of the door and should not exceed 10 mm. Twist is caused, particularly in panelled wood doors, where movement of the spiral grain, due to changes of moisture content, causes one free corner to move away from the frame. This should not exceed 10 mm.

Weather resistance

As a component part of an external wall, a door should serve to exclude wind and rain depending on the anticipated conditions of exposure described for windows. Careful detailing of the door jambs and inclusion of weather-stripping will help to reduce or eliminate draughts and assist with the conservation of heat. Laboratory tests on doors show that external doors, particularly those opening inwards, are more susceptible to water leakage than windows. It is difficult to design an inward-opening external door that will meet the same standards of watertightness that are expected of windows, without the protection of some form of porch or canopy. For maximum watertightness, a door will need effective weather-stripping, which will to an extent make opening more difficult, and a high or complex threshold, which may obstruct ease of access.

Fire safety

Doors may serve two functions in the event of fire in buildings: first as a barrier to limit the spread of smoke and fire and second to protect escape routes.

To limit the spread of fire, it is usual to divide larger buildings into compartments of restricted floor area by means of compartment floors and walls. Where doors are formed in compartment walls, the door must, when closed, act as a barrier to fire in the same way as the walls. For this purpose, doors must have a notional integrity, which is the period in minutes that they will resist the penetration of fire. Approved Document B gives provisions of tests for fire resistance of doors, with minimum requirements in minutes for the integrity of doors. These are usually stated as, for example, FD20, being a provision of 20 minutes minimum integrity for a fire door. FD20S denotes a door that will prevent the passage of smoke for 20 minutes.

There should be adequate means of escape from buildings in case of fire, to a place of safety outside, which is capable of being safely and effectively used at all times. To meet this basic requirement, it is usual to define escape routes from most buildings along corridors and stairways that are protected by fire barriers and doors from the effects of fire for defined periods. This latter function of a fire door is described as smoke control. A majority of doors along escape routes will need to serve as fire doors to resist the spread of fire and to control smoke.

Resistance to the passage of heat

Operation, strength, stability and security have always held more importance in the construction of a door than resistance to heat transfer. With increased thermal insulation of the external fabric, the importance of fitting thermally efficient door sets has grown. There are two considerations. First, the resistance to heat transfer by both door and frame, and second, the air leakage between door and frame, and between frame and wall. Most doors in current use are poor thermal insulators. A typical single-skin timber panel external door has a U-value of $3.0 \, W/m^2 K$. Houses that are classed as energy efficient include doors with U-values of $0.5 \, W/m^2 K$ or less. An example of door construction, which could have a U-value of less than $0.5 \, W/m^2 K$, is shown in Figure 8.2.

Airtightness

Like windows, doors are particularly susceptible to air leakage. All doors, whether unplasticated polyvinyl chloride (uPVC), metal or wood, expand and contract due to temperature differences between the internal and external environments and seasonal changes. Over time, the doors and frames distort, the fittings wear and the effectiveness of seals may be reduced. To maintain airtightness, weather-stripping that is capable of retaining a seal if the door distorts should be used.

Resistance to the passage of sound

Generally, the heavier a door (the greater its mass) the more effective a barrier it is in reducing sound transmission. A solid panel door is more effective than a hollow-core flush door. To be effective as a sound barrier, a door should be fitted with air seals all round as

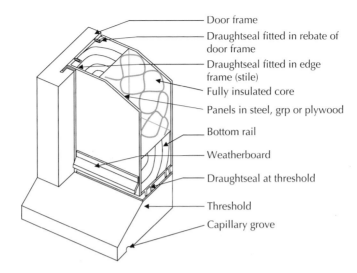

Figure 8.2 Typical construction of an insulated door.

a barrier against airborne sound. Figure 8.3 shows threshold seals that can be used to improve insulation against airborne sound.

Aesthetics

Given our high level of interaction with doors and door openings, it is important that the aesthetic qualities of the door are considered. This is very much a personal consideration and is not addressed in this book, other than to say that the choice of door and door furniture (handles, hinges, locks, etc.) will be influenced by personal taste as well as by functional and performance requirements.

8.2 Door types

Doors are supported in openings (doorways) on hinges as side hung, on pivots as double swing and on tracks as sliding or folding doors as illustrated in Figure 8.4. The side on which the door is hung is termed its hand or handing. Whether it opens into, or out of, a space or room is a matter of convenience in use. By convention, doors usually open into the room of which they are part of the enclosure. There have been systems of describing the hand of doors by reference to opening in or out and as either left-hand or right-hand, clockwise or anti-clockwise. These are of very little use because of the difficulty of clearly defining what is outside and what is inside. If a doorway is used as part of an escape route, the door should be hung so that it opens in the direction that people will egress the building.

❑   Hinged, single-swing, side-hung doors are for frequent use between rooms and between rooms and corridors or landings.

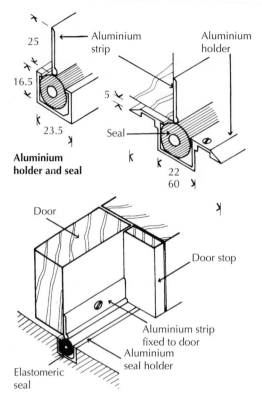

Figure 8.3 Threshold seal for internal doors.

❑ Hinged and pivoted side-hung, double-swing doors are for frequent use along corridors to accommodate two-way foot traffic.
❑ Sliding and sliding folding doors hung on overhead track are for occasional use in openings between rooms to convert single rooms into larger double rooms.
❑ Combinations of single-swing, double-swing and sliding and folding doors may be used for specific purposes.

Timber door classification

Timber doors may be classified as (Figure 8.5):

❑ Panelled doors
❑ Glazed doors
❑ Flush doors
❑ Matchboarded doors

The traditional door is formed from solid softwood or hardwood members framed around panels. This construction has been in use for centuries with little modification other than

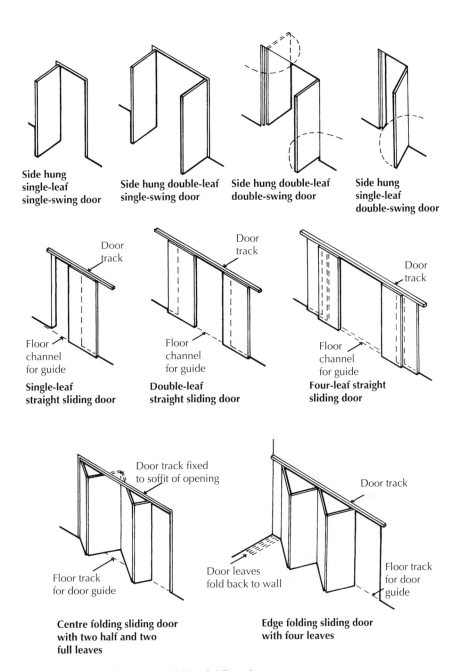

Figure 8.4 Hinged, sliding and sliding folding doors.

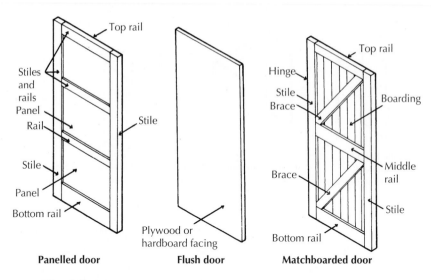

Figure 8.5   Wood doors.

in changes in jointing techniques due to machine assembly and the use of substitute materials for wood.

Panelled doors

Panelled doors are framed with stiles and rails around a panel or panels of wood or plywood. Because the door is hinged on one side to open, it tends to sink on the lock stile. The stiles and rails have to be joined to resist the tendency of the door to sink, and the two types of joint used are a mortice and tenon joint (Figure 8.6) and a dowelled joint (see Figure 8.8).

Mortice and tenon joint
This is a strong joint used to frame members at right angles in joinery work. Figure 8.6 shows the stiles and rails of a panelled door before they are put together and glued, wedged and cramped around the panels (which are not shown).

Haunched tenon
The tenons cut on the ends of the top rail cannot be as deep as the rail if they are to fit into enclosing mortices, so a tenon about 50 mm deep is cut. It is possible that the timber of the top rail may twist as it dries, so to prevent this from happening, a small projecting haunch is cut on top of the tenon, which fits into a groove in the stile.

Two tenons are cut on the ends of the middle and bottom rails. It would be possible to cut one tenon to the depth of the rails, but the wood around the mortice might bow out and so weaken the joint. Also, a tenon as deep as the rail might shrink and become loose in the mortice. To avoid this, a tenon should not be deeper than five times its thickness, hence the use of two tenons on the middle and bottom rails. Double tenons are sometimes

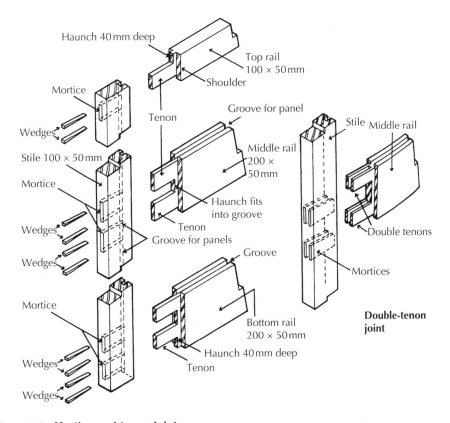

Figure 8.6 Mortice and tenon joint.

cut on the ends of the middle rail as illustrated in Figure 8.6. The purpose of double tenons is to provide a space into which a mortice lock can be fitted without damaging the tenons.

Glueing, wedging and cramping

The word cramp describes the operation of forcing the tenons tightly into mortices. The members of the door are cramped together with metal cramps, which bind the members together until the glue in the joints has hardened. Before the tenons are fitted into the mortices, both tenon and mortice are coated with glue. When the members of the door have been cramped together, small wood wedges are knocked into the mortices on the top and bottom of each tenon. When the glue has hardened, the cramps are released and the projecting ends of tenon and wedges are cut off flush with the edges of the stiles.

Mortice and tenon joints may, over time, become loose as the door timber shrinks, and the door will lose shape. To prevent this, panelled doors are sometimes put together with pinned mortice and tenon joints. The mortices and tenons are cut in the usual way and holes are cut through the tenons and the sides of the mortices, as illustrated in Figure 8.7. The tenons are fitted to the mortices, and wooded pins (dowels) of 13 mm diameter are

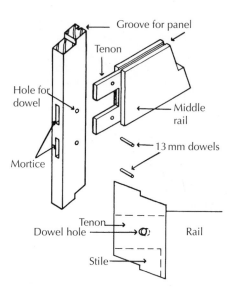

Figure 8.7 Pinned mortice and tenon.

driven through both mortice and tenon. Because the holes in the tenons and mortices are cut slightly off centre, the pins, as they are driven in, draw the tenons into the mortices. Pinned mortice and tenon joints are glued and wedged. This joint should be used for heavy-panelled doors.

The economic advantage of woodworking machinery cannot be exploited to the full in the cutting, shaping and assembly of mortice and tenon joints. It is practice, therefore, to use a jointing system better fitted to woodworking machine operations in the cutting and assembly of mass-produced doors.

Dowelled joints

With the dowelled joint, which is used to frame standard size panelled doors, continuous grooves are cut in the edges of the stiles and rails to take the panels and continuous, protruding haunches on the ends of rails as illustrated in Figure 8.8. The haunches on the ends of the rails fit into the grooves cut in the edges of stiles to secure and level the members. By this arrangement, the cutting of grooves and haunches are continuous operations suited to the use of woodworking machinery. Wood, plywood or wood particleboard, plain panels are cut to size to fit into the grooves cut in the stiles and rails. With improvements in glues, this type of joint will strongly frame the members of a panelled door.

Panels – avoiding shrinkage cracks

The comparatively thin wood from which panels are made will in time shrink due to loss of moisture, particularly in heated buildings. As drying shrinkage of wood occurs mainly across the long grain, panels may develop vertical cracks. To minimise shrinkage cracking of wood panels, it is practice to make panels that are more than 250 mm wide from boards

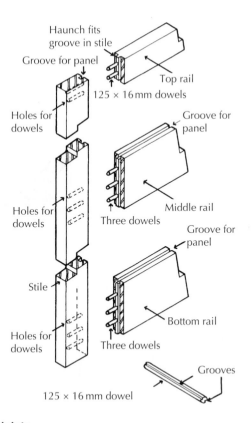

Figure 8.8   Dowelled joints.

that are tongued together. The term 'tongued' describes the operation of jointing boards by cutting grooves in their edges into which a thin tongue of wood is cramped and glued as illustrated in Figure 8.9. By using boards rather than one panel, shrinkage cracking may be avoided.

To avoid shrinkage of panels, plywood may be used. Plywood is made from three, five, seven or nine plies of thin sheets of wood firmly glued together so that the long grain of one ply is at right angles to the long grain of the plies to which it is bonded. The opposed long grain of the plies strongly resists shrinkage cracking. Three-plywood 5 or 6.5 mm thick is generally used for panels.

Securing panels
The traditional method of fixing and securing panels in doors is to set them into the grooves cut in the edges of the stiles and rails. To allow for drying shrinkage and any framing movements, there should be a clearance of 2 mm between the edges of the panels and the bottom of the grooves in which they are set. The advantages of setting panels in grooves in the framing members is that the panels are securely fixed in place and shrinkage and movement of the frame will not cause visible cracks to open up around the panels.

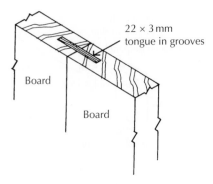

Figure 8.9   Boards through tongued to form panels.

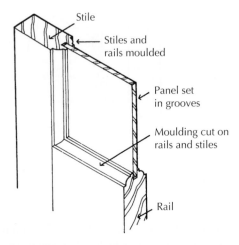

Figure 8.10   Framing moulded around panels.

To improve the appearance of a panelled door, mouldings are cut on the edges of the stiles and rails around panels as illustrated in Figure 8.10. For this finish, the ends of rails have to be scribed to fit around the moulding cut on the stiles. The term 'scribed' describes the operation of cutting the wood to shape to fit closely around the moulding, which is cut continuously down the length of the stile.

A cheaper method of giving the appearance of mouldings around panels is to nail moulded timber beads around each panel as illustrated in Figure 8.11. Owing to the drying shrinkage, cracks may open up between the beads and the framing and panels of doors.

Panelled doors

The traditional panelled door was constructed with four or six panels with central framing members, termed muntins, tenoned to rails as illustrated in Figure 8.12. The advantage of

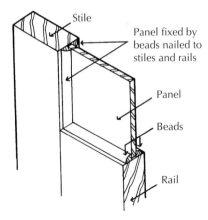

Figure 8.11   Planted moulding.

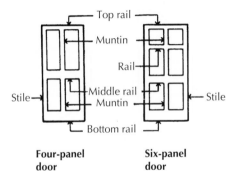

Figure 8.12   Traditional panelled doors.

this arrangement is that the width of the panels is limited to reduce the possibility of shrinkage cracks and the shape of the panels emphasises the verticality of the door.

For the sake of economy in using woodworking machinery and a change in fashion, standard, panelled wood doors are made without muntins with panels between the stiles of the framing as illustrated in Figure 8.13.

Doors with raised panels

Entrance doors and doors to principal rooms in both domestic and public buildings are often made more imposing and attractive by the use of panels that are raised so that the panel is thicker at the centre than at the edges. Such doors are often made of hardwood, which is finished to display the colour and grain of the wood by polish or French polish.

To avoid the ends of tenons showing on the edges of the door, it is practice to use stub tenons, which are secured with foxtail wedges, as illustrated in Figure 8.14. The foxtail wedges fit to saw cuts at the ends of the stub tenons so that when the tenon is cramped

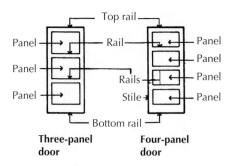

Figure 8.13   Standard interior panelled doors.

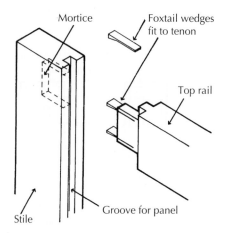

Figure 8.14   Foxtail wedges.

into the mortice the wedges spread the tenon to bind to the mortice. This type of joint, which has to be very accurately cut, makes a sound joint. As an alternative, dowelled joints may be used. Raised panels are either bevel raised, bevel raised and fielded, or square raised and fielded.

Bevel-raised panels
These are cut with four similar bevel faces each with a shallow rise from the edges of the panel to a point with square panels and ridge with rectangular panels as illustrated in Figure 8.15a.

Bevel-raised and fielded panels
These are cut with four similar bevel faces rising from the edges of the panel to a flat surface, termed the field, illustrated in Figure 8.15b. At the field, the panel is either as thick as or slightly less thick than the stiles. The proportion of the fielded surface to the whole panel is a matter of taste.

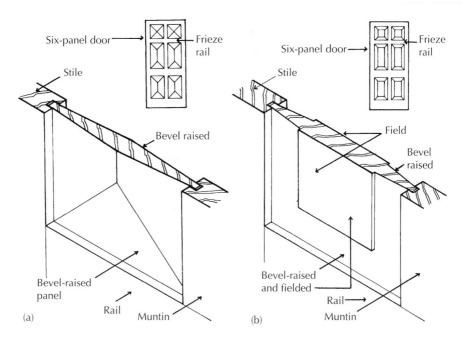

Figure 8.15   (a) Bevel-raised panels. (b) Bevel-raised and fielded panels.

Raised and fielded panel
The panel, which is of uniform thickness around the edges, is raised to a flat field at the centre with a shallow sinking as illustrated in Figure 8.16a, the field being square or rectangular depending on the shape of the panel. Panels may be raised on both sides, as shown in Figure 8.15a, or on one side only, as illustrated in Figure 8.16a.

Bolection moulding
A bolection moulding is planted (nailed) around the panels of a door for the sake of appearance. The moulding is cut so that when it is fixed it covers the edges of the stiles and rails around the panel, as illustrated in Figure 8.16b. This particular section of wood moulding may be used with both raised and fielded panels on one or both sides of a door.

Double-margin door
A wide-panelled door may be constructed as one door or as two doors hinged to meet in the middle. A single door would tend to look clumsy with oversize panels, whereas two doors would each be too narrow for comfortable access when only one leaf is opened. As a compromise such doors can be made up as if they were two doors with the two doors fixed together and acting as a single door as illustrated in Figure 8.17.

Solid panels – flush panels
Solid panel doors are constructed with panels as thick as the stiles and rails around them for strength, security, aesthetics or where the door acts as a fire check door.

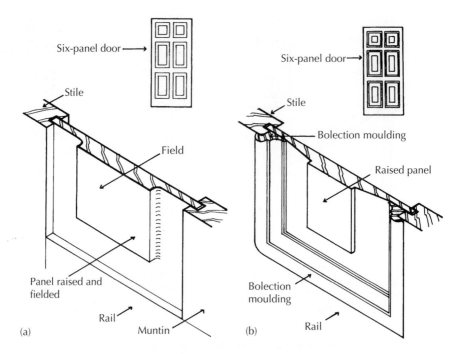

Figure 8.16   (a) Raised and fielded panels. (b) Bolection moulding.

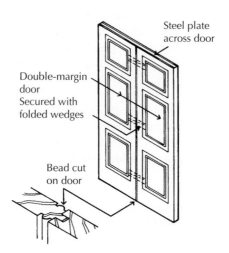

Figure 8.17   Double-margin door.

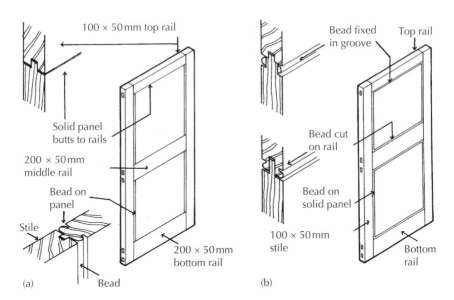

Figure 8.18 **(a) Door with solid panels bead butt. (b) Door with solid panels bead flush.**

These doors are usually constructed of hardwood, such as oak. The solid panels are tongued to grooves in the stiles and rails and are either cut with a bead on their vertical edges or with a bead all round each panel for appearance's sake. Timber shrinks more across than along its long grain, and because the long grain in these panels is arranged vertically the shrinkage at the sides of the panels will be more than at the top and bottom. For this reason, beads are cut on the vertical edges of the panels to mask any shrinkage cracks that might appear, as illustrated in Figure 8.18.

Double-swing doors

Double-swing doors are used at the end of and along corridors and as shop-entrance doors for the convenience of two-way foot traffic. As it is easier and quicker to push than pull a door open, this type of door is used for the convenience of passing through the doorway in either direction. To avoid the danger of the door being pushed simultaneously from both sides, these doors are constructed with either a glazed top panel or they are fully glazed.

To allow for the double-swing action of the door, the vertical edges of the door are rounded to rotate inside a rounded rebate in the timber door frame as illustrated in Figure 8.19. The door is either hung on double-action hinges designed to accommodate the double swing or supported by double-action floor springs and a top pivot as illustrated in Figure 8.19. The door is supported by a shoe that fits to the spindle of the double-action spring, which fits into a box set in the floor. A top plate, which is screwed to the spring box, finishes flush with the floor. A bearing plate, fixed to the top of the door, fits to a pin protruding from a plate fixed to the door frame. Because of the necessary clearance gaps around the door, it provides poor thermal and acoustic resistance.

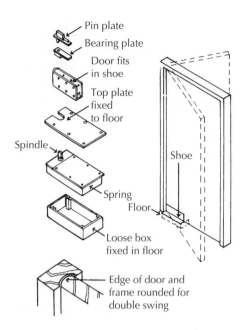

Figure 8.19   Double-swing door.

Sliding doors

Sliding doors are used either to provide a clear opening or to act as a barrier between adjacent rooms or spaces. Sliding doors can only be used where there is room for the door or doors to slide to one or both sides of the opening. These doors may slide as two leaves, one to each side or as one leaf, to one side of the opening.

Sliding folding doors

Sliding folding doors are used to separate large rooms or internal spaces into two separate spaces. Because of the folding sliding action the hinged leaves of the door can slide and fold back against a wall to occupy little space. Sliding and sliding folding doors should be hung on an overhead track. An inverted 'U'-shaped track is fixed to the overhead beam with brackets as illustrated in Figure 8.20. Hangers, in the form of a four-wheeled trolley, run in the overhead track and support the door leaves through brackets screwed to the doors.

At the bottom of the door a channel, set and fixed in the floor, acts as a guide to a pin or wheel fixed by brackets to the door. To maintain a reasonably easy movement of sliding doors, it is necessary to keep the bottom guide track free from dirt that might obstruct movement and to keep the hangers reasonably oiled.

The door leaves of both sliding and sliding folding doors may be of either panelled or flush construction. The lighter the doors, the easier their movement, and the heavier, the better they serve as a sound barrier.

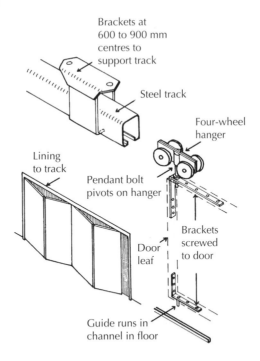

Figure 8.20 Sliding folding door.

Glazed doors

Doors with panels of glass, or fully glazed doors, are used to provide daylight into unlit spaces or to allow people to see through the door. A vision panel may prevent a person from opening a door into the path of a person on the other side. Fully glazed doors fixed in an external wall serve as both door and window.

Glazed entrance doors, which stand within a glazed screen, should be clearly differentiated from the screen. Approved Document M requires a high-contrast strip to be used on the top, bottom and sides of the door.

It is important that glazed doorways are easily recognisable since they can pose problems for people with impaired sight. Approved Document M requires that fully glazed doors and screens are defined with 'manifestations' on the glass at two levels, 850–1000 mm and 1400–1600 mm above the floor, which can be seen from both the inside and outside. A manifestation takes the form of a logo, sign, image or decorative feature. The manifestation should be at least 150 mm high, providing a contrast strip.

Where glass doors are capable of being held open, they must be protected by a guard to prevent the leading edge causing a hazard, thus preventing people from walking into the open door.

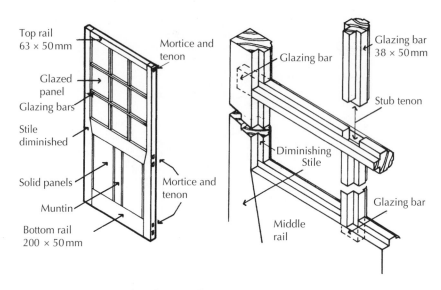

Figure 8.21   Glazed door with diminishing (gun stock) stiles.

Purpose-made glazed doors

A common form of traditional wood door for external use has glazed upper panels to admit daylight to halls. This door is framed as solid lower panels with upper glazed panels. The door may be framed as a normal panelled door with continuous width stiles or with diminishing stiles to provide a greater width for glazing, and this is illustrated in Figure 8.21.

French casement

The traditional form of first-floor window to many French and northern Mediterranean countries is in the form of a timber-framed door, fully glazed as illustrated in Figure 8.22. The door is made with vertical and horizontal glazing bars as part of the framing. These doors serve as windows to the first-floor rooms and as doors, opening in, for access to balconies and ventilation during summer. Louvred timber shutters are hung externally to close over the opening, providing shading from the sun while also providing natural ventilation.

Flush doors

A variety of flush doors are manufactured with plain flush faces both sides and fibreboard facings press-moulded, often with comparatively shallow sinkings, to resemble the appearance of panelled doors.

Cellular core flush doors are made with a cellular, fibreboard or paper core in a light softwood frame with lock and hinge blocks, covered with plywood or hardboard facings glued to the frame and core both sides as illustrated in Figure 8.23. These lightweight doors are for LD such as internal domestic doors. They do not withstand rough usage and provide poor acoustic privacy, security and fire resistance. They are mass-produced in a small range of standard sizes and are cheap.

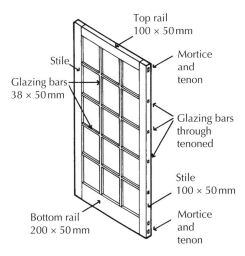

Figure 8.22   Casement door (French casement).

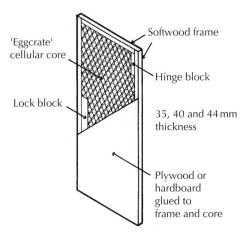

Figure 8.23   Cellular core flush door.

Skeleton core flush doors are made with a core of small-section timbers, as illustrated in Figure 8.24. The main members of this structural core are the stiles and rails, with intermediate rails as shown, as a base for the facing of plywood or hardboard. The framing core members are joined with glued, tongued and grooved joints. The door illustrated in Figure 8.24 has a skeleton core occupying from 30% to 40% of the core of the door. This is a LD door suitable for internal domestic use. A similar skeleton core flush door with more substantial intermediate rails at the core, where the core occupies from 50% to 60%, is an MD door suitable for use internally in domestic and public buildings and for external use in

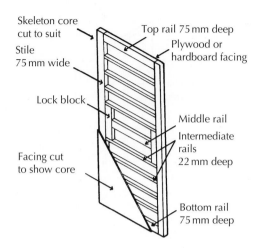

Figure 8.24   Skeleton core flush door.

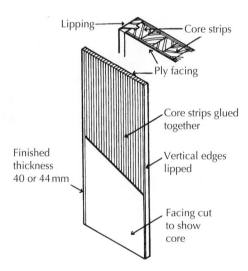

Figure 8.25   Solid core (laminate) flush door.

sheltered positions. This somewhat more substantial door will withstand normal use and maintain its shape stability better than a cellular core door.

Solid core flush doors are made with a core of timber, chipboard, flaxboard or compressed fibreboard strips. The solid core door illustrated in Figure 8.25 has a core of timber strips glued together, with plywood facings both sides glued to the solid core. The door is edged with vertical lipping to provide a neat finish. Because of the solid core, these doors have somewhat better shape and surface stability and acoustic resistance than the cellular or skeleton core flush doors.

The chipboard, flaxboard and compressed fibreboard strip core doors are made with a solid core enclosed in a light timber frame to which hardboard or plywood is fixed. These solid core doors are more expensive than cellular core or skeleton core doors. Solid core flush doors may be used as fire doors with an integrity rating of 20 or 30 minutes.

Fire doors

Fire doors serve to protect escape routes and the contents and structure of buildings by limiting the spread of smoke and fire. Fire doors that are fixed for smoke control are only capable of withstanding smoke at ambient (surrounding) temperatures and limited smoke at medium temperatures by self-closing devices and flexible seals.

Fire doors that are fixed as part of a fire compartment and as isolation of special risk areas should have a minimum fire resistance, for integrity only, of a period appropriate to that set out in Approved Document B.

Fire-resistance integrity

The current requirement for fire doors is that complete door assemblies be tested in accordance with BS 476 and certified as meeting the recommendations of performance for integrity and noted, for example, as FD20, FD30 as satisfying the requirements for 20 and 30 minutes integrity, respectively.

The performance test for fire doors that serve as barriers to the spread of fire is determined from the integrity of a door assembly or door set in its resistance to penetration by flame and hot gases. The test is carried out on a door assembly that includes all hardware, supports, fixings, door leaf and frame. Each face of the door assembly is exposed separately to prescribed heating conditions from a furnace, on a temperature–time relationship, to determine the time to failure of integrity. Failure of integrity occurs when flame or hot gases penetrate gaps or cracks in the door assembly and cause flaming of a cotton wool pad on the side of the assembly opposite to the furnace.

The door leafs illustrated in Figure 8.26 are constructed and faced to provide 30 and 60 minutes for integrity. The 30 minute skeleton core flush door is protected with plasterboard panels fixed to the skeleton core under the plywood or hardboard facings as illustrated in Figure 8.27. The 30 minute, solid core fire doors are protected with wood strips or high-density chipboard covered with plywood or hardboard facings. The 60 minute, solid core fire doors are protected with compressed mineral wool or high-density chipboard and hardboard or plywood facings. The flush, steel fire door to provide 60 minutes protection, shown in Figure 8.26 and illustrated in Figure 8.28, has welded sheet steel facings. The casing is pressed around steel stiffeners and a core of compressed mineral wool. The door is provided with intumescent seals and is hung on a pressed steel frame.

A fire door should be easy to operate, serve as an effective barrier to the spread of smoke and fire when closed, and be fitted with a self-closing device. For ease of operation, there must be clearance gaps around the door leaf. These clearance gaps are effectively sealed when a door leaf closes into and up to the rebate in a door frame. Where a door leaf has distorted in use and when the leaf is distorted by the heat of a fire, then the leaf will no longer fit tightly inside the rebate of the frame and smoke and flame can spread through the gaps around the door leaf. As a barrier to the spread of smoke, flexible seals should be

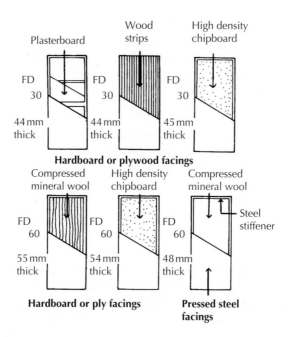

Figure 8.26 Fire doors.

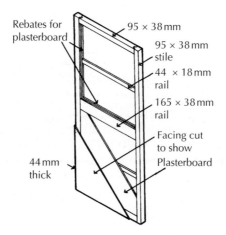

Figure 8.27 Standard half-hour fire door.

fixed between door leafs and frames and as a barrier to the spread of fire, heat-activated (intumescent) seals should be fitted.

Smoke-control door assemblies (FDS) that serve only as a barrier to the spread of smoke without any requirement for fire resistance, such as fire doors along an escape route, may be fitted in rebated frames or hung to open both ways. To provide an effective seal against the spread of smoke through gaps around these doors, flexible seals should be fitted.

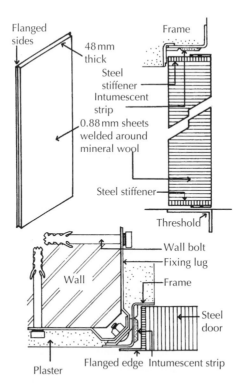

Figure 8.28   Flush steel fire door.

Smoke-control door assemblies that serve as a barrier to the spread of smoke and fire, such as doors leading to a protected escape route, should be hung in rebated frames and tested for a minimum integrity of 30 minutes against the spread of fire, and should be fitted with heat-activated (intumescent) seals and flexible edge seals against the likelihood of the door leaf deforming.

Fire door assemblies fixed in compartment walls and to enclosures to special risk areas should be hung in rebated frames and tested for integrity for not less than 30 minutes or such period as detailed in Approved Document B, and should be fitted with the intumescent seals. The currently accepted minimum size of a softwood door frame for a fire door is 70 × 30 mm exclusive of a planted stop.

Heat-activated intumescent seals

An intumescent seal is made of a material that swells by foaming and expanding at temperatures between 140° and 300°. The intumescent seals illustrated in Figure 8.29 consist of aluminium holders inside which the intumescent material is held. The aluminium holders are fixed in a rebate to the edges of doors or to the door frames. When the temperature rises sufficiently, the intumescent material inside the holders expands out through the vertical slots in the holder and effectively seals the door in the frame as a barrier to the spread of flame.

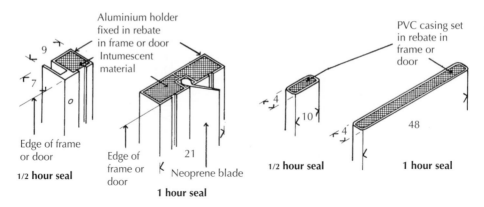

Figure 8.29   Intumescent fire seals.

The neoprene blade shown in the 1 hour seal acts as a seal against smoke that occurs in the early stages of a fire before the intumescent material is sufficiently hot to expand. The intumescent seals shown in Figure 8.29 comprise a polyvinyl chloride (PVC) casing in which the intumescent material is held. The seals are set in rebates at the edges of the door or in the rebate of the frame. As the temperature rises, the thermoplastic PVC casing gradually softens so that when the temperature has risen sufficiently, the intumescent material expands, ruptures the PVC casing and acts as a seal around the door.

For a fire door to be effective against smoke and the spread of fire, it should, when not in use, be positively closed to the frame by some self-closing device. The door closers that are used are overhead door closers or one of the floor springs (Figure 8.19).

The overhead door closers illustrated in Figure 8.30 consist of a hydraulically operated cylinder in a metal casing that is either screwed to the door face or set in a housing in the top of the door leaf for appearance sake. Pivoted arms, one to the housing and one attached to the door frame, act to automatically close the door to the frame, after the door has been opened. Because these door closers are fixed to doors that are generally in frequent use, they require regular maintenance if they are to serve their purpose.

Matchboarded doors

These simple doors are made with a facing of tongued, grooved and V-jointed boards that are nailed to horizontal ledges, braces between ledges or to a frame. These relatively crude doors are sometimes described as 'matchboarded' doors because of the comparatively thin boards from which they are made (also known as 'cottage doors').

Ledged matchboard doors
Ledged matchboard doors are made by nailing matchboards to horizontal ledges, as illustrated in Figure 8.31a. The nailing of the boards to the ledges does not strongly frame the door, which is liable to sink and lose shape. This door is used for narrow openings only.

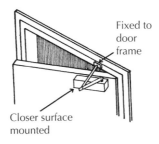

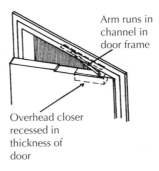

Figure 8.30   Overhead door closer.

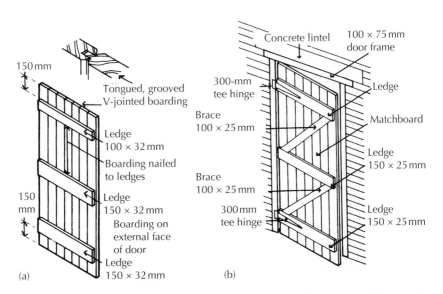

Figure 8.31   (a) Ledged matchboarded door. (b) Ledged and braced matchboarded door.

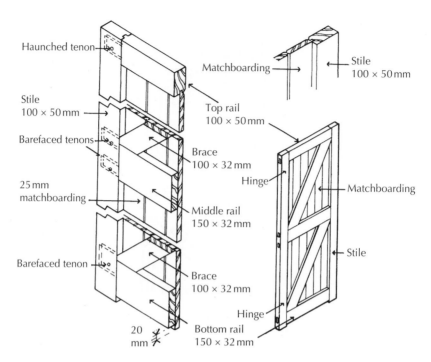

Figure 8.32   Framed and braced matchboarded door.

Ledged and braced matchboarded doors
Ledged and braced matchboarded doors are strengthened against sinking, with braces fixed between the rails at an angle to resist sinking on the lock edge (Figure 8.31b). The matchboarding is nailed to ledges and braces.

Framed and braced matchboarded doors
Framed and braced matchboarded doors are made by nailing matchboarding to a frame of stiles and rails that are framed with mortice and tenon joints with braces to strengthen the door against sinking, as illustrated in Figure 8.32. The boarding runs from the underside of the top rail to protect the end grain of the boards from rain and is carried down over both middle and bottom rails. To allow for the boards running over them, the middle and bottom rails are less thick than the stiles to which they are joined with a barefaced tenon joint (Figure 8.33). This joint is used instead of the normal joint with two shoulders so that the tenon is not too thin. These doors are used for large openings to garages, factories and for entrance gates.

8.3 Door frames and linings

A door frame is made of timbers of sufficient cross section to support the weight of a door and to serve as a surround to the door into which it closes. A majority of door frames are

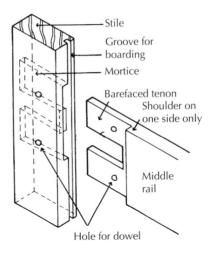

Stile

Groove for boarding

Mortice

Barefaced tenon
Shoulder on one side only

Middle rail

Hole for dowel

Figure 8.33   Barefaced tenon.

rebated to serve as a stop for one-way swing doors. The door frame is secured in the wall or partition opening to support external doors and heavier internal doors. The choice of frame or lining is to an extent a matter of appearance and convenience in fixing and the methods of masking the junction between plaster finishes and frames and linings.

Door linings are thin sections of wood or metal that are fixed securely in a doorway or opening as a lining around the reveal (thickness) of the wall or partition. A door lining which may not be substantial enough by itself to support the weight of a door will depend on its fixing to the wall or partition for support.

Door linings are generally used for internal doors in thin partitions where the width of the lining is the same as the thickness of the partition and wall plaster both sides. In this way, the junction of the plaster and lining can be masked by an architrave and the door opening emphasised by the lining and architrave. Figure 8.34 illustrates a door frame and a lining.

Door frames, commonly used for external doors and heavier internal doors, may not be as wide as the thickness of the wall in which they are fixed as illustrated in Figure 8.34. It is necessary, therefore, to run plaster finishes around the angle of the wall, into the reveal and up to the door frame. In time the junction between the plaster and frame will open up as an unsightly crack. A wood bead may be fixed to hide this potential crack. More substantial linings, or combinations of frames and linings, are used for panelled doors in walls that are one brick or more thick to combine the strength of a frame with the appearance of a lining.

Timber door frame

Timber door frames are assembled from three members for internal doors and four to most external doors. The members of the frame are two side posts, a head and a cill for external doors (where regulations do not prohibit an upstanding cill that would obstruct

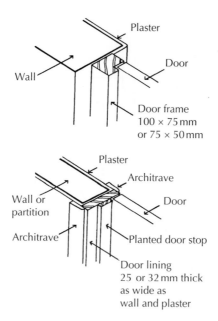

Figure 8.34 Door frame and door lining.

access). The members of the frame are usually cut with a rebate into which the door closes or a wood stop may be planted on a plain-faced timber, as a door stop. Because the frame is made to carry the weight of the door by itself, the members are joined with mortice and tenon joints to provide a rigid joint that will maintain the frame true square as illustrated in Figure 8.35a. The haunched mortice and tenon joints between the posts and head of the door frame are formed as illustrated in Figure 8.35a and in detail, in Figure 8.35b. The joint is formed by projecting the head of the frame some 100 mm each side of the posts as horns. These horns also provide a means of securing the frame by building them into surrounding brickwork where the frame is set at least ½ B back from the external face of the wall.

Where it is not convenient to build in horns when the door frame is fixed closed to the external face of a wall, the head is finished flush with the back of the posts. As it is not possible to form an enclosing mortice, a slot mortice and tenon joint is formed as illustrated in Figure 8.36. The tenon is secured in the slot mortice with a 16 mm dowel driven through the mortice and tenon.

Door frames to external walls may be built-in using fixing lugs. The frame is secured using fixing lugs that are built into horizontal courses of brickwork or blockwork as the walls are raised. The usual form of fixing lug used to secure wood window and door frames in solid walls is 'L' shaped, formed from galvanised steel. One tail of the lug is screwed to the back of the frame and the other tail is bedded in a horizontal brick or block course (Figure 6.35a). Three lugs are fixed to each post, and when solidly bedded in mortar they will secure the frame in position.

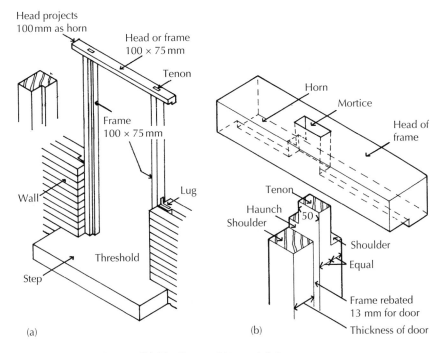

Figure 8.35   (a) Door frame. (b) Mortice and tenon joint.

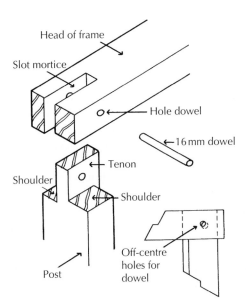

Figure 8.36   Slot mortice and tenon.

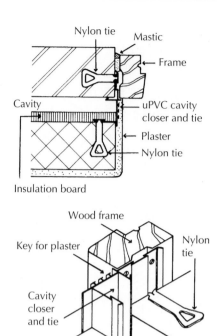

Figure 8.37   uPVC cavity closer and tie.

A uPVC cavity closer and ties, illustrated in Figure 8.37, is designed to close the cavity of a wall and allow the cavity insulation to be carried up to the back of the closer, so that there is no thermal bridge around the openings and to provide secure fixing for a door frame through nylon ties. The cavity closer is nailed to the back of the frame and the nylon ties are adjusted in slots for building into horizontal brick or block courses as the frame is built-in. Alternatively, the cavity closer can be installed after the brickwork has been raised. The cavity closer can simply be installed into the cavity and secured by fixing (plug and screwed) to the wall reveals.

Most door frames are installed after the brickwork has been raised. In such cases, the horns of the door frame will not be necessary and nylon or metal fixing lugs will be used to fix the frame in position.

Threshold

The threshold of a doorway or opening is the surface at the bottom of the opening that is level for internal doors and may be level or formed as a wood cill as part of the door frame. A level threshold may be formed for ease of wheelchair access. The disadvantage of this is that there is no positive check to wind-driven rain that runs down the door face, which can be blown-in under the closed door.

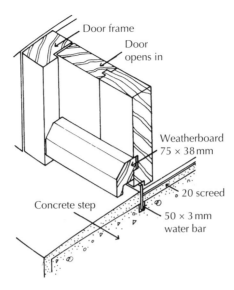

Figure 8.38   Water bar and weatherboard.

As a barrier to wind-driven rain penetrating under a door, a galvanised steel weather or water bar may be set in the threshold, so that when the door is closed it makes contact with the bar and forms a seal against the entry of rain as illustrated in Figure 8.38. To direct rainwater out from the door, a weatherboard of timber or metal is often fixed to the bottom of the door. The disadvantage of the water bar is that it is of small section and not always obvious to the unwary who may trip over it.

It is common to hang external doors to open inwards. The disadvantage is that the rebate in the door frame, into which the door fits when closed, does not so positively act as a check to wind and rain as it does with an outward-opening door. In exposed locations, wind pressure may force an inward-opening door away from the rebate to the extent that wind and rain may penetrate.

Traditionally, external door frames were made up with a wood, usually oak, cill as part of the framing with the posts of the frame tenoned to mortices in the cill. The oak cill, which is of a wider section than the posts, is weathered and cut with a drip on its lower edge to throw water off as illustrated in Figure 8.39. The cill may be rebated as housing for the door when closed or finished flush with the floor. A metal weather bar is set in the cill to fit into a rebate cut in the underside of the door as a weathercheck. A wood weatherboard was usually fixed to the bottom of the outside of the door to throw rain out from the bottom of the door. Unless the protective paint film is well maintained over the door surface, the wood weatherboard will before long become saturated and rot. A range of aluminium sections is available for use as weatherboards.

Weather-stripping

As a check to wind-driven rain and draughts of cold outside air that penetrate clearance gaps around external doors, it is necessary to fit weather-stripping around doors. The two

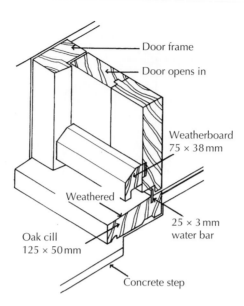

Figure 8.39 Oak cill.

systems of weather-stripping that are used are flexible seals fixed towards the inside face and compression seals fixed up to the outside face of doors.

Flexible seals are made from PVC or synthetic rubber in the form of a strip for housing in the frame, from which a flexible blade makes contact with the edges of the door. The seal may be fitted into an aluminium channel holder, which is fixed into a groove cut in the frame. The flexible seal does not make it difficult to open or close the door, but is sufficiently resilient to make positive contact with the closed door.

The members of the door frame and door illustrated in Figure 8.40 are cut with rebates to form a drainage channel up to the outer face of the flexible seals. This small channel serves the purpose of drainage channel and to reduce wind pressure on the flexible seal. The synthetic rubber seals are similar to those described and illustrated for use with wood casement windows.

Compression seals consist of hollow ball section strips of synthetic rubber in an aluminium alloy or plastic holder. The holders are screwed or pinned to the head and posts of the door frame so that the elastic seal presses against the closed outside of the door as a weather seal. The bottom edge of the door is sealed by a holder, fixed to the door, in which an elastic bulb presses on an aluminium alloy threshold strip fixed to the cill as illustrated in Figure 8.41. Unlike the flexible blade weather-stripping, these compression seals and their holders are visible, which may be unacceptable to some.

Both the flexible PVC blade strip and the compression seal bulb will after some years lose resilience and be less effective as seals. They should be replaced from time to time. When regular painting of doors and frame is carried out, care should be taken to avoid painting over the blade seal and the compression seal as a dried, painted film will make

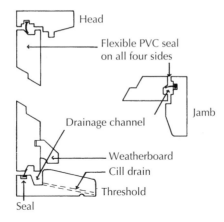

Figure 8.40   Inward-opening external door with weather strips.

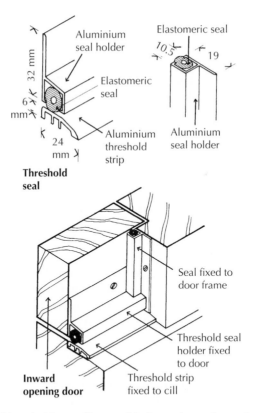

Figure 8.41   Threshold and side weather seal to inward-opening external door.

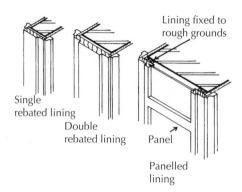

Figure 8.42   Door linings.

them less effective. As an alternative to the compression seals illustrated in Figure 8.41, one of the self-adhesive weather strips may be used.

Wood door linings

Wood door linings (door casings) may be plain and rebated or plain with planted stops, double rebated for appearance sake or panelled as illustrated in Figure 8.42. Plain linings with either a rebate or a planted stop are used for light doors in thin partitions, double-rebated linings for thicker brick or block partitions and panelled linings for heavier panelled doors in thicker walls or partitions.

Linings are fixed in position in the door opening before plastering to walls is carried out so that the finished plaster level is flush with the edges of the lining. The linings are nailed to rough wood grounds. Rough grounds are sections of plain sawn wood that are nailed to the surrounding brick or block partition to provide a level fixing to which the lining is nailed. The purpose of this is to avoid damaging the lining by driving large nails through the lining to find a fixing in the brick or block partition.

To provide a secure fixing for all but very narrow wood architraves around doors, it is practice to fix rough, wood grounds on each side of the lining. In some cases, grounds may be used by plasterers as a guide for the finish of the plaster. Linings are cut to the overall thickness of partitions and the thickness of plaster on both sides. Plain linings are usually cut from timber 47 or 54 mm thick for rebated linings and 31 or 38 mm thick for linings with planted (fixed) stops. Figure 8.43 is an illustration of a plain lining with planted stops fixed to a partition. The sides of the lining are jointed to the head with a tongued and grooved joint to secure the three sections in position.

Standard door frames and linings

Door manufacturers offer a range of standard frames and linings for standard size doors. The door frames are cut from sections of 102 × 64 mm and 89 × 64 mm rebated for doors and with sills for external doors. Door linings (or casings as they are sometimes called) are cut from sections 138 × 38, 138 × 32, 115 × 38 mm and 115 × 32 mm rebated for doors.

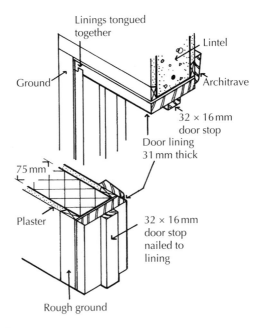

Linings tongued
together

Lintel

Ground

Architrave

32 × 16mm
door stop

Door lining
31 mm thick

75 mm

Plaster

32 × 16mm
door stop
nailed to
lining

Rough ground

Figure 8.43   Door linings.

The width of these linings, which is chosen to suit common partition and plaster thicknesses, may not match the overall thickness of some partitions and finishes.

Door sets

Door sets (door assemblies) are combinations of doors with door frames or linings and hardware such as hinges and furniture, prepared as a package ready for use on site. This plainly makes economic sense where many similar doors are to be used and packets of doors can be ordered and delivered instead of separately ordering doors, frames and hardware.

There is often inadequate fixing for a door frame or lining in a thin non-loadbearing partition so that the door, in use, may cause some movement in the frame or lining relative to the partition, to the extent that cracks in finishes around the frame or lining and particularly in the partition over the head of the door may appear. To provide a more secure fixing for doors in thin partitions, it is often practice to use storey-height frames that can be fixed at the floor and ceiling level.

These storey-height or floor-height frames are cut to line the reveal of the door opening and in that sense serve as linings, and are put together with floor-height posts, a head that can be fixed to the ceiling, a transom at the head of the door and also a cill for fixing to the floor, as illustrated in Figure 8.44. The frame sections may be rebated for the door or be plain with planted stops. The frame may be of uniform width for the full height with a panel fixed in the space over the door, or the width of the frame may be reduced over the

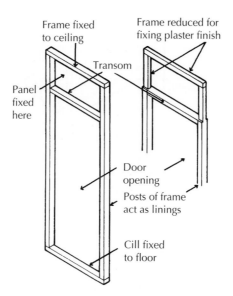

Figure 8.44   Storey-height door frame.

door so that finishes, such as plaster, may be run across the frame over the door. The cill of the storey-height door frame may be fixed to the floor so that floor boards can be fixed over it or finished flush with the floor finish for carpeted finishes. The advantage of these frames is that they provide a degree of stability to block non-loadbearing partitions.

Metal door frames

Metal door frames are manufactured from mild steel strip pressed into one of three standard profiles. The same profile is used for head and jambs of the frame. The three pressed steel members are welded together at angles. After manufacture, the frames are hot-dip galvanised to protect the steel against corrosion. Two loose pin-butt hinges are welded to one jamb of the frame and an adjustable, lock strike plate to the other. Two rubber buffers are fitted into the rebate of the jambs to which the door closes to cushion the impact of sound of the door closing. Figure 8.45 is an illustration of a standard metal door frame. The frames are made to suit standard door sizes. The frames are provided with steel, base ties welded across the foot of the posts of the frame to maintain the correct spacing of the posts.

Metal door frames are built-in and secured with adjustable, metal building-in lugs that are bedded in the horizontal joints of brick or blockwork, three to each side (Figure 8.45). The frames are bedded in mortar and filled with mortar. Alternatively, frames can be fitted-in after the brickwork has been raised. When the frame is fitted in the fixing, lugs are fitted to the internal leaf of the cavity wall.

These frames are made to suit standard external and internal wood doors. When used for internal doors in non-loadbearing partitions, a profile of metal frame is selected that is

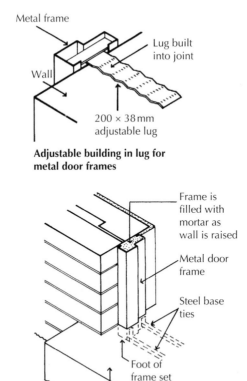

Figure 8.45   Metal door frame.

wider than the combined thickness of the partition and plaster both sides. In this way, the metal frame serve as a door lining that projects some 16 mm each side of the finished plaster. Plaster is run up to and under the lipped edges of the frame to avoid the necessity for an architrave to mask the junction between plaster and frame.

Glazed steel doors

Glazed steel doors are fabricated from the hot rolled steel sections used for windows. The sections are assembled with welded corner joints. The doors and frames, which are hot-dip galvanised after manufacture, may be finished with an organic powder coating. Single glass is either putty or clip-on aluminium bead glazed. Double glazing is bedded in mastic tape and secured with clip-on aluminium beads. Glazed steel doors, which have largely been superseded by aluminium doors, are mainly used for replacement work.

Flush steel doors (Figure 8.46) are manufactured from sheet steel, which is pressed to shape, often with lipped edges, hot-dip galvanised and either seam welded or joined with plastic, thermal break seals around a fibreboard, chipboard or foamed insulation core, generally with edge, wood inserts as framing and to facilitate the fixing of hardware. The

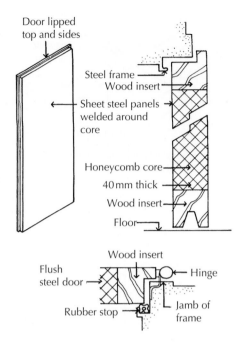

Door lipped top and sides

Steel frame

Wood insert

Sheet steel panels welded around core

Honeycomb core

40 mm thick

Wood insert

Floor

Wood insert

Flush steel door

Hinge

Rubber stop

Jamb of frame

Figure 8.46 Flush steel door.

sheet steel facings may be flush-faced or pressed to imitate wood panelling or with glazed panels. The exposed faces of the doors may be finished ready for painting or with a stoved-on organic powder or liquid coating.

Aluminium doors

An extensive range of partly glazed and fully glazed doors is manufactured from extruded aluminium sections. The slender sections possible with the material in framing the doors provide the maximum area of glass, and hence, good views out, for example to the garden. These glazed doors (commonly advertised as 'patio doors') are made as both single-and multi-leaf doors to hinge, slide or slide and fold to open. Glazed doors serve as a window by virtue of the large area of glass, and as doors by the facility to open them from the floor level. To minimise condensation on the inside faces of these double-glazed doors, it is practice to fabricate them as thermal break doors. The main framing sections of the doors, which are joined with corner cleats, are fixed to aluminium facings through plastic sections that act as a thermal break. Figure 8.47 is an illustration of an aluminium section glazed door designed to slide open. The doors are double glazed to reduce heat loss and have weather-stripping and drainage channels to protect against wind and rain.

Single-leaf aluminium doors can be manufactured to resemble traditional panelled wood doors. These are framed from extruded aluminium sections in the same way that windows and fully glazed doors are fabricated with the addition of a middle, horizontal

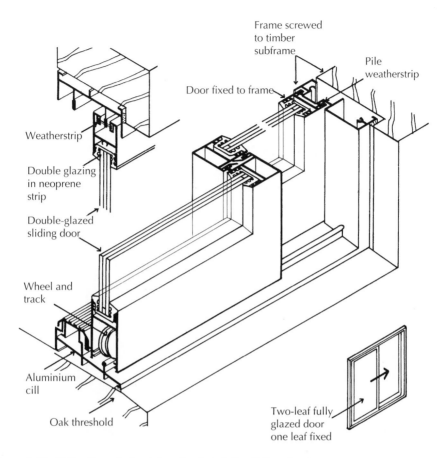

Frame screwed
to timber
subframe

Pile
weatherstrip

Door fixed to frame

Weatherstrip

Double glazing
in neoprene
strip

Double-glazed
sliding door

Wheel and
track

Aluminium
cill

Oak threshold

Two-leaf fully
glazed door
one leaf fixed

Figure 8.47 Fully glazed aluminium horizontally sliding door.

rail to imitate the middle or lock rail of a wood door. The sections are made to take either glazed or solid panels secured with internal pop-in glazing beads. The solid panels, which are fabricated from PVC or glass fibre reinforced plastic sheets around an insulating core, may be moulded to imitate traditional wood panels. An advantage is that they may be finished in a range of coloured powder or liquid coatings that do not require periodic painting for maintenance. These doors are sufficiently robust for use in domestic buildings and may be fabricated as thermal break construction to minimise condensation on the internal faces of the aluminium framing.

uPVC doors

Single-leaf uPVC doors are fabricated from a frame of comparatively bulky, extruded uPVC hollow sections similar in size to the stiles and rails of wood-framed doors. The hollow-framing sections are reinforced with galvanised steel or aluminium sections in the main cell of the hollow sections that are mitred and heat fusion welded at corners. A mid-rail

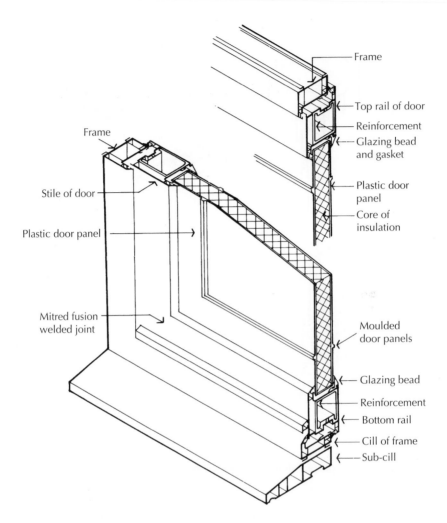

Figure 8.48 uPVC door and frame.

member is fitted to match the middle or lock rail of a wood door. These door leafs are hung to extruded, hollow section uPVC frames and thresholds as illustrated in Figure 8.48.

The uPVC door leafs, which are framed for glazing with double glazing secured with internal pop-in beads, are weathered with wedge and blade gaskets. As an alternative to glazed panels, a variety of plastic panels are produced from press-moulded acrylic, generally moulded to imitate wood door panels either as full-door height panels or as two panels fitted to a middle rail.

The hollow panels may have a core of some insulating material and a foil or thin sheet of aluminium as a barrier to breaking and entering by fracturing the panel. The majority of these doors are made as white or off-white impact modified uPVC to minimise the considerable thermal expansion that this material suffers as a result of solar radiation. Coloured and wood grain finishes are also supplied.

The advantage of uPVC doors is that they require little maintenance during their useful life, other than occasional washing. The disadvantage of these doors is that they may suffer thermal expansion causing them to jam, and knocks and indentations cannot be disguised by painting.

Garage doors

Up-and-over garage doors

Pressed metal doors are suited for use as garage doors because they are lightweight and have adequate stiffness and shape stability for a balanced 'up-and-over' opening action. The doors are manufactured from pressed steel or aluminium sheet, which is profiled to give the thin sheet material some stiffness. The sheet is welded or screwed to a light frame to give the door sufficient rigidity. Steel doors are hot-dip galvanised and primed for painting or coated with PVC, and aluminium doors are anodised. Figure 8.49 is an illustration of a steel up-and-over garage door.

To open, the door is lifted to slide on wheels in overhead tracks under the roof of the garage. Spring-loaded side-stays attached by pivots to the base of the door serve to steady the upward and downward movement of the door leaf and serve to balance the movement

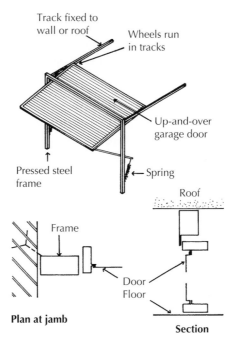

Figure 8.49   Steel 'up-and-over' garage door.

so that it may easily be raised and lowered by hand. The advantage of the overhead action is that the door does not obtrude on the outside.

Electronic opening devices may be fitted to garage doors that eliminate the need to lift the doors by hand. A sensor placed in the car will trigger the door to open when approaching the garage door. A wide range of electronic opening devices and associated doors are available.

8.4 Hardware

Hardware is the general term used for the hinges, locks, bolts, latches and handles for a door. Ironmongery was a term used when most of these were made of iron or steel. The term 'door furniture' is sometimes used to describe locks, handles and levers for doors. A wide variety of hardware is available to suit all types and styles of door. Materials used include steel, aluminium and brass with a variety of finishes. Locks and associated security devices should conform to rigorous security tests. Most insurance companies will insist on certain minimum security standards or offer a discount for high specification security measures. See also *Barry's Advanced Construction of Buildings* for further information on security.

Hinges

Pressed steel-butt hinge
Pressed steel-butt hinges are the cheapest and most commonly used hinges. They are made from steel strip, which is cut and pressed around a pin, as illustrated in Figure 8.50. They are used for hanging doors, casements and ventlights. The pin of the standard butt hinge is fixed inside the knuckle. These hinges are also made as loose pin-butt hinges, with the flap, which is screwed to the edge of the door, loose inside the knuckle so that the doors can be taken off by lifting. This avoids the necessity of unscrewing one flap of each hinge to take the door off for adjustment and for repair. These hinges are usually galvanised as protection against corrosion. The two flaps of the hinge are screwed in position into shallow sinkings cut in the wood frame and edge of the door, respectively, so that they are flush

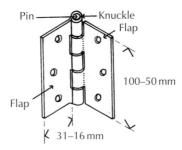

Figure 8.50   Pressed steel-butt hinges.

with the wood faces and with the knuckle of the hinge protruding from the face of the door.

Double-pressed steel-butt hinge

Double-pressed, heavy, steel-butt hinges are made of two strips of steel each folded back on itself as a flap and pressed and cut to form the knuckle around the pin as illustrated in Figure 8.51. Because of the double thickness of steel strip from which they are made, these hinges serve as HD to support larger, thicker doors.

Rising butt hinge

The bearing surfaces of the knuckle of both flaps of the rising butt hinge are cut on the skew so that as the hinge opens one flap rises, as illustrated in Figure 8.52. Because of the action of the hinge, as it opens, these hinges are generally described as 'rising butt hinges'.

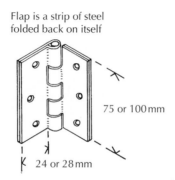

Flap is a strip of steel folded back on itself

75 or 100 mm

24 or 28 mm

Figure 8.51   Double-pressed heavy steel-butt hinge.

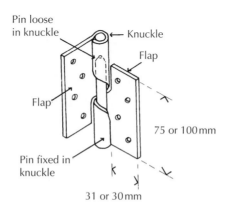

Pin loose in knuckle

Knuckle

Flap

Flap

Pin fixed in knuckle

75 or 100 mm

31 or 30 mm

Figure 8.52   Steel skew-butt hinge (rising butt hinge).

These hinges are used for hanging doors so that as the door opens it rises over and so clears such floor coverings as fitted carpets to reduce wear. The action of the skew butt will tend to make all but the lightest doors, self-closing.

Tee hinge

Steel tee hinges, illustrated in Figure 8.53, comprise a rectangular steel flap and a long tail, or hinge, which are pressed around a pin as knuckles. These hinges are made for use with matchboarded doors where the length of the long tail of the hinge will give support across the face of the door rather than at the edge. They are pressed from mild steel strip and are either galvanised or painted ready for fixing.

Hook-and-band hinge

Hook-and-band hinge, illustrated in Figure 8.54, is fixed in a housing in the wood door frame and to the face of the door. Hook-and-band hinges are made of more substantial

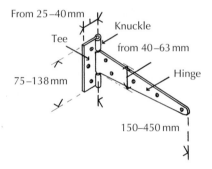

Figure 8.53   **Steel tee hinge.**

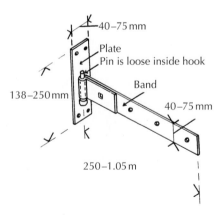

Figure 8.54   **Hook-and-band hinge.**

thicknesses of steel than tee hinges to support heavy wood doors, such as those to garages and workshops. The plate has a pin welded to it, around which the knuckle of the band fits as illustrated in Figure 8.54. The band is reinforced with a second plate of steel at the knuckle end. As a protection against rusting, the hinge is galvanised after manufacture.

To provide secure support for the doors, the band is holed towards its knuckle end for a coach bolt that is fitted to a hole in the stile of the door and bolted in place to provide solid support. For convenience in taking doors off for easing (trimming to fit), the hinge fits to a loose pin on the plate.

Latches and locks

The word latch is used to describe any wood or metal device that is attached to a door or window to keep it closed. These simple devices serve the purpose of keeping the door or window in the closed position. They do not lock the door.

A lock is any device of wood or metal which is attached to a door or window to keep it closed by the operation of a bolt that moves horizontally into a striking plate or staple fixed to the door or window frame.

Mortice lock

The mechanism most used today for doors is the mortice lock, so called because the metal case containing the operating parts is set into a mortice cut in the door. Locks for external doors and internal doors, where security is a consideration, consist of a latch bolt and a lock bolt. For internal doors in continuous use, the locks contain a single latch bolt to keep the door closed.

A mortice lock for an external door is set inside a mortice cut in the stile and middle rail of the door. The horizontal, two-bolt mortice lock, illustrated in Figure 8.55, consists of a case, a forend and a striking plate. The case fits into a mortice in the door through which holes have been drilled for a loose key and for the spindle for knobs or handles. The lock may be made with a forend plate that is screwed into position flush with the edge of the door or finished with a forend and cover plate of brass for appearance's sake.

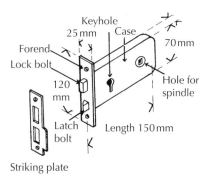

Figure 8.55   Horizontal two-bolt mortice lock.

A striking plate is fixed over mortices cut in the door frame to house the two bolts. This plate is termed a striking plate, as it serves the purpose of directing the shaped end of the latch bolt into the plate as the door is shut.

For flush doors and those without a middle rail, an upright two-bolt mortice lock is used. The lock, illustrated in Figure 8.56, is designed specifically for this use. Two-bolt mortice locks are supplied with two loose keys to operate the lock bolt.

Single-bolt mortice locks, which are supplied for internal doors, comprise a case, forend and striking plate with one latch bolt which is operated by knobs or lever handles and a spindle. As they do not lock the door, these devices should properly be called latches. LD internal doors, which are often too thin to accommodate a mortice lock, may be secured with a rim lock. These locks are designed for fixing to the face of doors (Figure 8.57).

Mortice dead lock

A mortice dead lock consists of a single bolt that is operated by a loose key. There is no latch bolt. It is a dead lock in the sense that once the bolt is shot, moved into the closed

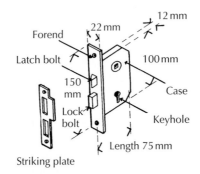

Figure 8.56   Upright two-bolt mortice lock.

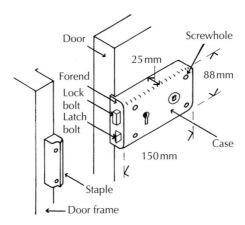

Figure 8.57   Horizontal two-bolt rim lock.

position and the key removed, it is dead to being operated. These locks are used in conjunction with and separate from a cylinder night latch for security. The upright case is housed in a mortice cut in the stile of the door and screwed through the forend to the edge of the door. The lock bolt is closed into a lock plate screwed over a mortice cut into the door frame. Figure 8.58 is an illustration of a typical dead lock.

The security of this locking device depends to an extent on the number of so called levers that are operated by the key. The greater the number of levers, the greater the security. This type of lock may be used for both wood doors and wood casement windows where the stile of the casement is wide enough to house the lock case.

Cylinder rim night latch (springlatch)

A cylinder rim night latch is designed to act as a latch from inside and a lock from outside for convenience in use on front doors. It is made as a rim latch for fixing to the inside face of doors (Figure 8.59). This type of latch offers poor security as it is fairly easy to push back the latch from outside by means of a piece of thin plastic or metal inserted between the door and the frame.

A more secure type of night latch is designed as a mortice lock which is opened as a latch from inside by means of a lever and from outside by a loose key. The lock has a

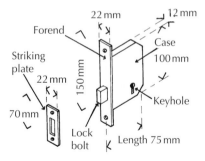

Figure 8.58   Mortice dead lock.

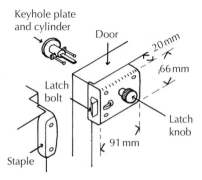

Figure 8.59   Cylinder rim night latch.

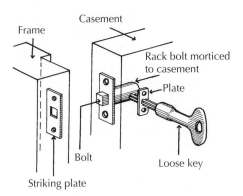

Figure 8.60   Rack bolt for security for wood windows and doors.

double-throw action which, by two turns of the key from outside, locks the latch in position so that it cannot be pushed back from outside. As it is the most convenient means of keeping external front doors closed by use of a latch bolt and offers some small security through its operation of opening from outside by a loose key, this lock is much favoured. It should be used in conjunction with a mortice lock to entrance doors to houses and flats.

Rack bolt

A rack bolt is a single-locking device used for locking wood doors and casement windows. It consists of a cylindrical case and bolt that is fitted into a mortice cut in the stile of doors and casements of windows as illustrated in Figure 8.60. A loose key from inside the door or window operates the bolt.

The bolt is fitted to a mortice in the door or window and the forend screwed to a shallow housing in the door or window edge. A striking plate is fixed over a mortice in the frame and screwed in position in a shallow housing. Two of these rack bolts fitted at the top and bottom to doors and casements or one to small casements serve as an economical and very effective means of locking shut.

Entry-control systems

Flats and residential developments, such as sheltered housing, need to provide entry to communal and private space. This has to be done without compromising the security of the residents. Entry-control systems that incorporate closed circuit television monitoring allow remote operation by individual tenants and are widely used.

9 Stairs and Ramps

For access between floors and different levels in buildings a ladder, stair or ramp is used. Lifts and escalators are also used for vertical circulation and are covered in *Barry's Advanced Construction of Buildings*, Chapter 9. Accessibility for all, regardless of disability, must be addressed when considering changes in level, no matter how small the difference in the height of finished floor surfaces.

Definitions

Ladders

A ladder is made as a series of narrow horizontal steps (rungs), fixed between two uprights of wood or metal, on which a person usually ascends (climbs up) or descends (climbs down) facing the ladder. A ladder may be fixed in an upright, vertical position or more usually at a slight angle to the vertical for ease of use (Figure 9.1a). Approved Document K recommends that a ladder should only be used for access to a loft conversion of one room, where there is not enough space for a stair, and that the ladder be fixed in position and fitted with handrails on both sides.

Stairs

A stair, or stairway, is the name given to a set of steps formed or constructed to make it possible to pass to another level on foot by putting one foot after the other on alternate steps to climb up or down the stair. A stair may be formed as a series of steps rising in one direction between floors as a straight flight of steps. More usually, a stair is formed as two or more straight flights of steps arranged to make a quarter or half turn at intermediate landings between floors. Using two flights helps to limit the number of steps in each flight making use of the stairway potentially safer. Because of the slope of the stair and the need to limit the number of steps in each flight, a typical half-turn stair occupies a considerable space in small houses as illustrated in Figure 9.1b.

A stair is the conventional means of vertical access between floors in buildings. It should be constructed to provide ready, easy, comfortable and safe access up and down with steps that are easy to climb, within a compact area, so as not to take up excessive floor space. Access for wheelchair users and less mobile persons may be facilitated by the inclusion of

Barry's Introduction to Construction of Buildings, Third Edition. Stephen Emmitt and Christopher A. Gorse.
© 2014 John Wiley & Sons, Ltd. Published 2014 by John Wiley & Sons, Ltd.

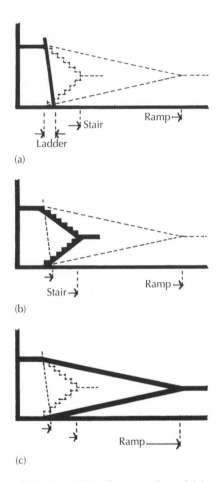

Figure 9.1   **Space used by (a) ladder, (b) half-turn stair and (c) ramp.**

chairlifts fixed to the side of the staircase. In such cases, adequate attention needs to be given to the stair width and overall design and installation of the stair and associated fittings (Figure 9.2).

Ramps

A ramp is a surface, sloping uniformly as an inclined plane, linking different levels. A ramp is formed or constructed at a slope of at least 1 in 20 (1 m rise vertically in 20 m horizontally). Because of the comparatively shallow slope of a ramp it occupies a considerable area and this must be considered early in the design process so that enough room is allowed (Figure 9.1c). The advantage of a ramp is that it allows relatively easy and safe access for wheelchairs and pushchairs and may be less daunting than a staircase to people with reduced mobility.

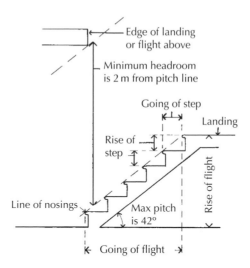

Figure 9.2 Rise, going and headroom for stairs (in a private dwelling).

9.1 Functional requirements

The primary functional requirement of a stair or ramp is:

❑ To allow vertical movement, safely, from one level to another

Additional functions and performance requirements include:

❑ Strength and stability
❑ Ease of use
❑ Fire safety
❑ Control of impact sound
❑ Safety in use
❑ Avoiding traps
❑ Durability
❑ Aesthetics

Strength and stability

A stair, or ramp, and its associated landings serves much the same function as a floor in the support of the occupants of a building, with the stepped inclined plane of flights, or shallow rise of the ramp, serving as support for movement between different levels. The requirements for strength and stability in supporting dead and imposed loads for floors apply equally to stairs and ramps.

Ease of use

Many different people will use the stairway. Care should be taken so that stairs are designed for users of all abilities and ages. Handrails can help all users, but are particularly important for those less able to move between levels. Landings may provide well-needed resting stages, or a platform that allows others to pass, enabling less mobile users to move at their own pace.

Fire safety

Stairs are one of the principal means of escape to the outside in case of fires in buildings. Stairs should be constructed of materials that are capable of maintaining strength and stability for a period of time sufficient for those in a building to escape to the outside or a protected area. Careful choice of materials is important, but so too is the protection of the stairs or ramp (the means of escape) from fire and smoke. The size of the steps and the width of the stair should be adequate for the safe escape of those in the building.

Whenever a stairway forms part of an escape route, attention should be given to the requirements imposed in Approved Document B Fire Safety. Stairways that form part of an escape route will need to be of sufficient width to allow the occupants to evacuate the building safely. Table 9.1 provides widths for vertical escape routes (protected stairways).

Where a stairway provides access to more than one dwelling, it should be situated in a fire-resisting enclosure (protected shaft). Protected stairways need to be relatively free of potential sources of fire and all escape routes should have adequate artificial lighting. Escape lighting that automatically illuminates if the main electrical supply fails should also be provided.

Impact sound

Consideration of materials used for the stair tread is an important factor in helping to keep impact sound from foot traffic to an acceptable minimum. For example, metal treads and

Table 9.1   Widths of vertical escape routes and exits (adapted from Approved Document B DETR, 2000)

Widths of vertical escape route and exits		
Situation of stair (or ramp)	Maximum number of people served	Minimum width (mm)
In an institutional building (unless only used by staff)	150	1000
In an assembly building and serving an area used for assembly purposes (unless less than 100 m³)	220	1100
In any other building and serving an area with an occupancy of more than 50	over 220	*
Fire-fighting stair		1100
Any stair not described above	50	800

*Tables 7 and 8 of Approved Document B provide widths for stairs used in simultaneous evacuation (stairs must allow all people to leave the building at once) and phased evacuation (parts of the building may be evacuated before others).

timber treads can be noisy and may be a nuisance to building users. This may not be a problem in a busy bar, but in a house the impact sound could prove a significant irritation to the building users. Direct impact sound can be reduced by using resilient packing between stairways and the building structure and/or with the use of dense material that absorbs sound energy, such as concrete. Carpet is widely used to absorb impact sound on stair treads.

Safety in use

A considerable number of accidents occur on stairs each year. While many of these are due to simple trips and falls, many are a result of damaged, broken or incorrectly fitted treads and handrails. It is essential that both the handrail and flight of stairs are securely fixed at the right height; an observation that also applies to the design of ramps. The guidance in Approved Document K is concerned with the safety of users in determining the rise, going and headroom of stairs and the dimensions of handrails and guarding. Guidance on ramps also covers the incline of the ramp and its uninterrupted length, positioning of landings, handrails and guarding.

Avoiding traps

Gaps in stairs can pose a problem to all users. Poorly positioned handrails, where the gap between the wall and the handrail is small or the space narrows such that fingers become trapped, can result in bruised, broken and even amputated fingers. Care should be taken to avoid gaps where a child's head may become stuck. Fixing brackets and variations in the walls can cause the gap to become a hazard. Sharp or protruding objects, such as exposed woodgrain, screws and bolts, can easily cut and injure hands as they move across the handrail and for obvious reasons such hazards must be corrected immediately.

Durability

The durability of a stair or ramp will be determined by the type of building it serves, the amount of use received in service and whether it is located within or outside the building envelope. Both internal and external stairs and ramps will need to remain functional and safe despite heavy use; therefore, the material used and any applied finishes must be relatively durable. Materials that wear easily and become loose or unstable are not suitable for stairways and ramps.

Aesthetics

In addition to their functional requirements, the stair or ramp may be designed to be a significant architectural feature within a building. Aesthetic requirements will determine the position and shape of the stairs or ramp as well as the materials used to construct it. Architectural fashion and increased attention to interior design have led to a big increase in the designs and materials used in stairs and ramps. In particular, stainless steel and structural glass have become very popular, with stairs and ramps forming an integral part of the building's character (Photograph 9.1).

(a) Stairway: glass cladding exposes stairway to the public

(b) Glass balustrades provide a feature, expose movement of people and reduce the risk of children climbing

(c) Open stairways with glass handrails, behind glass cladding combine internal and external features to enhance views from inside and outside the building

(d) The heavy cantilevered supporting steel structure has a visual impact. Open stair allows external light to penetrate gaps

Photograph 9.1   Stairway design and aesthetics.

9.2 Materials, terms and definitions

Internal stairs and ramps must be durable to everyday use, while external stairs and ramps must withstand weathering and be durable to everyday use. Timber, stone, concrete, metal, stainless steel and, more recently, structural glass are the main materials used in the construction of stairs and ramps (see further). Factory-made prefabricated staircases are now used in the majority of developments. Timber stairs are the most common, but steel and glass stairs are also manufactured off-site and delivered as discrete components.

Terms and definitions

There are a number of terms and definitions that are used to describe discrete elements of stairs and ramps. These are explained further.

Flight

The word flight describes an uninterrupted series of steps between floors or between floor and landing, or between landing and landing. A flight should have no more than 16 risers. Small changes in height are not always obvious to users and single steps should not be used because they represent a trip hazard. Where single steps exist (in existing buildings), it is necessary to provide visual warnings, for example, contrasting colours on the tread and riser, to warn people of the level change.

The rise and going of each step in one flight and in flights and landings between floors should be consistent. Variations in the rise of steps will interrupt the user's rhythm and might cause the user to trip and fall.

Approved Document K of the Building Regulations does not provide a specified flight width; however, if the stair is to form part of an escape route Approved Document B (Fire Safety) will apply (see Table 9.1). See also Approved Document M.

Treads and risers

The steps of a stair may be constructed as a series of horizontal open treads with a space between the treads or as enclosed steps with a vertical face between the treads, called a riser. The horizontal surface of a step is described as the tread, and the vertical or near vertical face as the riser. With enclosed steps, the treads usually project beyond the face of the riser as a nosing to provide as wide a surface of tread as practicable. Figure 9.3 illustrates the use of the terms tread, riser and nosing.

A stair, which is constructed with horizontal treads with a space between them, is described as an open-riser stair (Figure 9.4). When open risers are used and it is expected that children under the age of 5 will use the stairs, the stair shall be constructed so that any openings in the flight would not allow a 100 mm sphere to pass through it. Open treads should not be used when stairs provide disabled access. To help the visually impaired recognise and distinguish between steps, a 55 mm wide strip, which provides a visual contrast, should be permanently fixed to all nosings on both the tread and the riser.

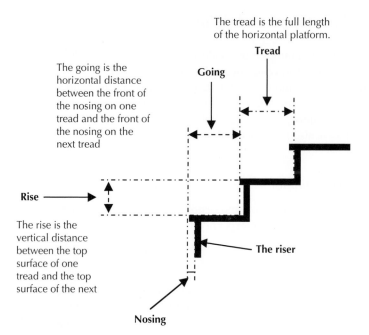

The tread is the full length of the horizontal platform.

Tread

The going is the horizontal distance between the front of the nosing on one tread and the front of the nosing on the next tread

Going

Rise

The rise is the vertical distance between the top surface of one tread and the top surface of the next

The riser

Nosing

Figure 9.3   Stair terminology.

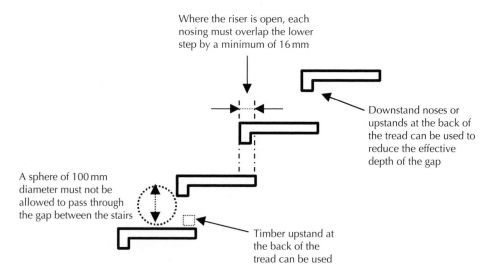

Where the riser is open, each nosing must overlap the lower step by a minimum of 16 mm

Downstand noses or upstands at the back of the tread can be used to reduce the effective depth of the gap

A sphere of 100 mm diameter must not be allowed to pass through the gap between the stairs

Timber upstand at the back of the tread can be used

Figure 9.4   Open-riser stairs.

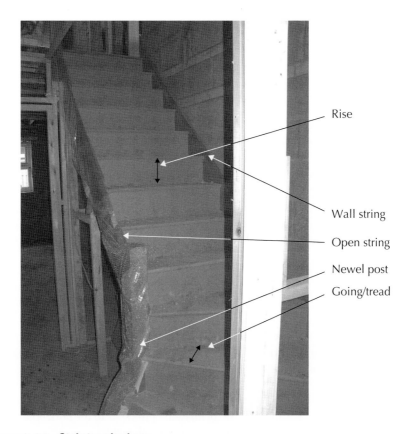

Rise

Wall string

Open string

Newel post

Going/tread

Photograph 9.2 Stair terminology.

The word rise describes the distance measured vertically from the surface of one tread to the surface of the next or the distance from the bottom to the top of a flight. The word going describes the distance, measured horizontally, from the face of the nosing of one riser to the face of the nosing of the next riser, as shown in Figure 9.3. The dimensions of the rise and going of steps determine whether a stair is steep or shallow (Photograph 9.2).

The recommended rise and going is given in Approved Document K1. A differentiation is made between stairs in a private dwelling, a utility stair and a stair for general access, with the minimum and maximum rise and going varying depending on use:

❑ Private stair (in a domestic dwelling) would have a pitch range between 150 mm and 220 mm, with a going dimension between 220 mm and 300 mm.
❑ Utility stair. This also has a minimum rise of 150 mm, but a shallower maximum rise of 190 mm to accommodate heavier use. The goings are larger compared to a private stair to facilitate ease of use, with a minimum of 250 mm to a maximum of 400 mm.

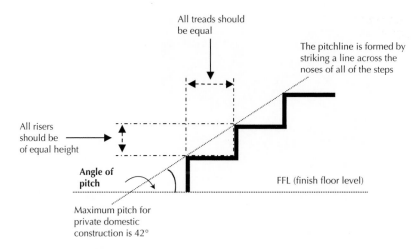

All treads should
be equal

The pitchline is formed by
striking a line across the
noses of all of the steps

All risers
should be
of equal height

Angle of
pitch

FFL (finish floor level)

Maximum pitch for
private domestic
construction is 42°

Figure 9.5   Angle of pitch.

❑   General access stair. As the use becomes more varied the maximum rise is reduced, with a minimum rise of 150 mm and a maximum of 170 mm. The going is the same range as the utility stair, from 250 mm to 400 mm.

The incline of a stair can be described either by the rise and going of the steps or as the pitch of the stair, which is the angle of inclination of the stair to the horizontal, as illustrated in Figure 9.5. Private stairs are pitched at not more than 42°.

Pitch
To set out a stair, it is necessary to select a suitable rise and adjust it, if necessary, to the height from floor to floor so that the rise of each step is the same, floor to floor, and then either select a suitable going or use the formula $2Rises + 1Going =$ between 550 and 700 mm, to determine the going.

Headroom
For people and for moving goods and furniture, a minimum headroom of 2 m, measured vertically, is recommended between the pitchline of the stair and the underside of the stairs, landings and floors above the stair as illustrated in Figure 9.6.

Handrails
The general requirement for handrails is that there should be at least one handrail on a stairway. If the stairway is greater than 1 m, two handrails are required. The height of the handrails should be between 900 and 1 m. Where the stair acts as a gangway greater than 1.8 m wide, a handrail should be used to divide the stairway (Figure 9.7 and Figure 9.8).

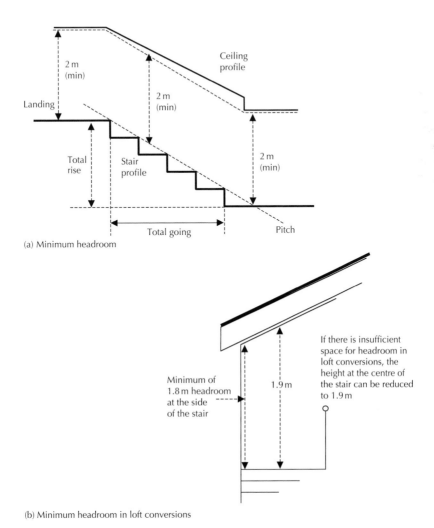

(a) Minimum headroom

(b) Minimum headroom in loft conversions

Figure 9.6 Minimum headroom.

Approved Document M provides guidance on the size and height of handrails for use by disabled people. The handrail should be 900–1100 mm above the landing height and 900 mm to 1 m above the pitchline of the stair; the handrails should also start 300 mm before the first step and end at the top step (Figure 9.9).

Landings
The position and general arrangement of landings to stairs are illustrated in Figure 9.10, Figure 9.11 and Figure 9.12. The maximum number of risers in any flight is 16 (12 for disabled use), after which a landing must be provided. The landing provides a position for the stair user to rest, if necessary, before continuing up or down the stairs and thus is an

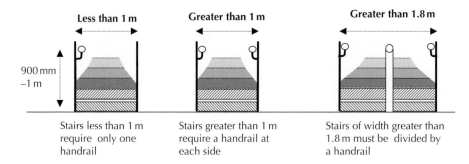

Less than 1 m

Greater than 1 m

Greater than 1.8 m

900 mm
–1 m

Stairs less than 1 m
require only one
handrail

Stairs greater than 1 m
require a handrail at
each side

Stairs of width greater than
1.8 m must be divided by
a handrail

Figure 9.7   Handrails and stair widths.

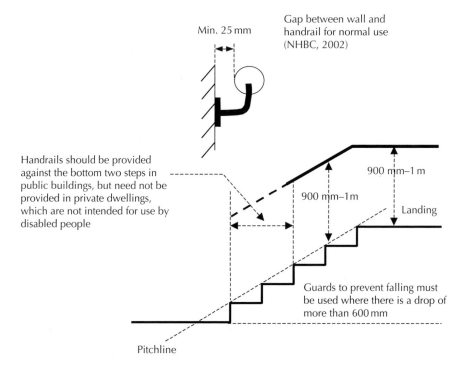

Min. 25 mm

Gap between wall and
handrail for normal use
(NHBC, 2002)

Handrails should be provided
against the bottom two steps in
public buildings, but need not be
provided in private dwellings,
which are not intended for use by
disabled people

900 mm–1 m

900 mm–1m

Landing

Guards to prevent falling must
be used where there is a drop of
more than 600 mm

Pitchline

Figure 9.8   Handrail height for general use.

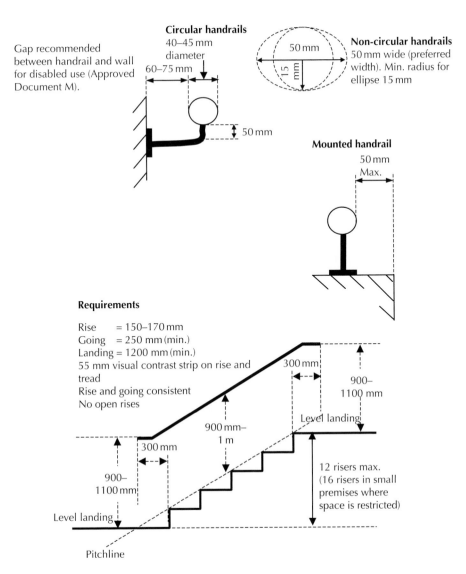

Circular handrails
40–45 mm diameter
60–75 mm

Gap recommended between handrail and wall for disabled use (Approved Document M).

50 mm

50 mm

Non-circular handrails
50 mm wide (preferred width). Min. radius for ellipse 15 mm

15 mm

Mounted handrail
50 mm Max.

Requirements
Rise   = 150–170 mm
Going   = 250 mm (min.)
Landing = 1200 mm (min.)
55 mm visual contrast strip on rise and tread
Rise and going consistent
No open rises

300 mm

900–1100 mm

Level landing

900 mm–1 m

300 mm

900–1100 mm

Level landing

12 risers max. (16 risers in small premises where space is restricted)

Pitchline

Figure 9.9 Handrails for the disabled.

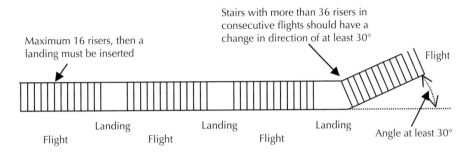

Maximum 16 risers, then a landing must be inserted

Stairs with more than 36 risers in consecutive flights should have a change in direction of at least 30°

Flight

Flight   Landing   Flight   Landing   Flight   Landing   Angle at least 30°

Figure 9.10 Change in direction.

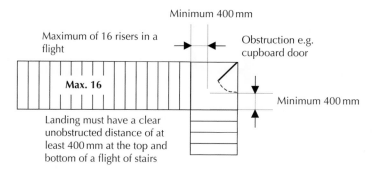

Figure 9.11   Maximum steps in flight – minimum unobstructed landing distance.

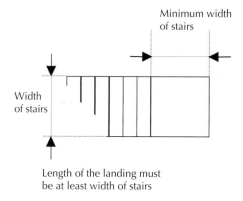

Figure 9.12   Length of landings.

important safety consideration. Landings should also be provided at changes of direction. When landings are designed for use by people with limited mobility, they should have an unobstructed length of 1200 mm.

9.3 Types of stair

The three basic ways in which stairs with parallel treads are planned are illustrated in Figure 9.13. These are a straight-flight stair, a quarter-turn stair and a half-turn stair.

Stairs with parallel treads

Straight-flight stair
A straight-flight stair rises from floor to floor in one direction with or without an intermediate landing, hence the name. It is usually the most economical use of a stair.

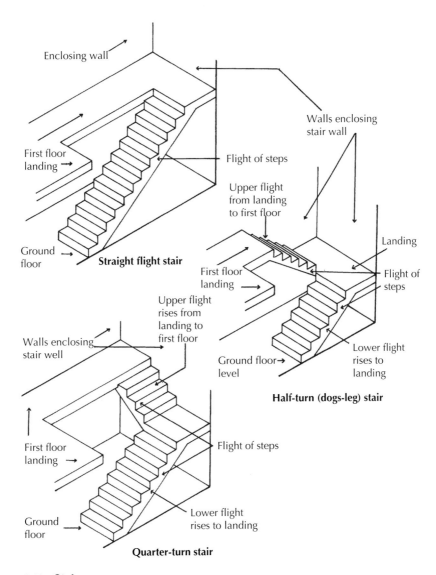

Figure 9.13   Stairs.

Quarter-turn stair

A quarter-turn stair rises to a landing between two floors, turns through 90° and then rises to the floor above, hence 'quarter turn'. The quarter-turn landing is sometimes replaced with winders for further economy in the use of space.

Half-turn or dog-leg stair

A half-turn stair rises to a landing between floors, turns through 180° and then rises parallel to the lower flight to the floor above, hence 'half turn'. The landing is described as a half-space or half-turn landing. A half-turn stair is often described as a 'dog-leg' stair because it looks somewhat like the hind leg of a dog. This, the most common arrangement of stairs, has the advantage in planning that it lands at, or roughly over, the starting point of the stair which can be constructed within the confines of a vertical stair well, as a means of access to and escape from similar floors.

Stair well

Stairs are sometimes described as 'open-well stairs'. The description refers to a space or well between flights. A half-turn or dog-leg stair can be arranged with no space between the flights or with a space or well between them, and this arrangement is sometimes described as an open-well stair. A quarter-turn stair can also be arranged with a space or well between the flights. As the term 'open well' does not describe the arrangement of the flights of steps in a stair, it should only be used in conjunction with the more precise description of straight-flight, quarter- or half-turn stair (e.g. half-turn stair with open well).

Geometrical stairs

Geometrical stairs are constructed with treads that are tapered on plan, with the tapered treads around a centre support as a spiral (helical) stair, an open-well circular stair or as an ellipse on part of an ellipse on plan, as illustrated in Figure 9.14.

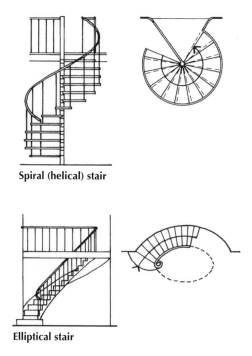

Spiral (helical) stair

Elliptical stair

Figure 9.14   Geometrical stairs.

A spiral (helical) stair with the treads tapering to a central, vertical support is the most economical way of planning a stair as it takes up little floor area. Because the treads taper sharply to the central post and one needs to make sharp turns up and down this type of stair, it is difficult to use and may be dangerous to young children and people with limited mobility. Spiral stairs, which form a helix around a central column or post, are used where space is limited for access to an intermediate floor of one room. A spiral stair is illustrated in Figure 9.14.

Circular or elliptical stairs are constructed around a generous open well with the treads having a shallow taper towards the well. These stairs, which are extravagant in the use of space, are used as a feature for grand means of access in large buildings. An elliptical stair is illustrated in Figure 9.14.

9.4 Timber staircases

A staircase, which is a stair with treads and risers constructed from timber boards put together in the same way as a box or case, hence the term staircase, is the traditional stair for houses of two or more floors where the need for resistance to fire does not dictate the use of concrete. Each flight of a staircase is made up (cased) in a joiner's shop as a complete flight of steps, joined to strings. Landings are constructed on site and the flight or flights are fixed in position between landings and floors.

Timber stair construction

The members of the staircase flight are string (or stringers), treads and risers. The treads and risers are joined to form the steps of the flight and are housed in or fixed to strings whose purpose is to support them. Because the members of the flight are put together like a box, the boards can be thin and yet strong enough to carry the loads normal to stairs. The members of the flight are usually cut from timbers of the following sizes: treads 32 or 38 mm, risers 19 or 25 mm and strings 38 or 44 mm. Figure 9.15 is an illustration of a flight of a staircase with some of the treads and risers taken away to show the housings in the string into which they fit and the construction of a landing.

Joining risers to treads

The usual method of joining risers to treads is to cut tongues on the edges of the risers and fit them to grooves cut in the treads, as illustrated in Figure 9.16. Another method is to butt the top of the riser under the tread with the joint between the two, which would otherwise be visible, masked by a moulded bead housed in the tread, as illustrated in Figure 9.16. The tread of the stair tends to bend under the weight of people using it. When a tread bends, the tongue on the bottom of the riser comes out of the groove in the tread and the staircase 'creaks'. To prevent this, it is common practice to secure the treads to the risers with screws (Figure 9.16). The nosing on treads usually projects 32 mm, or the thickness of the tread, from the face of the riser below. A greater projection than this would increase the likelihood of the nosing splitting away from the tread and a smaller projection would reduce the width of the tread. The nosing is rounded for appearance. Figure 9.16 illustrates the more usual finishes to nosings.

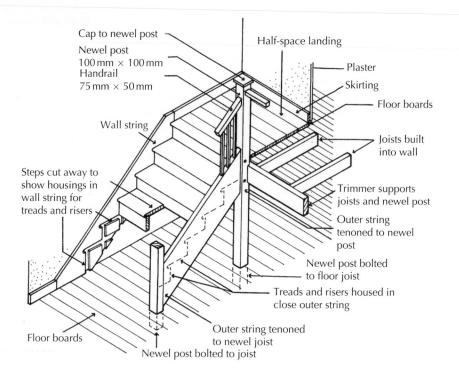

Figure 9.15   Lower flight of half-turn staircase.

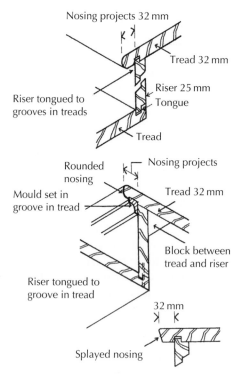

Figure 9.16   Method of jointing risers to treads.

Strings, treads and risers

Strings (stringers) are cut from boards 38 or 44 mm thick and of sufficient width to contain and support the treads and risers of a flight of steps. Staircases are usually enclosed in a stair well. The stair well is formed by an external wall or walls and partitions, to which the flights and landings are fixed. The string of a flight of steps, which is fixed against a wall or partition, is termed the wall string and the other string, if it is not fixed to a wall, is the outer string.

A string, which encloses the treads and risers it supports, is termed a close or closed string. It is made wide enough to enclose the treads and risers and its top edge projects some 50 or 63 mm above the line of the nosings of treads. The width of the string above the line of nosings is described as the margin. Figure 9.17 shows a closed string. A string 250 or 280 mm wide is generally sufficient to contain steps with any one of the dimensions of rise and going and a 50-mm margin. Wall strings are generally made as close strings so that wall plaster can be finished to them for a neat appearance. Outer strings can be made as closed strings or as open (cut) strings. The ends of the treads and risers are glued and wedged into shallow grooves cut in closed string. The grooves are cut 12 mm deep into strings and tapering slightly in width to accommodate treads, risers and the wedges, which are driven below them, as illustrated in Figure 9.17.

Angle blocks – glue blocks

After the treads and risers have been assembled, glued and wedged into the string, the angle blocks are glued in the internal angles between the underside of the treads and risers. Angle-glue blocks are triangular sections of softwood cut from say 50 mm square timber and each 120 mm long. Their purpose is to strengthen the right-angled joints between treads, risers and strings. Three or four blocks are used at each junction of the tread and riser and one at the junctions of treads, risers and string. Angle blocks are shown in Figure 9.18 (Photograph 9.3).

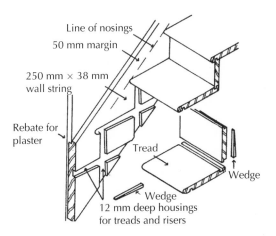

Figure 9.17 Housing treads and risers in close string.

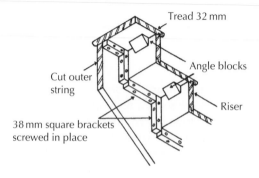

Figure 9.18 Underside of flight to show fixing of treads, and risers to cut outer string.

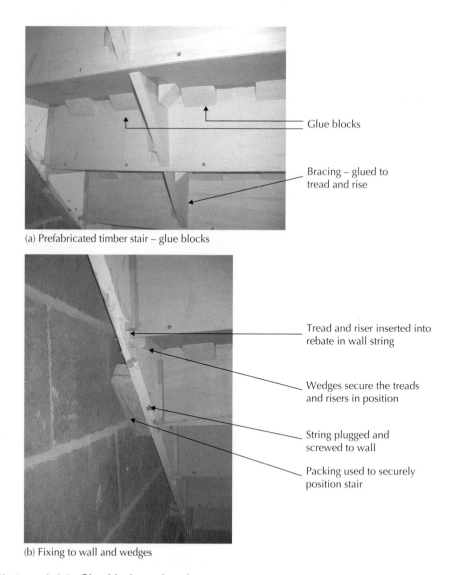

(a) Prefabricated timber stair – glue blocks

(b) Fixing to wall and wedges

Photograph 9.3 Glue blocks and wedges.

Cut or open string

A closed outer string does not show the profile of the treads and risers it encloses. Looking from the side of the staircase, only a single straight string with no steps in it can be seen. Alternatively, strings can be cut so that the profile of the steps can be seen. This type of string is termed a cut or open string. Because more labour is involved, a flight with a cut string is more expensive than one with closed strings. As the string is cut to the outline of the treads and risers, they cannot be supported in housings in the string and are secured to brackets screwed to both treads and risers and string, as illustrated in Figure 9.18. It is difficult to cut a neat nosing on the end grain of treads to overhang the cut string, so planted nosings are fitted as shown in Figure 9.19. The planted nosings are often secured to the ends of treads by slot screwing. This is a form of secret fixing used to avoid having the heads of screws exposed.

Landings – half-space (turn) landing

A half-space landing is constructed with a sawn softwood trimmer which supports sawn softwood landing joists or bearers and floorboards, as illustrated in Figure 9.20. As well as giving support to the joists of the half-turn landing, the trimmer also supports a newel or newel posts. Newel posts serve to support handrails and provide a means of fixing the ends of outer strings.

Newel posts

The newel posts are cut from 100 × 100 mm timbers and are notched and bolted to the trimmer. The outer string fits to mortices cut in the newel, as illustrated in Figure 9.20. For appearance, the lower end of the newel post is usually about 100 mm below the flights and moulded. As it projects below the stair it is called a drop newel.

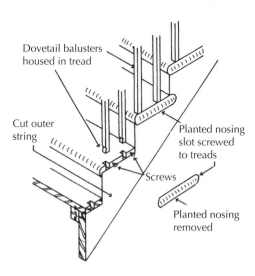

Figure 9.19   Cut string.

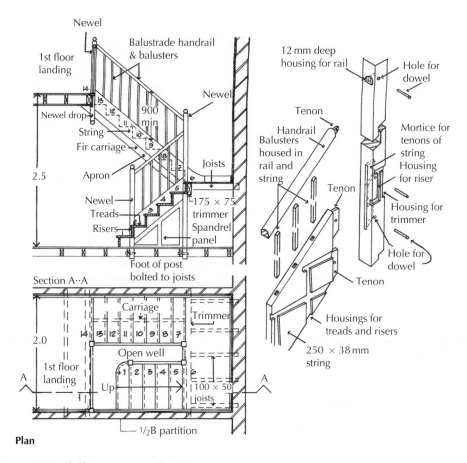

Figure 9.20 Half-turn open-well staircase.

Open balustrade

The traditional balustrade consists of newel posts, handrail and timber balusters, as illustrated in Figure 9.20. The newel posts at half-turn landings and at landings at first-floor level are housed and bolted to trimmers. These newels are fixed in position so that the faces of the risers at the foot and head of flights are in line with the centre line of the newel.

Handrail

The top of the handrail is usually fixed at a minimum height of 900 mm vertically from the pitchline to the top of the handrail and 900 mm above landings for domestic stairs in a single house and 900–1000 mm above the pitchline for other stairs. The handrail is cut from 75 × 50 mm timber, which is shaped and moulded. The ends of the handrail are tenoned to mortices in the newels.

Balusters

Balusters may be 25 or 19 mm square or moulded. They are either tenoned or housed in the underside of the handrail and tenoned into the top of closed strings or set into housings in the treads of flights with cut string, as shown in Figure 9.19. To prevent children under 5 years of age from becoming stuck between them, balusters should be so spaced that a 100 mm sphere cannot pass between them.

The traditional balustrade of vertical balusters, either plain or moulded, may not provide a satisfactory appearance for some situations, for example, enclosed glass-panelled balustrades with the glass set in a metal channel above the pitchline to concrete steps or stairs and set into a channel fixed under or below a handrail. An enclosed balustrade to a close string stair is illustrated in Figure 9.21, with a plywood panel set in a softwood frame fixed between the top of a close, outer string and the underside of the handrail.

Bullnose and curved steps

For the sake of appearance, the bottom step of a timber staircase may be framed to project beyond the newel post and be shaped as either a quarter or a half circle as illustrated in Figure 9.22. By projecting from the enclosing strings into the floor, the bottom step gives the sense of the stair belonging to the floor as well as the staircase. The bullnose, quarter circle and the rounded, half circle, end steps are made by cutting a riser made from a veneer

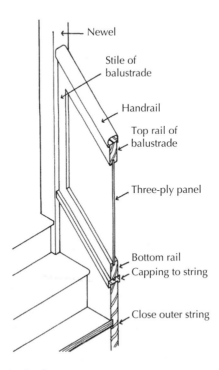

Figure 9.21   Enclosed balustrade.

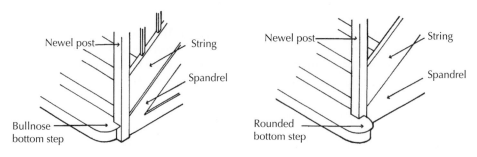

Figure 9.22 Shaped steps.

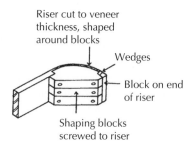

Figure 9.23 Bullnose step.

of thin wood which is shaped around three curved blocks as illustrated in Figure 9.23. The tread is cut with a projecting nosing to return around the shaped bottom step.

Spandrel

The triangular space between the underside of the lower flight of a stair and the floor is the spandrel. It may be left open and be a nuisance to keep clean or more usually it is enclosed with spandrel filling (boarded up) or closed with timber-framed panelling as illustrated in Figure 9.22. Where the length of the spandrel filling is sufficient, the panelling may be framed around a door so that the spandrel space below the stair may be used for storage.

Carriage

A sawn softwood carriage is usually fixed below the flights of a staircase to give support under the centre of each step. The fir (softwood) carriage illustrated in Figure 9.24 is $100 \times 6 \times 75\,mm$ in section and nailed to landing trimmers or joists for support with the top surface of the carriage directly under the angle of junction of treads and risers. Short off-cuts from $175 \times 25\,mm$ boards are nailed alternate sides of the carriage, so that the top edge of these brackets bears under treads to reduce creaking of the stair.

Where the soffit of flights of staircases is to be plastered, two additional fir carriages should be fixed, one next to the wall and the other next to the outer string as fixing for plasterboard.

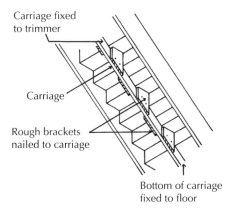

Figure 9.24   Underside of flight showing carriage and brackets.

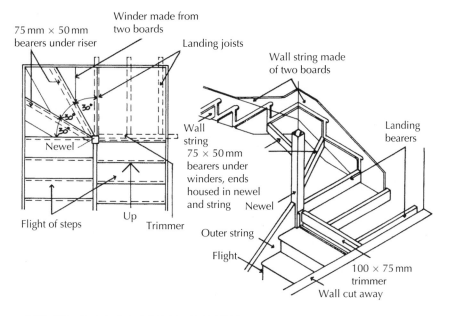

Figure 9.25   Winders and quarter-space landing.

Winders

Winders is the name given to tapered treads that wind round quarter- or half-turn stairs in place of landings to reduce the number of steps required in the rest of the stair and to economise on space. These winders may be used in domestic stairs, although they may present a hazard to the young and elderly and are not recommended for use in means of escape stairs or stairs in public buildings. The winders illustrated in Figure 9.25 are constructed as three taper treads at the quarter turn of a half-turn stair with a quarter-turn landing leading down to the lower flight.

The newel post may be continued down to and supported at the floor below so that it may support the trimmer for the quarter-space landing and the bearers for the winders. The treads of the winders are made of two boards tongued and grooved together. To support the edge of winders, 75 × 50 mm bearers are housed in the newel post and wall string. Because of the extra width of the tread of winders where they are housed in wall strings, the wall string has to be made of two boards into which both treads and risers are housed in 12 mm grooves and wedged and glued.

Open-riser or ladder stair

An open-riser or ladder stair consists of strings with treads and no risers so that there is a space between the treads, with treads overlapping each other by at least 16 mm (they are not suitable for disabled use). Various materials and forms of construction are used for exposed open-riser stairs, such as wood strings, treads and handrail, reinforced concrete strings and treads, reinforced concrete central carriage with cantilever treads, and steel strings and treads to a steel handrail supporting glass treads hung from the handrails.

More traditional open-riser stairs are illustrated in Figure 9.26. The strings may be either close or cut to the outline of the treads. The treads should be cut from 38 or 44 mm thick

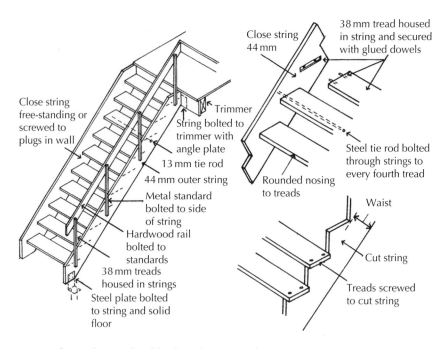

Figure 9.26 Open-riser stair with closed or cut string.

timbers, which are housed in closed strings and secured in position with glued wood dowels. To strengthen the fixing of the treads to the strings against shrinkage and twisting, 10 or 13 mm tie rods, one to every fourth tread, are bolted under the treads through the strings. The strings are fixed to the floor with steel plates, which are bolted to the sides of the timber strings and bolted to timber trimmers or cast into concrete floors.

Open-riser timber stair
Open-riser timber stairs are often constructed as straight-flight stairs between floors and there is no newel post to provide a fixing for the handrail. The handrail and balustrade are fixed to the sides of the strings, as illustrated in Figure 9.26.

Gaps in stairways
Where it is likely that children under the age of 5 are likely to use the stairway, any gap or opening must not allow a 100 mm sphere to pass through it. This is to prevent a child's head from becoming lodged between parts of a stairway. Thus, openings in open stairways and between balusters should be less than 100 mm.

Protection of stairs during construction
The importance of protecting stairs and ramps from damage during the construction period should not be underestimated. Stairs and ramps constructed in situ will be used by construction personnel and therefore must be protected to ensure that the surface finish is not damaged before the building is complete. Although prefabricated staircases can be installed much later in the construction process, it is common practice to install them as early as possible to provide safe access to the upper floor. Where stairs are used during the construction period, they should be protected with temporary treads. Temporary timber treads can be simply tacked (but securely fixed) in place until the construction period is finished. The whole stairway should also be safely covered with a protective membrane, such as building paper (Photograph 9.4).

9.5 Stone stairs

Stone stairs are constructed of steps of natural stone of rectangular or triangular section built into an enclosing wall so that each stone is bedded on the stone below in the form of a stair. Traditionally, the end of each stone was built into the wall of the stair well from which it cantilevered and took some bearing on the stone below in the form of a prop cantilever. The steps were either of uniform rectangular section with a stepped soffit or rectangular section cut to triangular section to form a flush soffit, as illustrated in Figure 9.27. These steps had splayed rebated joints and nosings cut on the edge of the tread surface. Landings were constructed with one or more large slabs of natural stone built into enclosing walls and bearing on the step below. Because of the cost of natural stone, this type of stairs is now made of cast stone or cast concrete which is usually reinforced and cast in the same sections as those illustrated for natural stone, or as a combined tread and riser with a rectangular end for building into walls and a stepped soffit.

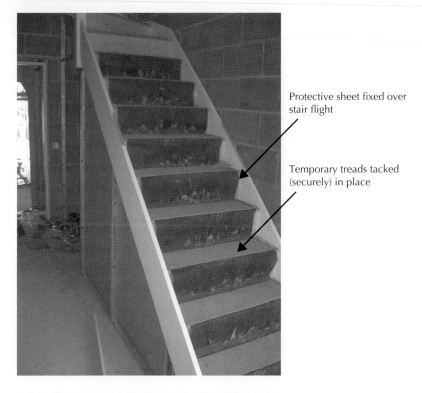

Photograph 9.4 Temporary protection of timber stairways.

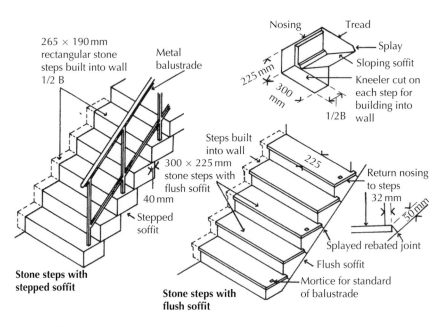

Figure 9.27 Stone steps.

9.6 Reinforced concrete stairs

A reinforced concrete stair has better resistance to damage by fire than a conventional timber staircase and is used for access and a means of escape stairs in most buildings of more than three storeys. The width, rise, going and headroom for these stairs and the arrangement of the flights of steps as straight-flight, quarter-turn, half-turn and geometrical stairs is the same as for timber stairs. The usual form of a reinforced concrete stair is as a half-turn (dog-leg) stair either with or without an open well. The construction of the stair depends on the structural form of the building and the convenience in casting the stair in situ or the use of reinforced concrete supports and precast steps.

Concrete stair construction

Concrete stairs which span between landings

Where there are loadbearing walls around the stair, it is generally economic to build the landings into the side walls as one-way spanning slabs and construct the flights as inclined slabs between the landings, as illustrated in Figure 9.28. This form of stair is of advantage where the enclosing walls are of brick or block as it would involve a great deal of wasteful cutting of bricks or blocks were the flights to be built into the walls and the bricks or blocks cut to fit to the steps.

Concrete stairs – cranked slab

As an alternative, the stair may be designed and constructed as a cranked (bent) slab spanning through landing, flight and next landing as one slab with no side support as shown in Figure 9.29a. This is a more costly construction than using the landings as slabs to

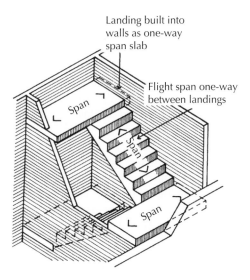

Figure 9.28 Inclined slab stair.

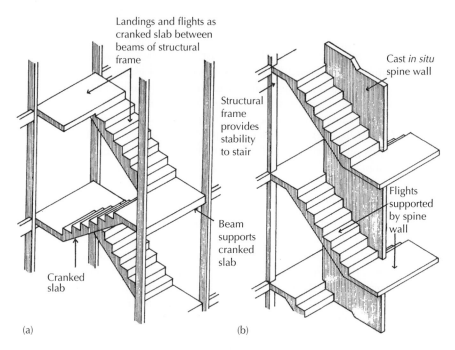

Figure 9.29   (a) Cranked slab stair. (b) Cantilevered spine wall stair.

support the flight. The flights have a greater span and therefore the cost of the stair is greater. This form of construction is used where the landings cannot gain support each side of the stair.

String and trimmer concrete stair
Another construction is to form a reinforced concrete frame of beams to landings support-ing inclined beams to flights, as illustrated in Figure 9.30. The landing beams are supported by side walls or the beams of a frame and in turn support inclined beams that support the steps. It is best suited to the use of precast concrete steps that bear on the inclined beam under the flight with step ends built into enclosing walls or on two inclined beams, and the use of precast landings.

Concrete stairs cantilevered from a central spine
Where a reinforced concrete half-turn stair is constructed around a reinforced concrete centre spine wall between the flights, the stair may be constructed to cantilever from this spine wall, as illustrated in Figure 9.29b, or partly cantilever from the spine wall and be supported by the enclosing frame or walls.

Reinforcement for concrete stairs
The reinforcement of a concrete stair depends on the system of construction adopted. The stair illustrated in Figure 9.31 is designed and built with the landings built into the enclos-ing walls as a two-way slab, and an inclined slab as flights spanning between landings

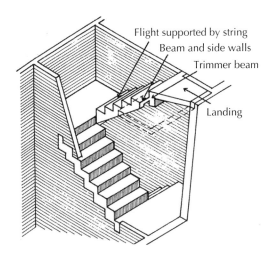

Figure 9.30 String and trimmer stair.

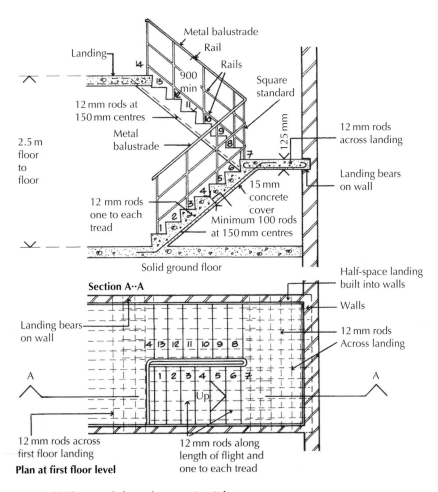

Figure 9.31 Half-turn reinforced concrete stair.

independent of the side walls. The main reinforcement of the landings is both ways across the bottom of the slab, and the main reinforcement of the flights is one way down the flights. The effective depth of the inclined slab that forms the flights is at the narrow waist formed in section by the junction of the tread and riser and the soffit of the flight. It is this thickness of the slab that has constructional strength and the steps play no part in supporting loads. The reinforcement has to have a cover of concrete around it to inhibit rust and protect steel rods against damage by fire.

Balustrade

The balustrade to a stone, cast stone or reinforced concrete stair is usually of metal, the uprights of which are either bolted to the sides of the flights to studs cast or grouted into the material or bolted through the material or set in mortices either cast or cut in the material. These vertical metal supports or standards in turn support rails as a balustrade for security and a handrail.

To provide maximum rigidity for the uprights that support a balustrade and handrail for stone or concrete stairs and landings, the uprights should be bolted through the thickness of the flights and landings as illustrated in Figure 9.32. The metal uprights are bolted

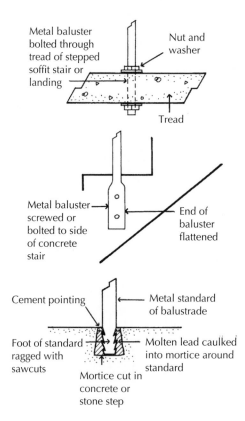

Figure 9.32   Setting metal standard in stone or concrete step.

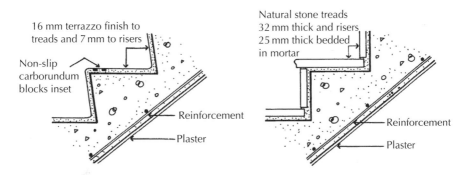

Figure 9.33   Finishes to concrete steps.

with nuts and washers through the depth of the stair. Uprights at some 900–1200 mm intervals will support a frame fixed just above pitchline and up to and including a handrail.

As an alternative fixing, for flush sloping soffit flights, for example, the uprights may be secured by expanding bolts to the side of a reinforced concrete stair as illustrated in Figure 9.32. The traditional method of fixing the metal uprights, standards or balusters to natural stone stairs and landings is by fixing the end of the verticals into a mortice cut in the stone. The ragged end of the verticals is then set in the mortice and molten lead is run into the mortice and caulked (rammed) to complete the joint. As this fixing for the uprights of a balustrade is not so rigid or secure as bolting through, the uprights have to be at fairly close intervals of 400–600 mm to support the balustrade frame and handrails.

Hard, durable, natural stone steps and landings may be left as a natural finish for the benefit of the appearance of the stone. Even the hardest natural stone will become scuffed and dirty in time and is laborious to keep clean. Usually, stone and concrete stairs are given an applied finish to create a surface that is easy to clean and is visually more attractive. In situ or precast terrazzo is often used for its appearance and ease of cleaning, with carborundum inserts as a non-slip surface, as illustrated in Figure 9.33. Wood treads of hardwood screwed to plugs in each step provide an attractive, durable and quiet-in-use surface. Stone treads and risers may be bedded as a surface finish for reinforced concrete stairs, as illustrated in Figure 9.33.

9.7 Structural glass stairs

Glass stairs, handrails and barriers have become increasingly common in recent years. This is partly down to architectural fashion, but mainly a result of considerable advances in the strength of glass used for structural purposes. Although aesthetics play a considerable part in the choice of a glass stairway, there are some functional reasons why glass may be used, especially when used in handrails and balustrades/guarding.

In many public buildings, such as airports and other large complexes, it has become increasingly important to be able to see the movement of people and their belongings.

Photograph 9.5   Examples of structural glazing used in stairs.

Balustrades made out of concrete, steel or timber may obscure vision and provide a place where people can hide, become trapped or encounter difficulty without others being aware. Glass balustrades help to improve safety and security (Photograph 9.5).

9.8 Ramps

Ramps should be designed to form as gentle an incline as possible, avoiding steep gradients that will create difficulties for users. Restricting the incline of the slope allows greater ease up the slope and controls the speed at which people descend. Landings may be required to allow the users and their helpers to rest during the ascent or descent of the ramp. The landing must allow adequate space for users to rest and also allow others to pass safely.

Internal ramps

Criteria for internal ramps (Figure 9.34) are:

❑   Slip-resistant surface, frictional resistance of ramp and landing must be similar
❑   Visual colour contrast between landing and ramp
❑   Surface width of at least 1.5 m

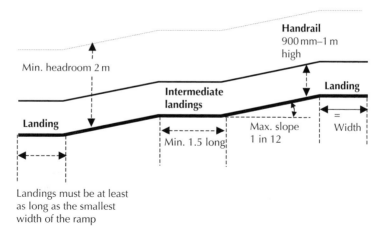

Figure 9.34   Internal ramps – positioning of landings.

Table 9.2   Limits of ramp gradient (Approved Document K)

Ramp gradients between landings		
Going of flight (m)	Maximum gradient	Maximum rise (mm)
10	1:20	500
5	1:15	333
2	1:12	166

❑   If the ramp provides means of escape, the number of persons expected to use the ramp and the type of building need to be considered to determine the width (see Table 9.1)

❑   Maximum gradient (see Table 9.2)

❑   No flight should have a going greater than 10 m, or a rise of more than 500 mm

❑   The ramp must have a top and bottom landing at least as long as the narrowest width of the ramp

❑   All ramps should have clear headroom of 2 m

❑   Handrails should be between 900 mm and 1 m high on the flight and 900 and 1100 mm on the landing

❑   Handrails should extend 300 mm beyond the top and bottom of the ramp

❑   Handrails should be on both sides

❑   Landings at the foot and head must be at least 1.2 m long

❑   Intermediate landings must be at a minimum length of 1.5 m and be clear of doorways

❑   Intermediate landings should be a minimum of 1.8 m wide and 1.8 m long, where they are required as passing places (when wheelchair users cannot see from one end of the ramp to the other, or the ramp has three flights).

External ramps providing access to a building

The requirements imposed by Approved Document K for external ramps are more stringent than for those used inside buildings; obviously, the exposure to the weather is a determining factor. The following provide a summary of the requirements imposed by Section K on ramps that are used in situations other than in buildings:

❏ Ramped access should be clearly signposted
❏ Signs are also required to warn users of hazard, when rise is greater than 300 mm
❏ Risk is greatest at the head of the ramp; signage should be sufficient to allow a person to stop before the ramp
❏ Max gradient as in Table 9.2
❏ No flight has a going greater than 10 m or rise more than 500 mm
❏ Alternative access must be provided if the total rise is greater than 2 m (e.g. a lift)
❏ Slip-resistant surface
❏ Visual colour contrast between landing and ramp
❏ Frictional characteristics of landing and flight must be similar
❏ Corduroy hazard-warning surface should be provided at the top and bottom
❏ Width of the ramp must be at least 1.5 m
❏ Must have top and bottom landing, minimum length 1.2 m, clear of obstructions
❏ Intermediate landings must be at a minimum length of 1.5 m, clear of obstructions
❏ Intermediate landings used as passing places must be a minimum of 1.8 m long and 1.8 m wide
❏ Landings are level, subject to a maximum gradient of 1:60 and a maximum crossfall of 1:40
❏ Handrails on both sides should extend 300 mm beyond the ramp
❏ Handrails should provide a visual contrast against the background and not be cold to the touch
❏ Handrail dimensions are as those for disabled stairs (see Figure 9.9).

10 Surface Finishes

Finishes form the interface between building users and the building and hence affect the way in which we interact and perceive our built environment. Surfaces are seen, touched and smelt by building users. Colour, or the lack of it, affects our psychology and the atmosphere of our buildings. Materials give off scent and this too will influence our internal environment and may affect our health. Given the wide variety of finishes available, the materials used to form a finish are quite extensive. In this volume, we have confined ourselves to the floor, wall and ceiling finishes most commonly found in domestic properties and small-office developments.

10.1 Functional requirements

The primary function of a surface finish is to provide a durable, visually attractive and low-maintenance surface to floors, walls and ceilings. There may be differences between the functional requirements of the finish depending on whether it is an external or internal finish as detailed further.

External finishes

External finishes are important in determining the aesthetic appeal of the building. The external finishes will also, in conjunction with the detailing and quality of the construction, determine how the building will weather over time. Thus, the quality of the materials used for external finishes and the manner in which they are applied will determine the durability of the building fabric. The functional requirements are:

❏ Aesthetic appeal
❏ Durability
❏ Strong mechanical or chemical bond to structural substrate
❏ Flexibility, the ability to withstand thermal and moisture movement (via control joints)
❏ Health and safety considerations

Barry's Introduction to Construction of Buildings, Third Edition. Stephen Emmitt and Christopher A. Gorse.
© 2014 John Wiley & Sons, Ltd. Published 2014 by John Wiley & Sons, Ltd.

Internal finishes

Internal finishes are important in creating a sense of place and in helping to ensure a healthy indoor environment. As we spend a great deal of time within buildings, the quality of the internal environment is particularly important in ensuring a sense of well-being and enjoyment. Materials will be touched, experienced visually and will give off scent, which, combined with furnishings and appliances, will influence our perception of the space in which we live or work and affect indoor air quality. The functional requirements are:

❏ Aesthetics
❏ Durability and flexible ability to withstand thermal and moisture movement
❏ Ease of maintenance and cleaning
❏ Strong mechanical or chemical bond to structural substrate
❏ Expel water from the surface that forms as condensation (particularly in kitchens and bathrooms)
❏ Prevent and resist mould growth or insect attack
❏ Provide visual finish, high levels of contrast (e.g. nosing on stairs)
❏ Tactile or touch-sensitive finish as an aid to those with visual difficulties
❏ Non-toxic

Types of finish

There are two different types of finish to the building fabric: those inherent in the material and those applied to a background.

Inherent finishes

Many materials such as timber, stone, brick and glass provide a natural finish without any need for further work – an inherent finish. Attention to joints, fixings and the quality of work is critical (because they will not be covered up). When brickwork and blockwork is to be left fairface (not plastered) it is important to specify this so that the joints are built to an appropriate quality. Carefully chosen, materials with an inherent finish may help to reduce construction time and initial construction costs. Furthermore, the use of materials with an inherent finish may be an important consideration when disassembling the building and reclaiming materials at a future date for reuse, since the material has not been compromised by the application of a finish.

Applied finishes: ecological consideration

Application of materials to existing backgrounds, such as plaster or render to a wall, or paint to timber, is an applied finish. The durability of the finish will depend upon the material properties of the finish and the material it is applied to, as well as the bond between the two materials. Ecological design goals aim to minimise the pollution from applied finishes. Petrochemical paints, stains and varnishes should be avoided and preference given to products with natural pigments that are not harmful to animals, plants or people. Frequency of maintenance and re-application of applied finishes needs to be considered in the overall life cycle of the building, and compared against inherent finishes.

10.2 Floor finishes

A floor finish should be level, reasonably resistant to the wear, be capable of being maintained and remain in a safe condition during its designated design life and capable of being easily cleaned. For specific areas of buildings, additional requirements may be that the surface should be non-slip, smooth for cleaning and polishing, resistant to liquids and chemical spillages, seamless for hygiene, etc. There is no one finish that will satisfy the possible range of general and specific requirements. There is a wide range of finishes available from which one may be selected as best suited to a particular function. For the small floor areas of rooms in houses and flats, the choice of floor finish is dictated largely by appearance and ease of cleaning. For the larger floor areas of offices, public and institutional buildings, ease of cleaning is a prime consideration where power-operated cleaning and polishing equipment is used.

Finishes to concrete floors

It is convenient to make a broad general classification of finishes to concrete floors (described in Chapter 4) as:

- ❑ Jointless
- ❑ Flexible thin sheet and tile
- ❑ Rigid tiles and stone slabs
- ❑ Wood and wood based

Jointless floor finishes

This group includes the cement- and resin-based screeds and mastic asphalt. These are laid while plastic and other than the provision of movement joints they provide a homogeneous surface.

Cement screeds
A cement- and sand-screed finish to a concrete floor may be an acceptable, low-cost finish to small area floors of garages, stores and outhouses where the small area does not justify the use of a power float and considerations of ease of cleaning are not of prime importance. Premixed, cement- and sand-screed material reinforced with polymer fibre is available; the fibre reinforces against drying, shrinkage and cracking. To produce improved surface resistance to wear and resistance to the penetration of oils and grease, a dry powder of titanium alloy with cements may be sprinkled on to the wet surface of concrete or screed and trowelled in.

Granolithic paving
Granolithic paving consists of a mixture of crushed granite, which has been carefully sieved so that the particles are graded from coarse to very fine in such proportions that the material, when mixed, will be particularly free of voids or small spaces, and when mixed with cement will be a dense mass. The usual proportions of the mix are 2½ of granite chippings to 1 of cement by volume. These materials are mixed with water and the wet mix is spread

uniformly and trowelled to a smooth flat surface. When this paving has dried and hardened, it is hard wearing. There are a number of additives variously described as 'sealers' or 'hardeners', that may be added to the granolithic mix to produce improved resistance to surface wear. This floor finish is used for factories, stores, garages and other large floor areas that have to withstand heavy wear.

Anhydrite floor finish

Premixed, dry bagged screed material of anhydrite and sand is used as a floor finish. Anhydrite is a mineral product of heating gypsum that will, when mixed with water, act as a cement to bind the grains into a solid mass as the material dries and hardens. The advantage of anhydrite is that it readily combines with water and does not shrink and crack as it dries out and hardens. The wet mix of anhydrite and sand may be pumped and spread over the concrete base as a self-levelling screed or spread and trowelled by hand. The material may be pigmented. A disadvantage of the material is that it fairly readily absorbs water and is not suited to use in damp situations.

Resin-based floor finish

A range of resin emulsion finishes is available for use where durability, chemical resistance and hygiene are required in laboratories, hospitals and food-preparation buildings. This specialist application finish is composed of epoxy resins as binders with cement, quartz, aggregates and pigments. The material is spread on a power-floated or cement-screed base by pumping and trowelling to a thickness of up to 12 mm. The aggregate may be exposed on the surface as a non-slip finish and as decoration. On larger floor areas, it is used for the advantage of a seamless finish that can be cleaned by a range of power-operated devices.

Polymer resin floor surface sealers

Polyester, epoxy or polyurethane resin floor sealers are specialist thin floor finishes used for their resistance to water, acids, oils, alkalis and some solvents. The materials are spread and levelled on a level power-floated or screed surface to provide a seamless finish to provide an easily cleaned surface. Polyester resin, the most expensive of the finishes, is spread to a finished thickness of 2–3 mm to provide the greatest resistance. Epoxy resin provides a less exacting resistance. It is sprayed or pumped to a self-levelling or trowelled thickness of 2–6 mm. Polyurethane resin, which has moderate resistance, can be spread on a somewhat uneven base by virtue of its possible thickness. It is pumped to be self-levelling for thin applications and trowelled for the thicker applications. Thicknesses of between 2 and 10 mm are used.

Mastic asphalt floor finish

Mastic asphalt serves both as a floor finish and a damp-proof membrane (dpm). It is a smooth, hard-wearing, dust-free finish, easy to clean but liable to be slippery when wet and less used since the advent of the thin plastic tiles and sheets.

Flexible thin sheet and tile

Linoleum is made from oxidised linseed oil, rosin, cork or wood flour, fillers and pigments compressed on a jute canvas backing. The sheets are made in 2 m widths, 9–27 m lengths and thicknesses of 2.0, 2.5, 3.2 and 4.5 mm in a variety of colours. The usual thickness of the sheet is 2.5 mm. Tiles 300 and 500 mm square are 3.2 and 4.5 mm thick. Linoleum should be laid on a firm level base of plywood or particle board on timber floors or on hardboard over timber-boarded floors and on a trowelled screed on concrete floors. The material is laid flat for 48 hours at room temperature and then laid on adhesive and rolled flat with butt joints between sheets. Linoleum has a semi-matt finish, is quiet and warm underfoot and has moderate resistance to wear for the usual 2.5 mm thick sheets and good resistance to wear for the thicker sheets and tiles. Linoleum has been used instead of vinyl for the advantage of the strong colours available in the form of sheets and also in the form of decorative patterns by combining a variety of colours in various designs from cut sheet material, a characteristic that has seen the material experience a return to fashion more recently.

Flexible vinyl sheet and tiles

Polyvinyl chloride (PVC), generally referred to as vinyl, is a thermoplastic used in the manufacture of flexible sheets and tiles as a floor finish. The material combines PVC as a binder with fillers, pigments and plasticisers to control flexibility. The resistance to wear and flexibility vary with the vinyl content, the greater the vinyl content the better the wear and the poorer the flexibility.

Vinyl sheet flooring has become the principal sheet flooring used where consideration of cost and ease of cleaning combine with moderate resistance to wear. Sheet thicknesses from 1.5 to 4.5 mm in widths from 1200 to 2100 mm are produced in lengths of up to 27 m. Foam-backed vinyl sheet is produced to provide a resilient surface with the advantages of resilience and being quiet underfoot but at the expense of the material being fairly easily punctured. The material is extensively used in domestic kitchens and bathrooms, and offices where a low-cost, easily cleaned surface is suited to moderate wear.

The thin sheet material should be laid on a smooth, level-screeded surface particularly free from protruding hard grains that might otherwise cause undue wear. The thicker, less flexible sheet may be laid on a power-floated concrete finish. The sheets are bonded on a thin bed of epoxy resin adhesive and rolled to ensure uniformity of adhesion. For large areas of flooring, the sheets may be heat welded to provide a seamless finish. A range of flexible vinyl tiles is produced in a variety of colours and textures in 225, 250 or 300 mm squares by 1.5–3 mm thicknesses. Various shapes of cut sheet may be used to provide single- or multicoloured designs.

Clay floor tiles

Natural clay floor tiles have been used for centuries as a hard, durable floor surface and finish for both domestic and agricultural ground floors. The two types of tiles may be distinguished as floor quarries and clay floor tiles. The word quarry is derived from the French *carr*, meaning square.

Floor quarries

Floor quarries are manufactured from natural clays. The clay is ground and mixed with water and then moulded in hand-operated presses. The moulded clay tile is then burnt in a kiln. Manufacturers grade tiles according to their hardness, shape and colour. The first or best quality of these clay floor quarries is so hard and dense that they will suffer the hardest wear without noticeably wearing. Because they are made from plastic clay, which readily absorbs moisture, quarries shrink appreciably when burnt, and there may be a noticeable difference in the size of individual tiles in any batch. The usual colours are red, black, buff and heather brown. Some common sizes are $100 \times 100 \times 12.5$ mm thick, $150 \times 150 \times 12.5$ mm thick and $229 \times 229 \times 32$ mm thick.

Clay floor tiles

Where finely ground clay is used, the finished tiles are very uniform in quality, and because little water is used in the moulding, very little shrinkage occurs during burning. The finished tiles are uniform in shape and size and have smooth faces. The tiles are manufactured in red, buff, black, chocolate and fawn. Because of their uniformity of shape, these tiles provide a level surface that is resistant to all but heavy wear, does not dust through abrasion, is easily cleaned with water and has a smooth, non-gloss finish which is reasonably non-slip when dry. They are used for kitchens, bathrooms and halls where durability and ease of cleaning are an advantage. Some common sizes are $300 \times 300 \times 15$ mm thick, $150 \times 150 \times 12$ mm thick and $100 \times 100 \times 9$ mm thick.

Vitreous floor tiles

Vitreous tiles are made from clay and feldspar, which gives the tile a semi-gloss finish. The tiles are uniform in shape and size and have a very smooth semi-gloss or gloss surface that does not absorb water or other liquids and can be easily cleaned by mopping with water. Both vitreous and fully vitreous tiles may be moulded with a textured finish to provide a moderately non-slip surface. The gloss finish is impervious to most liquids, dust-free and liable to be slippery, particularly when wet. Sizes are generally similar to those of plain colour tiles.

Laying clay floor tiles

The considerations that affect the choice of a method of laying floor tiles are:

❑ Tolerance: the fixing adhesive and tiles must be capable of accommodating any undulations and variations in the structure to which they are fixed so that the final tile finish is level
❑ Good adhesion to the base to provide solid support; this is particularly important for thin tiles if cracking of the tile is to be avoided
❑ Providing a means of accommodating relative structural, moisture and thermal movements between the base and the finish to prevent arching of the tile floor

Tiles are laid by the direct bedding method or the thin bed adhesive model.

Direct bedding method
The traditional method of laying tiles is to bed them on a layer of wet cement and sand spread over a screeded or level concrete floor with control joints at the manufacturer's recommended spacing.

Quarry tiles are laid and bedded in sharp sand and cement, 1 : 3 or 1 : 4 mix, spread to a level thickness of 15–20 mm, depending on the thickness of the tiles, on a fully dry concrete base. The cement and sand should be mixed with just sufficient water for workability and pressing the tiles into the bed. Too wet a mix will cause excess drying shrinkage. The main purpose of the bed is to accommodate the appreciable variations in thickness of the quarries to provide a reasonably level finish. The joints between the quarries will be up to 15 mm wide, to allow for variations in shape, and filled with cement and sand and finished level with the floor surface, or just below the surface, to emphasise the individual tiles.

The direct bedding method of laying is used for plain clay tiles on a bed some 10 mm thick and with joints between 5 and 10 mm wide, depending on variations in the size of the tiles and the need to adjust tile width to that of a whole number of tiles with joints to suit a particular floor size, thus avoiding the need to cut tiles on site.

Thin bed adhesive method
The majority of the thin, vitreous tiles that are used today is bedded and laid on an adhesive that is principally used as a bond between the tiles and the base, and to a lesser extent as a bed to allow for small variations in tile thickness. The adhesives that are used are rubber latex cement, bitumen emulsion and sand and epoxy resins. These adhesives are spread on a level power-floated concrete or a screed finish, to a thickness of 3–5 mm, combed to assist bedding and the tiles are then pressed and levelled in position. Where the thin bed epoxy resins are used as an adhesive for thin, vitreous tiles, there should be no large protruding particles of aggregate or sand in the floor surface over which the brittle tile could crack under load.

Tiles arching – control joints

The word arching is the effect of tiles rising above their bed (the structural floor) in the form of a shallow arch. Arching is caused by expansion of the tiles relative to their bed or contraction of the bed relative to the tiles. With most finishing materials, it is advisable to provide control joints which allow the materials to shrink without causing unsightly cracks and expand without separating from their substrata.

Where there is a realistic likelihood of arching, the tiles may be laid on a bed spread over a separating layer so that movement of either the tiles or the base will not affect the floor finish. A layer of polythene film, bitumen felt or building paper is spread with 100 mm lapped joints over the concrete floor. The tiles are then laid and bedded on a cement/sand mix spread and levelled to a thickness of 15–25 mm, depending on the thickness of the tiles, and jointed in the same way as for direct bedding.

Concrete tiles

Concrete tiles made of cement and sand, which is hydraulically pressed to shape as floor tiling, have been used as a substitute for quarry and plain colour clay tiles. The usual size of tiles is $300 \times 300 \times 25$ mm, $225 \times 225 \times 19$ mm and $150 \times 150 \times 16$ mm. The material may be pigmented or finished to expose aggregate. The density and resistance to wear depend on the quality control during manufacture and the nature of the materials used. They are laid on a level power-floated concrete or screed surface and jointed in the same way as quarries and plain colours.

Stone slabs

A wide range of natural stone slabs is used as a floor finish, from the very hard slabs of granite to the less dense soft marbles. Stone is selected principally for the decorative colour, variations in colour, grain and polished finish that is possible, and for durability. The method of bedding natural stone slabs as an internal floor finish varies with the thickness, size, nature and anticipated wear on the surface. Large, thick slabs of limestone, sandstone and slate up to 50 mm thick are laid on cement and sand with cement and sand joints. Thin slabs of granite and marble are laid by the thin bed adhesive method or the dry sand bed method, which is usually used for marble.

Joints

The width of the joints between tiles and slabs as an internal floor finish is determined by the uniformity of shape of the material used. For quarries, joints of up to 12 mm may be necessary to allow for the variations in size, and joints as little as 1 mm may be possible with very accurately cut and finished thin slabs of granite and marble. The disadvantage of wide joints is that the material used, such as cement and sand for quarries, will be more difficult to clean and will more readily stain than the floor material. Ideally, the jointing material should have roughly the same density, resistance to wear and ease of cleaning as the floor finish.

Control joints

The joints between tiles and slabs will serve the purpose of accommodating some movement of the floor finish. Some small expansion or contraction of the floor finish will be taken up in the joints through slight cracks or crushing of the very many joints.

It is good practice to form control joints to accommodate structural, moisture and thermal movement. These flexible joints should be continued through the rigid floor finishes as a flexible joint. Control joints should be formed around the perimeter of floors and against rigid abutments (e.g. columns) with an elastic sealant joint. It may also be necessary to divide large floor areas and rigid floor finishes into bays, separated by control joints. The disadvantages of these joints are that the joint material is necessarily softer than the surrounding surface, difficult to keep clean and will encourage wear of the edges of the finish next to the joint, thus careful consideration is needed when positioning control joints.

Timber floor finishes

Natural wood-floor finishes such as boards, strips and blocks are used for the advantage of the variety of colour, grain and texture of this natural material, which is warm, resilient and comparatively quiet underfoot.

Floorboards

Floorboards (described in Chapter 4) may be used as a floor surface to timber and to concrete floors. Either plain edge or tongued and grooved (T & G) boards are used. The boards are screwed or nailed to wood battens set in a screed or to battens secured in floor clips.

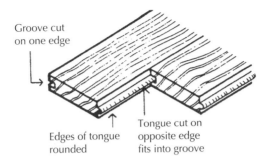

Figure 10.1   Tongued and grooved strip flooring.

Wood strip flooring

Strips of hardwood or softwood of good quality, specially selected so as to be particularly free of knots, are prepared in widths of 90 mm or less and 19, 21 or 28 mm in thickness. The type of wood chosen is one that is thought to have an attractive natural colour and decorative grain. The edges of the strip are cut so that one edge is grooved and the other edge tongued, so that when they are put together the tongue on one fits tightly into the groove in its neighbour, as illustrated in Figure 10.1. The main purpose of the T & G is to interlock the strips so that its neighbours resist any twisting within an individual strip of timber.

There is always some tendency for wood strips to twist out of true due to the wood drying out, and to resist this, the strips have to be securely nailed to wood battens which are secured to the concrete floor, either by means of plugging and screwing the battens to the structural concrete, using mechanically fired concrete nails ('Hilti' nails, which penetrate both the timber batten and concrete), galvanised metal floor clips, or casting the batten in a cement and sand screed. The illustration of part of a concrete floor finished with wood strips nailed to battens is shown in Figure 10.2.

Wood strip flooring can also be fixed by the thin bed adhesive method. Comparatively short lengths of wood strip, 300 mm long, with T & G edges and joints, are used to minimise drying deformation. The strips are bedded on an epoxy resin adhesive spread over a true-level screed on to which the strips are laid and pressed or rolled to make sound, adhesive contact. The strips are usually laid with staggered end joints. This is a perfectly satisfactory method of laying wood strip flooring as the narrow width and short length of strip is unlikely to suffer drying deformation which can tear strips away from the adhesive bed.

Proprietary systems are also available that employ secret fixing methods, their cost and durability determined by the type of timber used and its quality. Some of these systems use solid wood, although to reduce the cost of the flooring units it is common to apply a thin layer of timber to a solid backing. Ease of fixing and initial cost needs to be considered against maintenance and resistance to impact damage. Solid wood can be sanded, thin laminates cannot. Proprietary systems are usually laid directly on top of the floor screed and most can be used in conjunction with underfloor heating.

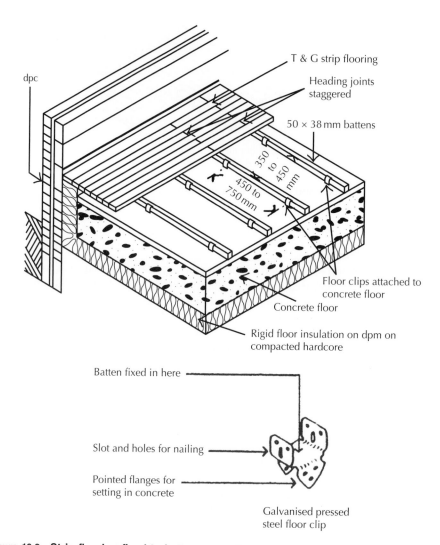

Figure 10.2 Strip flooring fixed to battens and clips.

Wood block floor finish

Blocks of wood are used as a floor finish where resistance to heavy wear is required, as in halls, corridors and schools, to provide a surface which is moderately resilient, warm and quiet underfoot. An advantage of the comparatively thick blocks is that after wear, the top surface may be sanded to reduce the block to a level surface. The blocks are usually 229–305 mm long by 75 mm wide by 21–40 mm thick, and are laid on the floor in a bonded, herringbone or basket weave pattern. The usual patterns are illustrated in Figure 10.3.

Wood blocks are laid on a thoroughly dry, clean, level cement and sand-screeded surface, which has been finished with a wood float to leave its surface rough textured. The tradi-

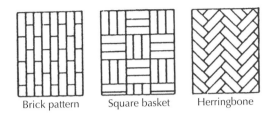

Brick pattern   Square basket   Herringbone

Figure 10.3   Wood blocks patterns.

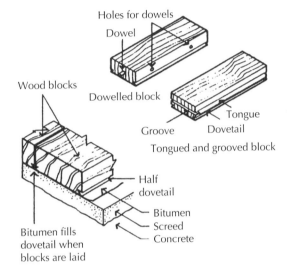

Figure 10.4   Joints for wood blocks.

tional method of laying blocks is to spread a thin layer of hot bitumen over the surface of the screed into which the blocks are pressed. The lower edges of the blocks of wood are usually cut with a half dovetail incision so that when the blocks are pressed into the bitumen, some bitumen squeezes up and fills these dovetail cuts and so assists in binding the blocks to the bitumen, as illustrated in Figure 10.4. After the surface has been sanded to provide a level finish, a wax polish or polyurethane seal is applied to provide an easily cleaned finish.

10.3 Wall and ceiling finishes

Fairface finishes

Fairface brickwork, blockwork, etc.
Fairface brickwork or blockwork is used to describe the higher standard of finish (work-manship and quality of bricks or blocks) that is required to provide an aesthetically pleasing

appearance. Particular attention is given to the joints; the perpendicular joints should line up vertically and all of the mortar joints should be of a consistent thickness. A neat joint should be formed and drips and snots of mortar should not be allowed to come into contact with the face of the brickwork. A wide variety of rough, smooth and polished facing blocks and bricks are available. The type of brick or block selected will depend on the performance requirements, for example, the performance requirements of fairface blockwork in industrial buildings, sports halls, showrooms and large arenas (which may accommodate concerts) are quite different.

Internal plastering

Plaster is the word used to describe the material that is spread (plastered) over irregular wall and ceiling surfaces to provide a smooth and level finish. The initially wet material is spread and levelled over uneven backgrounds such as brickwork, and over lath fixed to the underside of timber floor joists so that as it hardens and dries it forms a smooth, level wall and ceiling finish. The purpose of plaster is to provide a smooth, hard, level finish, which can be painted with emulsion paint, or to which wallpaper can be applied.

Plaster undercoats and finish coat

The finished surface of plaster should be flat and fine textured (smooth). It would seem logical, therefore, to spread some fine-grained material, such as lime or gypsum mixed with water, over the surface and trowel it smooth and level. The maximum thickness to which a wet, fine-grained material can be spread and levelled is about 3 mm. The irregularities in the surface of even the most accurately laid brick or blockwork are often more than 3 mm and it would be necessary to apply two coats to achieve a satisfactory finish. Instead of applying a fine-grained plaster in two coats, it is common practice to spread some cheaper, coarse-grained material that is easily spread as one or two coats to render the surface level and then finish this with a thin coat of fine-grained plaster to provide a smooth finish. The coarse-grained coat or coats of plaster are termed undercoats and the fine-grained final coat a finish or finishing coat.

Where some part of the surface of brick or block is particularly uneven, it is practice to fill the hollows with undercoat plaster up to the general level of the wall face. This practice is termed 'dubbing out', ready for a two-coat plaster finish. One-coat plaster is gypsum plaster, which combines the qualities of both undercoat and finish plaster in one product. The material, which is spread and built up by hand to a thickness of 11–13 mm on brick and block backgrounds, is progressively trowelled to a reasonably level, smooth matt finish ready for decoration. The material, which is mainly used as a do-it-yourself plaster, can be harder work to apply than a two-coat plaster.

Photograph 10.1a–c shows an internal wall where two-coat plaster has been applied. Prior to application, the blockwork was chased out so that a rebate could be formed to accommodate the services (Photograph 10.1a). The trunking, ducts and socket housings can be installed ready to receive the first fix services. Once installed, the wet plaster can then be applied.

Photograph 10.1 **(a) Services are chased and recessed into the walls ready for the wet plaster to be applied. (b) Coat and finish coat completed up to ceiling level – services protrude into ceiling void. (c) Finish coat applied with first fix services installed.**

Airtightness and wet plaster and dry lining methods

Brick and blockwalls on their own are not particularly airtight. Gaps between horizontal joints, weep holes and unsealed cavities provide an interconnected passage of air from the inside to the outside of the dwelling. Wet plaster methods have been found to greatly improve the airtightness of masonry walls. Dry lining does not have the same effect. If the plasterboard is applied using the dot and dab method, with plaster being spotted on the wall and the plasterboard pressed against the wet plaster dabs, a gap is left between the back of the plasterboard and the face of the wall. This gap then provides an additional air passage. Any gaps in the blockwork can then connect the cavity to the internal environment allowing air leakage. To increase airtightness when using plasterboard, it is suggested that a ribbon of plaster (continuous strip of plaster) should be placed around the perimeter of each board (Figure 10.5a and b). Another method to increase airtightness is to coat the surface of the blockwork with a thin layer of rough plaster (2–3 mm) prior to applying the plasterboard (Figure 10.5c).

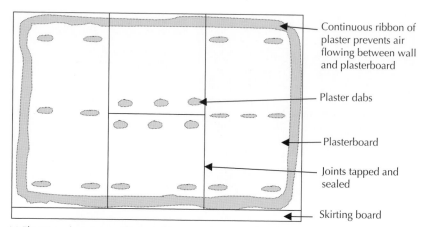

(a) Elevation showing continuous ribbon running around perimeter of the wall

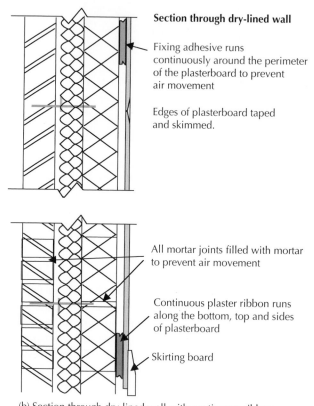

(b) Section through dry-lined wall with continuous ribbon

Figure 10.5   Sealing dry-lined walls.

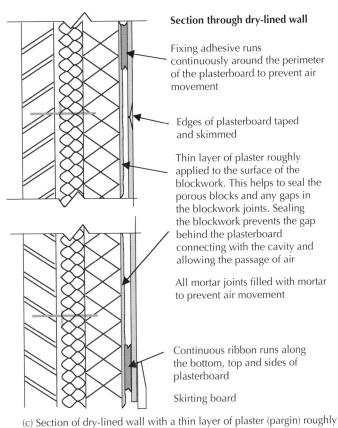

Section through dry-lined wall

Fixing adhesive runs continuously around the perimeter of the plasterboard to prevent air movement

Edges of plasterboard taped and skimmed

Thin layer of plaster roughly applied to the surface of the blockwork. This helps to seal the porous blocks and any gaps in the blockwork joints. Sealing the blockwork prevents the gap behind the plasterboard connecting with the cavity and allowing the passage of air

All mortar joints filled with mortar to prevent air movement

Continuous ribbon runs along the bottom, top and sides of plasterboard

Skirting board

(c) Section of dry-lined wall with a thin layer of plaster (pargin) roughly applied to seal surface of the blockwork

Figure 10.5   (Continued)

Materials used in plaster

Lime plaster

Lime plaster is used in the restoration and preservation of older buildings. Lime is mixed with sand and water in the proportion of 1 part of lime to 3 parts of sand by volume, with water for use as undercoat, and by itself mixed with water as a finish coat. As lime plaster dries and hardens, it shrinks and fine hair cracks may appear on the surface. To restrain shrinkage and to reinforce the plaster, long animal hair is included in the wet undercoat mix, 5 kg of hair being used for every square metre of the lime undercoat (coarse stuff). The resulting haired, coarse stuff is plastic and dries out and hardens without appreciable shrinkage and cracking.

Cement plaster

Cement is mixed with sand and water for use as an undercoat for application to brick and blockwalls and partitions. It is used on strong backgrounds as 1 part of cement to 3 or 4 parts of clean, washed sand by volume. A wet mix of cement and clean sand (sharp sand) is not plastic and requires a deal of labour to spread. It is usual, therefore, to add a plasticiser to the wet mix to produce a material that is at once plastic and sets and hardens to form a hard surface. Usual mixes are 1 of cement, ¼ lime to 3 of sand, 1 of cement, 1 of lime to 6 of sand or a mix of 1 of cement to 4 of sand, with a mortar plasticiser by volume. As cement plaster dries out, it shrinks fiercely and cracks may appear on the surface. In general, the more cement used, the greater the shrinkage. The extent of the cracking that may appear depends on the strength of the surface to which the plaster is applied and the extent to which the plaster binds to the surface.

Gypsum plaster

The advantage of gypsum plasters is that they expand very slightly on setting and drying and are not, therefore, likely to cause cracking of surfaces. Gypsum is a chalk-like mineral, being a crystalline combination of calcium sulphate and water ($CaSO_42H_2O$). It is available as both natural gypsum, which is mined in areas all over the world, and as a synthetic by-product of major industries such as fossil-fuelled power stations.

When powdered gypsum is heated to about 170°C, it loses about three-quarters of its combined water and the result is described as hemihydrate gypsum plaster ($CaSO_4¼H_2O$). This material is better known as plaster of Paris. When gypsum is heated to a considerably higher temperature than 170°C, it loses practically all its combined water and the result is anhydrous gypsum plaster. British Standard 1191 Parts 1 and 2 cover the manufacture of all traditional and modern gypsum-based plasters. It is convenient to categorise gypsum plasters relative to their use as wall and ceiling plasters as casting, undercoat, finish, one-coat and machine-applied plasters.

Casting plaster

Finely ground hemihydrate gypsum (plaster of Paris) when mixed with water sets and hardens so quickly (about 10 minutes) that it is unsuitable for use as a wall or ceiling plaster. It is ideal for making plaster casts for buildings. Wet plaster of Paris is brushed into moulds to provide cornices and other decorative plasterwork. The wet plaster is usually reinforced with open weave hessian and is generally referred to as fibrous cast plaster or fibrous work.

Retarded hemihydrate gypsum plaster

The gypsum used for undercoats is retarded hemihydrate gypsum in which a retarding agent is added to plaster of Paris to delay the setting time for 1.5–2 hours to allow time for spreading and levelling the wet material as undercoat. For general use as an undercoat, the retarded hemihydrate gypsum powder is mixed with lightweight aggregates, such as expanded perlite or vermiculite, as a dry mix powder, which is delivered to the site in bags as premix undercoat.

Premix gypsum undercoat

The advantage of this material is that the premix avoids the messy, wasteful operation of mixing dry powdered lime or cement with sand. The wet mix is comparatively easy to spread and level and the lightweight aggregate gives a small degree of thermal insulation. A disadvantage of the material is that the lightweight aggregate may not provide adequate resistance to damage by knocks. This material is applied as one undercoat to a finished thickness of 8–11 mm. In addition to the standard lightweight aggregate undercoat, other gypsum undercoat plasters are produced for specific backgrounds and also to suit specific performance criteria related to building use.

Bonding undercoat
Where the undercoat plaster is to be applied to a surface with particularly low suction, which does not readily absorb water, gypsum bonding undercoat is formulated to provide adequate adhesion.

High-impact undercoat
In some situations where it is anticipated that rough or careless usage may damage standard undercoat plaster, high-impact gypsum undercoat is used. A dense aggregate such as grains of silica are used in the premix in lieu of the usual lightweight aggregate to provide improved resistance to knocks.

Finish plaster
Finish plaster is powdered, retarded hemihydrate gypsum by itself for use as a thin finish coat for both gypsum undercoats and to plasterboards. Mixed with water, the plaster is spread and finished to a thickness of about 2–5 mm and sets in about 1–2 hours. This plaster, which is polished to a smooth surface, is also used as a finish to cement and sand undercoats.

Anhydrous gypsum plaster
Anhydrous gypsum plaster was commonly used as a thin finish coat to cement-based undercoats. The powdered gypsum is mixed with a mineral sulphate to accelerate its set, which otherwise would be so slow as to make it unsuitable for use as a finish plaster. A characteristic of this gypsum plaster is that it can be brought back (retempered) by sprinkling the stiff surface to make it plastic for trowelling smooth, although this makes it unsuitable for use in damp or moist situations.

One-coat gypsum plaster
One-coat gypsum plasters are retarded hemihydrate plasters, which combine the properties of an undercoat and a finish coat. One coat is applied to a thickness of up to 20 mm as an undercoat. As the plaster begins to set (stiffen), it is sponged with water and trowelled to bring the fine particles to the surface so that it may have a finish comparable to that of a separate finish coat. Because of the considerable labour required to build up and level the surface, this plaster is not extensively used.

Machine-applied gypsum plaster
Machine-applied or projection plasters are one-coat gypsum plasters designed to provide a longer setting time to allow for mixing, pumping, spreading and trowelling. The material is mixed with water, pumped and applied to the wall by a projection machine, which effectively halves the application time it would take to spread by hand. As the wet plaster covers the wall, it is treated in the same way as one-coat plaster by trowelling level and smooth.

Because of the additional labour in mixing and pumping and the necessary work of cleaning equipment after use, this type of plaster may be most economically used on large flat areas of the wall.

Phase changing materials: Wax microcapsules embedded in plaster

Phase changing materials (PCMs) are materials that change their form under certain known conditions. In this case, as wax capsules, which are embedded in materials, change from one form to another, they either take in or give out energy, in the form of latent heat helping to reduce fuel consumption.

The microcapsules, which measure less than 0.01 mm, make use of the latent heat required to change them from solid to liquid and liquid to solid. When the plaster gets hot, the wax within the plaster melts, and as it melts it uses the excess heat within the building to transform the capsules from solid to liquid. This transformation helps to keep the building cool, and captures and stores the latent heat (Figure 10.6). When the weather becomes cold and the internal climate changes, the capsules will solidify, releasing the latent heat. PCMs can be integrated in both solid and liquid materials; thus they can be used in paints, plasters and concrete.

The plaster, which incorporates one-third PCM, has the same storage capacity as a 230 mm brick wall. The wax capsules (PCM) help to equalise (stabilise) the temperature, making it cooler during hot periods and warmer during cold periods (Figure 10.6). The PCM material helps to optimise the room temperature. During warm periods, excess heat

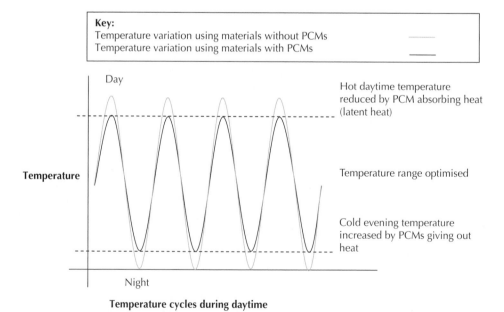

Temperature cycles during daytime

Figure 10.6   Graph showing room temperature adjustment using PCM (adapted from http://www.functionalpolymers.basf.com).

is taken from the internal environment and used as latent heat energy to transform the wax microcapsules from their solid form to liquid. The latent heat is stored and is released as the temperature drops and the material changes from liquid to solid, this time giving out heat energy as the material freezes.

Background surfaces for plaster

The surface of walls to be plastered will affect the type of plaster used and its application. The surface of rough-textured bricks and concrete blocks affords a 'key' for the mechanical adhesion of plaster to the background wall. As the wet plaster undercoat is spread and pressed into the surface, wet plaster fills the irregularities and as it hardens it forms a mechanical key to the background. With smooth-faced bricks or blocks, the mortar joints should be raked out some 12 mm, as the wall is built, to provide a key for the plaster. An advantage of the key is that it will restrain shrinkage of cement-based undercoats.

Suction

The word suction is used to describe the degree to which a surface will absorb water and so assist in the adhesion of plaster to a surface. Some lightweight concrete blocks readily absorb water and have high suction to the extent that wet plaster applied to them may lose so much water that it is difficult to spread and may not fully set due to loss of water. Suction may be reduced by spraying the surface with water prior to plastering or by the use of a liquid primer. The most straightforward way of testing the suction of a surface is to spray it with water to judge the degree of absorption of water. There are liquid pre-treatments that can be applied to control the suction of surfaces.

PVA bonding agent

There are two main types of treatments to improve the adhesion of plaster to surfaces with low suction, such as concrete, to avoid the laborious process of hacking the surface to provide a key. The first is based on polyvinyl acetate (PVA) that is brushed or sprayed on to the surface. The plaster is applied before the PVA has fully dried and is still tacky; it is the tackiness that provides the bond.

Polymer bonding agent

The second pre-treatment is polymer based and incorporates silica sand. Once the polymer is fully dry, the plaster is applied and gains bond through the silica grains and does not, therefore, depend on applying plaster as soon as the bonding agent is tacky. Where suction is high, as with lightweight blocks, the suction may be controlled by spraying with water or by spraying the surface with a liquid primer designed for the purpose.

Reinforcement for angles

A range of galvanised steel beads and stops is made for use with plaster and plasterboard as reinforcement to angles and stops at the junction of wall and ceiling plaster and plaster to other materials. An angle bead is pressed from a steel strip to form reinforcement to angles. The bead has expanded metal wings, as shown in Figure 10.7. The wings of the bead are bedded in plaster dabs on each side of the angle. The bead is then squared and plaster

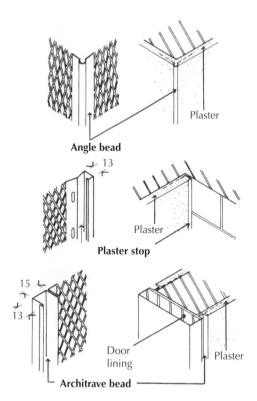

Figure 10.7   Metal beads and stops for plaster.

run up to it. A metal stop with an expanded metal wing is pressed from steel strip and used as a stop to make a neat finish at the junction of plaster with other materials at angles, skirtings and around doors and windows, as illustrated in Figure 10.7. The stop is either bedded in plaster or nailed to timber and the plaster is run up to the stop. These stops make a neat, positive break at junctions that would otherwise tend to crack or require some form of cover mould or bead to mask the joint. They are particularly useful at the junction of the plaster to another material or where the plaster has to stop. Another bead or stop is designed for use at the junction of the plaster and door and window frames to provide a definite break in surface between different materials, as illustrated in Figure 10.7. The stop is either bedded in plaster dabs or nailed to wood. The advantage of these beads is that they act as a break at the junction of dissimilar materials where they will mask any crack that may open. They are used instead of architraves.

Plaster finishes to timber joists and studs

Timber lath

The original method of preparing timber ceilings and timber-stud walls and partitions for plaster was to cover them with fir lath spaced about 7–10 mm apart to provide a key for

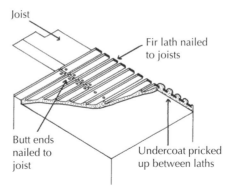

Figure 10.8   **Fir lath and plaster.**

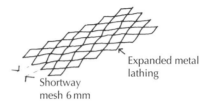

Figure 10.9   **Expanded metal lathing.**

the plaster. The usual size of each lath is 25 mm wide by 5–7 mm thick, in lengths of 900 mm. The softwood lath is either split or sawn from fir. Seasoned fir lath is nailed across the joists or timber studs. Obviously, the ends of the laths must be fixed to a joist or stud, as illustrated in Figure 10.8, and the butt-end joints of laths staggered to minimise the possibility of cracks in the plaster along the joints.

Fir lath is covered with three coats of plaster. The first coat is spread and forced between the laths so that it binds to it. This coat is described as pricking up. A second undercoat, termed the float coat, is spread and finished level and then covered with the finish or setting coat. Originally, the undercoats consisted of haired-coarse stuff (1 part lime to 3 parts sand, with hair) gauged with plaster of Paris, and the finishing coat of lime and water gauged with plaster of Paris. The purpose of the gauge (addition of a small amount) of plaster of Paris is to cause the material to harden more quickly so that vibration due to the applications of the next coat, or vibrations of the floor above, will not cause the plaster to come away from the lath before it is hard. It is now used for specialist restoration work.

Expanded metal lath (EML)

This lath is made by cutting thin sheets of steel so that they can be stretched into a diamond mesh of steel, as shown in Figure 10.9. This lath is described as expanded metal lath (EML). The thickness of the steel sheet, which is cut and expanded for plasterwork, is usually

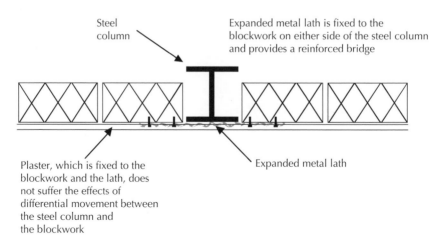

Figure 10.10   Plan view of expanded metal lathing used to bridge different substrates.

0.675 mm and the lath is described by its shortway mesh. A mesh of 6 mm shortway is generally used for the plaster. To prevent expanded steel lath from rusting, it is either coated with paint or galvanised. As a background for the plaster on timber joists and studs, the lath, which is supplied in sheets of 2438 × 686 mm, is fixed by nailing with galvanised clout nails or galvanised staples at intervals of about 100 mm along each joist or stud. During fixing, the sheet of lath should be stretched tightly across the joists. Edges of adjacent sheets of the lath should be lapped by at least 25 mm.

EML can be used where the substrate to the plaster temporarily changes. For example, if a blockwall butts up to either side of a steel column and the plaster finish is required to cover the wall and the steel, the EML can be used to provide a reinforced bridge as shown in Figure 10.10.

Gypsum plasterboard

Gypsum plasterboard consists of a core of set (hard) gypsum plaster enclosed in, and bonded to two sheets of heavy paper. The heavy paper protects and reinforces the gypsum plaster core, which otherwise would be too brittle to handle and fix without damage. Plasterboard is made in thicknesses of 9.5, 12.5, 15 and 19 mm, for use either as a dry lining or as a background for the plaster in boards of various sizes. Plasterboard is extensively used as a lining on the soffit (ceiling) of timber floors and roofs and on timber-stud partitions. The advantage of this material as a finish is that it provides a cheaper finish and can be fixed and plastered more speedily than lath and plaster. All gypsum plasterboards have inherently good fire-resisting properties due to the incombustible core. The disadvantages are that, because it is a fairly rigid material, it may crack due to vibration or movement in the joists to which it is fixed, and it is a poor sound insulator (although special plasterboards with acoustic properties are now available). Many types of gypsum plasterboards are made for specific applications as dry linings.

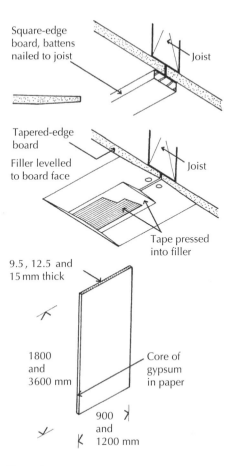

Figure 10.11   Gypsum wallboard.

Gypsum wallboard

Gypsum wallboard, which is the most commonly used board, is principally made for use as a dry lining wallboard to timber or metal-stud frames with the joints between the boards filled ready for direct decoration. The boards have one ivory-coloured surface for direct decoration and one grey face. The boards are 9.5, 12.5 and 15 mm thick, from 900 mm wide and up to 3600 mm long, as illustrated in Figure 10.11. The length of the boards is chosen as a multiple of standard timber joist or stud spacings such as 400 or 450 mm to minimise wasteful cutting of boards. The boards are made with two different edges as illustrated in Figure 10.11, tapered for smooth seamless jointing and square for cover-strip jointing.

Photograph 10.2a–e, shows wallboard being applied to internal walls. Once applied, a gun is used to apply the tape and filled over the joints. When the joints are filled, they are quickly skimmed level leaving a smooth surface. This process requires much less skill than wet plastering, is quicker and does not need as much time to dry out.

Photograph 10.2 (a) Photograph plasterboard applied to a blockwall with 'dot and dab' plasterboard adhesive. (b) Taping and skimming. (c) The joining gun or pogo can be used to apply filler and tape into the joint. (d) Tape and skim applied to vertical joints. (e) Once applied, the filler is quickly skimmed over and minor irregularities can be sanded out.

Wallboard is fixed to timber or metal supports with its length at right angles to the line of joists or studs. Timber or metal noggins are fixed between supports to provide support and fixing for the ends of boards, which do not coincide with a support. Noggins are short lengths of timber, 50 × 50 mm in section, nailed between supports. The boards are fixed with galvanised nails, 30 mm long for 9.5 mm and 40 mm long for 12.5 mm thick boards, or self-tapping screws for metal studs. Nails and screws should be driven home to leave a shallow depression ready for spot fitting. Square-edged boards are designed for use with a cover strip over all joints either for a panelled effect or for demountable partitions. Wood, metal or plastic strips are nailed or glued over joints. It is not uncommon for the wallboard to be used as a base for a thin skim coat of gypsum plaster, even though the smaller base-

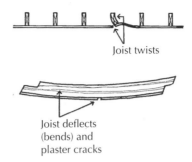

Joist twists

Joist deflects
(bends) and
plaster cracks

Figure 10.12   Cracking in plasterboard finishes.

board is more convenient to use, particularly to ceilings. Tapered-edge boards are made for jointing with a smooth, flush finish ready for direct decoration. The shallow depression at the joint between boards is first filled with joint filler made of gypsum and water, into which a 50 mm wide paper jointing tape is pressed. The joint is completed with filler, which is finished flush with the board as illustrated in Figure 10.11. Nail heads are covered with filler, finished smooth as spot filling.

The principal causes of cracking in these finishes are twisting and other moisture movements of timber joists or studs to which they are fixed and deflection of timber joists under load. New timber is often not as well seasoned as it should be and as the timbers dry out they tend to shrink and lose shape. Joists may wind (twist) and cause the rigid boards fixed to them to move and joints open up as illustrated in Figure 10.12. Strutting between joists will restrain movement and minimise cracking. Under the load of furniture and persons, timber floor joists bend slightly. The degree to which they bend is described as their deflection under load. Even with very small deflection under load, a large rigid sheet of plasterboard will bend and cracks appear at joints as illustrated in Figure 10.12. One way of minimising cracking with large boards is to use joists some 50 mm deeper than they need be to carry the anticipated loads. This additional depth of joist reduces deflection under load and the possibility of cracking. Baseboard, made specifically for a plaster finish, is smaller than a full-size wallboard and may, therefore, be less liable to show cracks due to shrinkage and movement cracks.

Gypsum baseboard

Gypsum baseboard is designed specifically for use as a base for gypsum plaster. The board is 9.5 mm thick, 900 mm wide and 1220 mm long, and used as ceiling lining for their manageable size. The boards have square edges as illustrated in Figure 10.13. The boards are fixed with 30 mm nails at 150 mm centres with a gap of about 3 mm at joints. The joints are filled with filler into which a reinforcing paper tape is pressed and the boards are covered with a board-finish gypsum plaster that is spread and trowelled smooth to a thickness of 2–5 mm, as illustrated in Figure 10.14.

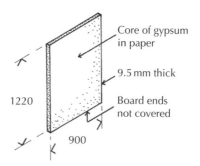

Figure 10.13   Gypsum baseboard.

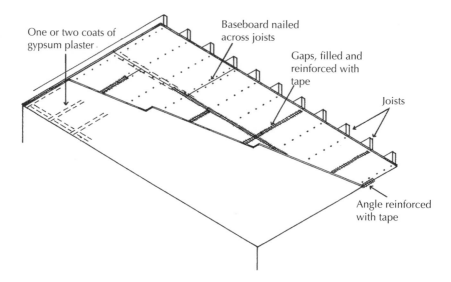

Figure 10.14   Fixing baseboard for plastering.

Gypsum plank

Gypsum plank is 19 mm thick, 600 mm wide and 2350 and 3000 mm long with either tapered edges for seamless jointing for direct decoration or square edges for plastering. This thicker board, which may be used for enhanced fire resistance or a small increase in sound insulation, may be fixed at 600 mm centres with 60 mm nails. For direct decoration, the tapered-edge boards are jointed and finished as described for the wallboard. For plastering, the square-edge boards are fixed with a gap of about 3 mm into which plaster or filler is run and reinforced with a 50 mm wide paper tape. Gypsum-finish plaster is then spread over the surface of the boards, levelled and trowelled to a smooth finish to a thickness of 2–5 mm.

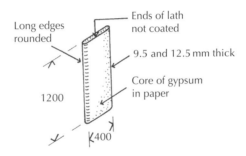

Long edges rounded

Ends of lath not coated

9.5 and 12.5 mm thick

Core of gypsum in paper

1200

400

Figure 10.15   Gypsum lath.

Gypsum lath

These comparatively small boards are made specifically as a base for ceiling lining for ease of holding and fixing, and as a base for plaster either as a thin skim coat or more particularly for a two-coat finish. The two-coat finish is preferred, as the undercoat will facilitate accurate levelling over the many joints. Lath is 400 mm wide, 1200 mm long and 9.5 and 12.5 mm thick. The long edges of the boards are rounded as illustrated in Figure 10.15. The lath is fixed to timber joists and ceiling with a gap of not more than 3 mm between boards. The joints do not have to be reinforced with tape. The lath is covered and finished with one finish coat to a thickness of up to 5 mm or more, usually with a gypsum undercoat 8 mm thick and a finish coat of about 2 mm thickness. The thicker board with a two-coat gypsum plaster finish will provide improved resistance to fire and some increase in sound insulation.

10.4 Skirtings and architraves

Skirting

A skirting (or skirting board) is a narrow band of material formed around the base (skirt) of walls at the intersection of the wall and the floor. Usually, the skirting sits proud of the wall surface. It serves to emphasise the junction of vertical and horizontal surfaces, the wall and the floor, and is made from material that is sufficiently hard to withstand knocks. The skirting protects the wall finish at a vulnerable point and also covers the joint between the wall and the floor. The materials commonly used for skirting are timber, medium-density fibreboard (MDF), plastic, metal and tile.

Timber-skirting board

Softwood boards, 19 or 25 mm thick, from 50 to 150 mm wide and rounded or moulded on one edge, are usually used. The skirting boards are nailed to plugs, grounds or concrete fixing blocks at the base of walls after plastering is completed. Figure 10.16 illustrates some typical sections of skirting board and the fixing of the board. Plugs are wedge-shaped pieces of timber that are driven into brick or block joints from which the mortar has been cut out

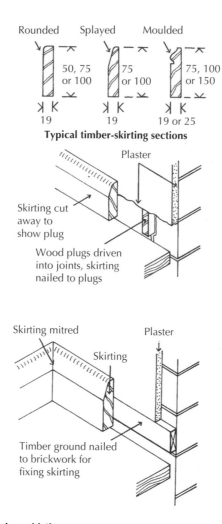

Figure 10.16   Fixing timber skirting.

as illustrated in Figure 10.16. This rough method of providing a fixing is usually unsatisfactory as it is not always possible to drive the wedge into a joint without damage and it is difficult to ensure that the face of the plug finishes at the required level. Nailing into the end grain of the wood plug may not provide a secure hold. An alternative is to nail soldiers to the wall. Soldiers are short offcuts of sawn softwood timber 38 or 50 mm wide and the same thickness as finished plaster. The soldiers are fixed vertically at intervals of 300–400 mm apart as a fixing for the skirting board. Soldiers, which provide a more secure fixing than plugs, are laborious to fix.

Grounds are small-section lengths of sawn softwood timber, 38 or 50 mm wide and as thick as the plaster on the wall. These timber grounds are nailed horizontally to the wall as a background to which the skirting can be nailed, as illustrated in Figure 10.16. Grounds are generally fixed before plastering is commenced so that the plaster can be finished down on to and level with them. Concrete fixing blocks are either purpose-made or cut from lightweight concrete building blocks and built into brick walls at intervals of 300–400 mm as a fixing for skirtings. Another method of fixing wood skirtings is to run wall plasters down to the floor level and fix the skirting with an adhesive directly to the plaster. The contact adhesive is spread at the back of the skirting and the plaster from a hand-operated cartridge gun. These adhesives provide a secure bond to the plaster. With long runs of skirting, it is wise to supplement the adhesive bond with two or more screws, driven through the skirting into plugs in the wall, to resist shrinkage twisting of the wood skirting, which would otherwise pull the skirting from its adhesive fixing.

Medium density fibreboard

MDF is increasingly used as a timber-based material into which grooves, rebates and complicated shapes can be cut. The material, which is made from fine particles of wood compressed and glued together, is less expensive than other timber boards, easy to work and finish and is more flexible than most natural wood boards. Where timber finishes are to be painted rather than varnished, MDF may offer an acceptable alternative to timber.

Metal skirting

The traditional wood skirting was initially used to mask the joint between wall plaster and timber floor finishes where it was impossible to make a neat joint between lime plaster and boarded floors. The plasters used today are less liable to shrinkage and cracking, and may be finished directly on to solid floor finishes or down on to pressed metal skirtings. A range of pressed steel skirtings is manufactured for fixing either before or after plastering. The skirting is pressed from a mild steel strip and is supplied, painted with one coat of red oxide priming paint. Figure 10.17 illustrates these sections and their use. The skirting is fixed by nailing it directly to lightweight blocks or to plugs in brick and block joints or to a timber ground. Special corner pieces to finish these skirtings at internal and external angles are supplied.

Tile skirting

The manufacturers of most floor tiles also make skirting to match the colour and size of their products. The skirting tiles have rounded top edges and a cove base to provide an easily cleaned rounded internal angle between skirting and floor. The skirting tiles are first thoroughly soaked in water and then bedded in sand and cement against walls and partitions as the floor finish is laid. Special internal and external angle fittings are also manufactured. Figure 10.18 illustrates the use of these skirting tiles. Skirting tiles make a particularly hard-wearing, easily cleaned finish at the junction of floor and walls and are commonly used with quarry and clay-tile finishes to solid floors in rooms and places where the conditions are wet and humid, such as kitchens, bathing areas and laundries. To avoid excessive condensation on smooth, hard surfaces such as tiles, they should be applied to floor and external wall surfaces that are adequately insulated.

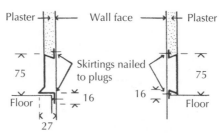

Type A for fixing before plaster

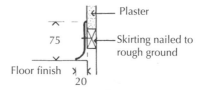

Type B for fixing after plastering

Figure 10.17   Metal skirtings.

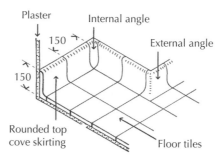

Figure 10.18   Tile skirting.

Architrave

The word architrave describes a decorative moulding fixed or cut around doors and windows to emphasise and decorate the opening. An architrave can be cut or moulded on blocks of stone, concrete or clay, built around openings externally. Internal architraves usually consist of lengths of moulded wood nailed around doors and windows. An internal architrave serves two purposes: to emphasise the opening and to mask the junction of wall plaster and timber door or window frame (because a crack usually opens up at this junction).

A timber architrave is usually 19 or 25 mm thick and from 50 to 100 mm wide. It may be finished with rounded edges, or splayed into the door or decorated with some moulding. The usual practice is to fix architraves so that they diminish in section towards the door or window. Narrow architraves can be fixed by nailing them to the frame or lining of the

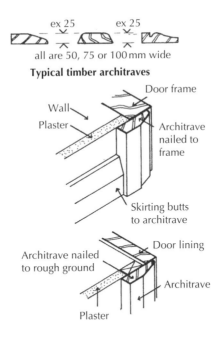

Figure 10.19   Architraves.

door or window. Wide architraves are usually fixed to sawn softwood grounds nailed to the wall around the frame or lining as a background to which the architrave can be securely nailed. Architraves are mitre cut (458 cut) at angles. Figure 10.19 illustrates some typical sections and fixing of architraves.

10.5 Painting and decorating

The term 'painting and decorating' is used to encompass a wide range of paints, wood stains, varnishes and oils for internal and external use, as well as the covering of internal walls with decorative wallpaper. The function of paints, stains, varnishes and oils is twofold: to provide a protective surface to materials, e.g. timber, and to provide a visually attractive finish to the material. Colour and surface texture does affect our mood and can be used to create a strong or subtle visual effect. The function of wallpaper is primarily to provide an applied decorative finish to internal (plastered) walls.

Colour

An important factor in the description and specification of paints, stains and varnishes is colour. Colour for building purposes is usually defined by British Standard (described later). Other systems such as Colour Dimension, Munsell, Pantone and RAL are also used in the construction sector to specify colours. This allows the contractor to purchase paint from any of the paint manufacturers, safe in the knowledge that the colour will be as specified. Paint manufactured for the domestic market (purchased and applied by householders) tends to be described by a wide range of exotic names, some of which relate directly to British Standard or RAL; others are specific to a paint producer. Paint manufacturers produce colour cards and small sample pots of paint (which cover approximately 2 m² of wall) in an attempt to help people make a choice from the vast range of colours available. Colour appears to change in different sized rooms and against other colours. Natural light also varies in its quality through the day and this too will influence the appearance of a particular paint colour.

Patterned and textured wallpaper, by its very nature, has to be specified by manufacturer, pattern name and reference number. Rolls carry a pattern reference number and a batch number. Rolls of wallpaper with the same batch number should provide the best match for colour and texture.

British standard

Under the British Standard system, a specific colour is defined by a three-part code. For example, Magnolia is defined as 08 B 15, Black as 00 E 53. The first part of the code defines the hue (ranging from 00 to 24), the second part the greyness (A to E) and the third part the weight (a subjective scale from 01 to 58, which incorporates reflectivity to incident light and greyness). To specify a paint colour, it is only necessary to specify the relevant reference, e.g. 08 B 15, although it is common practice to give the paint name (in this case Magnolia) as well.

Required finish

The type of paint finish required will also influence the visual appearance of paint colour. These range from flat, full gloss, gloss, low sheen, matt and semi-gloss and must be specified along with the appropriate reference number. Manufacturers may produce some colours in one finish only, while other paints are manufactured in several different finishes. For example, a paint finish to an old property would most likely be in a flat finish to echo the type of paint finish available at the time. Paints may also be formulated for particular situations or uses. For example, bathroom and kitchen paints provide a robust finish to walls that can be regularly wiped down to remove surface condensation and splashes without damaging the surface finish. By comparison, ceiling finishes do not need to be so robust and therefore special 'ceiling' paints are available for the purpose. A number of 'effects' paints are also available, e.g. water-based paints for interior use that give a metallic finish, and acrylic-based paints for exterior effects.

Paint systems

Paint is composed of a number of ingredients, each with a specific purpose. Typically paint will include a binder, driers, solvents, pigments and a base material. The medium, or binder, solidifies after application to create the protective paint film. Alkyd resins and vinyl or acrylic resins have replaced linseed oil as the medium for the majority of paints. Driers are added to induce the polymerisation of the binder to ensure rapid drying. The solvent, either water or an organic material, helps to create fluidity to the paint to facilitate painting with a brush and/or roller and spraying. Water-based paints have better environmental and health and safety credentials than those made from organic solvents. Colour is added with organic and inorganic dyes and pigments. Opacity is achieved through the addition of a base material, usually titanium dioxide, and other inert extenders such as silica.

The finished paint system is made up of a series of 'coats' or layers, each coat of paint performing a specific task. For internal emulsion paint, it is common practice to apply a 'mist coat' to the plastered wall, the substrate. This is emulsion thinned with water to act as an undercoat, followed by two finishing coats of the emulsion. For external and internal gloss paint, the base material, the substrate, requires a coat of primer, followed by an undercoat and one or two finishing coats. Paint manufacturers provide extensive guidance on the different types of paint that they manufacture and typical specifications for application to a wide range of materials. Higher specification paints are manufactured for external use in exposed conditions and marine/coastal climates.

Substrate

The surface of the material to be painted is known as the substrate. The type of material and its condition will determine the type of paint to be used. Preparation of the substrate is an important factor in achieving a durable paint finish, and preparation can account for as much as 50% of the cost of painting a surface.

Primers

The purpose of the primer is to adhere to the substrate, to provide protection from corrosion or deterioration and to offer a good base for the undercoat. The substrate material must be free of all loose material, be clean and, in the case of timber, have low moisture content at the time of application. Different primers are available for different materials and circumstances.

Undercoats

Undercoats provide an opaque cover and a good base for the finishing coat. It is usual practice to paint the undercoat in a different colour to the top, finishing coats to help the decorator monitor progress more easily. Undercoats are normally based on acrylic emulsions or alkyd resins.

Finishing coats

The finishing coat provides the protective and decorative surface finish. The exact colour and type of finish must be specified. The finishing coat should be inspected regularly for signs of damage and promptly repaired by removing loose paint and repainting to ensure protection to the substrate.

Application of paints

Paints and stains can be applied by brush, roller or application pads. For special effects, cotton rags and sponges may be used. All paints should be applied in accordance with manufacturers' instructions and in accordance with prevailing health and safety legislation and guidance. Low-odour paints should be specified where possible. Internal rooms should be well ventilated to avoid a build-up of fumes. Where paints are sprayed, protective breathing apparatus and protective clothing must be worn. The temperature of the surface to be painted is also a determining factor in achieving a durable finish. Most paint manufacturers recommend application at a surface temperature of 10°C or above and recommend against application of paint in extremely hot or cold weather.

Storage and disposal

Storage of paint presents a fire safety hazard and it must be stored in accordance with current health and safety regulations. Disposal of paints should also be undertaken in accordance with the manufacturer's instructions and due regard for environmental laws.

Special paints

Specially formulated paints have a wide variety of uses; the more common ones are described here.

Masonry paints

External walls of fair face brick, block, stone, concrete and render may be painted with masonry paint. These are predominantly water-based acrylic resin-based products, although solvent-based systems and mineral silicate paints are also produced. The paint usually contains fungicides to prevent, or at least delay, the growth of moulds and algae on the paint surface. Masonry paint is available in a range of colours and is produced to provide a smooth finish or a sand textured finish. The sand textured finish is more suited to covering fine cracks in render than the smooth finish. Larger cracks should be repaired with flexible filler prior to decoration.

Fungicide paints

In internal areas where mould growth is a problem (e.g. kitchens and bathrooms), it is common to paint walls with fungicide paint (sometimes marketed as mould protection paint). This acrylic paint contains a mix of fungicides to resist mould growth and is both washable and durable. The paint is formulated to release the fungicide gradually over a long period. Being a cheap and effective solution, efforts should also be made to improve ventilation to the problem area, to determine the cause of the mould growth and to fix the problem (if possible) prior to decoration.

Multicolour (fleck) paints

Multicolour paints, applied by roller or spray system, incorporate coloured flecks to help disguise unsightly surfaces (and cover graffiti) and provide some deterrent to new graffiti. Some of these systems are designed to make the removal of graffiti easier compared with other paint systems.

Water repellent paints

Water repellent paints may be applied to porous surfaces to prevent water penetrating the wall, yet still allowing the evaporation of water from the masonry. These silicone-based paints can be applied to brickwork, concrete, stone and render to provide a moderate degree of protection to the building fabric.

Waterproofing paints

Epoxy waterproofing paints provide an impervious surface finish to surfaces and tend to be used where high humidity or wet operations would cause damage to normal paint finishes. Epoxy ester paint systems are also highly resistant to spillages of oil and some chemicals. In industrial buildings, these paints may be used to provide a surface finish to a 'wet room' for washing equipment, etc. In existing buildings, epoxy waterproofing systems may be used to improve the wall finish to damp basement walls, provided that a good bond can be achieved between the wall and the epoxy system. Bituminous paints serve a similar function, forming a waterproof finish to masonry and metal.

Heat-resisting paints

Heat-resisting paints are formulated to resist high temperatures. For example, aluminium paint is resistant to temperatures of up to around 250°C.

Flame-retardant paints

A flame-retardant paint will give off non-combustible gases when subject to fire, thus retarding the surface spread of flame.

Intumescent coatings

Fire protection to structural steel is achieved through a thin coat (1–2 mm) of intumescent paint, usually applied by spray guns in the steel fabricator's paint shop to control paint thickness. When subjected to fire, the paint expands to form a thick layer of insulating foam. Intumescent coatings are applied to give fire protection times of 30, 60 and 120 minutes. Intumescent emulsion paints and clear varnishes are also available for use on timber.

Wood stains

Wood stains are applied by brush and penetrate the surface of the timber, creating a permeable and water repellent sheen finish. Wood stains for exterior use are water-based or solvent-based and include a preservative basecoat, which helps to control rot and mould growth. Stains for interior use are also available. Wood stains are commonly specified as low-, medium- or high-build systems. For example, timber cladding and rough sawn timber would be applied with a low-build stain system, while external timber joinery, such as a smooth window cill, would benefit from a medium- or high-build system. The stain is usually applied in two coats. The first coat will penetrate the surface of the timber and also adhere to the surface. The second coat creates a microporous surface finish that is water repellent. The finished stain is permeable to moisture movement, thus allowing the timber to 'breathe' (unlike a gloss paint finish), which may help to reduce moisture-induced movement of the timber. The finished stain will fade on exposure to weather, and therefore regular application of new top coats will be required to maintain an attractive finish and adequate weather protection to the substrate.

Varnishes for timber

Varnishes are applied by brush to create a protective surface film to timber. Polyurethane varnishes are either water-based or solvent-based systems and are available in gloss, satin or matt finishes. The solvent-based varnishes produce the harder and more durable coatings, e.g. 'yacht varnish'. Varnishes are available in a clear finish, to maintain the natural wood colour, are manufactured to enhance the natural colour of the wood (usually by darkening a light timber) and can be used to add some colour to the timber. For most applications, two coats of varnish provide adequate protection.

Oils and waxes for timber

Oils and waxes are mostly used for internal applications to provide a protective finish to timber furniture. Natural oils, such as linseed oil, are liberally applied to timber with a soft cloth and any excess wiped off. Waxes, such as beeswax-based polishes, are applied in a similar manner but the timber is polished with a soft cloth to bring out the natural grain of the timber and also to provide a gloss finish. Oils and waxes need to be applied on a regular basis to maintain appearance. Oils, such as teak oil, are also used externally to provide weather protection to, for example, timber garden tables and chairs. The oils for external use are formulated to produce an ultraviolet-resistant and microporous finish. Regular application will be required to maintain an attractive appearance and weather protection in situations where the furniture is exposed to sunlight and rain. A wide variety of finishes is available, ranging from a transparent ('natural') finish to an opaque one.

Wallpaper

Wallpaper is the term used to cover a wide variety of sheet papers produced for use on internal walls. The papers are mainly used in domestic premises, applied by professional decorators or by householders. Where papers are applied in commercial premises, it is important to check that the application of a paper does not compromise fire safety (surface spread of flame).

Papers adhere to the wall by means of a water-based adhesive paste. The paste may be mixed with water and applied to the wall or more commonly to the back of the paper. The sheet of pasted paper is then 'hung' on the wall. The paper is carefully positioned and then the surface is brushed over with a soft brush to apply enough pressure to remove any air bubbles that have formed under the paper. Once positioned, the sheet of wallpaper can be trimmed at the ceiling and skirting with a sharp knife. Some papers are manufactured with an adhesive back which is activated by soaking the paper in water for a few minutes or by brushing water to the back of the paper before it is applied to the wall. Some skill is required in keeping the sheets of paper plumb and patterns lined up at the join (especially if the wall is slightly out of true).

Lining paper

Lining paper is a relatively dense plain paper with a smooth finish that provides a good key for emulsion paints. The paper is applied to walls that are slightly uneven to provide a relatively smooth surface on which to apply emulsion paint.

Plain textured papers

In situations where internal walls are in poor condition, i.e. cracked and uneven, a heavily textured paper may be used to disguise the poor wall finish. The texture is created when the paper is rolled or created by adding texture to the paper, e.g. very small wood chips. These textured papers provide a good key for the application of emulsion paint. This is a cheaper solution than applying a skim coat of finishing plaster (which provides a smooth and level finish) but the textured finish may not suit all tastes. Woodchip paper tends to be used on walls where the plaster is in very poor condition and/or the walls are badly cracked, something worth remembering when inspecting a property.

Patterned papers
Patterned papers are applied to walls to create an attractive internal finish. The vast majority of wallpaper is machine printed and mass produced, the popular ranges being the cheapest. Hand-painted wallpaper may occasionally be commissioned for prestigious refurbishment projects. These papers are expensive and require special skills to apply them correctly. The choice of wallpaper is a personal one.

Thin strips of paper are also produced to provide a border, e.g. at the junction of ceiling and wall, in a range to complement the chosen pattern.

Patterned textured papers
Wide ranges of papers are also available that have both a pattern and a textured surface.

Washable papers
Ranges of washable vinyl papers are manufactured for use in rooms where they may need to be wiped down. Examples may be in a kitchen, bathroom or a child's playroom.

10.6 External rendering

Owing to their colour and texture, common bricks, concrete and clay blocks do not provide what is commonly considered to be an attractive external finish for buildings. The external faces of walls built with these materials are often rendered with two or three coats of cement and lime mixed with natural aggregate and finished either smooth or textured (Photograph 10.3a–c). Because an external rendering generally improves the resistance of a wall to rain penetration, the walls of buildings on the coast and on high ground are often rendered externally to provide additional weather protection. External paints specially formulated for rendered surfaces may provide additional protection, while also helping to improve the appearance of the surface with the addition of some colour. In certain parts of the UK, render is the traditional finish to all buildings, old or new. Render is also used as a relatively cheap and convenient way of improving the weather resistance of old, less durable walls.

Renders depend on a strong bond to the background wall, on the mix used in the rendering material and on the surface finish of the background. The rendering should have a strong bond or key to the background wall as a mechanical bond between the rendering and the wall so that the bond resists the drying shrinkage inevitable in any wet-applied mix of rendering. The surface of the background wall should provide a strong mechanical key for the rendering by the use of keyed flettons, raking out the mortar joints, hacking or scoring otherwise dense concrete surfaces and hacking smooth stone surfaces. If there is not a strong bond of rendering to background walls, the rendering may shrink, crack and

Photograph 10.3   (a) Renovation project – smooth finish render being applied to a block of flats. (b) The smooth render coat applied to insulation with fabric reinforcement. (c) Externally insulated building with smooth render finish.

come away from the background and water will enter the cracks and saturate the background from which it will not readily evaporate. As a general rule, the richer the mix of cement in the rendering material the stronger should be the background material and key.

The mixes for renderings depend on the background wall, lean mixes of cement and lime being used for soft porous materials and the richer cement and lime mixes for the denser backgrounds so that the density and porosity of the rendering corresponds roughly to that of the background. The types of external rendering used are smooth (wood-float finish), scraped finish, textured finish, pebbledash (drydash), roughcast (wetdash) and machine-applied finish.

Types of render

Smooth or wood-float finish

Smooth (wood-float finish) rendering is usually applied in two coats. The first coat is spread by trowel and struck off level to a thickness of about 11 mm. The surface of the first coat is scratched before it dries to provide key for the next coat. The first coat should be allowed to dry out. The next coat is spread by trowel and finished smooth and level to a thickness of about 8 mm. The surface of smooth renderings should be finished with a wood float rather than a steel trowel. A steel trowel brings water and the finer particles of cement and lime to the surface, which, on drying out, shrink and cause surface cracks. A wood float (trowel) leaves the surface coarse textured and less liable to surface cracks. Three-coat rendering is used mostly in exposed positions to provide a thick protective coating to walls. The two undercoats are spread, scratched for key and allowed to dry out to a thickness of about 10 mm for each coat; the third or finishing coat is spread and finished smooth to a thickness of 6–10 mm (Photograph 10.3a–c).

Spatterdash

Smooth, dense wall surfaces such as dense brick and situ-cast concrete afford a poor key and little suction for renderings. Such surfaces can be prepared for rendering by the application of a spatterdash of wet cement and sand. A wet mix of cement and clean sand (mix 1:2, by volume) is thrown on to the surface and left to harden without being trowelled smooth. When dry, it provides a surface suitable for the rendering which is applied in the normal way.

Scraped-finish rendering

An undercoat and finish coat are spread as for a smooth finish and the finished-level surface, when it has set, is scraped with a steel-straight edge or saw blade to remove some 2 mm from the surface to produce a coarse-textured finish.

Textured finish

Textured rendering is usually applied in two coats. The first coat is spread and allowed to dry as previously described. The second coat is then spread by trowel and finished level. When this second coat is sufficiently hard, but still wet, its surface is textured with wood combs, brushes, sacking, wire mesh or old saw blades. A variety of effects can be obtained by varying the way in which the surface is textured. An advantage of textured rendering is that the surface scraping removes any scum of water, cement and lime that may have been brought to the surface by trowelling and which might otherwise have caused surface cracking.

Pebbledash (drydash) finish

Pebbledash finish is produced by throwing dry pebbles, shingle or crushed stone on to, and lightly pressed into, the freshly applied finish coat of rendering, so that the pebbles adhere to the rendering but are mostly left exposed as a surface of pebbles. Pebbles of from 6 to 13 gauge are used. The undercoat and finish coat are of a mix suited to the background and are trowelled and finished level. The advantage of this finish is that the pebbledash masks any hair cracks that may open due to the drying shrinkage of the rendering. Photograph 10.4a–e shows a pebbledashed render applied to an existing house. Photograph 10.5 shows the pebbledash being applied by hand to the trowelled surface.

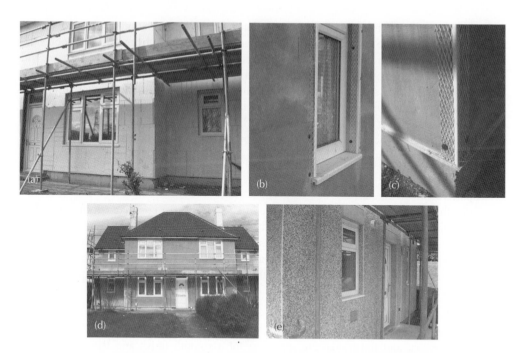

Photograph 10.4   (a) External insulation applied to existing structure. **(b)** Expanded metal reinforcement is used to make a strong bond around the window frame. **(c)** External beading used to ensure a good finish at the corner. **(d)** Rendered house ready for pebbledash. **(e)** Pebbledash finish applied.

Photograph 10.5   Pebbledash applied to render on an externally insulated wall (courtesy of Weber Saint Gobain).

Roughcast (wetdash) finish

A wet mix of rendering is thrown on to the matured undercoat by hand to a thickness of from 6 to 13 mm to produce a rough irregular textured finish. The gauge of the aggregate used in the wet mix determines the finish.

Machine-applied finish

A wet mix of rendering is thrown on to a matured undercoat by machine to produce a regular coarse-textured finish. The texture of the finish is determined by the gauge of the aggregate used in the final wetdash finish, which may have the natural colour of the materials or be coloured to produce what are called Tyrolean finishes.

11 Internal Environment and Energy Supply

The design of the services provision should be integral to the design of the building and its fabric, thus allowing greater ease of constructability, better access for routine repair and maintenance, and ease of recovery at the end of the building's life. Heating, cooling, ventilation and electrical services allow the building to be used, while influencing the well-being of the building users. Indeed, it is the combination of heating, cooling, ventilation and lighting that contributes to the building users' perceived level of comfort. And this requires energy. The European Union (EU) directive on energy use of buildings is concerned with reducing the energy use of the overall building fabric, not just U-values, thus helping to reinforce the need for an integrated and sustainable approach to the design, construction and use of buildings.

11.1 Functional requirements

Services design must form an integral part of the design philosophy for the building. Input from appropriate specialist engineers and suppliers should be made early in the design process to ensure compatibility and ease of constructability. Heating, cooling and services equipment will require regular maintenance and will have a much shorter service life than the main building structure; thus ease of access for repair, maintenance and upgrading is an important consideration for all building types.

Functional requirements will vary depending on the type of building and the anticipated use of space within the building. Generic functional requirements are:

- ❑ Comfort and well-being of building users
- ❑ Easy and safe access for installation, maintenance and replacement
- ❑ Low (preferably zero) energy consumption during operation
- ❑ Low (preferably zero) emissions
- ❑ Ease of control by building users (manual and/or by smart technologies)

Barry's Introduction to Construction of Buildings, Third Edition. Stephen Emmitt and Christopher A. Gorse.
© 2014 John Wiley & Sons, Ltd. Published 2014 by John Wiley & Sons, Ltd.

11.2 Indoor climate control

Technological innovations in the materials used for the building fabric (the building envelope) can play a significant role in the creation of a comfortable internal environment that is also energy efficient. The manner in which the building envelope responds to solar radiation, how heat loss (and gain) is controlled, how ventilation is provided and how noise is excluded are key factors. So too are the materials and finishes with which building users interact. Careful consideration should be given to the use of natural materials and the elimination of materials that give off chemical odours and/or attract dust and other sources of irritants. When we fail to create a healthy and enjoyable indoor environment the occupants may experience what is known as sick building syndrome (SBS). Individuals feel unwell when they are in the building; the discomfort caused by a combination of physical and psychological factors.

Innovation in building services provision since the Industrial Revolution has primarily focused on the control of the internal environment through mechanical and electrical systems that consume large amounts of energy. Internal spaces such as office buildings, retail units and leisure buildings are largely isolated from changes in climate, with a constant temperature and humidity maintained by sophisticated services equipment. Residential developments also consume large amounts of energy and legislation and guidance is now in place to try and reduce this unsustainable demand. Attention has turned to the specification of more energy-efficient mechanical systems and also to passive systems that consume little or no energy.

Comfort

A large number of variables, both physical and psychological, influence our feeling of comfort within a building. These include:

- ❏ Air quality (and level of pollutants and irritants)
- ❏ Air movement (controlled air changes)
- ❏ Presence of odours
- ❏ Ambient water vapour pressure (humidity)
- ❏ Radiant temperature
- ❏ Air temperature (and temperature gradients within the space)
- ❏ Activity level of users
- ❏ Type of clothing worn by users
- ❏ Natural and artificial lighting levels
- ❏ Background noise levels
- ❏ Interaction with other users
- ❏ Degree of individual control over the internal environment

Our comfort levels are subjective and individual. Discomfort will be felt if there is a large variation in one or more of the conditions, e.g. the air movement being too high (perceived as a draught and hence uncomfortable). The degree of control an individual has over his or her internal environment is an important consideration in the design of appropriate systems for shared spaces, e.g. office buildings. This too may influence our perception of comfort.

Heating

Space heating may be provided by a number of energy sources, the most common of which are described briefly here. Efficiency of the system is dependent upon the heat source and the design of pipe or duct runs (and associated maintenance). Choice of heat emitter must be made to suit the construction of the building and the needs of the users. Consideration should be given to:

❏ Thermal response of building fabric
❏ Pattern of use of internal building spaces
❏ Appearance (available space)
❏ Individual level of control
❏ Installation and maintenance costs

The type of thermal response is usually defined as rapid, intermediate or slow. If the building is used sporadically, it may be beneficial to have a rapid thermal response. Conversely if the building were used continually throughout the heating period, a slow thermal response would normally be a better choice. Some of the more common systems are outlined here.

Rapid – warm air systems

Warm air systems include electric fan convectors (fixed or portable) and forced warm air systems. These provide the building user with the ability to quickly respond to changes in external air temperature, which is good for lightweight building fabric. Warm air systems can be designed to include a mechanical ventilation heat recovery unit. Ductwork to carry the warm air is typically run in floor and roof voids. Outlets tend to be positioned in the floor and/or at a low level in the internal walls. Heating of the room is rapid but so too is cooling.

Intermediate – water-filled radiators

Hydronic heating systems with a gas or oil-fired energy source are the most common systems installed in the UK. Pipework is run to wall-mounted panel radiators, which come in a wide range of sizes and designs. The boiler will have a thermostatic device for general control of output. A thermostatic valve connected to each radiator controls individual room temperature locally. The thermostatic valve is a relatively simple valve that is adjusted by hand to control the amount of hot water flowing through the radiator, and hence the temperature of the radiator and surrounding space.

Slow – underfloor heating and storage radiators

Two systems that are slow to respond to changes in external temperature are electric storage radiators and underfloor heating systems. Both are good for buildings with a high thermal mass and which are occupied on a continual basis, such as houses. Storage radiators contain material that retains heat and radiates it over a long period. Electric storage heaters are usually designed to use off peak electricity (or more recently electricity generated from photovoltaic (PV) systems, building up heat overnight and releasing it during the daytime. Underfloor heating is more energy efficient than radiators because the heat is more evenly distributed around the internal space. The absence of radiators also allows greater flexibility in the use of wall and floor space. Underfloor heating is particularly efficient when used with ceramic and stone tiles as a floor finish but can equally be used with timber and carpeted floor finishes. There are two systems, either 'wet' or 'dry'.

The wet system comprises hot water running through plastic pipes, fed from a gas-fired boiler or a solar collector system. The water temperature in the pipes is less than that required for radiators, hence making the system suitable for use with solar collectors. The dry system comprises electric cables placed on a reflective sheet placed on top of the floor insulation. Both systems warm the room from the floor up, creating a decreasing heat gradient from the floor upward. Although slow to warm the air temperature compared with other systems, the heat retained in the floor structure will ensure a relatively stable internal air temperature throughout the heating season. Systems can be zoned (by room or area) to provide individual control to minimise energy consumption. Access for repair and/or replacement can be disruptive since the floor finish will need to be lifted and replaced.

Cooling

Rising concern over climate change has resulted in greater interest in mechanical cooling systems for residential and commercial premises. Although many mechanical heating systems can be switched to cooling mode, it is not unusual to see individual cooling and air conditioning units installed within buildings. These systems use energy and there is a move to design and construct buildings that use natural ventilation and shading devices, and hence less energy. A well-designed and constructed domestic building anywhere in the UK should have no need for mechanical cooling. Unfortunately many existing buildings are not so well designed or constructed to respond to their immediate climate, and so it is inevitable that some form of mechanical cooling may be necessary during the warmer months to keep occupants cool.

Shading devices (with PV cells)

Shading to windows can be provided internally by curtains, blinds and shutters. Shading can also be provided by physical shading devices fixed to, or built into, the external face of the building. Traditional designs often featured deep window reveals that provided shading from unwanted solar glare and solar gain during the summer months. The trend for flat façades to buildings means that the shading has to be incorporated into the façade design, with blinds and solar control glazing. Vertical fins can be bolted to the façade to provide shading and also to provide some relief to an otherwise 'flat' building.

Ventilation

Naturally ventilated buildings are a feature of vernacular architecture, with carefully positioned opening windows and louvred vents providing sufficient air flow through the building. Vernacular buildings were designed and built to respect their local climate and use it to maximise internal comfort. Mechanical ventilation systems are a feature of more recent, highly serviced buildings, which are sealed to the outside environment. In many modern buildings, the local climate is largely ignored and the internal environment controlled solely by mechanical and electrical equipment.

Naturally ventilated spaces

The use of passive ventilation is gaining widespread acceptance as an environmentally responsible way in which to ventilate buildings. The art is still being rediscovered and is set to evolve further over the coming years as mechanical systems lose favour. Natural ventilation is, in principle, a rather simple concept, although expert advice is required to design an effective naturally ventilated building. Naturally ventilated buildings are usually ventilated from one side, or cross ventilated, and also rely on the stack effect for effective ventilation (illustrated simply in Figure 11.1). Because it relies on natural forces, natural ventilation is less predictable than mechanical systems but, when implemented successfully, it is possible to make large savings on capital and running costs, while providing a healthy interior environment.

Assisted natural ventilation systems

Assisted natural ventilation systems offer the benefit of knowing that the system will be able to cope with peak demands. Simple ceiling-mounted fans may be sufficient for the purpose. Alternatively, systems are available that incorporate fans and heat recovery devices.

Mechanical ventilation systems

Mechanical ventilation with ductwork systems is widely used in commercial buildings. Mechanical air handling plant takes up a lot of space (usually positioned within the basement and/or on the roof) and consumes energy in use. Ductwork tends to be quite large in cross section, and necessitates the use of raised floors and suspended ceilings to conceal the ducts. Thus the design of the building must take into account the quite considerable space requirements of mechanical plant. This is reflected in the initial cost, and complexity, of the building. Although advances have been made to reduce the energy requirements of mechanical ventilation systems (and hence the associated pollution), the majority of systems are expensive to run and maintain.

11.3 Energy sources

The majority of the energy produced in the UK comes from the combustion of fossil fuels. This process produces carbon dioxide (CO_2). Buildings use energy, and our building stock accounts for approximately 50% of all CO_2 emissions. Approximately two-thirds of these building-related emissions are from the domestic sector and the remaining third from the commercial sector. The biggest energy use is for heating internal space (typically somewhere around 50%), followed by lighting (around 15%). Building design and the care with which buildings are constructed will influence energy consumption and hence the CO_2 emissions for the building as a whole. For example, poor work may result in unnecessary

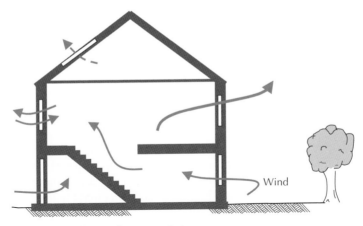

Section: showing stack effect and cross ventilation

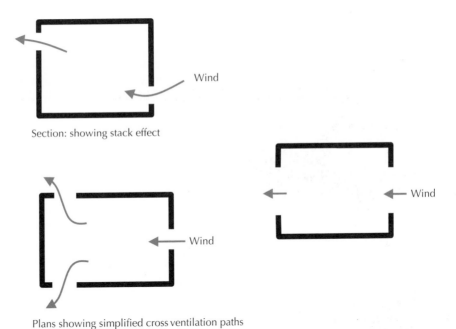

Section: showing stack effect

Plans showing simplified cross ventilation paths

Figure 11.1   Stack effect and cross ventilation.

air leakage. Reducing energy use can be addressed relatively easily with new buildings but is more problematic for the majority of the existing building stock.

Emphasis should be given on reducing the amount of energy consumed by the building in use and on the provision of clean and sustainable energy sources. Increasing the thermal insulation of the building fabric and eliminating unwanted air leakage are obvious areas to address. Similarly, installing or upgrading to more energy-efficient space heating systems will help to reduce energy usage, and the associated pollution. There is a growing number of zero energy demonstration buildings around the UK and some good examples in Europe from which to take inspiration. Some of these buildings also incorporate passive ventilation principles.

Energy is either finite in supply or from renewable sources.

Finite sources

The solid fuels available are bituminous coal (house coal), anthracite, smokeless fuels, coke and peat. Solid fuel tends to be used for burning in fireplaces and stoves for radiant heat to supplement the central heating, or for occasional use in the spring and autumn when the central heating is not on. There is an increasing trend to use fires and stoves as architectural features and a wide range of styles are available from which to choose. Wood-burning stoves represent the most environmentally friendly solution, burning wood (biomass) from sustainable sources.

Gas is primarily used for gas-fired boilers to work central heating systems and hot water supplies. Gas is also used for cooking. Oil is an alternative in more remote areas that do not have gas mains in close proximity.

Electricity can be generated from finite materials, such as coal and gas, and also from renewable sources, such as hydroelectric and wind power as well as through PV cells. In the UK, the majority of buildings are connected to the national grid, which generates electricity from a variety of sources, including nuclear energy. An alternative to the grid is to generate electricity local to the building, on or within its grounds, from renewable sources. Electricity is used to power electric central heating and hot water supplies as well as for cooking, lighting and general power for electrical appliances.

The benefits of electricity over gas for space heating are mainly associated with ease of installation and adaptability, as well as reduced installation costs and annual maintenance costs. Running costs may be lower than gas if off-peak electricity is used; however, we would urge readers to carry out a comprehensive life cycle economic evaluation of all systems prior to installation.

Renewable sources

Primary renewable sources of energy are from the sun and from the wind. Solar energy can be harnessed by PVs to produce electricity and by solar thermal panels to produce hot water (Photograph 11.1a and b and Photograph 11.2). Wind energy is harnessed through wind turbines (Photograph 11.3a and b). Hydroelectric power is an option, provided that there is a reliable source of fast-flowing water. Tidal movement, with the vast oceans moving due to the gravitational pull of the moon and the sun, offers immense power; this source of energy has yet to be properly harnessed.

Photovoltaics

PV materials convert sunlight directly into electricity. Sunlight is a widely available resource and, although it varies with location, season and the time of day, PVs can capture energy from the sun, even in climates such as that of the UK. PV cells produce direct current (dc)

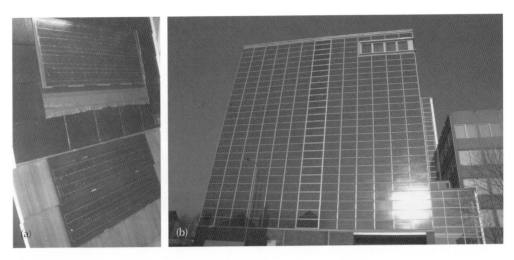

Photograph 11.1   (a) Photovoltaics incorporated in roof tiles. (b) Wall of photovoltaics.

Photograph 11.2   Solar power.

Photograph 11.3 (a) 1.5 MW wind turbine – hub height 67 m. (b) 1.5 M saving up to 3161 tonnes of carbon dioxide per year.

electricity that can be used immediately, transferred to the national grid or stored in batteries for later use (since the cells can only create electricity when the sun is shining). During their operation, they give off no atmospheric or water pollutants; the only negative environmental impact is associated with the use of potentially toxic chemicals in their manufacture. PV cells are made of a semi-conducting material, such as crystalline silicon, which convert up to 23% of sunlight into electricity. Alternative materials known as 'thin films' (e.g. gallium arsenide, copper indium diselenide and titanium dioxide) are also available and have potential as architectural features. Research and development is continuing to improve the efficiency of PV technologies and reduce the costs involved in their manufacture. Their use in buildings has considerable potential and can help in the drive to reduce carbon dioxide emissions. PV cells are incorporated into proprietary roof tiles, cladding panels and glazing (Photograph 11.1a and b, Photograph 11.4 and Photograph 11.5).

The efficiency of PV panels is constantly improving, although performance will vary between manufacturers and, of course, positioning. Typically, roof-based PV systems are capable of generating:

❑ 1 kW from an 8 m² system, can generate an estimated output 850 kWh (depending on orientation and shading)
❑ 4 kW from a 14 m² system, can generate an estimated output 1700 kWh (depending on orientation and shading)
❑ 4 kW from a 28 m² system, can generate an estimated output 3400 kWh (depending on orientation and shading)

Solar farms are starting to be developed in the UK, comprising large arrays of solar panels. These farms cover large expanses of land, but generate significant quantities of electricity.

Photograph 11.4   Photovoltaic acts as both roof covering (roof tiles) and solar energy receptor.

Photograph 11.5   Photovoltaic integrated with roof tiles.

With the installation of solar trackers the panels can be orientated to gain maximum benefit from the sun. The energy conversion ratio of solar panel is typically somewhere in the region of 12–19%, although some manufacturers are reporting much higher figures as the technologies continue to improve and energy losses associated with inverter and other electrical resistances are addressed.

Solar thermal panels

Solar thermal panels are usually positioned on the roof of a building to convert solar radiation into hot water. Collectors comprise a black surface to absorb incoming radiation and are glazed and insulated to help reduce heat losses. Collectors are positioned to optimise the amount of energy falling on them, usually on south-facing roof slopes. The heat generated in the collector is circulated to a storage device (hot water cylinder) by a water circuit. A back-up system (e.g. electrical immersion heater) provides heat during periods when solar heating is insufficient. Recent technological advances have combined PV technology with active solar heating systems. Typically, a solar thermal panel system can provide around 70% of domestic hot water and 40–60% of the home's space heat requirements (typically used for underfloor heating). Systems are relatively low cost to install and require little maintenance, having a service life of around 35 years. The disadvantage is that a larger than average hot water storage tank is required, typically around 300 L in capacity, together with a buffer tank approximately twice the size to store heat. Photograph 11.6 and Photograph 11.7 show solar thermal panels.

Photograph 11.6   Solar thermal panels fixed to the roof.

Photograph 11.7 Solar thermal panels.

Wind turbines

Wind turbines are a proven method of generating electricity from a renewable source with minimal impact on the environment, although their visual impact on the environment remains a cause for concern for many people. Large wind farms are used to contribute electricity to the national grid, usually located in remote or sparsely populated areas or offshore. Smaller, local wind farms are being introduced to supply renewable energy to urban and semi-urban areas (often as part of a larger regeneration initiative). The energy co-operatives are owned and run by local people, helping to improve social, economic and environmental quality. Concerns over visual impact and noise pollution have to be balanced against clean and sustainable sources of energy. Large wind turbines such as those shown in (Photograph 11.3a and b) are far more effective than those that are being offered for domestic use. The 400 W dc household turbines, at best, are currently only capable of powering the household lighting, if the house is fitted with fluorescent energy-saving light bulbs, whereas the large 2 MW wind turbines could power an office, and the 5 MW turbines could start to produce the equivalent of small power stations, possibly saving up to 10,000 tonnes of CO_2 per day. The larger wind turbines have diameters of up to 150 m with blades up to 75 m long. Large and more exposed wind turbines are more likely to prove to be the most effective and efficient.

Small wind turbines that generate 1 kW in optimum conditions are available for domestic properties. However, the actual output is usually less than 1 kW since optimum conditions only occur around 20% of the time, and this may be when electricity is not needed. Care should be taken when attaching a wind turbine to the structure of a building since the energy generated can cause structural damage. Ideally, the turbine should be mast-mounted, although this may not be possible due to local town planning restrictions.

Earth to air heat exchangers

Earth to air heat exchanges work on the principle that the temperature below the surface of the ground is more constant than the air temperature; also at considerable depth, the temperature below ground increases. By taking advantage of the warmer or cooler temperatures below ground, a heat exchange system can be used to keep dwellings warmer in the summer and cooler in the winter. The heat exchange pipes can be run horizontally at shallow depths or vertically, sunk in bored holes, at much greater depths (Figure 11.2). Systems are normally closed, meaning that water, mixed with an antifreeze liquid, is contained in the pipes and is continually pumped around a complete circuit. Alternatively, if there are deep wells or other water sources nearby, the system can simply draw off the water,

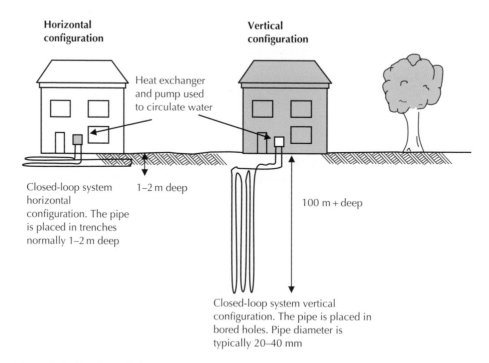

Figure 11.2 Earth to air heat exchanger.

feed it around the building, through the heat exchanger, and back to the original water source. Open systems are usually used for cooling purposes.

Biomass

Timber logs can be burnt in a wood-burning stove to heat a room and wood pellets can be used (via an automated pellet feeder) to fire a biomass central heating and hot water boiler. Biomass is regarded as being carbon neutral, since the carbon released during burning is the same as when the wood decays naturally.

11.4 Solid fuel-burning appliances

Open fire

The traditional open fire consists of a grate inset in a fireplace recess formed in a brick chimney breast, as illustrated in Figure 11.3. An open fire is clearly visible and this is its chief attraction, usually provided as a means of additional or alternative warmth to a central heating system.

Formed at floor level on a solid hearth, the fire will radiate much of its heat at floor level. So that an open fire will radiate heat more uniformly into the body of a room, the hearth may be positioned above the floor level, as illustrated in Figure 11.4. In larger living areas,

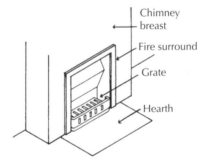

Figure 11.3 Open fire inset in recess.

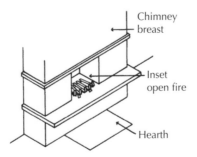

Figure 11.4 Open fire inset above floor in brick or block chimney breast.

the fire may be built as a free-standing structure (Figure 11.5) or fitted where it is visible from one or more sides and where the heat generated by the fire will radiate from all four sides more generally to the living space.

An inset open fire is formed in a recess inside a chimney breast or wall under a flue built in the chimney breast. The burning fuel transfers heat to the room by radiation and convection. Inset open fires for the normal domestic room are generally for burning coal (and smokeless fuels and logs) on an iron grate to contain the burning coals and provide some draught of air under and through the coals. The traditional grate is set in a fireback. Figure 11.6 is an illustration of a typical fireback, grate and fret. The fireback is made from fire

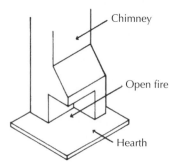

Figure 11.5   Free-standing open fire with brick or block chimney.

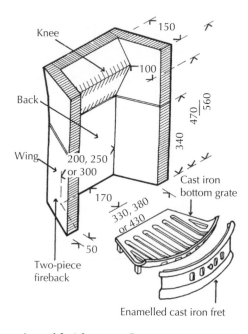

Figure 11.6   Fireback, grate and fret for open fire.

clays, which are clays that contain a high proportion of sand with some alumina and can withstand considerable heat without damage. The fireback may be in one, two, four or six pieces, the four-and six-piece backs being made to facilitate replacing an existing damaged fireback.

Wood-burning and multi-fuel stoves

Wood-burning stoves and multi-fuel stoves are often provided as an additional source of heat to a central heating system. These use wood, or recycled wood products (artificial logs); or a variety of fuels such as wood and coal as a fuel for multi-fuel stoves. There is a wide range of styles and sizes available to suit taste and room size. The fuel is burnt in a chamber which is accessed by a heat-resistant hinged glazed door. This prevents hot timber or coals from falling out and on to the surrounding floor, while also providing a view of the fire.

The advantages of stoves over open fires are that air intake and combustion can be controlled via a series of control vents, ensuring a greater degree of control compared to an open fire. The whole of the surface area of the stove is used to heat the room by radiation and convection of air around the heater. The stove is set on a hearth in front of the chimney breast with a short length of flue pipe running from the back of the heater through a hole in the plate that seals the fireplace recess.

Fire safety

Fireplace recesses

Heat-producing appliances must be installed and fireplaces and chimneys constructed so as to reduce to a reasonable level the risk of the building catching fire. Approved Document J gives recommendations for the position and thickness of solid, non-combustible materials to be used in surrounds to open fireplaces and the thickness and dimensions of solid, non-combustible materials in hearths under open fireplaces in, and in front of, open fireplace recesses.

The solid, non-combustible materials to enclose an open fire recess are brick, concrete block or concrete. The least thickness of these materials at the back of a fireplace recess should be at least 200 mm of solid walling or for each leaf of a cavity wall, at least 100 mm, and where a fireplace is built back to back at least 100 mm, as illustrated in Figure 11.7. In an external wall with no combustible cladding, there should be at least 100 mm of solid backing. The least width of the jambs at the sides of a fireplace opening is 200 mm. This dimension applies whether the jambs project from a wall to support a projecting chimney breast or are part of the wall into which the recess is formed with the chimney breast or chimney projecting from the opposite side of the wall.

There should be an area of solid, non-combustible material in and projecting from fireplace recesses and under stoves, room heaters and other heat-producing appliances. Where the floor under the fireplace recess and under heat-producing appliances is at least a 125 mm thickness of solid concrete, the floor may be accepted as a hearth. The minimum area and thickness of solid, non-combustible hearths in and in front of fireplaces and free-standing hearths are illustrated in Figure 11.8 and Figure 11.9.

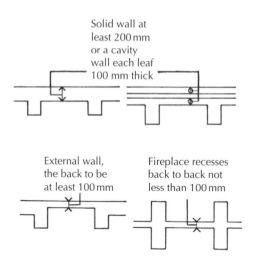

Figure 11.7   Thickness at back of fireplace recess.

Combustible material should not be used under a hearth unless it is to support the edges of a hearth or there is an air space of at least 50 mm between the material and the underside of the hearth, or there is a distance of at least 250 mm between the material and the top of the hearth. A free-standing hearth should extend at least 840 mm around the back and sides of an enclosed heat-producing appliance.

A heat-producing appliance should be separated from combustible materials by some solid non-combustible material 200 mm thick if the appliance is 50 mm or less from the non-combustible material, and 75 mm thick if it is between 50 and 150 mm from the non-combustible material. The non-combustible material should extend at least 300 mm above the top of the appliance.

Flues

A flue is a shaft or pipe, usually vertical, to induce an adequate flow of combustion air to a fire and to remove the products of combustion to the outside air. The material that encloses the flue, e.g. brick, block, stone or metal, is termed a chimney. A chimney may take the form of a pipe run to the outside or can be constructed of solid brick, block or stone, either free-standing or as part of the construction of a partition or wall (Figure 11.12).

Approved Document J provides guidance on the efficient working of flues and chimneys. Section 2 of the guidance, which applies to solid fuel-burning appliances with a rated output up to 45 kW, requires a ventilation opening direct to external air of at least 50% of the appliance throat-opening area, for open appliances, and at least 550 mm for each kilowatt of rated output.

The requirements for an air supply for combustion, like the requirements for room ventilation, are dictated by the trend over recent years to air-sealed windows and doors to

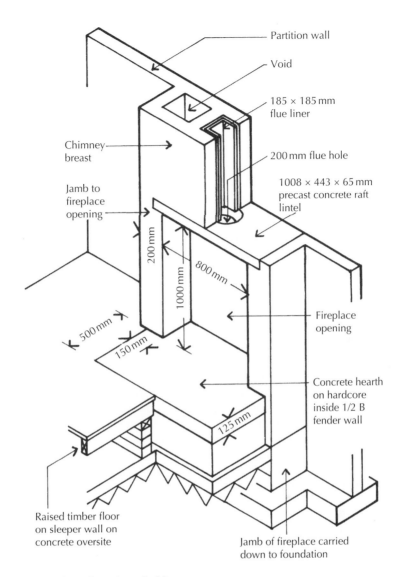

Partition wall

Void

185 × 185 mm
flue liner

Chimney
breast

200 mm flue hole

1008 × 443 × 65 mm
precast concrete raft
lintel

Jamb to
fireplace
opening

200 mm

1000 mm

800 mm

Fireplace
opening

500 mm

150 mm

Concrete hearth
on hardcore
inside 1/2 B
fender wall

125 mm

Raised timber floor
on sleeper wall on
concrete oversite

Jamb of fireplace carried
down to foundation

Figure 11.8 Fireplace, hearth and chimney.

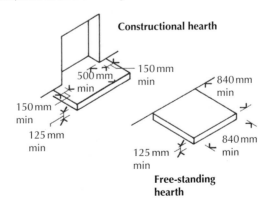

Constructional hearth

500 mm
min

150 mm
min

840 mm
min

150 mm
min

125 mm
min

840 mm
min

125 mm
min

**Free-standing
hearth**

Figure 11.9 Constructional hearths.

Table 11.1   Size of flues

Installation	Minimum flue size
Fireplace recess up to 500 × 550 mm	200 mm diameter or square of equivalent area
Inglenook recess appliances	Free area 15% of area of the recess opening
Open fire	200 mm diameter or square of equivalent area
Closed appliance up to 20 kW output burning bituminous coal	150 mm diameter or square of equivalent area
Closed appliance up to 20 kW output	125 mm diameter or square of equivalent area
Closed appliance above 20 kW and up to 30 kW output	150 mm diameter or square of equivalent area
Closed appliance above 30 kW and up to 45 kW output	175 mm diameter or square of equivalent area

Source: Reprinted from Approved Document J, The Building Regulations 1991, HMSO.

contain heat and avoid draughts. Thus a supply of outside air, via an airbrick or flue, to the room in which the fire is burning is required for combustion, unless the appliance is fitted with a separate air intake. The practical guidance in Approved Document J to the Building Regulations sets out the minimum flue size shown in Table 11.1.

For maximum efficiency, the flue should be straight and vertical without offsets. As the heated products of combustion pass up the flue, they cool and tend to condense on the surface of the flue in the form of small droplets. This condensate will combine with brick, block or stone work surrounding the flue to form water-soluble crystals which expand as they absorb water and may cause damage to the chimney and finishes such as plaster and paint. To protect the chimney from possible damage from the condensate, to encourage a free flow of air up the flue and to facilitate cleaning the flue, flue liners are built into flues.

Flue liners are made of burnt clay or concrete. Clay flue liners are round or square with rounded corners in section and have rebated ends (Figure 11.10). These liners are built in as the chimney is raised and supported on raft lintels over fireplace openings, and the liners are surrounded with mortar and set in place with the liner socket uppermost so that condensate cannot run down through the joint into the surrounding chimney. Bends are made of the same cross sections for use where flues offset. Concrete flue liners are made of high alumina cement and an aggregate of fired diatomaceous brick or pumice cast in round sections with rebated ends. Liners should be jointed with fireproof mortar and spaces between the liners and the brickwork of flues should be filled with a weak mortar or insulating concrete.

To match the dimensions and bonding of concrete blocks, a range of purpose-made precast concrete flue blocks is made for building into concrete walls as flue and liner. Flue blocks are made of expanded clay aggregate concrete with a flue lining of high alumina cement and are rebated and socketed as illustrated in Figure 11.11. Both straight and offset blocks are produced to suit bends in the flue. At the junction of the chimney and the roof, a corbel block is used to provide support for the chimney blocks and the brick facing to the chimney above roof level. The brick facing is used as protection against rain penetration and for appearance sake. A precast concrete coping block caps the chimney and provides a bed for the flue terminal.

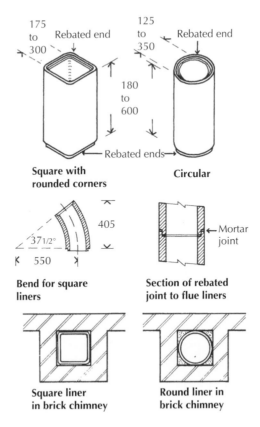

Figure 11.10 Clay flue liners.

When gas-burning open fires and gas-fired appliances are connected to existing unlined brick-built flues, it is necessary to line the flue with flexible stainless steel liners that are drawn up the flue and connected to the appliance and a terminal.

Chimneys

Some of the heat of combustion of fires and stoves will be transferred to the chimney, which heats the structure and air surrounding it (Figure 11.12). To take the maximum advantage of this heat, the best position for a chimney is as a free-standing structure in the centre of a room or building where it is surrounded by inside air. As an alternative, the chimney may be built as part of an internal partition. Where buildings are constructed with a common separating or party wall, it is convenient to construct chimneys back to back on each side of the separating wall. Chimneys constructed as part of an external wall suffer the disadvantage that some of the heat will be lost to the outside, but this can be minimised by continuing the cavity of an external wall and cavity insulation behind the chimney. The four positions for a chimney are illustrated in Figure 11.13.

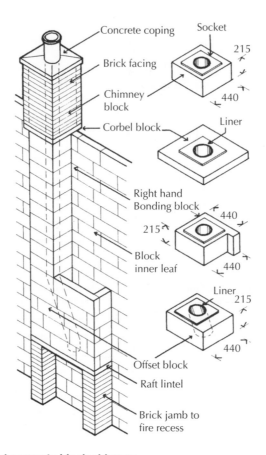

Concrete coping   Socket

Brick facing

215

440

Chimney
block

Corbel block

Liner

Right hand
Bonding block
440

215

Block
inner leaf
440

440

Liner
215

440

Offset block

Raft lintel

Brick jamb to
fire recess

Figure 11.11   Precast concrete block chimney.

It is usual to construct chimney breasts and chimneys as projections into rooms, with the chimney breast projecting into the ground floor room and the chimney projecting into the room above. The advantages of this arrangement are that the heated surfaces of the breast and chimney will transfer some heat to the inside by radiation and convection and that the projecting breast will give some emphasis to the comparatively small fireplace openings. To support the brick or blockwork of the chimney breast over the fireplace opening, two piers are built on either side of the recess. These piers, which are described as jambs (legs), support a brick arch or reinforced concrete lintel that supports the chimney breast. It is common to use a reinforced concrete raft lintel over the fireplace recess to support the chimney breast (Figure 11.8).

Factory-made insulated chimneys
Factory-made chimneys are designed to be free-standing inside buildings as a feature. Their advantages are their comparatively small cross section, ease of installation, high thermal insulation and smooth faces to encourage draught and facilitate cleaning. Flue sections

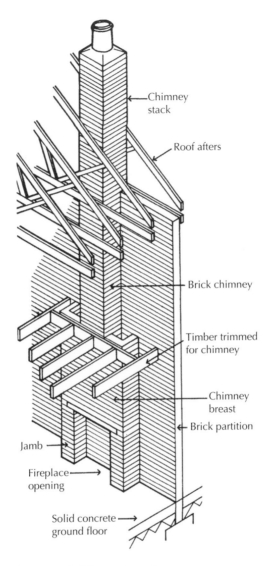

Figure 11.12   Brick chimney and chimney breast.

require some support at roof and intermediate floor levels, and have stainless steel inner linings and either stainless steel or galvanised steel outer linings around a core of mineral insulation that conserves heat and prevents condensation in the flue. Because of the insulation and construction of these flue sections, structural timbers may be as close as 50 mm to the outside of the flue, which avoids the need for trimming timbers and facilitates supporting the chimney at floor and roof level. The cylindrical flue sections are joined with socket and spigot ends that are locked together with a bayonet locking joint (Figure 11.14).

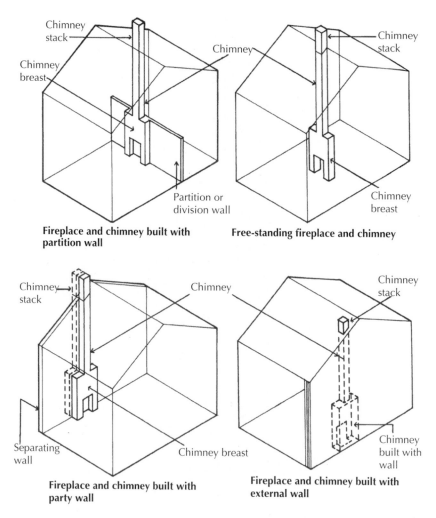

Figure 11.13 The four positions of fireplace and chimney.

Where the chimney passes through timber floor and roofs, metal fire stop plates that fit around the chimney are nailed to the underside and top of the joists around the flue and to timber dust stops nailed between the joists. Mineral wool fibre is packed around the flue sections and the joists and dust stops. At roof level, a lead flashing dressed under the covering and around the flue fits into the spigot end of a weather sleeve section on to which a coping cap is fitted.

Proximity of combustible materials to chimneys

Combustible materials such as timber floor joists and timber rafters should be separated from flues in brick or blockwork chimneys and fireplace recesses to minimise the possibility

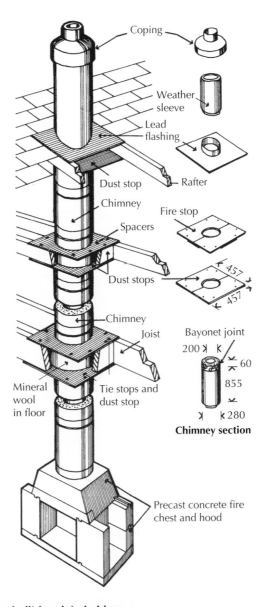

Figure 11.14 Factory-built insulated chimney.

of them becoming so hot as to catch fire. Approved Document J recommends that timber be at least 200 mm from a flue and 40 mm from the outer surface of a brick or blockwork chimney or fireplace recess unless it is a floorboard, skirting, dado or picture rail, mantelshelf or architrave. Metal fixings in contact with combustible materials should be at least 50 mm from a flue. So that no combustible material is closer than 200 mm from a flue it is

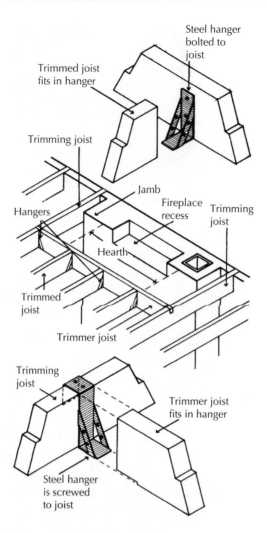

Steel hanger
bolted to
joist

Trimmed joist
fits in hanger

Trimming joist

Hangers

Jamb

Fireplace
recess

Trimming
joist

Hearth

Trimmed
joist

Trimmer joist

Trimming
joist

Trimmer joist
fits in hanger

Steel hanger
is screwed
to joist

Figure 11.15   Trimming floor with metal hanger fixings.

necessary to build in the ends of trimming joists each side of the chimney breast as illustrated in Figure 11.15.

Chimney stacks above roof

Chimney stacks are raised above roofs to encourage the products of combustion to rise from the flue to the open air by avoiding down draught. The practical guidance in Advisory Document J to the Building Regulations sets minimum dimensions for the outlet of flues above roof level as illustrated in Figure 11.16. For roofs pitched at less than 10°, the outlet of any flue in a chimney or flue pipe should be at least 1 m above the highest point of contact

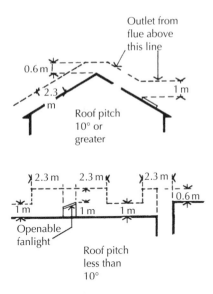

Figure 11.16   Outlets from flues.

between the chimney/flue pipe, and for roofs pitched at 10° or more, should be 2.3 m measured horizontally from the roof surface.

The outlet of any flue in a chimney or flue pipe should be at least 1 m above the top of any openable part of a window or skylight or any ventilator or similar opening which is in the roof or external wall and is not more than 2.3 m horizontally from the top of the chimney or flue pipe. The outlet of any flue in a chimney, or any flue pipe, should be at least 600 mm above the top of any part of an adjoining building, which is not more than 2.3 m horizontally from the top of the chimney or flue pipe. For the sake of strength and stability, the practical guidance in Approved Document A sets the height of masonry chimneys at 4.5 times the least width of chimneys measured from the highest point of intersection of roof and chimney to the top of chimney pots or terminals.

Chimney pots, terminals

The traditional method of finishing the top of a brick chimney stack above the roof is with a chimney pot. The purpose of the chimney pot is to provide a smooth-sided outlet to encourage the outflow of combustion gases to outside air and to provide a neat terminal to a stack which, with cement and sand flaunching around the pots, will provide resistance to weather. Some typical terracotta (burnt earth) chimney pots are illustrated in Figure 11.17. The internal dimensions of these pots should be similar to that of the flue and the pots should be at least 150 mm high. The pots are bedded in cement and sand on the stack and flaunched around in coarse sand and cement and weathered to slope out as illustrated in Figure 11.18. A variety of chimney pots and galvanised steel terminals are produced with louvred sides, horizontal outlets and metal terminals designed to rotate with changes of wind direction to minimise down draught in flues.

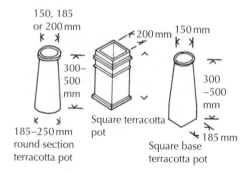

Figure 11.17   **Flue terminals (chimney pots).**

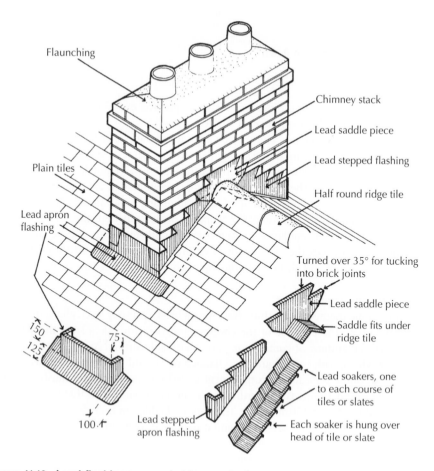

Figure 11.18   **Lead flashings around chimney stack.**

Weathering around chimneys

At the junction of roof coverings and chimney stacks, it is necessary to provide some form of weathering to prevent the penetration of rain into the roof; this is achieved by forming a system of flashings. The flashing should also be able to accommodate that differential movement between wooded roof and masonry stack. Where the pitched roof covering abuts the brick or stone chimney stacks, it is weathered (sealed) by stepped lead flashings. Horizontal cuts, 25–35 mm deep, are made into the masonry joints. The inserts are positioned below the damp-proof course (dpc) so that any water from the dpc is cast to the outside of the wall and is carried away by the flashings. The flashings are inserted into the horizontal joints of the brickwork and dressed over the roof tiles, as illustrated in Figure 11.18 and Figure 11.19. Soakers, which lap under the flashing and over the tiles, may be

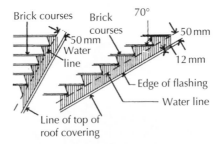

Setting out stepped apron flashing

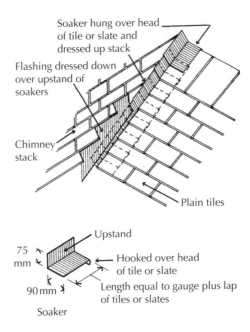

Figure 11.19 Stepped flashing and soakers.

used where it is difficult to dress the flashings over the tiles. The stepped apron flashing overlaps the upstand of soakers by 50 mm and soakers overlap one another by the gauge of the roof covering. The edge of each step of the flashing is wedged into a raked out horizontal brick joint and pointed in mortar.

At the junction of the ridge of the roof and a stack, a saddle piece is used and at the front of a stack, which is at right angles to the slope of the roof, a front apron flashing is used as illustrated in Figure 11.18. These flashings are tucked into horizontal brick joints, wedged and pointed. At the junction of side lap roof tiles and a chimney stack, the stepped lead flashing is dressed down over the tiles to exclude rainwater as illustrated in Figure 11.20. Here, there is no need for soakers, as the flashing dressed over the roll will suffice.

At the junction of a roof slope down towards a chimney stack, it is necessary to form a lead back gutter to collect water running down the slope and divert it to run each side of the stack. The back gutter is shaped out of one sheet of lead to form an upstand, gutter bed and apron to dress under the roof covering as illustrated in Figure 11.20.

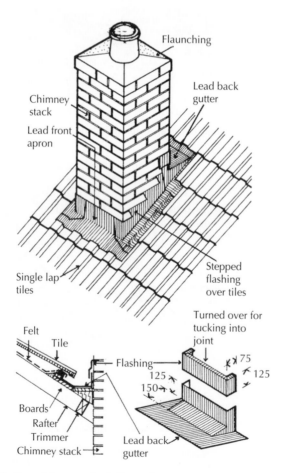

Figure 11.20   Single-flue chimney stack through one slope of roof.

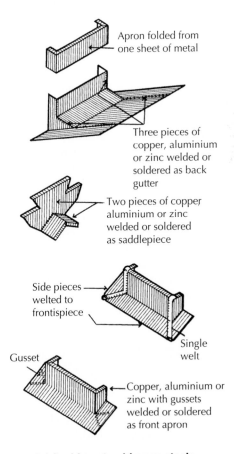

Apron folded from
one sheet of metal

Three pieces of
copper, aluminium
or zinc welded or
soldered as back
gutter

Two pieces of copper
aluminium or zinc
welded or soldered
as saddlepiece

Side pieces
welted to
frontispiece

Single
welt

Gusset

Copper, aluminium or
zinc with gussets
welded or soldered
as front apron

Figure 11.21   Non-ferrous metal flashings to chimney stack.

Lead sheet, thickness code 5 or 4, is best suited to use as weatherings to stacks and roofs. The other metals in sheet form that may be used are copper, aluminium or zinc (Figure 11.21). Copper, which is moderately malleable, suffers the disadvantage that it forms a blue oxide of copper in combination with water, which will be washed on to roof coverings to form an unsightly blue stain. Aluminium is moderately malleable and unlikely to cause stains. Zinc sheet is stiff and difficult to bend and shape, compared with other metals. Its low cost is outweighed by the difficulty in shaping the sheet.

11.5 Domestic gas installations

Natural gas is a non-toxic, finite resource. Because it is odourless, distinctive smelling additives provide warning of leaks.

For the ignition and subsequent combustion of natural gas, a supply of oxygen and a gas temperature of about 700°C is necessary. In gas-fired cookers, fires and boilers oxygen is taken from an intake of air, which is drawn into the combustion chamber, mixes with gas, is ignited and burns to complete combustion. The products of the complete combustion of gas are carbon dioxide and water vapour, which are expelled, either due to the pressure of combustion and the intake of air, to outside air by flues, or by convection. Where there is an insufficient intake of air to provide oxygen for the complete combustion of gas, the by-product will contain carbon monoxide, a gas that in very small quantities is lethal and can cause death in a few minutes. It is important, therefore, to ensure that all gas appliances are installed and serviced on a regular basis (annually) by an authorised Gas Safe registered person. On 1 April 2009 the Gas Safe Register took over the statutory gas registration scheme in Great Britain from Council for Registered Gas Installers (CORGI).

Gas-fired boilers (combustion devices) are used, primarily, to heat water, which is then pumped around interconnecting pipework with radiators to give off heat. Stop valves and thermostats help to provide local control. Some warm air systems are also in use, with the air warmed by the boiler and driven around a duct system with fans. A detailed description of these systems is outside the scope of this book; however, it is important to know what type of system is going to be used early in the construction so that the building can be planned and subsequently constructed to enable economic pipe and duct runs. Boilers are generally described as 'conventional', 'condensing' or high efficiency. To conserve energy, condensing boilers and high-efficiency boilers should be installed in preference to conventional boilers.

Gas supply

Gas is supplied under pressure through the gas main. A branch service pipe is run underground from the main to buildings. The service pipe is laid to fall towards the main so that any condensate runs back to the main, where it is collected. Where the consumption of gas is high, as in commercial and other large buildings, a valve is fitted to the service pipe just inside the boundary of the site to give the supplier control of the supply, e.g. in the case of fire. Domestic service pipes are run directly into the building without a valve at the boundary, and the meter valve or cock controls the supply.

The gas service pipe must not enter a building under the foundation of a wall or load-bearing partition, to avoid the possibility of damage to the pipe by settlement of the foundations. Gas service pipes running through walls and solid floors must pass through a sleeve so that settlement or movement does not damage the pipe. The sleeve is usually cut from a length of steel or plastic pipe larger than the service pipe, which is bedded in mastic to make a watertight joint.

Gas meter and meter box

The service pipe connects to the supply pipe through a gas meter installation which comprises a cock or valve governor, filter and a meter, for domestic premises; with the addition of a thermal cut-out and non-return valve for larger installations. Domestic meters should be positioned in a suitable housing (metal meter box) on an outside wall or within a recess on the outside wall. Thus the meter is naturally ventilated and in a position where the meter reader can gain access when the occupier is out. In this position, the meter control cock or valve is more available for access in emergencies such as fire, than it is if it were indoors.

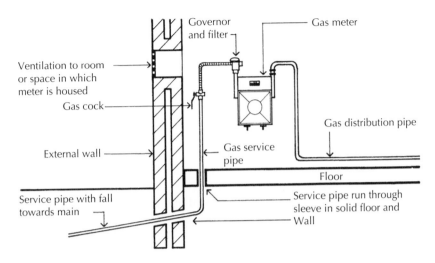

Figure 11.22   Gas supply.

Meter installations to large premises are often positioned in a purpose-built, ventilated, meter house.

Gas cock (valve)

A gas cock to control the supply from the service pipe to the governor and meter consists of a solid plug that in the shut position fills the bore of the cock, and in the open position only partly obstructs the flow of gas. A hand lever, as illustrated in Figure 11.22, operates the gas cock. The connection of the gas cock to the pressure governor is made with a short length of semi-rigid stainless steel tube that can accommodate any movement between the service pipe and meter, which might otherwise damage the meter.

Pressure governor and filter

For domestic installations, a combined pressure governor and filter is fitted at the connection of the service pipe to the meter, as illustrated in Figure 11.22, and a separate governor and filter are used for larger installations. The fine mesh filter is fitted to collect pulverised particles of rust and metal that are carried along the main by the gas. The governor is a spring-loaded diaphragm valve, the function of which is to reduce the pressure of gas in the main to a pressure suited to gas-burning appliances. The governor reduces mains gas pressure to 20–25 mbar standing pressure.

Pipework

The gas pipes, which are run from the meter to supply the various appliances, are described as supply pipes. Commonly, the system of pipes is described as a gas carcass and the work of running the pipes as gas carcassing. Copper tubulars are used for most small gas installations. Capillary or compression fittings at connections are used in the same way that water pipework is run. For safety, the size or bore of pipe used for gas carcassing should be adequate to deliver the necessary pressure of gas to each appliance. If the pipe is undersize, too low a pressure of gas at appliances may result in incomplete combustion and development of carbon monoxide, or there may be a flashback into pipework and an explosion may occur. The required size of pipe from the meter and the branches to each fitting, such as cooker, boiler and gas fire, depends on the volume and pressure necessary and the frictional resistance of the pipes and fittings to flow.

Testing pipework

Newly installed gas carcassing must be tested for leaks, which is usually done after the gas-burning appliances are connected and their supply shut off by closing gas cocks on each appliance. The pipework may be tested either before or after the meter has been installed. The procedure for testing after the meter has been installed is to turn off the main gas cock and all gas appliances and pilot lights. The screw of the test nipple, on the outlet of the meter, is removed and a pressure gauge or manometer is connected to the nipple. The main gas cock is slowly opened to let gas into the pipework until a standing pressure of 20–25 mbar is indicated on the manometer. The gas is turned off at the main cock and, after a wait of 1 minute, over the next 2 minutes the pressure indicated on the manometer should show no drop in pressure greater than 4 mbar, for domestic installations. If the pressure drops below the 4 mbar limit there is a leak, which can be indicated by leak detection liquid which, when sprayed around a leak, will bubble to indicate the position of the leak.

Purging is the operation of evacuating all air that has entered when the pipelines are first installed and when the gas is turned off for maintenance and repairs. The purpose of purging gas pipelines of air is to avoid the possibility of gas mixing with air in the pipelines, which could lead to an explosion. Air is purged by opening the main control cock and the cock to an appliance to cause the gas entering under pressure to force out all air. The gas is allowed to flow until there is a distinct smell of gas when the appliance or burner cock is closed. During this operation, windows are opened and naked flames and electric sparks are avoided until the escaped gas is dispersed.

Gas-fired boilers

Conventional gas-fired boilers have the lowest initial cost due to simplicity of design and relatively limited insulation levels to the boiler. The boiler heats water to feed hot water supply (either direct or indirect) and water-filled central heating systems, such as wall-mounted radiators and underfloor pipes. High-efficiency boilers have a higher initial cost and have more efficient heat exchangers and higher levels of insulation to the boiler casing. Condensing boilers have an enlarged, or additional, heat exchanger and are the most efficient of the three boilers. Under all conditions, the condensing boiler will recover heat from the flue gases and will also recover heat from the condensation of water vapour in the flue gases under suitable conditions. High-efficiency and condensing boilers should be specified since they are the most efficient, and the reduced energy usage is accompanied by a reduction in CO_2 emissions. Manufacturers of gas boilers have made significant advances in their product range in recent years; thus specifiers are urged to consult manufacturers for technical advice and guidance on initial cost, cost in use, maintenance requirements and emissions of CO_2, nitric oxide and nitrogen dioxide (collectively known as NO_x), and sulphur dioxide and sulphur trioxide (collectively known as SO_x). Regular maintenance by a Gas Safe (previously CORGI) engineer is essential to help maintain an efficient and safe boiler.

In many rural areas where there is no main gas supply it is common to install oil-fired boilers.

Flues to gas-burning appliances

The by-product of the combustion of gas and air is heated gases that will rise by natural convection. The purpose of a flue or ventilation system is to encourage the heated gases to rise, by convection or by force of a fan to outside air.

Open-flue appliances depend for their operation on an intake of air directly from the room, or enclosure, in which they are fixed; thus permanent ventilation to the outside air is essential. Manufacturers of gas appliances provide ventilation rates.

Natural convection flues depend for their operation on the natural draught or pull of heated gases rising from heating appliances. It is the function of flues to encourage heated gases to rise vertically from heating appliances to outside air. In general, the higher a flue rises, the greater will be the draught or pull of hot gases.

There is an optimum size of flue to provide the best draught. The best cross-sectional dimensions of a flue depend on the output from a heating appliance and the volume of air required for combustion of a particular appliance. Manufacturers provide detailed advice.

Existing flues

Where gas-burning appliances are fitted in existing buildings, it is common to use an existing open fire brick flue as the flue to the new appliance. The flue is first swept to clear loose, friable material and the draught up the flue is tested with a smoke-producing device. This will test for both the draught up the chimney flue and for an adequate intake of air by natural ventilation of air into the room. Where flues are built into an external brick or block wall, it may be necessary to line the flue to minimise condensation of flue gases inside the flue, caused by too rapid a cooling of the rising gases. To provide the best section of unobstructed flue inside an existing brick or block chimney, it may be necessary to line the flue with a flexible stainless steel liner which is pulled up the flue and sealed to a flue terminal and the gas appliance at the base. Flues built into new brick or block buildings are lined with clay pipes or purpose-made flue blocks.

Flue pipes

The flue pipe to a heating appliance may be independent of, or secured to, a wall. Stainless steel, enamelled steel or aluminium is used for both single-walled and double-walled pipes. Double-walled flue pipes consist of two concentric pipes separated by an insulating material. This insulating material maintains the heat and convection draught of flue gases and prevents the pipe, where it is exposed, from becoming too hot.

Flue terminal

If a flue is to be effective in discharging flue gases to the open air, it should rise above adjacent roofs and structures so that air currents blowing towards the flue terminal do not cause down draughts or suction currents as the blown wind gusts are deflected by surrounding buildings. A terminal is fixed to the top of the flue to deflect gusts of wind.

Room-sealed appliances

Room-sealed gas-burning appliances are those that draw their combustion air intake directly from outside air instead of from inside air, as is the case with open-flue appliances. The majority of small capacity boilers used for space and water heating for domestic premises such as flats and houses are specifically designed to be small and compact to fit into a kitchen or bathroom. These boilers operate through a terminal fixed to the external face of a wall, into which air is drawn and from which combustion gases escape either by natural convection or by fan-assisted operation. With natural convection, they rely on natural air movements and use a balanced flue to control intake and expulsion of gases. Alternatively, an electrically operated fan to assist extraction of combustion gases can be used. All these appliances are sealed so that no part of the intake air or exhaust gases enter the room in which the appliance is fixed. In operation, both the balanced flue terminal and the smaller fan draught terminal will adequately disperse flue gases with extension to flues or flue pipes.

11.6 Domestic electrical supply and installations

Electricity is supplied to the majority of premises via the national grid. Electricity companies charge consumers on the units of electricity consumed, which is measured on a meter. In isolated areas, alternative forms of generation are required, e.g. oil-fired generators. With continual innovations, it is now feasible and cost-effective to generate electricity from alternative sources such as PV cells and local wind turbines. This provides dc, which then needs to be converted to alternating current (ac).

Electrical safety in dwellings is covered by Approved Document P, with P1 covering the design and installation of electrical installations. The aim of Approved Document P is to protect persons operating, maintaining or altering electrical installations from injury or fire. The fundamental principles of Part P can be satisfied by following BS 7671:2001 and guidance given in manuals that is consistent with BS 7671:2001, such as the publications by the Institution of Electrical Engineers (IEE). The majority of electrical installation work to new and existing domestic buildings has to be notified to building control bodies or be carried out by a person registered with an electrical self-certification scheme. Exceptions are listed in Approved Document P. Where electrical installation work is to be carried out professionally, it is necessary to comply with the Electricity at Work Regulations 1989. Once complete, an electrical installation certificate should be issued by someone qualified to do so.

This section provides a brief overview of electrical provision in domestic properties. Readers should note that the regulations relating to electricity and wiring are frequently updated. Electrical standards and approved codes of practice are available on the Health and Safety Executive web page (http://www.hse.gov.uk); new rules are provided within the planning portal (http://www.planningportal.gov.uk and the electrical safety council (http://www.esc.org.uk).

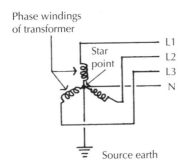

Figure 11.23 Phase conductors (L1, L2, L3) and neutral for TN-C-S supply.

Electrical circuits

To affect the transfer of energy from a source of potential electrical power, such as a generator or transformer, a complete circuit of some conductive material is necessary to provide a path of low resistance back to the source, so that the maximum energy is available around the circuit for conversion to lighting and heating and for motive power to rotary equipment. The material most often used as a conductor is metal in the form of a copper or aluminium wire.

With ac three-phase supplies, there are three separate conductors, one to each of the three separate phase windings of the generator or voltage reduction transformer, and a fourth conductor serving as neutral back to the source to complete the circuit.

❏ Live. The three separate wires, which are termed phase or phase wires, are usually called live or live wires.
❏ Neutral. The conductor that serves to complete the circuit is termed neutral.
❏ Earth. This may be a separate conductor in some cables or, for economy, may be combined in other cables with the neutral conductor. Figure 11.23 shows the terms used. The earth conductor serves as a protective and safety device by acting as part of a conductive circuit with the earth.

Electrical distribution

Consumer's installation
A consumer's electrical installation for small premises, such as a house, begins at the connection of the meter. From the meter, a low-voltage single phase, 240 V supply is run to the consumer's installation, which includes a consumer's unit and the necessary separate circuits for lighting, heating and power. A consumer's unit combines in one factory-made unit the necessary functions of a switch for isolation, circuit-breaking devices (fuses) and distribution of supply to the various circuits.

Most modern units now also have an integral earth bar, which can accommodate all of the circuits' earths or circuit-protective conductors (CPCs), the main earthing conductor and the main equipotential bonding conductors. Even with this arrangement, it is still

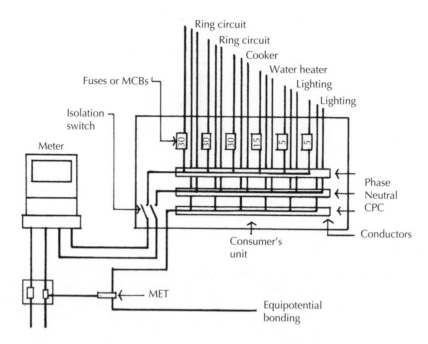

Figure 11.24   Connections to consumer's unit and circuit distribution.

convenient for testing to have a separate main earthing terminal (MET) connected to the consumer unit (Figure 11.24).

Earth system

The three distribution conductors housed in the consumer's unit are connected separately to the phase and neutral of the supply and to the earth conductor made through the earth connection block. From the three distribution conductors, the supply to each separate circuit is run from the phase conductor through an overcurrent protective device (fuse) and directly from the neutral and earth conductor to each circuit.

The UK electricity system is an earth system. This means that the star point or neutral point is connected to the general mass of earth; thus the star point is maintained at or about 0 V. This means that any person or livestock that comes into contact with a live wire and the earth is at risk of an electric shock (Figure 11.25). This can occur by direct contact, that is, touching live wires or parts of a system, or indirect contact, where a conductor, e.g. a metal casing, has become live due to a fault. The generally accepted effects of current passing through a human body are:

1–2 mA	Barely perceptible, no harmful effects
5–10 mA	Thrown off, quite a painful sensation
10–15 mA	Body suffers muscular contraction, person shocked, cannot let go
50 mA and above	Ventricular fibrillation and death

(Scaddan, 2003)

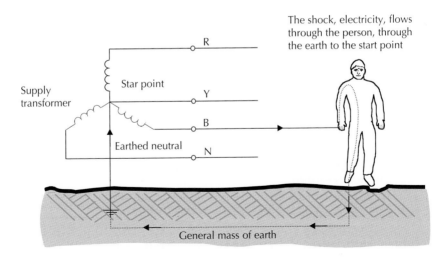

The shock, electricity, flows through the person, through the earth to the start point

Figure 11.25 **Risk of electric shock.**

Protection against direct contact can be achieved by:

❑ Insulating any live wires or parts
❑ Ensuring that all live parts and systems are housed behind protective barriers or suitable enclosures

Residual current devices (RCDs) cannot prevent direct contact, but they can reduce the degree of exposure and possible effects. Residual current breakers (RCBs), residual current circuit breakers (RCCBs) and RCDs are terms used for the same thing. Modern circuit breakers (CBs) now make it easy to protect individual circuits with a combined CB/RCD [residual-current circuit breaker with overload protection (RCBO)]. It should be noted that RCDs can malfunction and should not be the sole method of protection; however, they do provide effective back-up.

The most common protection against indirect contact, through contact with live or extraneous conductive parts while touching the earth is by earthed equipotential bonding and automatic disconnection of supply (EEBADS).

All of the extraneous conductive parts are joined together with a main equipotential bonding conductor, which is connected to the MET, and all of the exposed conductive parts are connected to the MET by the CPCs. This reduces the potential of electricity passing through a human body as the protective conductors provide an alternative path with less resistance. In addition to this, the overcurrent protection should operate fast enough when faults occur such that the risk of electric shock is significantly reduced.

Distribution and earthing systems

The IEE Regulations use the letters T, N, C and S as a classification system to identify the type of system used:

T = terre (French for earth); it indicates that there is a direct connection to the earth
N = neutral
C = combined
S = separate

When the letters are grouped together, the following system applies:

First letter. The first letter identifies how the supply is earthed.
    T = at least one point of the supply is directly connected to the earth
Second letter. The second letter shows the earthing arrangement, how the metalwork of
    an installation is earthed.
    T = all exposed conductive metalwork is directly earthed
    N = all exposed conductive metalwork is connected to an earth, which is provided by
        the supply company
Third and fourth letters. The third and fourth letters indicate the function of neutral and
    protective conductors.
    S = the earth and neutral conductors are separate
    C = the earth and neutral conductors are combined

Common supply and earthing arrangements are (Figure 11.26):

TT. Mostly used in rural areas where the supply is overhead. An example is an overhead
    supply with earth electrodes, and the mass of earth as a return path. The system has a
    direct connection of the supply source to earth and a direct connection of the installation
    metalwork to earth.
TN-S. This type of arrangement is used widely in the UK. The supply company provides
    an earth terminal with the intake cable. Normally, the supply cable has a metal (lead)
    sheath around the cable (supply protective conductor) which connects back to the star
    point at the area transformer, where it is earthed. Thus, the system has the supply source
    directly connected to the earth. The installation metalwork, which is connected via the
    lead sheath of the supply cable and the neutral and protective conductors, performs
    separate functions throughout the whole system.
TN-C-S. In this earthing system, the supply cable sheath forms the combined function of
    earth and neutral conductor; this is known as protective earth neutral. At the consumer's
    installation, it changes to the TN-C-S system. The earth and neutral are separate. This
    is also known as a protective multiple earthing.

Main intake in domestic installations

Unless the domestic premises are large, it is unlikely that a three-phase supply would be
needed. Single-phase systems are generally adequate for domestic properties.

Many TT systems have been installed and protected by a single 30 mA RCD; this does
not conform to the current IEE Regulations. Under the regulations, each circuit that needs
to be controlled separately (lighting and power) should be able to remain in operation in
the event that the other circuit fails. A failure on a socket outlet should not cause the lights
to be cut off.

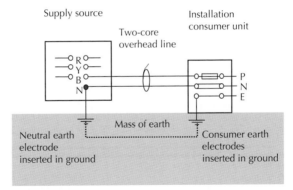

TT system

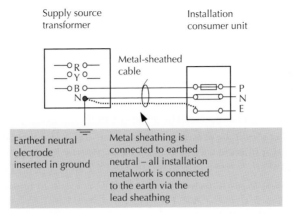

TN-S system

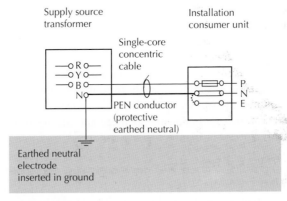

TN-C-S system

Figure 11.26   **Earthing systems.**

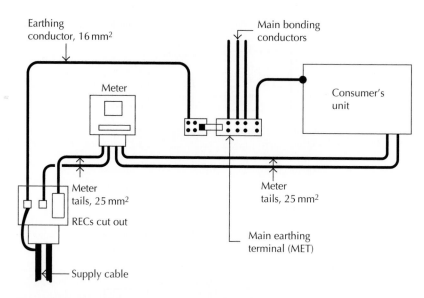

Figure 11.27 Supply connections for a TN-C-S system.

An installation can be protected by a 100 mA RCD and using a split-load consumer unit; the socket outlet should be protected with a 30 mA RCD. However, modern RCBOs, which combine RCDs/miniature circuit breakers (MCBs), make the use of 'split-load' boards unnecessary. Where socket outlets are intended for connection to portable appliances for external use, such as lawn mowers, a 30 mA RCD is required to protect the circuit.

MET and integral earth bars

To provide a means of connecting the earthing conductor of the TN-C-S supply cable to the means of earthing for the consumer's installation and to the main equipotential bonding, an MET is included as part of the consumer's installation. The MET is made as two blocks of metal with a disconnectable link, as illustrated in Figure 11.27. The disconnectable link is provided as a means of testing.

A conductor is run from the supplier's common earth and neutral conductor to the smaller MET block, and a conductor is run from the main MET block to the earth distribution in the consumer's unit. This provides the necessary earth connection to the various CPCs to protect the installation against damage and danger of fire or shock to persons. Where there is a failure of insulation to a live conductor, and contact between the conductor and a conductive part of the installation, the CPC provides an alternative path of least resistance for unpredicted currents to flow to earth. One or more earth conductors are run from the main MET block to the extraneous conductors, such as metal water, gas, oil and heating pipes, as main equipotential bonding. The conductors connect the gas and water services together, and maintain such services within the premises at or about earth potential (about 0 V). It must be remembered that bonding the installation earthing to gas, water and oil services is not done to gain an earth – many services are now made from non-metallic materials.

Most modern units have an integral earth bar, which can accommodate all of the circuit's earths or CPCs, the main earthing conductor and the main equipotential bonding conductors. Even with this arrangement, it is still convenient to have a separate MET, which is connected to the earthing conductor from the consumer unit. Such an arrangement is useful for installation testing.

Main isolation

Normally, the consumer unit has a single switch that provides the means to isolate the whole supply. There is a requirement that isolation switches should be accessible at all times. They should not be positioned in storage cupboards where they are likely to become obstructed.

Short circuit

A short circuit is a fault in a circuit where a live conductor comes into contact with another. A short circuit may cause overheating of conductors and breakdown of insulation and so damage the electrical installations. Overcurrent protective devices, such as fuses and MCBs, are fitted to limit the occurrence of a short circuit.

Overcurrent protective devices

In the phase (live) conductor to each circuit cable run from the distribution conductor, is a fuse or CB as protection against a current greater than that which the circuit can tolerate. The purpose of these devices is to cause a break in a circuit as protection against damage to conductors and insulation by overheating caused by excessive currents.

The main priority of a fuse or MCB is to protect the circuit conductor, not the appliances or user. It is therefore essential that calculation of cable size involves the correct selection of protective device. There are many types of fuses and CBs; however, the IEE Regulations only refer to the following:

❑ Fuses – BS 3036, BS 88, BS 1361 and BS 1362
❑ CBs – BS 3871 and BS EN 60898

Fuses

The fuse is an overcurrent protective device that is manufactured to 'break' at a predetermined maximum current, and so break the circuit. The three types of fuse in use are as follows.

Semi-enclosed rewirable fuse (BS 3036)
The semi-enclosed, rewirable fuse (Figure 11.28) consists of a porcelain fuse holder through which the fuse wire is threaded and connected to the two brass terminals. This fuse has lost favour to the cartridge fuse and the CB, which are easier to replace or reset. The fuse wire is usually rated at 5, 10, 15 and 30 A.

This type of fuse has disadvantages:

❑ Because it is repairable, the wrong size of fuse wire can be used.
❑ After long periods of use, the elements become weak and may break under normal conditions.

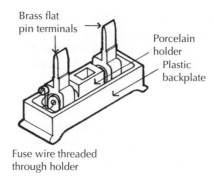

Figure 11.28   Rewirable fuse.

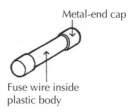

Figure 11.29   Cartridge fuse.

❏   Normal starting conditions which result in current surges may overload and break the circuit.
❏   The holder can become damaged as a result of a short circuit.

Cartridge fuse (BS 1361 and BS 1362)
A cartridge fuse consists of a fuse wire in a tube with metal-end caps to which the wire is connected. The fuse wire is surrounded by closely packed granular filler. Figure 11.29 shows a cartridge fuse. These fuses are cheap, easy to replace by pressing into place between terminals and do not deteriorate over time. Cartridge fuses are made with ratings from 2 to 15 A.

High-breaking capacity (HBC) Fuse
The HBC cartridge fuse consists of a ceramic tube with brass-end caps and copper-connecting tags to which the silver elements inside the tube are connected. The elements are surrounded by granulated silica filling to absorb the heat generated when the elements overheat, rupture and break circuit. These more sophisticated and expensive fuses are used for the more heavily loaded installations.

Circuit breakers (CBs)
A CB is a thermal-magnetic, magnetic-hydraulic or assisted bimetal tripping mechanism designed to operate on overload to break the connected circuit. MCBs are extensively used as protection against damage or danger resulting from current overload and short circuit

in final circuits in buildings. MCBs are factory moulded, sealed units available with terminals for plugging in or bolting to metal conductors in consumers' units. A range of ratings are provided to suit the necessary overload current ratings for particular final circuits.

Final circuits
Final circuits are the circuits of a consumer's installation that complete a circuit for the flow of current back to the supply neutral. For the low voltage, single-phase ac that is usual for small premises, a small compact consumer's unit will provide distribution terminals for the number of separate final circuits used. Typical circuits for a house could be one or more ring main circuits and one cooker circuit, each with a 30 A breaking capacity fuse or MCB and two or more lighting circuits each with a 5 A fuse or MCB. The purpose of segregating the various circuits is to afford economy in the cost of cables, the cross section of those with 5 A fuses being smaller than those with a 30 A rating. Figure 11.24 shows a typical consumer's unit layout. The length of each circuit is limited to provide an economical section of cable and to minimise the electrical resistance of the cable and the number of connections to each circuit, to limit overload current. The two types of circuit that are used are the ring and the radial.

Ring circuit
These are generally referred to as ring mains and are the most common method of supplying socket outlets in domestic installations in the UK. Ring circuits are commonly used for socket outlets that provide electrical supply to portable equipment. The circuit (Figure 11.30) makes a big loop or ring from one outlet to the next, round all the outlets and back to the consumer's unit, as the most economic and convenient cable layout. The recommended maximum length of cable run depends on the cross-sectional area of the

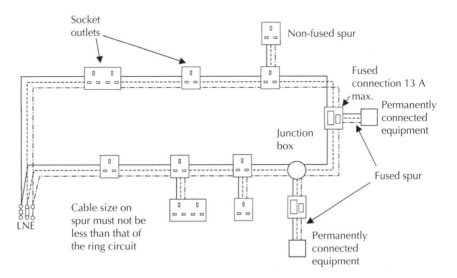

Figure 11.30   Ring circuit.

conductor chosen and the type of protective device used. While there is no recommended limit to the number of outlets served, it is recommended that a maximum of 100 m² of floor area be served by a ring circuit, protected by a 30-A fuse or MCB in domestic premises. The cable generally used for ring circuits is 2.5 m² conductor, twin and earth, polyvinyl chloride (PVC)-insulated and PVC-sheathed cable, or 1.5 m² mineral-insulated copper conductors (MICCs).

Spur outlets from a ring main
Where a socket outlet cannot be conveniently fed by a ring circuit, a spur outlet may be run from the ring circuit. A spur outlet is connected through a joint box or spur box, as shown in Figure 11.30. The number of fused spurs is unlimited, and there may be as many non-fused spurs as there are points on the ring main. Each non-fused spur may supply one single-socket outlet, one double-socket outlet or one item of permanently connected equipment.

Radial circuits
These are fed from the consumer unit and run either in a loop-type formation (lights) or go directly from the source (consumer unit) to a single item of equipment (e.g. a cooker). Typical domestic radial circuits are used for lighting, water, heating, storage heating and cooking.

The loop-in system (lighting circuits)
The circuit arrangement for lighting is generally in the form of radial circuits, each of which runs from the consumer's unit to its light fittings and back to the consumer's unit. In each radial circuit, the phase conductor runs in the form of a loop from the circuit through a switch back to the light fitting or fittings that the switch controls; the circuit cables 'loop' in and out of each lighting point (Figure 11.31). [For the sake of clarity, the CPC cable (earth conductor) is not shown in Figure 11.31]. By this arrangement, the switch controls the fittings allocated to it without controlling the rest of the light fittings connected to the circuit. Where one switch is used to control several light fittings, the circuit may be run as a ring circuit off a radial circuit, or as a ring circuit by itself, whichever is the most con-venient. A 5 A fuse or MCB is connected as an overcurrent protective device to the phase wire run out from the consumer's unit. The cables are run in 1.5 mm² conductor, twin and earth, PVC-insulated and PVC-sheathed cable. To limit current flow to suit the fuse or

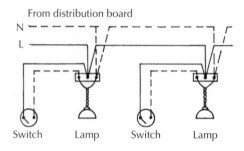

Figure 11.31   Radial lighting circuit.

MCB and cable size, there are usually two or more separate lighting circuits to most small installations.

Earthing

The Wiring Regulations (BS 7671:2001) include recommendations for protection against the dangers of electric shock to persons or livestock, damage to installations and the danger of fire from overcurrents and earth faults. As described earlier, the earth conductor in cables and the earthing connections to conductive metals provide an alternative path for unplanned flows of electrical energy.

The CPC, often called the earth cable, is connected to metal components of electrical appliances so that, in the event that a fault occurs, unpredicted currents have a direct flow to earth through the CPC, earth cable, rather than through a person.

With the generally used TN-C-S system of supply cables, the earth (CPC) conductor of an installation is connected to the combined neutral earth of the supply cable, which may not provide a wholly satisfactory path to earth. It is generally necessary, therefore, to provide other earth paths through earth electrodes connected to the installation's earthing block.

Main equipotential bonding

BS 7671 requires that all extraneous conductive metal parts, which are not part of an electrical installation, be connected to the MET. This earth bonding is provided as protection against the possibility of conductive metal outside (extraneous) the electrical installation becoming live due, for example, to failure of insulation of a cable run close to a heating pipe, making a live connection to the pipe. The conductive metals included are water, gas, oil, heating and hot water pipes, radiators, air conditioning and ventilation ducts, which should be earth bonded as illustrated in Figure 11.32.

Most modern units have an integral earth bar, which can accommodate all of the circuits' earths or CPCs, the main earthing conductor and the main equipotential bonding conductors.

Cross bonding

The purpose of equipotential bonding is to co-ordinate the characteristics of protective devices with earthing and the impedance (resistance) of the circuit to limit touch voltages until the circuit-protective devices cause disconnection. Because of the introduction of plastic pipes, connections and plastic coatings to metal, it may be necessary to provide earth-bonding connections across any plastic parts.

Bathrooms

Special consideration needs to be given to the position of electrical sockets and appliances in bathrooms. Different considerations are given to zones that are in close proximity to wet appliances and those that are further away from the wet appliances. Bathrooms are divided into four zones: 0, 1, 2 and 3. Zone 0 is the interior of the bath tub or shower basin; if the shower flows into a wet room with no shower tray, the zone is applied to a radius of 600 mm from a fixed head shower or 1200 mm radius for demountable head showers, at a depth of 50 mm above the floor (Figure 11.33).

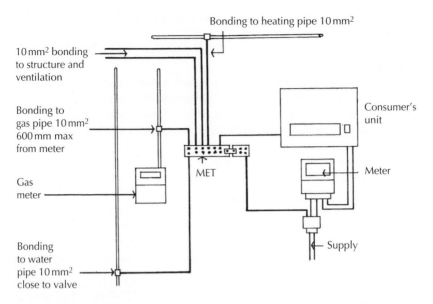

Figure 11.32   Equipotential bonding.

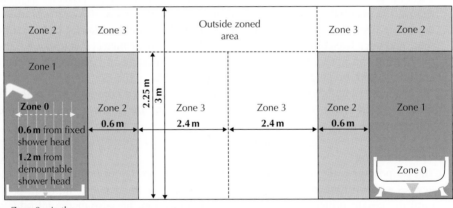

Figure 11.33   Bathroom zones.

The factors and considerations associated with this zone are:

❑ Only separated extra low voltage (SELV) 12 V, or ripple-free dc may be used as a measure against electric shock, the safety source being outside Zones 0, 1 and 2.
❑ No switchgear or accessories are permitted, other than equipment that is specifically designed for use in the zone. Equipment must be at least IP X7 (index of protection

code). IP 7X means that the installation should be protected against immersion in water. It must not be possible for water to enter the enclosure.

❏ Only wiring associated with specifically designed equipment may be installed.

Zone 1

This is the area that is above and below the Zone 0 area. Zone 1 extends to 2.25 m above the floor where Zone 0 area applies. The factors and considerations associated with this zone are:

❏ Only SELV switches and switches and controls specifically designed for use in this area are permitted.
❏ Water heaters and shower pumps can only be used provided they are specifically designed and suitably fixed for use in this zone.
❏ All equipment installed must be suitable for this zone and must be protected by an RCD (rated tripping current of 30 mA or less).
❏ Only wiring that is suitable for use in Zones 0 and 1 should be installed.
❏ Equipment designed for use in Zone 1 must be at least IP X4 (protection against splashing; liquid which is splashed from any direction should have no harmful effect) or IP X5. (The installation should be protected against water jets, which may be projected from any angle. The water jets should have no harmful effect.)

Zone 2

Zone 2 extends 600 mm horizontally beyond Zone 1 and covers the area above Zone 1 to a height of 3 m above the floor level. Thus, Zone 2 is the area above Zone 1 from 2.25 to 3 m above the floor level. The factors and considerations associated with this zone are:

❏ Only switches and socket outlets of SELV circuits and shaver units to BS EN 60742, and equipment specifically designed for use in this zone may be installed.
❏ Only water heaters, shower pumps, lights (luminaries), etc., designed for use in this zone should be installed.
❏ Insulating pull cords may hang down into Zones 1 and 2 but the main body of the switch must be in Zone 3 or outside the zoned areas.
❏ Equipment designed for use in Zone 2 must be at least IP X4 (protection against splashing; liquid which is splashed from any direction should have no harmful effect) or IP X5. (The installation should be protected against water jets, which may be projected from any angle. The water jets should have no harmful effect.)

Zone 3

Zone 3 extends 2.4 m beyond Zone 2 and is extended to a height of 2.25 m above the floor level. The factors and considerations associated with this zone are:

❏ Only SELV sockets or shaver units to BS EN 60742 may be installed in this zone. Equipment must be specifically designed for use in this zone, or must be protected by an RCD (rated tripping current 30 mA or less).
❏ Any equipment designed for use in this zone must be at least IP X5, if water jets are likely to be used. IP X5 means that the installation should be protected against water

jets, which may be projected from any angle. The water jets should have no harmful effect on the installation.

Supplementary equipotential bonding

It is a requirement of BS 7671 that supplementary equipotential bonding is provided to all simultaneously accessible, exposed conductive parts in special conditions of wet activities and high humidity, such as bathrooms and swimming pools, where the moisture in such conditions provides additional risks. The term 'simultaneously accessible' in relation to exposed conductive parts means the possibility of someone, e.g. in a bath of water, reaching out to touch a heated, metal towel rail and so providing a conductive path from the metal bath to the towel rail, through their body. The purpose of supplementary bondings is to spread the voltage potential across and between other adjacent bonded metal parts to equalise and so limit the voltage potential to that least likely to cause injury or death in the few seconds before CBs come into operation.

A bath is the appliance most requiring supplementary bonding between metal pipes and a metal waste pipe, as shown in Figure 11.34. Bonding to a basin is required between metal hot and cold water pipes and a metal waste pipe. Similarly, a metal supply pipe to the cistern of a water closet (WC) should be bonded to a metal waste pipe. To complete the supplementary bonding between all simultaneously accessible exposed conductive parts, there should be bonding between the cross bonding to a bath, basin, WC and a radiator. This connecting bonding may be affected by a conductive pipe that provides a link to sanitary fittings. Where there is no satisfactory conductive link by pipework, conductive bonding should be connected to all bonded fittings and connected to the radiator as a terminal.

To affect supplementary bonding, a conductor should be connected between simultaneously accessible, exposed conductive parts by means of pipe clamps, through which the conductor should run continuously. The bonding conductor may be of bare copper or insulated copper, the lowest cross-sectional area of which should be $4\,\text{mm}^2$. Because it is a requirement that supplementary bonds be accessible for inspection and testing, there may well be some untidy exposed conductors and pipe clamps that cannot be hidden behind removable bath panels or basin pedestals. Supplementary bonding must be carried out with equipment and any extraneous conductive part within Zones 1, 2 and 3.

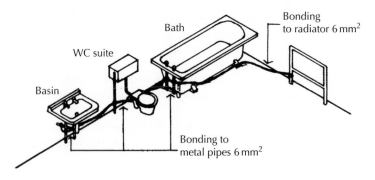

Figure 11.34   Supplementary equipotential bonding.

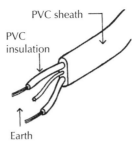

PVC sheath

PVC insulation

Earth

Figure 11.35   PVC-insulated and PVC-sheathed cable.

Cables and conduits

PVC cable – twin and earth

The cable most used for final circuits to 240 V, single-phase supplies is PVC-insulated, PVC-sheathed cable (commonly described as PVC twin and earth, describing the phase and neutral insulated conductors and the earth in the sheath), as shown in Figure 11.35. The size of the cable is defined by the cross-sectional area of the copper wires: 1.5 and 2.5 mm^2 for lighting and socket outlet circuits, respectively. For 415 V, three-phase supplies, a three-core PVC-insulated and PVC-sheathed cable is used.

Protection

The Wiring Regulations provide extensive recommendations on measures to be taken to minimise damage to cable during installation and in use. These may be broadly grouped under the headings mechanical damage, temperature, water and materials in contact including corrosive materials. Mechanical damage includes precautions to avoid damage during installation, such as drilling holes in the centre of the depth of floor joist for cable runs rather than using notches in the top or bottom of joists where subsequent nailing might damage the cable. At temperatures above 70°C, plastic will appreciably soften, and below freezing will become noticeably brittle and crack. Water, both as liquid and in vapour form, in contact particularly with terminals, may act as a conductor between live and earth conductors. To protect cables from damage by corrosive materials such as cement, cables run in plaster, concrete and floor screeds should be protected by channels or conduit.

Mineral-insulated metal-sheathed cable

This type of cable consists of single-stranded wires tightly compressed in magnesium oxide granules, enclosed in a seamless metal sheath of copper or aluminium, as shown in Figure 11.36. The combination of the insulation of the magnesium oxide and the metal sheath gives this cable an indefinite life and reasonable resistance to mechanical damage. For added protection, the metal sheath may be protected with a PVC sheath. Because of the high initial cost of the cable and the various fittings and seals necessary at bends, junctions and terminations, the use of this cable is confined to commercial and industrial installations where the cost may be justified. The metal sheath may serve as the earth (CPC) conductor.

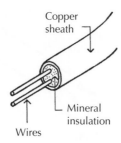

Figure 11.36   Mineral-insulated copper-sheathed cable.

Channels

Channels of plastic or galvanised steel, generally in the form of a hat section, are used to provide PVC cable, which is to be buried in plaster, with protection and to secure the cable in position while the plaster is being spread. The brim of the hat section is tacked to the wall surface around the cable. These sections, which are used solely as protection, do not provide a ready means of pulling cable through when replacing cable.

Conduit

Round or oval-section plastic or galvanised steel sections are used as protection where PVC cable is to be run in plaster, solid floors, walls and roofs. Conduit is used to provide all-round protection, particularly where the cheaper PVC-insulated cable is used, and to provide a means of renewing and replacing cable by drawing cable through the conduit. Because of the bulk of the conduit, it is generally necessary to form or cut chases (grooves) in wall surfaces and solid floors and roofs where screeds are insufficiently thick to accommodate the conduits.

Plastic conduit

Plastic conduit, which is considerably cheaper than metal conduit, is made in round or oval sections. Round-section conduit, which can be buried in thick screeds, has to be fixed in a chase or groove in walls. Oval-section conduit can be buried in thick plaster finishes and thin screeds. Lengths of conduit are joined by couplers, such as the one shown in Figure 11.37, into which conduit ends fit. A limited range of elbows and junctions is made for solvent welding to conduit. The conduit is secured with clips that are tacked in position. There is some facility for pulling through replacement cable, but not as much as with metal conduit. Because it is made from plastic, the conduit will not serve as an earth conductor.

Metal conduit

Metal conduit is manufactured as steel tubes, couplings and bends which are either coated with black enamel or galvanised. The cheaper black enamel conduit is used for cables run in hollow floors and roofs and other dry situations. The more expensive galvanised conduit is used where it is buried in concrete floors, roofs and screeds, and for chases in walls below plaster finishes where wet finishes might cause rusting if black enamel conduit were used.

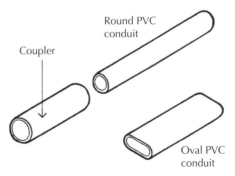

Figure 11.37   PVC conduit.

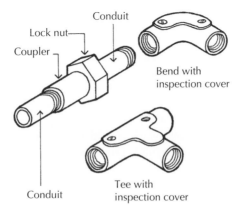

Figure 11.38   Steel conduit.

Metal conduit is produced with a variety of fittings designed to facilitate pulling out old cable and replacing it with new ones. The access or inspection covers to the bend and tee, shown in Figure 11.38, are for pulling through. Because the conduit is of conductive metal, it may serve as the earth (CPC) conductor, and single PVC-insulated cable may be used as the conduit provides protection against damage.

Trunking
Where there are extensive electrical installations and several comparatively heavy cables follow similar routes, with standard conduit too small to provide protection, it is practice to run cables inside metal or plastic trunking to provide both protection and support for the cables. Trunking is fixed and supported horizontally or vertically to wall, floor or ceiling structures, or above false suspended ceilings, with purpose-made straps secured to the structure. Trunking is also run inside ducts for vertical runs and above false ceilings where appearance is a consideration, or it is exposed in industrial premises. Trunking may be hollow, square, rectangular, circular or oval in section. Figure 11.39 is an illustration of square trunking with access for inspection and any necessary renewal or alterations.

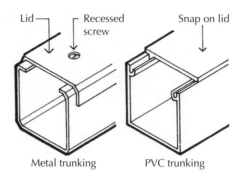

Figure 11.39 **Metal trunking.**

Socket outlets

Socket outlets comprise sockets in a wall-mounted front plate into which the terminals of a loose plug top fit to supply electricity to a wide range of moveable, portable and handheld electrical equipment. Wall-mounted socket outlets and switches should be positioned so that they are easy to reach, to comply with Approved Document M. Socket outlets are usually connected to a ring final circuit, there being no limit to the number of sockets connected to each ring circuit, which should not serve an area greater than 100 m² for domestic installations. A ring circuit is protected by a 30-A fuse or MCB at the consumer unit.

A socket outlet consists of a galvanised steel box to which a front plate is screwed after the terminals of the front plate have been connected to the electric supply cable. The galvanised, pressed steel boxes are made with circular knockouts, which can be removed for cable entry, and lugs for screws to secure the plastic front plate. Steel boxes are usually made for recessing into a wall and plaster and others for surface fixings. Figure 11.40a shows a recessed steel box for a two-gang, two-plug top outlet. Steel boxes for surface fixing have a smooth-faced finish with the knockouts for cable entry and holes for fixing screws at the back of the box. Figure 11.40b shows a steel box for surface mounting. Front plates to socket outlets are moulded from plastic to suit single, double or multi-gang plug top outlets, complete with sockets for square three-pin plug tops. Brass terminals for the phase, neutral and earth conductor cables are fixed to the back of the front plate, and it is holed for two screws for fixing to the steel box. Socket outlets may be supplied with a switch and a pilot light to indicate the 'on' position (Figure 11.40c).

Many socket outlets are connected to a ring circuit, which is protected with a 30-A fuse or MCB at the consumer unit. It is necessary to fit a cartridge fuse to the plug top to provide current overload protection to the flexible cord and appliance connected to the outlet. Fuses to plug tops to outlets vary from 2 or 4 A for lighting to 13 A for electric fires. Where there is a spur branch to a ring circuit, a spur box is fitted to provide a connection for the spur cables and to protect and control the spur outlet. A spur box consists of a steel box and plastic front plate in which there is a fuse and a switch (Figure 11.41). The terminals at the back of the front plate serve to make connection of the spur cables.

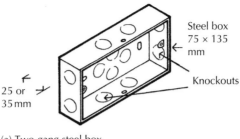

(a) Two-gang steel box

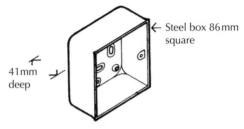

(b) Steel box for surface mounting

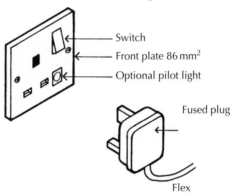

(c) 13 A outlet and plug

Figure 11.40   Socket outlets.

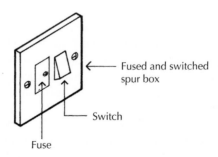

Figure 11.41   Spur box.

The electrical supply to an electric cooker with hobs, grill and oven requires a high fuse rating and cables. A separate radial circuit is run from the consumer's unit to the electric cooker, and protected by a 30, 40 or 45 A fuse or MCB in the consumer's unit. A wall-mounted cooker control unit is fixed close to the cooker. This control unit consists of a recessed steel box and plastic front plate in which a single switch is provided to isolate the appliance.

11.7 Artificial lighting

Artificial lighting enables us to carry out tasks safely and provides a means of creating ambience within a room. To minimise the consumption of electricity, low-energy or energy-efficient light fittings must be fitted to new developments. Sensors can also be used to switch off electric lighting during periods when the room is not being used. The drive to minimise energy use has now become mandatory through the most recent Approved Document Part L. Non-domestic buildings are required to provide lighting systems that are energy efficient. Part L gives guidance on how this can be achieved. The regulations apply to new dwellings and also to dwellings where there has been a material change of use.

The requirements in Approved Document L2 are much more detailed. They aim to make greater use of natural light, reduce the lighting of areas that are not in use and increase the use of energy-efficient lighting without endangering the occupants using the buildings. Switching devices that can help achieve these requirements include:

❑ Local manual controls so that lights can be turned off when the area is not in use
❑ Absence detection (device turns off when the room is not in use)
❑ Timed-off control (provides light for a set time then turns off – could be set to minutes or hours)
❑ Photoelectric dimming (lights dim when the natural light increases)
❑ Photoelectric switching (lights turn off when the natural light is sufficient)

By using a combination of energy-efficient lamps and controls, the requirements of the Building Regulations can be met.

Energy-efficient lighting

Traditional light bulbs, tungsten filament bulbs, are only able to convert around 5% of the electricity they use into visible light. As the filament is heated it starts to glow, giving off a yellow/white light. These bulbs do not have a long service life because the filament gradually evaporates. These bulbs are not energy efficient and should not be used. Bulbs with better energy efficiency include:

❑ Compact fluorescents (CFLs) are glass tubes that contain a gas. As the gas is charged with electricity it starts to glow, causing a coating on the inside of the tube to fluoresce, giving off light. These are available in a wide range of sizes and shapes, from small fittings through to longer 'strip' lights. There is also a wide range of colours of light available, ranging from a warm yellow light through to a cool pure white light for task lighting.

❑ Halogen lamps. These are very similar to tungsten filament lamps but run at a higher temperature and therefore use slightly less electricity. These are commonly used in spotlight downlighters. These are slowly being replaced by light-emitting diode (LED) fittings.

❑ LEDs use very little energy and have a very long service life compared with halogens. LED is a solid state device that emits light as electricity flows through it. These bulbs are expensive compared to halogen lamps, but given their much improved energy efficiency and their long service life they should be a cheaper option when viewed over a number of years.

External lighting

Where external lighting is fixed to a building, such as lamps over porches, reasonable provision should be made to enable effective control and efficient use of lamps (Approved Document L1). One method of complying with this regulation is to install systems that:

❑ Automatically extinguish when there is enough daylight, and when not required at night (e.g. on a timer), or

❑ Have sockets that can only be used with lamps having an efficacy greater than 40 lumens/circuit (e.g. CFL lamps).

For lighting paths, driveways and other external areas, low-voltage systems are available. Free-standing lights operated by solar cells (with a back-up battery) may be an energy-saving (and cheaper) alternative to hard wired systems. The length of time the solar light emits light will depend on how much solar radiation it has been exposed to through the day, thus they tend to glow for longer in the summer months compared to the shorter days of the winter months when the need for lighting is greater.

Lighting outlets

Common for high ceilings is a ceiling rose screwed to the ceiling, from which drops a pendant cable supporting a lampholder fixed at a convenient height. The ceiling rose is connected to a radial circuit protected by a 5 A fuse in the consumer unit, with a loop down to a wall switch that controls the ceiling light. The three cables, phase, neutral and earth, are connected to the terminals in the base, which is screwed to the ceiling. The lampholder is screwed to the base and the lampholder cover screwed around the lampholder. For low ceilings it is common to use recessed light fittings with LEDs.

More intense lighting systems used for display purposes in shops and showrooms have been used in the home for kitchen and general lighting. Systems of 'downlighting' have been used, consisting of a number of lights recessed into ceilings and systems of 'spotlighting', using lamps fixed to tracks in the ceiling, which may be adjusted to light a particular area. There have been rapid advances in the range of affordable halogen and LED light fittings suitable for all types of building use.

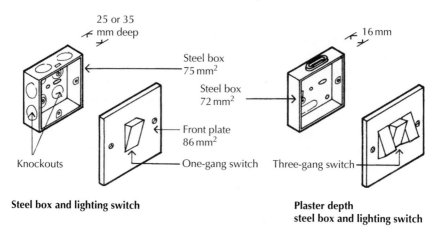

Figure 11.42   Wall switches.

Wall switches

A wall switch is generally recessed in a wall, partition or plaster depth for appearance sake. A galvanised, pressed steel box is set into a recess in the wall after knockouts have been removed for the cable entry. The three conductors, phase, neutral and earth, are connected to the terminals set into the back of the plastic front plate, which is then screwed to lugs in the steel box. Figure 11.42 shows a single gang switch, with steel box and plastic front plate. For small rooms, a single switch inside the access door or opening is generally sufficient for ceiling and wall lighting. For large area rooms, it is often convenient and economical to use two or more circuits, each with its own wall switch, so that a part of, or the whole of, the room may be illuminated. For staircases, a system of two-way or three-way switching to a circuit is used to control the lighting on upper and lower floors as necessary. The system of two-way or three-way switching involves additional wiring to provide the means of switching from two places. Figure 11.42 is an illustration of a steel box and plastic front plate for a three-gang switch.

11.8 Electronic communication systems: Broadband

The majority of building users require cable-based electronic communication systems, such as the high-speed Internet access 'broadband'. Installation to existing buildings is not always easy due to the unavailability of ready access for routing cables. The provisions in Approved Document Q of the Building Regulations aim to ensure that electronic communication services can be installed without inconvenience to the building owners and users, and without disruption to the building fabric and surrounding ground. In practical terms, this means that a duct should be laid from the site boundary to the building, into which appropriate cables may be easily installed and, if necessary, removed. A terminal chamber is required at the site boundary, and terminal boxes will be required externally and internally to the building (see Figure 11.43). Care is required to ensure that the external terminal box

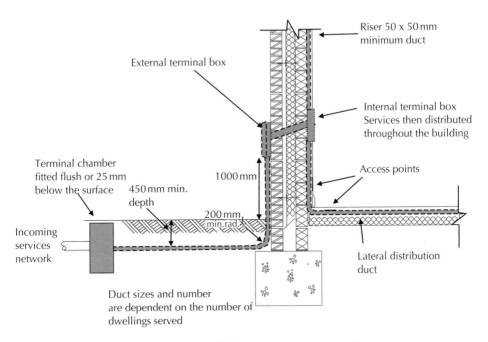

Riser 50 x 50 mm minimum duct

External terminal box

Internal terminal box
Services then distributed
throughout the building

Terminal chamber
fitted flush or 25 mm
below the surface   450 mm min.
depth

Access points

1000 mm

200 mm min.rad.

Incoming
services
network

Lateral distribution
duct

Duct sizes and number
are dependent on the number of
dwellings served

Figure 11.43   Means of supply to the building.

is positioned so as not to be too unsightly. Similarly, internal terminal boxes should be as small as practicably possible and positioned to suit anticipated room usage.

Installation

Commercial suppliers carry out installation of cable television and broadband service supply. The main supply will run under the pavement, with connections made to terminal chambers at the boundary of individual buildings. Connection will be made on payment of a subscription fee to the supplier. For new building work, the terminal chamber, ducts and terminal boxes should be installed as part of the general contract. This enables the service provider to add the cable to the duct and make the necessary connections quickly and easily.

Terminal chambers

The terminal chamber should allow for the connection of at least three electronic communication services' networks. The chamber should also allow for the connection of one or more customer ducts – see Figure 11.44. It is common practice to install the terminal chamber into a hole in the ground so that the top is flush with the finished ground level, or located 25 mm below a grassed or soiled area (but not covered over). Thus the terminal should be designed and constructed to be able to withstand a reasonable superimposed load. The chamber lid must be marked to clearly identify the purpose and contents of the terminal, i.e. electronic communications services. Chambers tend to be manufactured from plastics and are often identified by words such as 'Cable TV' or 'Broadband' on the lid.

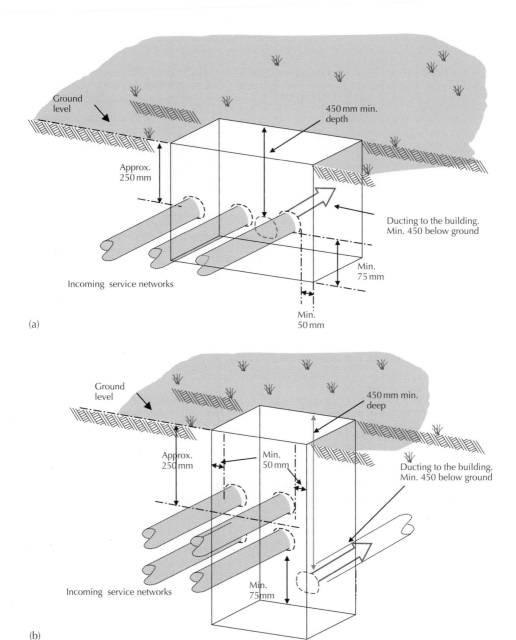

(a)

(b)

Figure 11.44   Terminal chamber configurations.

Ducts

The size and number of ducts will be dependent on the number of dwellings served. One dwelling would typically be served by one duct with an internal diameter of 50 mm. A 90 mm diameter duct is used for provision to more than one dwelling, with the number of ducts increasing to suit the number of dwellings. Guidance is provided in Approved Document Q. The duct should be installed a minimum of 450 mm below finished ground level and laid on a bed of at least 25 mm well-compacted fine fill material. The duct should be surrounded with well-compacted fine fill material, to a minimum depth of 50 mm above the duct – see Figure 11.45. The duct should be swept up to the external wall, or alternatively the internal face of the wall to the terminal box. For refurbishment work, it is most likely that the terminal box will be placed externally (to avoid disturbing the

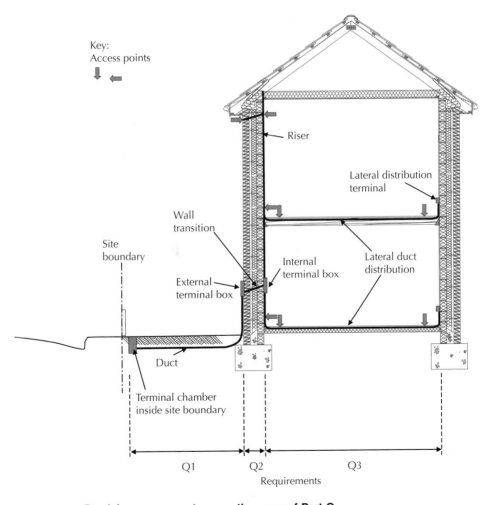

Figure 11.45   Provisions common to more than one of Part Q.

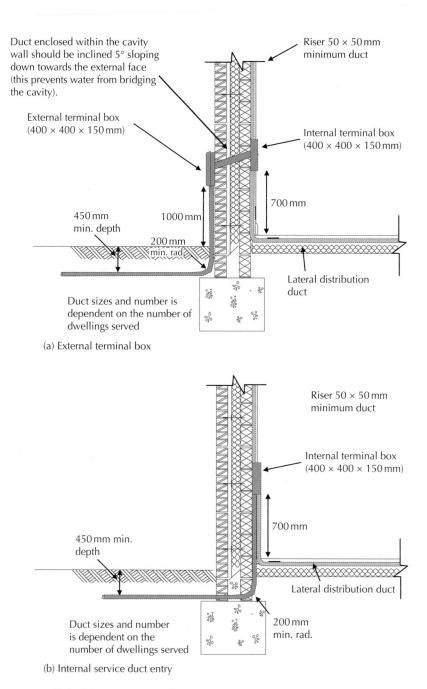

Duct enclosed within the cavity wall should be inclined 5° sloping down towards the external face (this prevents water from bridging the cavity).

Riser 50 × 50 mm minimum duct

External terminal box (400 × 400 × 150 mm)

Internal terminal box (400 × 400 × 150 mm)

700 mm

450 mm min. depth   1000 mm

200 mm min. rad.

Lateral distribution duct

Duct sizes and number is dependent on the number of dwellings served

(a) External terminal box

Riser 50 × 50 mm minimum duct

Internal terminal box (400 × 400 × 150 mm)

700 mm

450 mm min. depth

Lateral distribution duct

Duct sizes and number is dependent on the number of dwellings served

200 mm min. rad.

(b) Internal service duct entry

Figure 11.46 **Alternative arrangements for duct entry and distribution.**

foundations and floor construction). For new building work, it is preferable to install the terminal box on the inside face of the external wall. Where the duct passes under foundations, it should be protected.

External and internal terminal boxes

An external terminal box and the duct leading to it may look unsightly, so consideration should be given to positioning on the wall. The external box will need to be connected to an internal terminal box (as illustrated in Figure 11.46a) via a conduit or duct passing

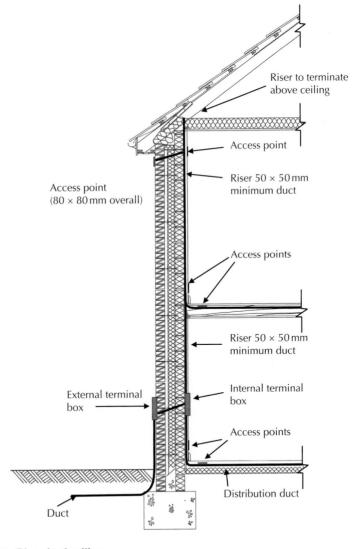

Figure 11.47   Riser in dwelling.

through the external wall construction. Installing an internal box (Figure 11.46b) is a neater alternative. The sizes recommended for the wall-mounted terminal boxes in Approved Document Q are quite large (e.g. 400 × 400 × 150 mm deep) and so some thought needs to be given to the position of the internal terminal box.

Distribution within the building

From the internal box will run the lateral distribution duct (20 × 40 mm). The duct should have access at a maximum spacing of 20 m. Access to vertical risers should be provided at each floor level. For dwelling houses, Approved Document Q recommends that the distribution duct should be positioned to allow access at each floor level and at least one habitable room at each floor level in dwellings. Ducting should be spatially separated from mains electricity cables. The riser, lateral distribution ducts and access points are illustrated in Figure 11.47. Lateral distribution ducts can be run within timber floor cavities, within raised floors and/or around the perimeter of the room (e.g. fixed to or behind a suitable skirting board). Lateral ducts should terminate with a flush-mounted box, the position of which should be so as to satisfy the provisions in Approved Document M. Additional guidance for flats and maisonettes can be found in Approved Document Q.

12 Water Supply and Sanitation

To ensure that good standards of hygiene are maintained in and around buildings, it is essential that care is taken when designing, installing and commissioning the services that will be used to supply cold and hot water and remove our waste (foul) water.

12.1 Cold water supply and distribution

Cold water supply

Mains water suppliers, the water undertakers, have an obligation to provide a constant, potable (drinkable) supply of water, for which service, either a water rate is levied or a charge by meter is made based on the volume of water used. All new premises and an increasing number of existing houses have a water meter fitted to measure water consumption. Water is supplied, under pressure, through pipes laid under streets, roads or pavements. In urban areas, duplicate trunk mains feed street mains. By closing valves, the individual lengths of mains may be isolated for repairs and renewals without interrupting the supply.

Functional requirements
The functions of water supply are to:

- ❏ Deliver clean water suitable for human consumption (Fluid Category 1)
- ❏ Provide a continuous supply of water at a constant pressure

Water categories and prevention of cross connection to unwholesome water
The quality of water is determined by five fluid categories, the highest Category 1 being the cleanest, 'wholesome' water supplied by a water undertaker and suitable for drinking, through to Fluid Category 5, which represents a serious health hazard.

All water fittings that convey unwholesome water, e.g. rainwater, recycled water or any fluid other than water supplied by a water undertaker, must be clearly identified so as to be easily distinguished from any supply pipe or distributing pipe. This is to avoid any contamination of the water supply, and is particularly important where systems are designed to use 'grey' water (e.g. for flushing toilets).

Barry's Introduction to Construction of Buildings, Third Edition. Stephen Emmitt and Christopher A. Gorse.
© 2014 John Wiley & Sons, Ltd. Published 2014 by John Wiley & Sons, Ltd.

Connections to the water main

Service pipe

The service pipe connection is made to the main and is run to a stop valve near to the site boundary of the building. Connection and maintenance is the responsibility of the water undertaker. A stop valve is situated either immediately outside or inside the boundary. The purpose of the stop valve is to enable the water undertaker to disconnect the water supply if there is a leak or for non-payment of the water bill. In situations where the supply is provided to two or more premises from the same supply or distributing pipe, then it should be fitted with a stop valve to which each occupier of those premises can have access.

Supply pipe

A supply pipe is any pipe, maintained by the consumer, which is subject to water pressure from the authority's mains. The supply pipe runs from the stop valve to, and into, the building; maintenance is the responsibility of the land/building owner. For convenience, it is usual to run the supply pipe into the building through drainpipes to facilitate renewal of the pipe if need be. At the point where the supply pipe enters the building, there should be a stop valve (Figure 12.1) to disconnect the supply for repair and maintenance purposes. To reduce the risk of freezing, the supply pipe should be laid at least 750 mm below the finished ground level. If the supply pipe enters the building and rises closer than 750 mm to the outside face of a wall, it should be thermally insulated from where it enters the building and up to the level of the ground floor.

Distributing pipe

A distributing pipe is any pipe (other than an overflow or flush pipe) that conveys water from a storage cistern or from hot water apparatus supplied from a feed cistern and under pressure from that cistern.

The Water Supply (Water Fittings) Regulations 1999

These Regulations apply to any water fitting installed and used in premises to which a water undertaker supplies water. The Act deals with the installation and enforcement of water supply and water fittings, providing guidance on materials and substances in contact with

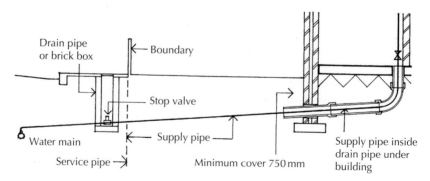

Figure 12.1 Service pipe and supply pipe.

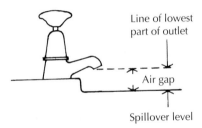

Figure 12.2   Air gap to taps to fittings.

water, workmanship, water system design and installation and provision against backflow (and hence potential contamination) of the water supply. The Act also sets out clear advice about the way in which water fittings are installed within a building, clearly stating that no water fitting shall be embedded in any wall or solid floor, thus allowing access for maintenance and repair.

The principal concern of the by-laws is the prevention of contamination of mains supplied water by the flow of potentially polluted water from a supply or distributing system back into the mains water supply. Backflow can occur when there is loss of pressure in the mains due to failure of pumps, or repair and maintenance on the mains, and also where a pumped supply in a building creates pressure greater than that in the mains. As a guard against backflow, all draw-off taps to baths, washbasins and sinks must either be fixed so that there is an air gap between the spillover level of the fittings and the outlet of the tap, as illustrated in Figure 12.2, or a doublecheck valve assembly must be fitted to the supply pipe to each draw-off tap. These requirements apply equally to taps connected to mains pressure supply pipes and to distributing pipes from a cistern. The Act also requires water supply for outside use to comply with provisions to save water and prevent contamination; this includes drinking water for animals and supply to ponds and pools.

Water system design and testing

The complete water system (all pipes and relevant fittings) must be designed to withstand an internal water pressure of not less than one and a half times the maximum pressure to which the installation is designed to be subject to when in operation. The Water Supply (Water Fittings) Regulations provide advice on the type of testing required, which differs for systems that include pipes made of plastics and those that do not. The whole system should be subjected to test pressure by pumping and subsequent maintenance of test pressure for a stipulated period (e.g. 1 hour in the case of systems that do not contain plastic pipes) to check for any visible leakage. After testing, every water system should be flushed and, where necessary, disinfected before it is first used.

Cold water supply systems

The two systems of cold water supply that are used are indirect, from a cold water storage cistern and direct, from the mains water pressure. The advantages and disadvantages of the systems are as follows.

Cistern feed – indirect supply

Advantages:

- ❏ The reserve of water in the cistern that may be called on against interruption of supply to provide regular flow
- ❏ The air gap between the supply pipe and the water level in the cistern acts as an effective barrier to backflow into the mains supply, which can cause contamination

Disadvantages:

- ❏ The considerable weight of a filled cistern has to be supported at high level
- ❏ The inconvenience of access to the cistern for inspection and maintenance
- ❏ The possibility of the cistern overflowing
- ❏ The need to insulate effectively the cistern and its associated pipework against freezing
- ❏ Animals and insects may enter the cistern and contaminate the water held in the open cistern

Mains pressure – direct supply

Advantage:

- ❏ A uniformly high-pressure supply to all hot and cold water outlets some distance below the pumped head of the supply

Disadvantages:

- ❏ Discontinuity of supply to all hot and cold water outlets if mains supply is interrupted
- ❏ The need for comparatively frequent inspection, maintenance and repair of many valves and controls to the system

Cistern feed cold water supply

A cistern is a liquid storage container that is open to the air and in which the liquid is at normal atmospheric pressure. Figure 12.3 illustrates the cold water supply to a two-storey house. The supply pipe rises through a stop valve and draw-off or drain tap to the ball valve that fills the cistern. A stop valve is fitted close to the cistern to shut off the supply for maintenance and repairs.

Small-capacity cisterns are used as header or feed tanks to provide the necessary head of cold water supply to boilers where an open-vented pipework system is used. The header or feed cistern or tank is fixed in a roof space or in some position above the boiler to provide the necessary head of water. An expansion pipe runs up from the boiler, to discharge over the header tank, so that water under pressure from overheating may discharge into the header tank. The majority of cisterns are manufactured in plastic materials, which do not rust and are easy to handle when empty (Figure 12.4). Many existing buildings still have

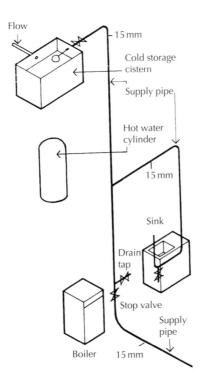

Figure 12.3 Cistern feed cold water supply.

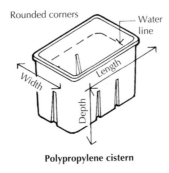

Polypropylene cistern

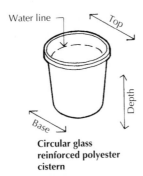

**Circular glass
reinforced polyester
cistern**

Figure 12.4 Plastic cisterns.

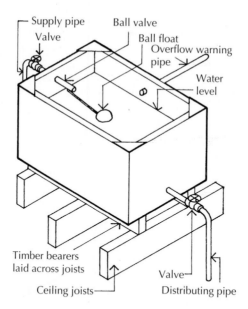

Figure 12.5 Support for cistern.

galvanised steel cisterns, which are prone to rust, and which should be checked regularly and replaced where necessary.

The most convenient place for a cold water storage cistern is in a roof space below a pitched roof, or in a tank room on a flat roof or at some high level. In whichever position the cistern is fixed, there must be space around and over the cistern for maintenance or replacement of the ball valve. The cold water storage cistern must be fixed not less than 2 m above the highest fitting it is to supply with water. This distance represents the least head of water necessary to provide an adequate flow.

Because of the weight of water they are designed to contain, all cisterns must be adequately supported on timber or steel spreaders to transfer the load over a sufficient area of timber or concrete ceilings and roofs, as illustrated in Figure 12.5.

The capacity of cold water storage cisterns is usually 114 L per dwelling where the cistern supplies cold water outlets and water closet (WC) cisterns, and 227 L per dwelling where the cistern supplies both cold water to outlets and domestic hot water cylinders. The alternative is to calculate the capacity of the storage cistern at 90 L per resident. Cisterns are holed for the connection of the cold water supply, the distributing pipes to cold water outlets and for the pipe to a hot water cylinder and the overflow pipe.

Ball valve

The water supply to the cistern is controlled by a ball valve that is fixed to the cistern, above the water line, and connected to the cold water supply through a valve as illustrated in Figure 12.5. A hollow copper or plastic ball, fixed to an arm, floats on the water in the cistern; as water is drawn from the cistern, the ball falls and the arm activates the ball valve

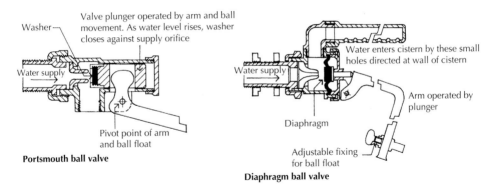

Figure 12.6   Ball valves.

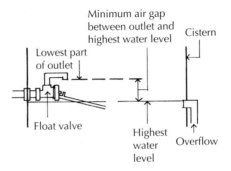

Figure 12.7   Air gap to ball valve.

that opens to let water into the cistern. When the water has risen to the water line marked on the cistern wall, the ball and arm rise to close the valve. The two types of valve in use are the Portsmouth valve and, for most new installations, the diaphragm valve, see Figure 12.6.

An air gap between the outlet of float valves to cisterns and the highest water level in the cistern is required to prevent backflow. This air gap is related to the bore of the supply pipe or the outlet and is taken as the minimum distance between the outlet of the float valve and the highest water level when the overflow pipe is passing the maximum rate of inflow to the cistern (Figure 12.7).

Overflow warning pipe

As a precaution against failure of the valve and consequent overflow of the cistern, most water undertakers require an overflow warning pipe to be connected to the cistern above the water line and carried out of the building to discharge where the overflow of water will give obvious warning. The overflow pipe should be of larger bore than the supply pipe to the cistern and preferably twice the bore of the supply pipe and not less than 19 mm bore.

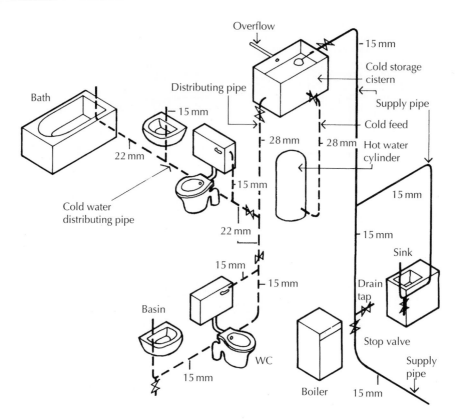

Figure 12.8   Cold water distributing pipe system.

Cistern feed cold water distributing pipe system

Figure 12.8 illustrates the cold water distributing pipe system for a two-storey house. The distributing pipe is connected to the cistern some 50 mm above the bottom of the cistern to prevent any sediment that may have collected from entering the pipe. A stop valve is fitted to the pipe adjacent to the cistern, isolating the whole system from the cistern in the event of repairs and renewals. The distributing pipe is carried down inside the building with horizontal branches to the first-floor and ground-floor fittings, as shown in Figure 12.8. The aim in the layout of the pipework is economy in the length of pipe runs, and on this depends a sensible layout of sanitary fittings. In Figure 12.8, it will be seen that one horizontal branch serves bath, basin and WC. For rewashering taps, stop valves are fitted to branches as shown. Where one branch serves three fittings as shown in Figure 12.8, one stop valve will serve to isolate all three fittings. Drain or draw-off taps should be provided where pipework cannot be drained to taps so that the whole distributing system may be drained for renewal or repair.

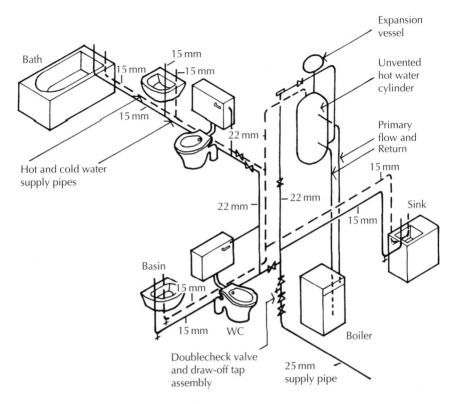

Figure 12.9 Cold and hot water supply.

Mains pressure cold water supply

Figure 12.9 illustrates the cold water, mains pressure pipe system to a two-storey house. Here, the first line of defence against contamination of the mains supply is a doublecheck valve. A doublecheck valve assembly is a combination of two non-return (check) valves with a test cock between them.

A non-return valve is a simple, spring-loaded valve that is designed to open one way to the pressure of water against the valve and to close when that pressure is reduced or stops (Figure 12.10). It prevents backflow from the supply pipe system into the mains.

A test cock is a simple shut-off device with a solid plug that, when rotated through 90°, either opens or closes. In use, a check valve may become coated with sediment, particularly in hard water areas, and not operate as it should. To test that the check valves are working and acting as non-return valves, the stop valve is closed to test one check valve and opened to test the other with the test cock open.

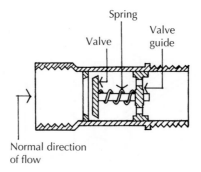

Figure 12.10   **Spring-loaded check valve, non-return valve.**

Recycled water systems

Up to 40% of household water consumption is, quite literally, flushed down the toilet. This is an unnecessary waste of treated water intended for drinking purposes. We also receive a great amount of 'free' water that falls as rain on our roofs throughout the year, and this too we throw away to soakaways or the main drainage system. The philosophy and practice of conserving water and recycling rainwater (known as 'rainwater harvesting') is well established in some countries, but has been slow to establish itself in the UK. Hot summers and water shortages are rare, and treated water is relatively cheap. There is, however, growing pressure on the mains water supply as the population of the UK grows, lifestyles change and demand increases. Add to this a growing awareness of environmental issues and the practicalities of recycling water start to become a practical alternative.

Recycled water systems are designed to use recycled water, primarily rainwater, for tasks such as flushing the toilet and irrigating the garden. The term 'grey water' is also used to describe the reuse of water from domestic appliances such as washing machines, and these systems are usually combined with recycled rainwater. In a recycled water system, the rainwater is collected from the roof in copper gutters and discharged to polypropylene containers, where it is stored. This water is then used to flush toilets and for other non-consumptive tasks such as bathing and irrigating the garden. Some systems are designed to also filter and purify the rainwater to provide drinking water (although this tends to be a relatively small amount of the total recycled water usage, approximately 5–10%).

To avoid contamination with domestic mains supply, the recycled water system is designed as a separate system to the main cold water distribution pipework. Pipework should be identified accordingly. There will be some duplication of pipework and, depending on how the recycled water is stored, some need for additional pumps. However, the initial capital cost of the system may be recovered quite quickly given the savings in mains water consumption.

Ideally the rainwater should be stored at roof level. In this approach, the toilet flushing system and other appropriate outlets are fed by gravity. Where the rainwater is stored at a lower level, e.g. in a basement, then the recycled water supply will need to be pumped to the outlets. In a pitched roof design, there may be enough space for water storage vessels to be placed within unused roof space. Care should be taken to ensure that the additional loading from the water can be accommodated, i.e. the structure should be designed accordingly. The storage vessels should also be protected from frost and inspected on a regular basis. Filters are required to stop leaves and associated debris from getting into the water storage vessels. These filters require regular maintenance and cleaning to avoid blockages.

The storage system must be designed with an overflow system, to prevent flooding once the storage vessels become full, e.g. in a particularly wet period. Normally the overflow will run to a water storage system for irrigating the garden. Alternatively the overflow may run to the mains drainage system or a soakaway. Similarly, the recycled water system will need to be connected to a mains water feed to provide back-up water in particularly dry periods. The usual precautions are required to prevent contamination of the mains supply.

Personnel with experience of these systems must carry out the design of an installation of recycled water systems. Building Regulation approval will be required for installation in new and existing properties.

12.2 Hot water supply and distribution

There are two hot water supply systems, the local and the central (Figure 12.11). In some buildings, it may be economic to use a combination of central and local hot water systems.

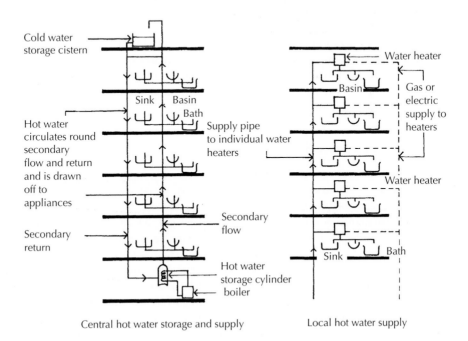

Central hot water storage and supply   Local hot water supply

Figure 12.11   Hot water supply systems.

Local hot water supply

A water heater, adjacent to the fittings to be supplied, is fired by gas or electricity. The water is either heated and stored locally or heated instantaneously as it flows through the heater. The advantages of this system are that there is a minimum of distributing pipework, initial outlay is comparatively low, and the control and payment for fuel can be local, an advantage, for example, to the landlord of residential flats. The disadvantage is that local heaters are more expensive to run and maintain than one central system.

Central hot water supply

Water is heated and stored in a central cylinder from which a pump circulates it around a distributing pipe system from which hot water is drawn. In the two-storey house used to illustrate hot water distribution or supply pipe systems, the hot water was drawn directly from single branches. In a small building, such as a house, where the sanitary fittings are compactly sited close to the cylinder, the slight inconvenience of running off the cooled water in the single branches before hot water is discharged is acceptable. In larger buildings, the inconvenience of running off cooled water from long single branches (known as 'dead legs') is a waste of water and energy. The length of these dead legs is limited to 20 m for 12 mm pipes ranging to 3 m for pipes more than 28 mm, unless the pipes are adequately insulated against loss of heat.

The storage cylinder contains hot water sufficient for both anticipated peak demand and demands during the recharge period. The system is therefore designed to supply hot water on demand at all times. The one disadvantage of the system is that there is some loss of heat from the distributing pipes no matter how adequately they are insulated. This is out-weighed by the economy and convenience of one central heat source that can be fired by the cheapest fuel available, and one hot water source to install, supply and maintain hot water being at hand constantly by simply turning a tap. Where a mains pressure supply system is used, the supply pipe connects to the unvented hot water storage cylinder from which supply pipes connect to the fittings, and there is no roof-level storage system.

Instantaneous water heaters

These water heaters operate by running cold water around a heat exchanger so that the water is heated as it flows. The heat exchanger only operates when water is flowing, hence the name instantaneous water heater.

Most instantaneous water heaters are fired by gas, which is ignited by a pilot light; immediately water flows to provide hot water instantaneously. Cold water running through a coil of pipework, wrapped around a combustion chamber and heat exchanger over a gas burner, is heated by the time it reaches the outlet. The cold water supply valve controls these heaters. When the valve is opened, the flow of water opens a gas valve to ignite the burners to heat water. A single-point, gas, instantaneous water heater is designed to supply hot water to single fittings such as a basin or sink. These heaters are usually fixed above the fitting to be supplied, and the hot water is delivered through a swivel outlet. Because of their small output and limited use, the air intake from the room and the exhaust outlet to the room is acceptable. Where effective draught seals are fitted to windows to rooms in which these heaters are fired, there should be permanent ventilation to the open air.

Multi-point instantaneous gas water heater

These heaters can supply hot water to a sink, basin and bath through dead-leg draw-off pipes to fittings, hence the name multi-point. When a tap over one of the fittings is opened, the flow of cold water through the coils of pipe around the heat exchanger is heated to deliver hot water. The initial flow of cold water opens a gas valve and the pilot light ignites the gas burners to provide heat. When all the taps to fittings are shut and there is no water flow, the gas valve shuts.

The rate of flow of hot water from these heaters is limited by the need for sufficient time to allow an adequate exchange of heat to the water coils in the heat exchanger. When more than one tap is opened, there will be a restricted rate of flow of hot water, and filling a bath can be a somewhat lengthy process. For these reasons these heaters are much less used than they used to be. The comparatively large output from these heaters necessitates a flue to open air to exhaust combustion gases, and also an adequate intake of air for efficient and safe combustion of gases. A flue and permanent air vents or a balanced flue are necessary. The gas valves in these heaters will only operate when there is comparatively high water pressure, such as that from a main supply, or a good head of water from a cistern.

Instantaneous electric water heater

Water is heated as it flows through coiled heating elements immersed in a compact, sealed tank. A flow switch in the cold water supply inlet operates the electric supply. The control valve on the cold water supply pipe is used to adjust the temperature of the hot water outlet. These heaters are fixed over the single appliance to be supplied, with hot water delivered through a swivel outlet discharging over the appliance. The output of hot water from these heaters is limited by the rate of exchange of heat from the heating elements to cold water. In hard water areas, limescale will coat the heating element and appreciably reduce the efficiency of these heaters. The heat exchange tank, which is heavily insulated, is housed in a glazed enamel casing for appearance sake. The advantage of these heaters is that they are compact, require only one visible supply pipe and may be fixed in internal, unventilated toilets as they have no need for air intake or a flue.

Hot water storage heaters

The larger water storage heaters are used to supply hot water to ranges of fittings such as basins, showers and baths used in communal changing rooms of sports pavilions and washrooms of students' hostels.

Large gas-fired water storage heaters consists of a water storage cylinder through which a heat exchanger rises to a flue from a combustion chamber. A thermostat in the water storage chamber controls the operation of the gas burners in the combustion chamber, cutting in to fire when the temperature of the water falls. A cold water supply pipe is connected to the base of the storage cylinder. Hot water is drawn off either through a dead-leg draw-off pipe where pipe runs are short, or by a circulating secondary pipe system where runs are long. The storage heater is heavily insulated to conserve energy. The size of the heater is determined by the anticipated use of hot water at times of peak use.

Electric water storage heaters may be used in communal washrooms of students' hostels and residential schools where peak demand for hot water in bulk is generally confined to mornings and evenings, between which times the heater automatically reheats the water. These heaters, which are heavily insulated to conserve energy, are housed in a separate enclosure away from the wet activities they serve, for safety reasons. Hot water for basins is drawn from the top of the cylinder, which is heated by an upper immersion heater. Hot water in bulk for baths is boosted by the operation of both the upper and lower immersion heaters. A thermostat and/or a timed switch operate the lower immersion heater. An advantage of the electric storage heater is that it does not have to be fixed close to an outside wall, which the gas heater does because of its flue.

Small, electric, single-point water heaters are designed to heat and store a small volume of water for the supply to single basins. The hot water storage cylinder and electric immersion heater, which are heavily insulated to conserve energy, are housed in a glazed enamel metal casing for appearance sake. A thermostat controls the electric supply to the immersion heater, cutting in to reheat the water as it is drawn off. These heaters are used for basins in single toilets where it is convenient to run water and an electrical supply for the occasional use of hot water.

Mains pressure hot water supply

The hot water supply system for a small two-storey house is illustrated in Figure 12.12. From the doublecheck valve assembly, the supply pipe rises to fill the unvented hot water cylinder with cold water. The cold water in the cylinder is heated by a heat exchanger that exchanges heat from the boiler.

Unvented hot water cylinder

The difference between the traditional vented hot water system and the unvented system is that in the vented system the expansion of hot water is accommodated by the vented expansion pipe that will discharge an excess of expansion water to the cistern, and in the unvented system, expansion of hot water is relieved by an expansion vessel which contains a cushion of gas or air sufficient to take up the expansion by compression of the gas. The advantages of the unvented over the vented system lie in the improved flow rates from showers and taps, reduction in noise caused by the filling of storage cisterns and virtually no risk of frost damage. The saving in eliminating the cistern, feed and expansion pipes is offset by the additional cost of the expansion vessel and temperature and expansion control valves, and the necessary, comparatively frequent, maintenance of these controls.

Figure 12.13 illustrates a low-pressure, unvented hot water cylinder. The pressure-reducing valve is fitted to the feed pipe where low-pressure systems are used and is provided to reduce mains pressure to a level that the cylinder can safely withstand. Where high-pressure systems are used and the cylinder is designed to stand high pressure, the pressure relief valve is omitted.

Expansion vessel

An expansion vessel is a sealed container in which a flexible diaphragm separates water from the cylinder and the air or gas, which the sealed vessel contains. As the water in the

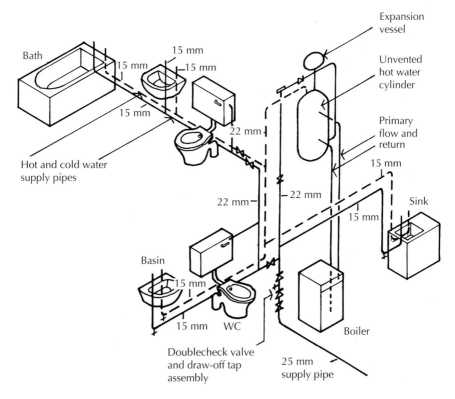

Figure 12.12   Unvented hot water supply.

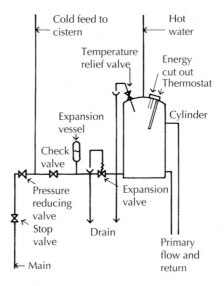

Figure 12.13   Diagram of low-pressure unvented hot water system.

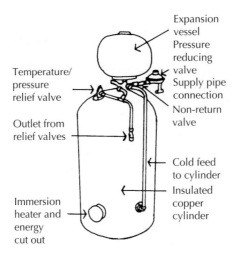

Temperature/
pressure
relief valve

Outlet from
relief valves

Immersion
heater and
energy
cut out

Expansion
vessel
Pressure
reducing
valve
Supply pipe
connection
Non-return
valve

Cold feed
to cylinder

Insulated
copper
cylinder

Figure 12.14   Unvented hot water storage cylinder.

cylinder heats, it will expand against the diaphragm, compress the air and gas and so relieve the water expansion. In this way, the expansion vessel acts in the same way as a vented expansion pipe to indirect systems. The pressure relief, or reducing valve, is a safety device against the expansion vessel being unable to take up the whole of the expansion of heated water. For control against overheating, there is a thermostat on the immersion heater and also a temperature-limiting cut-out that operates on the electricity supply, and a temperature-operated relief valve to discharge if the other controls fail. Figure 12.14 is an illustration of an unvented copper indirect cylinder prepared ready for installation as a packaged unit complete with factory applied insulation, expansion vessel and the necessary valves and controls.

Hot water supply pipework
The hot water supply pipework is run from the top of the cylinder with horizontal branches to the hot water outlets to fittings on each floor. Stop valves and draw-off taps are fitted to control and drain the pipe system as necessary for maintenance and repair.

Cistern feed hot water supply

Figure 12.15 illustrates the hot water distributing pipe system for a two-storey house. The hot water is drawn from a cylinder, which is fed by cold water drawn from the cold water storage cistern in the roof. The cold water in the hot water cylinder is heated by a heat exchanger in the cylinder through which hot water circulates from the boiler. The cold water feed to the cylinder is run through a stop valve to the bottom of the cylinder. In accordance with the Water Supply (Water Fittings) Regulations, hot water systems should be fitted with appropriate vent pipes, temperature control devices and relief valves.

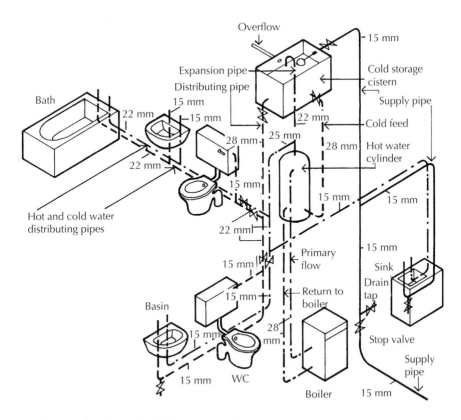

Figure 12.15   Hot water distributing pipe system.

Hot water storage cylinder

Hot water storage cylinders are designed to contain water under pressure of the head of water from the cold water storage cistern. Most hot water storage containers are cylindrical and are fixed vertically to encourage cold water fed into the lower part of the cylinder to rise, as it is heated by the heat exchanger, to the top of the cylinder from which hot water is drawn, and so minimise the mixing of cold and hot water. The cold feed pipe to the cylinder is run from the cold water storage cistern and connected through an isolating stop valve to the base of the cylinder. The hot water distribution pipe is run from the top of the cylinder to the draw-off branches to sanitary appliances and is carried up to discharge over the cold cistern as an expansion pipe, in case of overheating, as illustrated in Figure 12.15.

Vented hot water system

Because the expansion pipe discharges through the open end of the pipe from the cold water storage cistern (if the hot water expands due to the system overheating), this arrangement is described as a vented hot water system and the cylinder as a vented or open-vented system. This is to distinguish it from the unvented system used with the mains pressure system. The required storage capacity of the cylinder depends on the number of sanitary

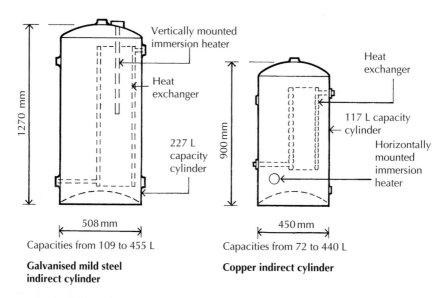

Figure 12.16 Indirect hot water storage cylinders.

appliances to be served and the estimated demand. The interval between times of maximum demand on domestic hot water is longer than the recovery period required to reheat water in storage systems, and it is reasonable, therefore, to provide hot water storage capacity of 50–60 L per person. Storage cylinders are made either of galvanised sheet steel or sheet copper.

Indirect cylinder

The hot water storage cylinders illustrated in Figure 12.16 are indirect cylinders, so called because the primary hot water from the boiler exchanges its heat indirectly through a heat exchanger to the hot water supply, there being no connection between the water from the boiler and the hot water supply. The purpose of this indirect transfer of heat is to avoid drawing hot water directly from the water system of the boiler. Where hot water is drawn directly from a boiler, it has to be replaced and in hard water areas, each fresh charge of water will deposit scale inside the boiler and its pipework and in time the build-up of scale will reduce the efficiency of the boiler and the bore to its pipes. With an indirect cylinder, there is no replacement of water to the boiler and its primary pipes and therefore no build-up of scale. Scale formation is proportional to water temperature. There is less build-up of scale in the secondary hot water circulation because of the lower water temperature in the system. Indirect cylinders are also a protection against the possibility of drawing scalding water directly from the boiler.

Heat exchanger

The heat exchanger, which is fixed inside the cylinder and immersed in the water to be heated, takes the form of a coil of pipes or annulus designed to provide the maximum surface area for heat exchange.

Primary flow

The system of pipes that carries hot water from the boiler through the heat exchanger and back to the boiler is described as a primary flow system.

Hot water boiler or heater

For a small building, such as the two-storey house, it is general practice to use one boiler for both space heating and hot water, to economise in the initial outlay on heating equipment and pipework and to make maximum use of space. The temperature of the water heated by the boiler is controlled by a thermostat, which can be set to water temperatures in the range 55–85°C, the boiler firing and cutting out as the water temperature falls and then rises to the selected temperature. Water heated in the boiler rises in the primary flow pipe to the heat exchange coil or container inside the hot water storage cylinder and, as it exchanges its heat through the exchanger to the water in the cylinder, it cools and returns through the primary return pipe back to the boiler for reheating. There is a gravity circulation of water in the primary pipe system. The primary flow and return pipes should be as short as practicable, i.e. the cylinder should be near the boiler to minimise loss of heat from the pipes. The circulating pipes from the boiler through the heat exchanger are termed primary flow and return as they convey the primary source of heat in the hot water system. Small, compact, boilers require a forced flow of water around the primary circulation system by an electrically operated pump. A small feed cistern (not shown in Figure 12.15) provides the head of water required for the boiler and its pipe system.

Immersion heater

An electric immersion heater is fitted to the hot water storage cylinder to provide hot water when the boiler is not used for space heating. An electric immersion heater is an inefficient and relatively expensive means of heating water; however, when powered by photovoltaic (PV) cells it becomes a more feasible option.

12.3 Water services to multi-storey buildings

Mains water is supplied under pressure from the head of water from a reservoir, or a pumped head of water, or a combination of both. The level to which mains water will rise in a building depends on the level of the building relative to that of the reservoir from which the mains water is drawn, or relative to the artificial head of water created by pumps. In built-up areas there will, at times such as early morning, be a peak demand on the mains supply, resulting in reduced pressure available from the water main. It is the pressure available at peak demand times that will determine whether or not mains supply pressure is sufficient to feed cold water to the higher outlets. Water pressure varies from place to place depending on natural or artificial water pressure, intensity of demand on the main at peak demand time and the relative level of the building to the available supply pressure.

Mains pressure supply

Figure 12.17 is a diagram of an eight-storey building where the mains pressure at peak demand times is sufficient to supply all cold and hot water outlets to all floors. Two supply

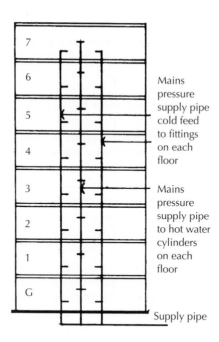

Figure 12.17   Mains pressure supply.

pipe risers branch to provide cold water to ranges of sanitary fittings in male and female toilets on each floor, and another riser branches to feed hot water storage cylinders on each floor for the toilets. This is the most economic arrangement of pipework where sanitary fittings are grouped on each floor, one above the other. To provide reasonable equality of flow from outlets on each floor, it is usual to reduce pipe sizes. In the arrangement shown in Figure 12.17, the bore of the risers will be gradually reduced down the height of the building to provide a reasonable flow from all outlets to compensate for reduced flows as pumped head pressure increases down the height of the building. Where the mains pressure is sufficient to provide a supply to a multi-storey building, it is not always possible to provide a reasonable equality of flow to fittings on each floor by varying pipe sizes alone. An increase in pipe size will provide a little reduction in pressure loss from the frictional resistance to flow of larger bore pipes and fittings. There is a limit to the resistance to flow that can be effected, because of the limited range of pipe sizes, without using uneconomic pipe sizes. Another method of providing equalisation of rate of flow from outlets floor by floor in multi-storey buildings is by the use of pressure-reducing valves at each flow level.

Multiple risers

Another method of equalising flow, being used experimentally, is the use of multiple rising pipes as illustrated in Figure 12.18. One rising supply pipe branches to supply the three lower floors and then rises to supply the three floors above, where it is joined and its supply reinforced by the second rising supply and then up to the top three floors where it is joined

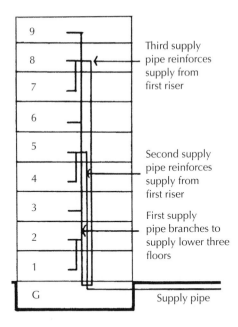

Figure 12.18 Multiple risers.

and reinforced by the third rising supply pipe. The logic in the use of this arrangement is that in multi-storey buildings in multiple occupations, such as a hotel, there will be unpredictable short periods of peak use. During this period, there may be heavy call on water use on one floor, which may cause unacceptable starvation of water supply. By providing alternative reinforcing sources of supply and judicious arrangements of pipe sizes, a reduction of flow rate may be avoided or at least smoothed out.

It is good practice in the design of pipework layout to make an assumption of frequency of use of draw-off water to sanitary appliances. The assumption is based on an estimate of peak period use, which does not allow for unpredictable heavy use. To provide for possible maximum use would involve uneconomic pipework. Where mains pressure is insufficient at peak demand times, it is necessary to install pumps in the building to raise water to the higher water outlets. In this situation, it is usual to supply from the supply pipe those outlets that the mains pressure will reach, and those above by the pumped supply, to limit the load on the pumps as illustrated in Figure 12.19.

Two mains pressure supply pipes rise to supply the lower five floors, one with branches to each floor level to cold water outlets and the other with branches at each floor level to hot water storage cylinders. The five upper floors are supplied by two rising supply pipes under the pressure of the pump in the basement. At each floor level, branches supply cold water to sanitary fittings and there are branches to hot water cylinders. There are two pumps, one operating and the other as standby in case of failure and to operate during maintenance. The pumps are supplied by the mains through a doublecheck valve assembly to prevent contamination of the supply by backflow should the pumps fail.

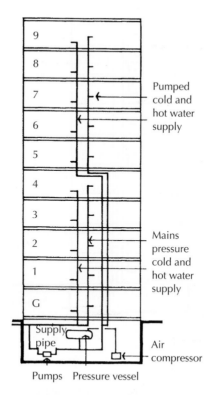

Figure 12.19   **Pumped and mains pressure supply.**

Auto-pneumatic pressure vessel

The auto-pneumatic pressure vessel indicated in Figure 12.19 and Figure 12.20 is a sealed cylinder in which air in the upper part of the cylinder is under pressure from the water pumped into the lower part of the cylinder. The cushion of air under pressure serves to force water up the supply pipe to feed upper-level outlets as illustrated. Water is drawn from the auto-pneumatic pressure vessels as water is drawn from the upper-level outlets so that when the water level in the pressure vessel falls to a predetermined level, the float switch operates the pump to recharge the pressure vessel with water. Thus the cushion of air in the pressure vessel and its float switch control and limit the number of pump operations. In time, air inside the pressure vessel becomes mixed with water and is replaced automatically by the air compressor.

Gravity feed cold water supply

In multi-storey buildings, where mains pressure is insufficient to raise water to roof level, cistern feed to cold water outlets may be used with a covered drinking water storage vessel or cistern. Figure 12.21 is a diagram of a 10-storey building in which the drinking water outlets to the lower five floors are fed by the mains supply. Because the mains pressure is

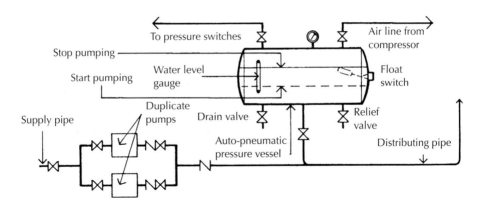

Figure 12.20   Auto-pneumatic pressure vessel.

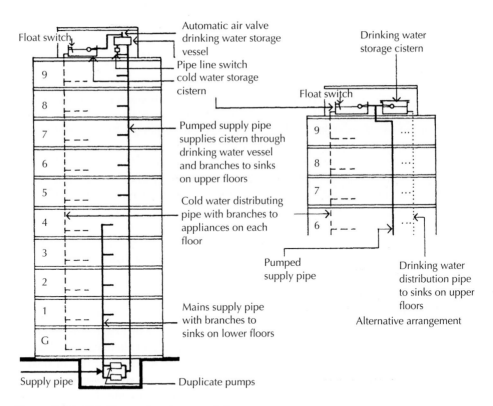

Figure 12.21   Drinking water storage cistern.

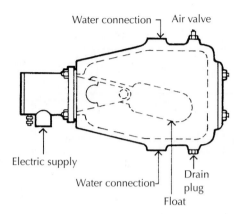

Figure 12.22   Pipeline switch.

insufficient to raise water to the roof level, the upper five floors are supplied with drinking water from a covered drinking water storage vessel or cistern. The roof-level cold water storage cistern, the drinking water vessel or cistern and the drinking water outlets to the top five floors are fed by a pumped supply. The cold water storage cistern supplies cold water to all cold water outlets, other than the sink, and to hot water cylinders on each floor.

There is a duplication of rising pipework to the five lower floors, which is considered a worthwhile outlay in reducing the load and wear on the pumps. The pumped supply feeds both the higher drinking water outlets and cold water storage cistern through a drinking water storage vessel. The sealed drinking water storage vessel and the cold water storage cistern are filled through the pumped supply, in which a pipeline switch is fitted. A pipeline switch, illustrated in Figure 12.22, is used to limit pump operations. As water is drawn from the drinking water vessel, the water level falls until the float in the pipeline switch falls and starts the pump. The cold water storage cistern is supplied through the drinking water vessel, from which it can draw water to limit pump operations. When the water level in the cold water storage cistern falls to a predetermined level, a float switch starts the pump to refill the cistern through the drinking water vessel.

Drinking water storage cistern

As an alternative to a sealed drinking water storage vessel, a drinking water storage cistern may be used (Figure 12.21). The cistern has a sealed cover and filtered air vent and overflow to exclude dirt and dust, as illustrated in Figure 12.23. A pump cut-out, or float switch, controls the pump operations at a predetermined level. A float switch controls pump operations to fill the cold water storage cistern. The pumped service pipe feeds both cisterns so that whichever switch operates, the pump operates to fill both cisterns. The advantage of the sealed drinking water storage vessel is that it requires less maintenance than the drinking water cistern whose ball valve and filters require periodic maintenance. But the roof-level switches have to be wired through a control box down to the basement-level pumps, and the switches will require regular maintenance in a position difficult to access. As a

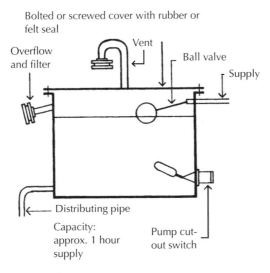

Figure 12.23   Drinking water storage cistern.

check to the possibility of backflow from a pumped supply into the main, and as a reservoir against interruption of the mains supply, it has been a practice to use a low-level cistern as feed to the pumped supply to a roof-level cistern, so that the air gap between the inlet and the water level in the cistern acts as a check to possible contamination of the mains supply.

Low-level cistern

Where the mains pressure is insufficient to supply cold water outlets on upper floors of multi-storey buildings, and a separate drinking water supply is required, a cistern feed supply to cold water outlets may be used in combination with a pumped supply to upper-level drinking water outlets, as illustrated in Figure 12.24. Here a low-level cistern is fitted as supply to the pumps. A low-level cistern is used as a form of check against contamination of the supply and as a standby against interruption of the mains supply. The covered low-level cistern serves to supply the upper-level drinking water outlets and the roof-level storage cistern. The operation of the pumps is controlled at low level through a pressure vessel similar to that illustrated in Figure 12.20.

Drinking water outlets to the lower floors are connected to the mains supply pipe, which in turn feeds a low-level storage cistern from which a supply is pumped to the upper-level drinking water outlets and the roof-level storage cistern. An auto-pneumatic pressure vessel, illustrated in Figure 12.20, and a delayed action ball valve to the roof-level cistern limit pump operations. As the low-level cistern supplies both drinking and cold water outlets, it has to be sealed to maintain the purity of the drinking supply. A screened air inlet maintains the cistern at atmospheric pressure. The pumped supply pipe shown in Figure 12.24 feeds the roof-level cold water cistern, which is fitted with a delayed action ball valve.

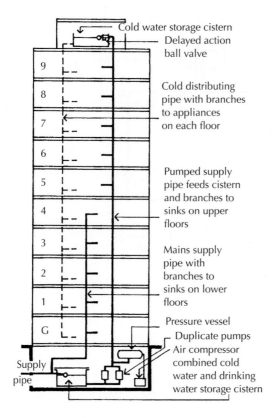

Figure 12.24   Low-level storage cistern.

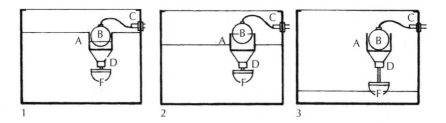

Figure 12.25   Delayed action ball valve.

Delayed action ball valve

A delayed action ball valve, illustrated in Figure 12.25, is fitted to the roof-level cistern to control and reduce pump operations. (1) The delayed action ball valve consists of a metal cylinder (A) that fills with water when the cistern is full and in which a ball (B) floats to operate the valve (C) to shut off the supply. (2) Water is drawn from the cistern and, as the

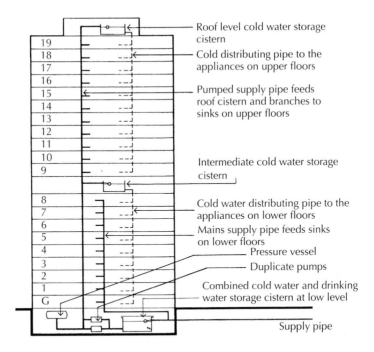

Figure 12.26   Intermediate-level storage cistern.

water level falls, the float (F) falls and opens the valve (D) to discharge the water from the cylinder (A). The ball (B) falls (3) and opens the valve (C) to refill the cistern through the pump, so limiting the number of pump operations.

Intermediate water storage cistern

Current practice in multi-storey buildings of more than about 10 storeys is to use a roof-level and one or more intermediate water storage cisterns to supply outlets other than drinking water taps. The intermediate cisterns spread the very considerable load of water storage and also serve to reduce the pressure in distributing pipes, for which reason they are sometimes termed 'break pressure cisterns'. Figure 12.26 shows a ground-level storage cistern supplying a pumped supply to an intermediate and roof-level cistern from which distributing pipes supply sanitary appliances on the lower and upper floors respectively, thus spreading the weight of water storage and limiting pressure in distributing pipes. The pumped supply also feeds drinking water outlets to upper floors. A float switch in the pressure vessel and delayed action ball valves in the cisterns limit pump operations. Intermediate cisterns are used at about every tenth floor.

Zoned supply system

The supply to the 22-storey building illustrated in Figure 12.27 is divided into three zones in order to help equalise pressure and uniformity of flow from outlets on all floors. The

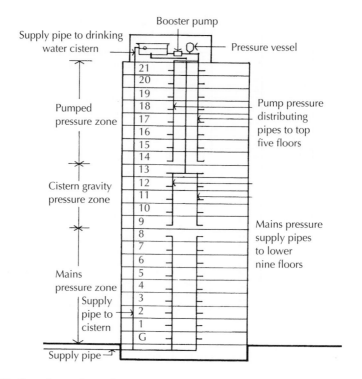

Figure 12.27 Zoned supply system.

supply to the lower nine floors is through two rising supply pipes taken directly from the mains supply. The supply to the eight top floors is from a pumped supply through a roof-level pump and pressure vessel, and the supply to the intermediate five floors is by gravity from a sealed drinking water cistern at roof level. In this way the loss of residual head to the lower and upper floors is limited, and the loss of head from the cistern is limited by feeding intermediate floors only. By dividing the building into three zones of supply, loss of head is limited in the main supply and distribution pipes, and by reduction in pipe diameter in each zone, reasonable equalisation of flow from taps on each floor is possible.

12.4 Pipes (tubulars) for water supply

Insulation

Cold and hot water supply and distributing pipes should be fixed inside buildings, preferably away from external walls to avoid the possibility of water freezing, expanding and rupturing pipes and joints. Where pipes are run inside rooftop tank rooms and in unheated roof spaces, they should be insulated. Similarly, roof-level and roof-space storage cisterns should be fitted with an insulation lining to all sides, the base and to the top of the cisterns.

The Water Supply Regulations prohibit the use of lead for either new or replacement pipework for water supply. The Regulations also prohibit the use of lead solders for joining copper pipes, and so tin/silver solders are used. Lead pipework may be found when working on existing buildings and should be replaced with a suitable material. The materials used for pipework for water supplies are copper, galvanised mild steel and plastic.

Copper pipe (copper tubular)

The comparatively high strength of copper facilitates the use of thin-walled light-gauge pipes or tubes for most hot and cold water services. The ductility of copper facilitates cold bending, the thin walls make for a lightweight material and the smooth surface of the pipe provides low resistance to the flow of water. Light gauge copper tube, size 6–159 mm (outside diameter) is manufactured; the sizes most used in buildings are 12, 15, 18, 22, 28, 35, 42 and 54 mm.

Jointing

Copper pipes are joined with capillary or compression joints and welding, for the larger bore pipe for drains above the ground where the pipework is prefabricated in the plumber's shop.

Capillary joint
A capillary joint is made by fitting plain ends of a pipe into a shouldered brass socket. Molten solder is then run into the joint, or internal solder is melted by application of heat. Pipe ends and fittings must be clean; otherwise, the solder will not adhere firmly to the pipe and socket. Figure 12.28 illustrates typical capillary joints. This is a compact, neat joint. Capillary joints are used for most small bore pipework, particularly where there are a lot of joints.

Compression joint
Compression fittings are either non-manipulative or manipulative. With the former, plain pipe ends are gripped by pressure from shaped copper rings; in the latter, the ends of the

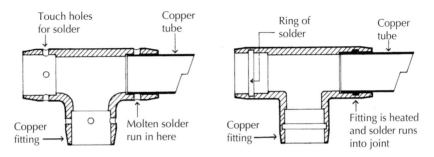

Figure 12.28 Soft solder capillary fittings.

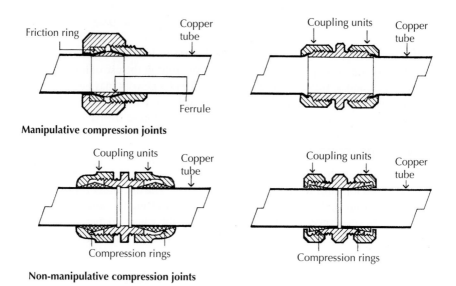

Figure 12.29   Compression joints.

pipes are shaped to the fitting, as illustrated in Figure 12.29. The manipulative fitting is used for long pipe runs and where pipework is not readily accessible, as the shaped pipe ends are more firmly secured than with the non-manipulative or capillary joint. Compression joints are often combined with capillary joints at the ends of pipe runs to facilitate repairs or alterations, because of the ease of disconnecting these joints. Thin-walled copper tubulars used for water services have poor mechanical strength in supporting the weight of a filled pipe. The pipework should, therefore, be supported at comparatively close intervals by pipe clips screwed to plugs in walls.

Galvanised mild steel (low carbon steel)

Galvanised mild steel tubulars are used for service, supply and distributing pipework, 40 mm bore and over, particularly where there are long straight runs of pipework, because of the economy of material and labour. For water services, tubulars should be galvanised to resist corrosion.

Plastic pipe or tube

Polythene and unplasticised polyvinyl chloride (uPVC) tube is used respectively for use underground and above the ground for water services. Polythene (polyethylene) is flexible and used for water services underground, while the more rigid polyvinyl chloride (PVC) is used for services above the ground. Both materials are lightweight, cheap, tough, do not corrode and are easily joined. These materials soften at comparatively low temperatures and are used principally for cold water services and drains above and below the ground. The pipes are manufactured in sizes from 17 to 609 mm, the sizes most used being 21.2,

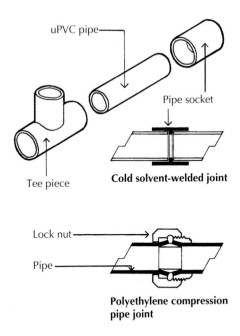

uPVC pipe

Pipe socket

Tee piece

Cold solvent-welded joint

Lock nut

Pipe

Polyethylene compression pipe joint

Figure 12.30   PVC pipe.

26.6, 33.4, 42.1 and 60.2 mm (outside diameter). uPVC and acrylonitrile butadine styrene (ABS) pipes are used primarily for waste pipes. Polythene tube is jointed with gunmetal fittings as illustrated in Figure 12.30. A copper compression ring is fitted into the pipe ends to prevent the wall of the pipe ends collapsing as the coupling is connected by tightening the nut. This type of joint is used to withstand high pressures, as in mains pressure supply pipework.

Solvent-welded joint

Solvent weld cement is applied to the ends of the pipes to be joined and the ends of the pipes are fitted into the socket or other fittings (Figure 12.30). The solvent cement dissolves the surfaces of the pipe end and fitting so that as the solvent hardens it welds the joined plastic surfaces together. The joint will set in 5–10 minutes, but will require 12–24 hours to become fully hardened. This type of joint is commonly used with uPVC pipework, such as low-pressure distribution pipework and more particularly waste branch pipes. Pipe runs in plastic are supported by clips at 225–500 mm horizontally and 350–900 mm vertically.

Valves and taps

The two general terms used to describe fittings designed to regulate or shut the flow of water along a pipeline, are valve and cock. A valve is a fitting that can be adjusted either to cause a gradual restriction in flow or a cessation of flow. For this reason, valves are also known as stop valves. A cock (or stop cock) is a fitting that consists of a solid cylindrical

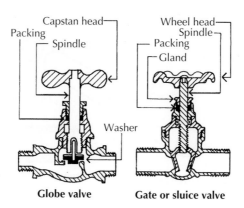

Figure 12.31   Valves.

plug, fixed with its long axis at right angles to flow. When shut, the plug fills the pipeline. A 90°, quarter turn of the plug brings a round port or hole in the plug into line with the flow, and so opens the plug. There is little control of flow with a plug that is designed to be either shut or open.

Valves

The two valves in common use are the globe valve and the gate valve (Figure 12.31). Globe valves control or shut flow through a disc that is lowered slowly by turning a screwdown spindle to a seating (they are also described as screwdown valves) and are commonly used in high pressure and hot water pipework. A gate valve operates by raising or lowering a metal gate into, or out of, the line of the pipework as the spindle is screwed down or up. This valve is sometimes referred to as a fullway gate valve as when it is fully open it does not restrict flow along the pipeline, unlike the globe valve. For this reason, the gate valve is used where there is low-pressure flow in the pipeline, such as that from cistern feed systems.

Taps

The word tap is used in the sense of drawing from or tapping into, as these fittings are designed to draw hot or cold water from the pipework. They are sometimes described as draw-off taps. There are three types of draw-off taps: the bib tap, the pillar tap (including mixer taps which mix hot and cold water) and the draw-off tap, which is used to draw-off water to drainpipe systems.

The bib tap (Figure 12.32) operates in the same way as a globe valve through a disc which is screwed down to close and up to open, except that the tap is at the end of a vertical pipeline to open to discharge water. A washer fixed to the base of the disc seals the tap when shut. A draw-off tap is similar to a bib tap except that it is operated by a loose wheel or capstan head and the outlet discharge is serrated to take a hose connection.

The pillar tap (Figure 12.32) is designed to be fitted to the bath, basin or sink it serves, with the supply pipe connected vertically to the base of the tap. The tap operates in the same way as a bib tap, through a screwdown spindle, which opens or closes a washer.

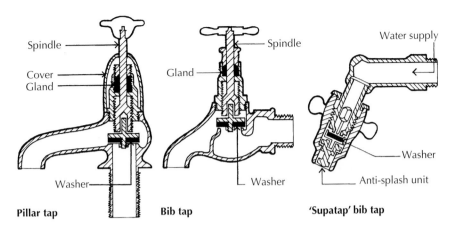

Figure 12.32   Taps.

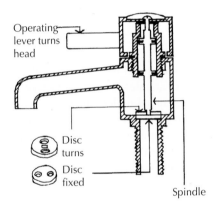

Figure 12.33   Quarter turn tap.

The 'supatap', illustrated in Figure 12.32 is designed so that the washer may be replaced without turning off the water supply, through an automatic closing valve that interrupts the supply.

Quarter turn taps
Many modern taps use a system of ceramic discs to open and close the supply. The lower disc is fixed and the top disc can be turned through 90°. When the two ports or holes in the top disc coincide with those in the bottom-fixed disc, the water flows. A lever, when turned or depressed, operates a spindle to effect the necessary quarter turn to open the tap for water to flow. The simple operation of these taps makes it possible to design plain, elegant taps that consist of a plain steel cylinder on which a lever operates to open and shut the tap and can be used to operate a hot and cold water mixer. Figure 12.33 is an illustration of a quarter turn tap with ceramic disc washers.

12.5 Sanitary appliances

Sanitary appliances, sometimes termed sanitary fittings, include all fixed appliances in which water is used either for flushing foul matter away or in which water is used for cleaning, culinary and drinking purposes. Two terms are used, soil appliances and waste appliances. Soil appliances include WCs and urinals, the discharge from which is described as soil, soiled or foul water. Waste appliances include washbasins, baths, showers, sinks and bidets, the discharge from which is described as wastewater.

Functional requirements

The primary functional requirements of sanitary appliances are:

❑ The safe and hygienic disposal of waste
❑ Durable and easy to clean/maintain

Soil appliances

WC suite

WCs are sold as a matched set of WC pan, seat, flushing appliance and any necessary flush pipe, which together are described as a WC suite. The letters WC stand for water closet, the word closet referring to the small room in which the early water pans were enclosed.

A WC pan is a ceramic or stainless steel bowl to take solid and liquid excrement, with an inlet for flushing and a trapped outlet. The seat is secured to the back of the pan. The flushing appliance is a cistern designed to discharge water rapidly into the pan through a flush pipe, for cleaning and disposal of contents. The flushing cistern may be fixed high above, near to or closely coupled to the pan, the three arrangements being described as high-level, low-level (low-down) or close-coupled WC suites, as illustrated in Figure 12.34.

Most WC pans are of the pedestal type, the base or pedestal being made integral with the pan, which is secured to the floor with screws through holes in the pedestal base to fixing plugs in solid floors or directly to timber floors (Figure 12.35). The flushing rim is designed to spread the water, which discharges through the flush outlet, around the pan to wash down the sides of the bowl. Most WC pans are made of vitreous china, which, after firing, has an impermeable body and a hard, smooth, glazed finish that is readily cleaned. The glazed finish to pans is generally white but the pan may be finished in various colour glazes. The flushing cistern body and cover to close-coupled WC suites are also made of vitreous china. Stainless steel is often used in public areas as a more durable alternative to vitreous china. WC pans have integral traps to contain a water seal against odours from the drainpipes or drains.

Washdown and syphonic pans are distinguished by the operation of the flush water in cleansing and discharging the contents of the pan. In the washdown pan (Figure 12.36), the flush water runs around the rim to wash down the bowl and then overturns the water seal to discharge the contents. In the syphonic pan (Figure 12.37 and Figure 12.38), the flush water washes the sides of the bowl and also causes a water trap or traps to overturn and create a syphonic action, which discharges the contents. The purpose of this arrangement is to effect a comparatively quiet flush and discharge of contents.

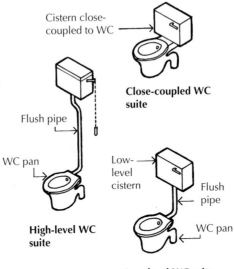

Cistern close-coupled to WC

Close-coupled WC suite

Flush pipe

WC pan

High-level WC suite

Low-level cistern

Flush pipe

WC pan

Low-level WC suite

Figure 12.34   WC suites.

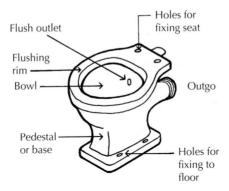

Flush outlet

Flushing rim

Bowl

Holes for fixing seat

Outgo

Pedestal or base

Holes for fixing to floor

Figure 12.35   Pedestal WC pan.

Discharge outgo

The majority of washdown WC pans are made with an outgo that is near horizontal, with a small slope down as illustrated in Figure 12.36. This standard arrangement is used for simplicity in production and consequent economy. This type of outgo, described as a P trap outgo (Figure 12.39), suits most situations as drain fittings are available to provide a connection to soil pipes relative to the position of the WC. In some situations, the WC pan may discharge to a drain below the floor level and it is convenient to have a vertical outgo. This type of outgo is described as an S trap (Figure 12.39). P traps and S traps are so named

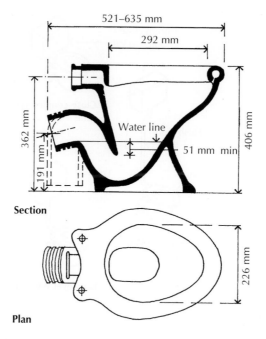

Section

Plan

Figure 12.36 Washdown WC pan.

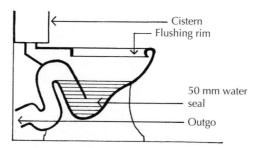

Figure 12.37 Single-seal syphon WC pan.

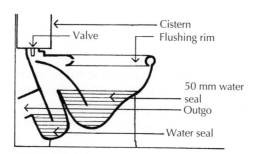

Figure 12.38 Double-seal syphonic WC pan.

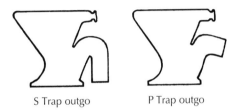

S Trap outgo   P Trap outgo

Figure 12.39   WC pan outgoes.

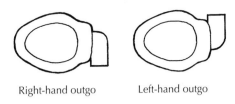

Right-hand outgo   Left-hand outgo

Figure 12.40   WC pan outgoes (plan).

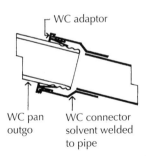

WC adaptor

WC pan   WC connector
outgo   solvent welded
                 to pipe

Figure 12.41   WC to plastic pipe.

for their sectional similarity to the letters. Where a WC pan has to discharge to one side of its position, rather than straight back through a wall, it may be convenient to have a left-hand or right-hand outgo rather than several unsightly drain fittings (Figure 12.40). The hand, either right or left, is indicated by facing the front of the pan.

The connection of the outgo of a ceramic pan to the branch drainpipe is a common site of leaks because of the difficulty of maintaining a watertight joint and making some allowance for inevitable movement between the pan and the drain branch. Soil pipes and branches are run in plastic pipe sections. The connection of the ceramic pan outgo to the plastic branch is effected by a plastic connector which is solvent welded to the soil pipe branch and whose socket end fits around the pan spigot. The seal is made with a tight fitting plastic adaptor that fits around the pan spigot outgo and adaptor as shown in Figure 12.41.

Flushing cisterns

No flushing device should exceed 6 L in a single flush and, where the flushing device is designed to give the option of a reduced flush, this should consume at least one-third less water, i.e. a maximum of 4 L of water per flush. Alternatives to flushing treated water are systems designed to use recycled water, 'grey water' (e.g. rainwater collected from the roof and stored for use) or compost toilets, which are based on a dry system of disposal.

Small bore macerator sanitary system

The Building Regulations permit the use of small bore pipe discharges from WCs. Their use depends on the macerator and pump fitted to the outlet of WC pans. The electrically powered macerator and pump comes into operation as the normal flush of a WC pan, by a conventional cistern, fills the pan. A macerator is a rotary shredder that reduces solid matter to pulp, which is, with the flush water, then pumped along a small (18–22 mm) pipe to the discharge stack. The macerator and pump unit fits conveniently behind a WC pan (Figure 12.42). The unit is connected to the horizontal outlet of a BS 5503 pan and a small bore outlet pipe. The macerator and pump are connected to a fixed, fused electrical outlet.

The particular advantage of the small bore system is in fitting a WC in either an existing or a new building some distance from the nearest foul water drainage stack, with a small bore (18–22 mm) pipe that can be run in floors or can be easily boxed in. In addition, because of the pumped discharge, the small bore branch discharge pipe can carry the dis-

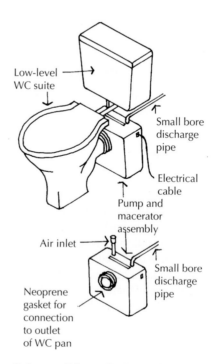

Low-level WC suite

Small bore discharge pipe

Electrical cable

Pump and macerator assembly

Air inlet

Small bore discharge pipe

Neoprene gasket for connection to outlet of WC pan

Figure 12.42 Macerator unit for small bore discharge for WCs.

charge for up to 20 m with a minimum fall of 1 in 180 and can also pump the discharge vertically up to 4 m, with a reduced horizontal limit, which is of considerable advantage in fitting WCs in basements below drain levels. The macerator and pump unit can also be used to boost the discharge from other fittings such as baths, basins, sinks, bidets and urinals along small bore runs, with a minimum fall, and over considerable runs not suited to normal gravity discharge. The small bore discharge system is not a substitute for the normal short-run branch discharge pipe system for fittings grouped closely around a vertical foul water drainage system, because of the additional cost of the macerator and pump unit and the need for frequent periodic maintenance of the unit.

Urinals

The three types of urinal in general use are the stall urinal, slab urinal and bowl urinal as illustrated in Figure 12.43. Materials used are stoneware with salt-glazed or white-glazed finish, or stainless steel, which tends to be more durable in use. Consideration should be given to ease of cleaning and maintenance and resistance to vandalism.

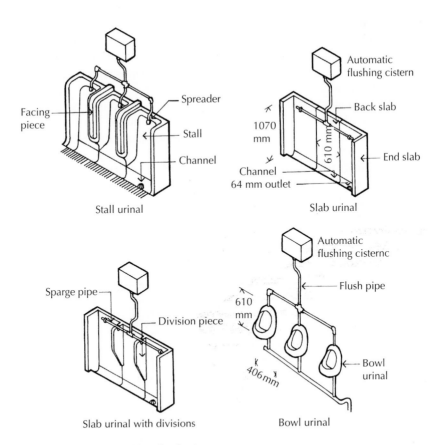

Figure 12.43   Stall, slab and bowl urinals.

Urinals are flushed by automatic flushing cisterns fixed above the urinal and discharging through a flush pipe, spreaders or sparge pipe. The automatic flushing cistern must not use more than 10 L of water per hour for a single urinal or 7.5 L/h per urinal bowl and is triggered by a sensor (thus only flushing when necessary). One outlet to the trap and branch discharge pipe is used for up to six stalls or slab units, the outlet being in the channel to the slab or in the channel of one of the stall units. The outlet, 40 mm minimum diameter, is covered with a domed, gunmetal grating and the outlet connected to a trap and waste. To accommodate the channel of urinals in a floor, a step is often formed.

Wastewater appliances

Washbasins

Washbasins, designed for washing the upper part of the body, are supported by wall brackets or by a pedestal secured to the floor, as shown in Figure 12.44. The standard washbasin consists of a bowl, soap tray, outlet, water overflow connected to the outlet and holes for fixing taps. The wall-mounted basin is fixed on brackets screwed to plugs in the wall. The pedestal basin consists of a basin and a separate vitreous china pedestal that is screwed to the floor and on which the basin is mounted. The purpose of the pedestal is to hide the trap, waste and hot and cold service pipes. Either the whole or a large part of the weight of the basin is supported by the pedestal. A resilient pad is fitted between the bottom of the basin and the top of the pedestal as the two separately made fittings rarely make a close fit. The majority of washbasins are made of vitreous china, although stainless steel is a popular alternative.

Hot and cold taps, connected to 12 or 15 mm hot and cold distributing or supply pipes are fixed to washbasins. An overflow is usually formed during the manufacture of basins, which consists of a hole in the top of the basin, which can drain to the outlet, as shown in Figure 12.45.

A waste outgo, with slot to drain the weir overflow, is formed in the basin. A waste is fitted to the outgo, bedded in setting compound and secured with a back nut, as shown in Figure 12.45. A plastic P trap with a 75 mm water seal is then connected to the waste outgo and the waste pipe.

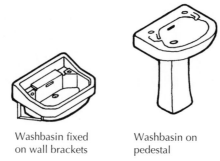

Washbasin fixed
on wall brackets

Washbasin on
pedestal

Figure 12.44   Washbasins.

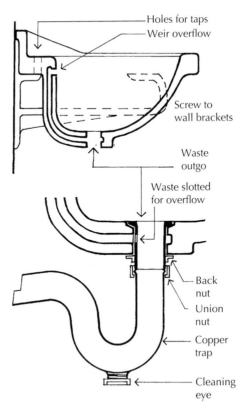

Holes for taps
Weir overflow
Screw to wall brackets
Waste outgo
Waste slotted for overflow
Back nut
Union nut
Copper trap
Cleaning eye

Figure 12.45   Washbasin waste and trap.

Baths

A wide range of styles and sizes of baths are available, with materials used ranging from porcelain-enamelled cast iron, enamelled pressed sheet steel, to plastic. Baths are fixed on adjustable feet, usually positioned against a wall or in a recess with side and end panels of pre-formed board, plastic or metal secured to wood or metal brackets or frames. A 75 mm seal trap is fitted to the bath waste and connected to the 40 mm waste. An overflow pipe is connected to the bath and either run through an outside wall as an overflow warning pipe or connected to the bath outlet or trap, as illustrated in Figure 12.46.

The 18 mm cold and hot distributing or supply pipes are connected to either individual pillar taps, a mixer with taps or a shower fitting, or both, with an air gap between the outlet and spillover level of the bath, similar to that for basins. Where shower fittings are provided, the wall or walls over baths should be finished in some impermeable material such as tile, and a waterproof screen or curtain should be provided.

Showers

This consists of a shower tray of glazed ceramic, enamelled cast iron or plastic to collect and discharge water, with a fixed or handheld shower head or rose and mixing valves. The

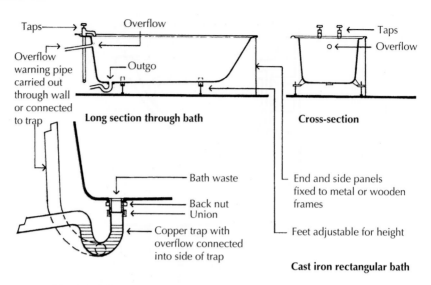

Figure 12.46   Plumbing to bath.

shower is either fixed in a wall recess or may be a free-standing cubicle. The walls around fixed showers are lined with some impermeable material such as tile, and the open side is fitted with a waterproof curtain or screen. Figure 12.47 shows some shower trays. The tray with a waste and no overflow is for use as a shower only. The tray with a waste plug and an overflow is for use either as a shower or a foot bath.

A shower compartment is often surrounded with an upstand curb or may be sunk into the floor to contain the shower water that would otherwise spill over the surrounding floor. The shower tray may be fixed on to or recessed into the floor. A 75 mm seal trap is fitted to the tray and connected to a 40 mm waste.

Bidets

A bidet consists of a glazed ceramic pedestal bowl, which is secured to the floor, usually backing on to a wall or partition. The shallow bowl has a flushing rim, a weir overflow connected to the waste and an inlet for a spray. An optional handheld spray may be fitted to the hot and cold supply. The bidet operates through the discharge of water around the flushing rim, and a spray of water that rises from the bowl or a handheld spray. The bowl may be filled with water and drained by the operation of a pop-up waste control. Figure 12.48 shows a bidet with a 75 mm trap from the waste outgo to a 40 mm waste pipe. As a precaution against the possibility of contamination of the mains supply from a bidet, particularly through the submerged spray, a separate cistern feed to a bidet or other effective device to sprays and handheld showers, such as doublecheck valve assemblies, is required.

Kitchen sinks

The traditional kitchen sink was made of glazed stoneware, usually white-glazed inside the bowl and salt-glazed outside, commonly known as Belfast sinks (Figure 12.49), fitted under

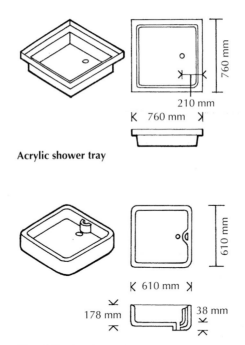

Acrylic shower tray

Glazed fireclay shower tray

Figure 12.47 Shower trays.

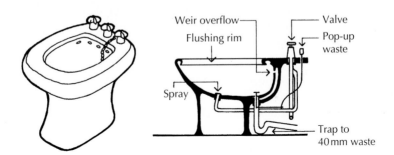

Figure 12.48 Bidet.

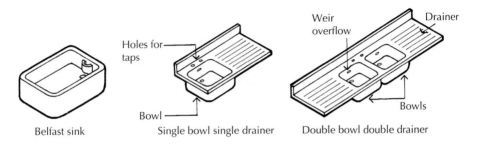

Figure 12.49 Range of sinks.

natural teak draining boards. A range of stainless steel sinks, designed to fit into kitchen and utility units, is made with single bowl and drainer, and double bowl and double drainer (Figure 12.49). The sinks are finished in the natural colour of the stainless steel from which they are pressed. Most sinks have weir overflows connected to the waste outlet and are holed for fitting hot and cold taps or mixers. A 75 mm seal trap is connected to the sink waste, which is connected to a 40 mm copper, or plastic waste pipe. Because of the flat base of the bowl a sink waste is unlikely to run full bore and cause self-syphonage.

Washing machines, dishwashers and other appliances
All appliances should be economical in the use of water, and manufacturers now rate their appliances for water consumption, energy use and noise, with 'A' being the best grade and provide typical consumption figures for typical cycles. Water fittings to appliances should be readily accessible to allow access for routine maintenance and replacement.

Waste traps

A trap or waste trap is a copper or plastic fitting formed as a bend in pipework to contain a seal of water as a barrier to odours rising from sanitary pipework and drains into rooms. The traps are formed as P or S traps to accommodate the position of the waste pipe relative to the sanitary fitting outlet. At the bottom of the trap is a cleaning eye which can be unscrewed to clear blockages. Figure 12.50 shows P and S traps.

The depth of the water seal is measured from the top of the first bend and the bottom of the second. The traps shown in Figure 12.50 are 75 mm deep seal traps, the depth of seal required for all sanitary fittings connected to single-stack systems of drainage, except for WCs that have an integral 50 mm seal.

Sanitary pipework

Ventilated system
Where pressure fluctuations in the stack and the branch wastes cannot be limited to prevent self-syphonage, induced syphonage and back pressure, e.g. in multi-storey buildings, a

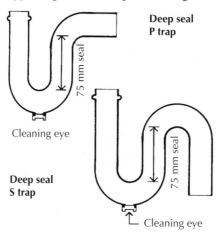

Figure 12.50   Deep seal S and P traps.

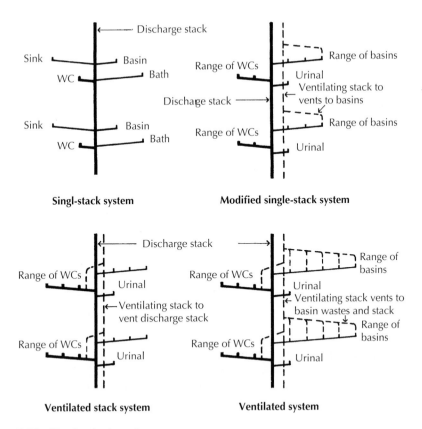

Figure 12.51 Single-stack systems.

ventilated system is used, as shown in Figure 12.51. The single vertical pipe collecting discharges from all sanitary appliances is the discharge stack and the pipes from all appliances to the stack are discharge pipes. The single vertical ventilating pipe is the ventilating or vent stack and the branches from it to the discharge stack and discharge pipes are ventilating or vent pipes.

Figure 12.52 illustrates the application of the single-stack system to a five-floor residential building with one group of appliances on each floor. The discharge pipes are arranged within the limitations set out in Figure 12.53. The 100 mm single stack (shown in Figure 12.51) can also be used to take the discharge from two groups of appliances per floor for up to five floors. The single-stack system of sanitary pipework was originally developed for houses. It has since been developed for use in multi-storey buildings, such as flats, where sanitary fittings are closely grouped, floor over floor, so that short branch discharge pipes connect to a common single stack for economy in drain runs.

Traps

Where sanitary appliances discharge foul water to the sanitary pipework system, there should be a water seal, provided by means of a trap, to prevent foul air from entering the

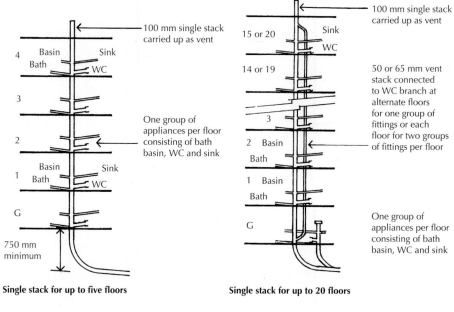

Single stack for up to five floors

Single stack for up to 20 floors

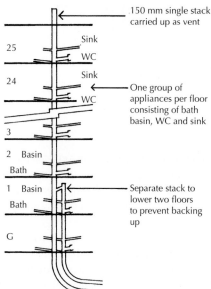

Single stack for up to 25 floors

Figure 12.52 Applications of the single-stack system.

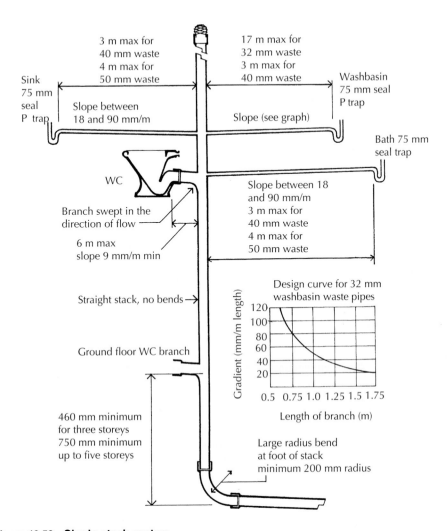

Figure 12.53 **Single-stack system.**

building under working conditions. There is a water seal trap to each of the appliances. The minimum size and depth of water seal for these traps are set out in Table 12.1. WC pans have a water seal trap that is integral with the pan in the form of a single or double water seal (as illustrated in Figure 12.37 and Figure 12.38).

Baths, bidets, sinks and washbasins have a trap that is fitted to the appliance and connected to the branch discharge pipe. To facilitate clearing blockages there should be a cleaning eye or the trap should be removable, as shown in Figure 12.50. To prevent the water seal in traps being broken by the pressures that can develop in a sanitary pipe system, the length and gradient of branch discharge pipes should be limited to those set out in Figure 12.53, or a system of ventilating pipes should be used.

Table 12.1   Minimum trap sizes and seal depths

Appliance	Diameter of trap (mm)	Depth of seal (mm)
Washbasin	32	75
Bidet		
Sink		
Bath		
Shower		
Food waste disposal unit	40	75
Urinal bowl		
WC pan		
(Syphonic only)	75	50

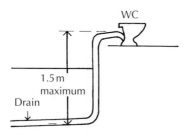

Figure 12.54   Direct connection of ground floor WC to drain.

Branch pipes

The discharge of foul water from sanitary appliances is carried to the vertical discharge stack by branch pipes, as shown in Figure 12.53. All branch discharge pipes should discharge into a discharge stack except those to appliances on the ground floor. Ground floor sinks, baths and washbasins may discharge to a gully, and WCs, bidets and urinals to a drain. A branch pipe from a ground floor WC should only discharge directly to a drain if the drop is less than 1.5 m, as shown in Figure 12.54. A branch pipe should not discharge into a stack lower than 450 mm above the invert of the tail of the bend at the foot of the stack for single dwellings up to three storeys, and 750 mm for buildings up to five storeys high. The branch pipes from more than one ground floor appliance may discharge to an unvented stub stack, with the stub stack connected to a ventilated discharge stack or a drain, provided no branch is more than 2 m above the invert of the connection to the drain and branches from a closet more than 1.5 m from the crown of the closet trap, as shown in Figure 12.55. Branch pipes from wastewater fittings such as sinks, baths and basins on the ground floor should discharge to a gully between the grating and the top level of the water seal. To avoid cross flow, small similar-sized connections not directly opposite should be offset by 110 mm on a 100 mm stack and 250 mm on a 150 mm stack. A wastewater branch should not enter the stack within 200 mm below a WC connection as shown in Figure 12.56. Pipes serving single appliances should be at least the same diameter as the appliance trap and should be the diameter shown in Table 12.2 if the trap serves more than one appliance and is unventilated.

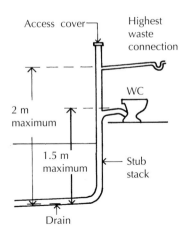

Figure 12.55   Stub stack ground floor appliances.

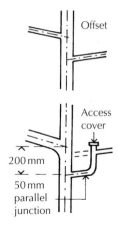

Figure 12.56   Branch connections to stack.

Table 12.2   Common branch discharge pipes (unvented)

Appliance	Maximum number to be connected	Maximum length of branch (m)	Minimum size of pipe (mm)	Gradient limits (fall per metre)	
				min (mm)	max (mm)
WCs	8	15	100	9	90
Urinals: bowls	5	*	50	18	90
Stalls	7	*	65	18	90
Washbasins	4	4 (no bends)	50	18	45

*No limitation as regards venting but should be as short as possible.

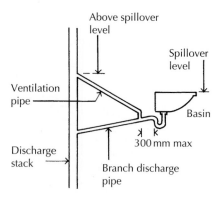

Figure 12.57   Branch ventilation pipe.

Bends in branch pipes, which should be avoided if possible, should have a radius as large as possible and a centre line radius of at least 75 mm for pipes of 65 mm or less in diameter. Junctions on branch pipes should be made with a sweep of 25 mm radius or at an angle of 45°, and connections to the stack of branch pipes of 75 mm diameter or more should be made with a sweep of 50 mm minimum radius or 45°.

It is not necessary to provide ventilation to branch pipes whose length and slope is limited to the figures given in Figure 12.53, or to the common branch discharge pipes set out in Table 12.2. Where the length or slope is greater than these limits, the branch pipes should be ventilated by a branch ventilation pipe to external air, to a discharge stack or to a ventilating stack where the number of ventilating pipes and their distance to a discharge stack are large. Branch ventilating pipes should be connected to the discharge pipe within 300 mm of the trap and should not connect to the stack below the spillover level of the highest appliance served, as illustrated in Figure 12.57. Branch ventilating pipes to branch pipes serving one appliance should be 25 mm in diameter or where the branch is longer than 15 m or has more than five bends, 32 mm in diameter. The discharge pipe is usually 100 mm in diameter.

Pipe materials

uPVC is the most commonly used material for discharge stack and branch pipe systems, because of its low cost, ease of cutting, speedily made joints and the range of fittings available. The pipework is secured with loose brackets that are nailed or screwed to plugs in walls. A variety of fittings are manufactured to suit the various branch waste connections for single-stack systems of sanitary pipework. The plastic pipework is usually jointed by means of an elastomeric ring seal joint. The synthetic rubber ring forms an effective seal as it is compressed between the spigot and socket ends of the pipe as the spigot end is pushed into the socket. Connections of uPVC branch pipes to outlets of appliances are made with rubber compression rings that are hand-tightened by a nut to a copper liner. Figure 12.58 is an illustration of an uPVC discharge stack to a house.

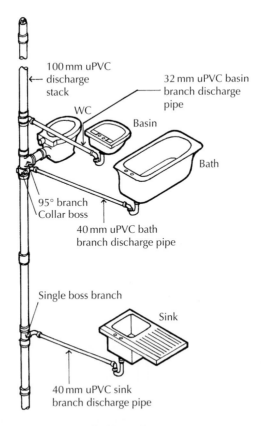

100 mm uPVC
← discharge
stack

WC

32 mm uPVC basin
branch discharge
pipe

Basin

Bath

95° branch
Collar boss

40 mm uPVC bath
branch discharge pipe

Single boss branch

Sink

40 mm uPVC sink
branch discharge pipe

Figure 12.58   uPVC discharge stack and branches.

Soundness test

Air test

The accepted method of testing the soundness of discharge stacks and pipes above the ground is the air test. A sound pipe system will contain air under pressure for a few minutes as an indication of its capacity to contain the flow of liquid in conditions normal to a discharge pipe system. The air test is carried out to the whole discharge pipework above the ground in one operation or, where the pipework is extensive, in two or more operations. The traps of all sanitary appliances are filled with water and the open ends of pipes are sealed with expanding drain plugs or bag stoppers. Air is pumped into the pipework through the WC pan trap and the air pressure is measured in a U-tube water gauge or manometer. A pressure equal to 38 mm water gauge should be maintained for at least 3 minutes if the pipework is sound. Figure 12.59 illustrates the equipment used for the air test. If the air pressure is not maintained for 3 minutes, leaks may be traced by spreading a soap solution around joints, with the pipework under air pressure; bubbles in the soap

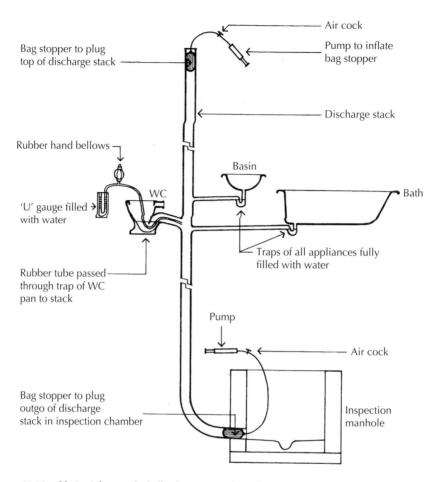

Figure 12.59 Air test for sealed discharge stack and branches.

solution will indicate leaks. Alternatively, smoke is pumped into the pipework from a smoke machine and the escape of smoke will indicate leaks. Leaking joints are made good and the air test applied to test for soundness.

In-use test

To test the performance of a discharge pipe system in use, a group or groups of appliances are discharged simultaneously to cause conditions most likely to produce maximum pressure fluctuations. In buildings with up to nine appliances of each kind to a stack, the top-floor sink and washbasin are filled to overflowing and the plugs pulled simultaneous to a normal discharge of the top-floor WC. After this test, a minimum of 25 mm water seal should be retained in every trap. With more than nine appliances of each kind to a stack, two or more WCs, basins and sinks are discharged simultaneously on the top floors for the performance in-use test.

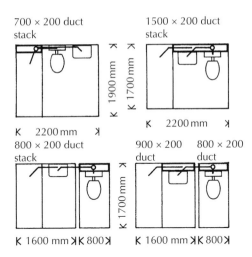

Figure 12.60   Pipe ducts.

Pipe ducts

In large buildings, all discharge pipework, together with hot and cold water services, are run internally in ducts for ease of access. Where there are internal bathrooms and WCs, to economise in pipework and to avoid overlarge ducts it is necessary to group sanitary appliances. Some compact groups of sanitary appliances and ducts and pipework are illustrated in Figure 12.60, which illustrates ducts to a bathroom with a WC and a separate bathroom and WC. Where there are two or more bathrooms and WCs on each floor of a multi-storey building, they will be grouped around a common duct.

Ventilation of internal WCs and bathrooms

Bathrooms and WCs are often sited internally in modern buildings, such as flats, and refurbishment schemes to maximise the use of space. It is necessary to provide means of extract ventilation to internal bathrooms and WCs, to dilute pollutants and moisture vapour by air changes (Figure 12.61). Extract ventilation may be provided by mechanical extract of air or by passive stack ventilation. Passive stack ventilation is a ventilation system that uses ducts from the ceilings of rooms to terminals on the roof to operate through the natural stack effect of warm air rising, as in a chimney stack. Mechanical ventilation is effected by an electrically operated extract fan, designed to evacuate air through a duct to outside air (Figure 12.62). The Building Regulations recommend mechanical extract ventilation of 60 L/s for bathrooms and sanitary accommodation located internally. The extract fan should be controlled by the operation of the light switch to the room and have a 15 minute overrun after the light is turned off. There should be a 10 mm gap under the door to bathrooms and sanitary accommodation through which replacement air can enter.

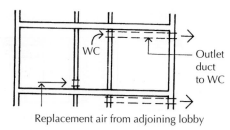

Horizontal outlet duct

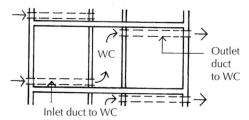

Horizontal inlet and outlet ducts

Figure 12.61 Ducts.

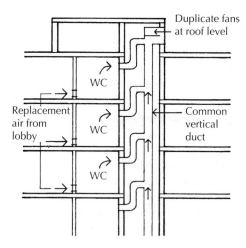

Figure 12.62 Mechanical extract system.

12.6 Foul drainage

Functional requirement

The primary function of a foul drainage system is to facilitate the transfer of foul water and matter quickly, economically and safely from a building to the mains drainage or collecting vessel.

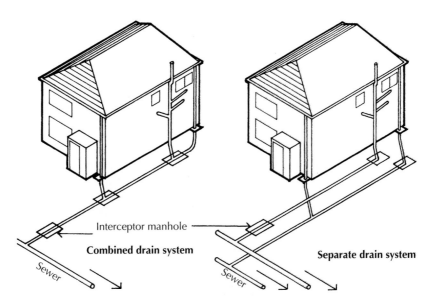

Interceptor manhole

Combined drain system

Separate drain system

Sewer

Sewer

Figure 12.63   Combined and separate drain systems.

Drainage layout

The layout of foul drains depends on whether foul water and rainwater are discharged to a common drain system (combined) or to separate drain systems (Figure 12.63), which in turn depends on whether there is one sewer carrying both foul and rainwater or separate sewers for foul and rainwater. Foul drainage systems should be kept simple. Fittings that discharge foul water should be grouped together on each floor to economise on pipe runs; discharge pipe branches should run to a common waste stack and groups of fittings on each floor should be positioned one over the other to minimise pipework. Rainwater pipes from roof gutters and gullies to collect water from paved areas should also be positioned to simplify drain runs. Wherever practicable, changes of direction and gradient should be limited to minimise the number of access points. Because the drains for foul water and rainwater will generally run across each other at some point, it is necessary to adjust the level or gradient (slope) of the drains to accommodate this.

A combined drainage system has to be ventilated throughout to conduct foul air discharge to the open air. In this system, it is necessary to fit trapped gullies with a water seal. With separate systems of foul and rainwater drains, it is necessary to fit trapped gullies only to the discharge of foul water fittings, where the discharge pipe does not serve as a ventilation pipe.

Foul water drainage systems should be ventilated by a flow of air, with a ventilating pipe to the head of each main drain and any branch drain that is more than 6 m long serving a single appliance or 12 m long serving a group of appliances. Ventilated discharge pipes such as discharge stacks, discharging directly to the drain, are commonly used for ventilation of drains.

Drain runs should be laid in straight lines wherever possible to encourage the free flow of discharge water by gravity, with gentle curves in drain runs only where straight runs are not practicable. Bends in drain runs should be limited to positions close to or inside inspection chambers and to the foot of discharge pipes and should have as large a radius as practicable. Where drain runs are near to or under a building, precautions should be taken to accommodate the effects of settlement without damage to the drain.

Drainpipes

Drainpipes are classified as being either rigid or flexible. Rigid pipes are those that fail by brittle fracture before they suffer appreciable deformation, and these include clay, concrete, cement and cast iron. Flexible pipes are those that suffer appreciable deformation before they fracture, and these include uPVC. The deformation of flexible and rigid pipes is controlled by the use of a granular bed, and limitations of load by the design of the trench and its backfilling.

Clay pipes

Clay pipes are mass-produced in highly automated plants. The selected clays are ground to a fine powder and water is added for moulding. Simple fittings are formed by extrusion and the parts of junctions are extruded and then cut and joined by hand. The moulded clay pipes and fittings are then dried in kilns to encourage uniform loss of water, to avoid loss of shape and then fired in continuous tunnel kilns. Pipes and fittings that are to be glazed have a ceramic glaze or slip sprayed on, or they are dipped in the slip before they enter the kiln.

The nominal bore (inside diameter) of clay pipes for drains is from 75 to 900 mm in increments of 25 mm between 75 and 250 mm, one increment of 50–300 mm and then increments of 75 mm from 300 to 900 mm, as illustrated in Figure 12.64. A wide range of more than 250 fittings is made for clay drains such as bends, junctions, channels and gullies, some of which are shown in Figure 12.65. The advantage of the clay pipe for drains is that the comparatively short length of the pipe and the wide range of fittings are adaptable both to the straightforward and the more complex drain layouts and the pipes themselves are inert to all normal effluents.

Jointing

Flexible joints should be used with rigid clay drainpipes so that the drain lines can accommodate earth movements under, around or over the pipeline, within the joint. The flexible joint can be made in all weather conditions and, once made and tested, the trench can be backfilled to protect the pipeline from damage. There are two types of flexible joint in use: the socket joint for socket and spigot pipes and the sleeve joint for plain-ended clay pipes. Typical joints are shown in Figure 12.66 and Figure 12.67. These flexible joint seals are made from natural rubber, chloroprene rubber, butyl rubber or styrene-butadiene rubber. The flexible socket joint is made with plastic fairings cast on the spigot and socket ends of the pipe to provide a simple push fit joint. It suffers the disadvantages that the joint may be damaged in handling and cut pipe lengths present difficulties on site. The flexible coupling joint is made with a close-fitting, flexible plastic sleeve with rubber sealing rings. The

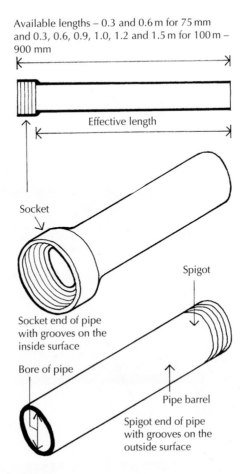

Available lengths – 0.3 and 0.6 m for 75 mm
and 0.3, 0.6, 0.9, 1.0, 1.2 and 1.5 m for 100 m –
900 mm

Effective length

Socket

Socket end of pipe
with grooves on the
inside surface

Bore of pipe

Spigot

Pipe barrel

Spigot end of pipe
with grooves on the
outside surface

Figure 12.64   Clay drainpipes.

Figure 12.65   Clay drainpipe fittings.

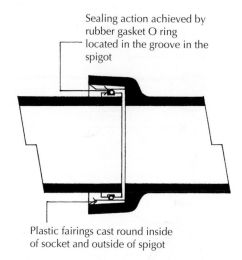

Sealing action achieved by
rubber gasket O ring
located in the groove in the
spigot

Plastic fairings cast round inside
of socket and outside of spigot

Figure 12.66   Section at junction of socket and spigot clay pipes.

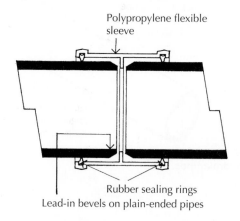

Polypropylene flexible
sleeve

Rubber sealing rings
Lead-in bevels on plain-ended pipes

Figure 12.67   Section at junction of plain-ended clay pipes showing flexible coupling.

ends of the plain-ended pipes are bevelled to assist in forcing pipe ends into the sleeve to
make a close watertight seal.

Concrete pipes

Spun concrete pipes are often used for major infrastructure works. The normal range of
the pipes is from 300 mm to 1.8 m. The standard pipe is usually around 2.5 m long and a
6000 mm diameter pipe weighs approximately 1200 kg. These pipes can only be lifted and
manoeuvred into position using mechanical plant; the arm of a backactor of a 360-tracked
excavator is usually used to lift, pull and push the pipes into position. Most concrete pipes

come fitted with gaskets, and once lubricant has been applied to the barrel, it can be pushed into the socket for a watertight seal.

Cast iron pipes

Cast iron pipes are used for drains because of their superior strength where there is unstable or made up ground, in shallow trenches, under buildings, for drains suspended under floors of buildings, in heavily waterlogged ground and where sewage is under pressure from pumping. Cast iron and ductile iron pipes are made with socket and spigot ends or with plain ends, as shown in Figure 12.68. All pipes are hot dip coated with either a bituminous or tar coating inside and out. Pipes are manufactured with bores of 75, 100, 150 and 225 mm and in lengths of 1.83, 2.74 and 3.66 m, together with a wide range of fittings.

Jointing

A flexible push fit joint is used for cast iron and ductile iron pipes. A rubber gaskin, fitted inside the socket of pipes, comprises a heel of hardened rubber that aligns the pipes, and

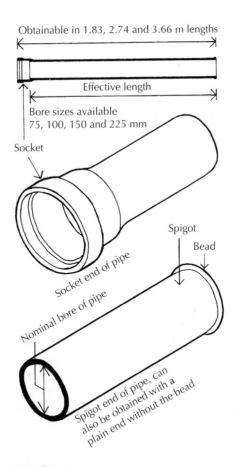

Figure 12.68 Cast iron drainpipes.

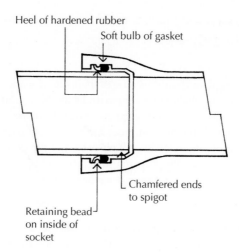

Heel of hardened rubber

Soft bulb of gasket

Chamfered ends
to spigot

Retaining bead
on inside of
socket

Figure 12.69   Flexible push fit joint for cast iron.

a bulb of softer rubber that makes the joint, as shown in Figure 12.69. The pipe ends must be clean and are lubricated and joined by leverage from a crowbar for small pipes, or a fork tool for larger pipes. The joint is flexible and will accommodate longitudinal and axial movements. The joint is rapidly made in any conditions of weather, the pipeline may be tested immediately and the trench can be backfilled to protect the pipeline.

uPVC pipes

uPVC pipe is extensively used because of its ease of handling, cutting and jointing and the low cost. The pipe is light in weight and flexible as it can, to some extent, deform under load without fracture. Deformation should be limited to an increase in horizontal diameter of not more than 5% to avoid blockages in pipelines or breaks to joints. Pipe sizes are described by the outside diameter of the pipe as 110, 160 and 200 mm, and lengths are 1.0, 3.0 and 6.0 mm (Figure 12.70). Because of the comparatively long lengths in which this pipe material is made, with consequently few joints necessary, the material lends itself to assembly at the ground level from where it can be lowered into narrow trenches.

Jointing

Socket and spigot push fit joints with rubber rings depend on a shaped socket end of pipe, as shown in Figure 12.71, or a shaped coupling designed to fit around rubber sealing rings that fit between the socket or coupling and spigot end to make a watertight seal. Because the rubber ring seal has to be a tight fit between the pipes, it requires careful manipulation to fit the spigot end into the socket end of the pipe or coupling while making certain the rubber seal takes up the correct position in the joint. Because of the considerable thermal expansion and contraction of this material, it is necessary to provide for movement along the pipe lengths through the allowance between the spigot end of pipes and the shoulder on the socket end of the pipe or coupling, as illustrated in Figure 12.71. Several proprietary systems of push fit joints are available.

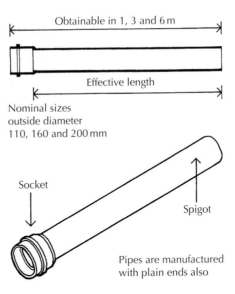

Obtainable in 1, 3 and 6 m

Effective length

Nominal sizes
outside diameter
110, 160 and 200 mm

Socket

Spigot

Pipes are manufactured
with plain ends also

Figure 12.70   PVC drainpipes.

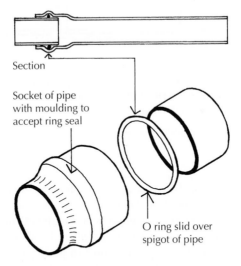

Section

Socket of pipe
with moulding to
accept ring seal

O ring slid over
spigot of pipe

Figure 12.71   PVC socket and spigot push fit joint.

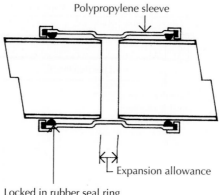

Polypropylene sleeve

Expansion allowance

Locked in rubber seal ring

Figure 12.72   PVC sleeve joint.

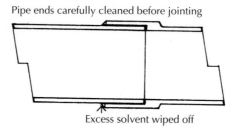

Pipe ends carefully cleaned before jointing

Excess solvent wiped off

Welded joint achieved by bringing
together spigot and socket of pipes which
have been brushed at the ends with
solvent – PVC solution and methyl chloride

Figure 12.73   PVC solvent-welded joint.

Sleeve joints of polypropylene with rubber sealing rings to plain-ended pipes depend on a separate plastic coupling sleeve that fits over the spigot ends of plain pipe lengths. Rubber seal rings in the ends of the sleeve compress on the spigot ends of the pipes being joined, as illustrated in Figure 12.72. To provide a tight fit between the coupling sleeve and the pipe ends, it is necessary to apply some force to push the pipe ends into the sleeve, for which cramps or jacks are available. The ends of the pipes being joined fit to shoulders in the sleeve to provide the expansion allowance illustrated in Figure 12.72. Sleeve jointing is generally favoured over the socket and spigot joint.

Solvent-welded joints (Figure 12.73) are available for short lengths of drainpipe. This type of joint is not used for long lengths of drain, as the rigid joint makes no allowance for expansion. The joint is made by bringing together the spigot and socket ends of the pipe after they have been brushed with a PVC solution and methyl chloride. The solvent dissolves the PVC, which fuses together after some time, but takes a while to harden fully.

Drainpipes made of uPVC are commonly used with the proprietary drain systems, which comprise a package of plastic connections, clearing eyes and access bowls for inspection, to combine economy and speed of assembly and laying. Because of the comparatively extensive range of fittings manufactured in clay, it is not uncommon for clay gullies, for example, to be used in conjunction with uPVC pipes.

Drain laying

Drains laid underground should be of sufficient diameter to carry the anticipated flow, and should be laid to a regular fall or gradient to carry the foul water and its content to the outfall.

Pipe gradient or fall

Research into the flow load of drains in use has shown that the size of the drainpipe and its gradient should be related to the anticipated flow rate in litres per second, so that the discharge entering a drain will determine the necessary size and gradient of the drain. For short drain runs, the invert level at the top and bottom of the run can be pegged out at the correct level, and either sight lines or string lines used to plot the gradient of the pipe. Pipe lasers are more commonly used today. The laser sits at one end of the pipe run and the target is positioned in the end of the pipe being levelled. When the laser strikes the centre of the target, the pipe is at the correct level (Figure 12.74 and Figure 12.75 and Photograph 12.1).

Drains are designed to collect and discharge foul water and rainwater by the flow of water under gravity. Drains are, therefore, laid to a regular fall (slope) towards the sewer or outflow. The necessary least gradient or fall of a drain depends on the anticipated flow of water through it and the necessary size of drain to carry that flow. Table 12.3 gives recommended minimum gradients for foul drains.

Approved Document H1 recommends that a drain carrying only wastewater should have a diameter of at least 75 mm, and a drain carrying soilwater at least 100 mm. The term wastewater is generally used to include the discharge from baths, basins and sinks, and soilwater includes the discharge from WCs.

Figure 12.76, from Approved Document H1, shows the relationship of flow rate to gradient for three pipe sizes with drains running three quarters of proportional depth. The rate of flow in drains with gradients as flat as 1:200 is given. Drains are not commonly run with gradients below 1:80 because the degree of accuracy necessary in setting out and laying drains required for shallow falls is beyond the skills of most building contractors.

Drains should be laid at a depth sufficient to provide cover for their protection and the excavation should be as narrow as practical for bedding and laying the drain lines. The greater the width of the trenches at the crown of the pipe, the greater the surcharge loads on the pipe. It is advantageous, therefore, to bed the drain in a narrow trench, which may be increased in width above the level of the crown of the drain for ease of working. With modern excavating machinery, flexible joint pipelines may be assembled above the ground and then lowered into and bedded in comparatively narrow trenches, so saving labour and cost in excavation, improving health and safety and providing the best conditions for the least loads on the pipeline.

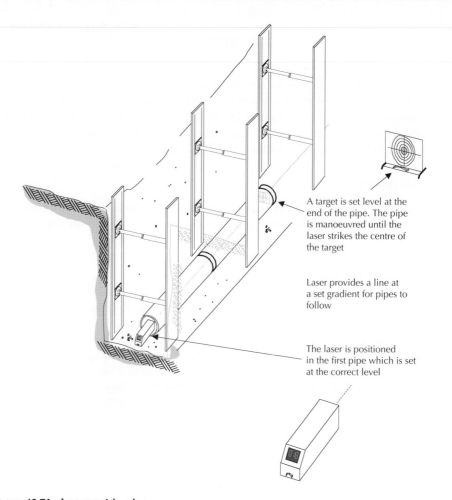

Figure 12.74 Laser set in pipe.

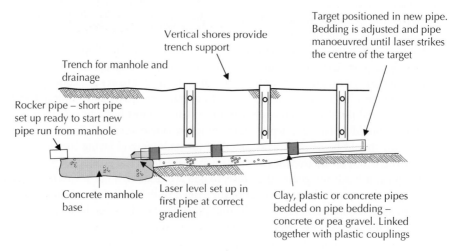

Figure 12.75 Section showing positioning of pipes using laser level.

(a) Laser set up in manhole. Once set at the correct gradient, pipes can be laid to the correct level

(b) The laser can be used to obtain a level for the pipe bedding; the drainage is placed onto the bedding and positioned to correct fall

(c) Once the pipe is correctly positioned, the pipe bedding covers the whole pipe

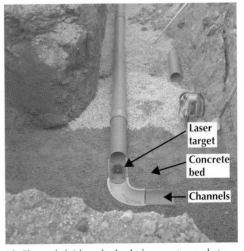

(d) Channels laid on the bed of concrete, ready to form a new manhole

Photograph 12.1 Laying drains.

The depth of the cover to drainpipes depends on the depth at which connections are made to the drain, the gradient at which the pipes are laid and ground levels. Depth of cover is taken as the level of finished ground or paving above the top of a drainpipe. A minimum depth of cover is necessary to provide protection to the pipe against damage, and a maximum depth to avoid damage to the drain by the weight of the backfilling of the drain trenches. Minimum and maximum cover for rigid pipes is set out in Table 12.4.

Table 12.3 Recommended minimum gradients for foul drains

Peak flow (L/s)	Pipe size (mm)	Minimum gradient (1:. . .)	Maximum capacity (L/s)
<1	75	1:40	4.1
	100	1:40	9.2
>1	75	1:80	2.8
	100	1:80*	6.0
	150	1:150†	15.0

*Minimum of one WC.
†Minimum of five WCs.

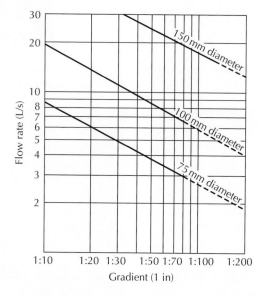

Figure 12.76 Discharge capacities of foul drains running 0.75 proportional depth.

Table 12.4 Limits of cover for standard strength rigid pipes in any width of trench

Pipe bore	Bedding class	Fields and gardens		Light traffic roads		Heavy traffic roads	
		Min	Max	Min	Max	Min	Max
	D or N	0.4	4.2	0.7	4.1	0.7	3.7
100	F	0.3	5.8	0.5	5.8	0.5	5.5
	B	0.3	7.4	0.4	7.4	0.4	7.2
	D or N	0.6	2.7	1.1	2.5	–	–
150	F	0.6	3.9	0.7	3.8	0.7	3.3
	B	0.6	5.0	0.6	5.0	0.6	4.6

Flexible pipes should have a minimum of 0.6 m of cover under fields and gardens and 0.9 m under roads.

Bedding flexible pipes

Flexible pipes should be laid in a narrow trench, on granular material such as clean, natural aggregate. This granular material is spread in the base of a trench that has been excavated and roughly levelled to the gradient or fall of the drains. The granular material is spread and finished to a thickness of 100 mm, to the drain gradient. Lengths of drainpipe are then lowered into the trench and set in position by scooping out the granular bed from under the collars of pipe ends, to set the pipeline in place at the centre of the trench. Further granular material is then spread and lightly packed on each side of the pipeline, to support it against deformation under load, as illustrated in Figure 12.77. A layer of granular material or selected fill, taken from the excavated material, is then spread over the pipeline to a depth of 100 mm. A further layer of selected fill, free from stones, lumps of clay or other material larger than 40 mm, is spread in the trench to a thickness of 200 mm. The trench is then backfilled with excavated material up to the ground level, and consolidated. This drain-laying operation requires care and some skill to bed the drain line correctly, and further care in backfilling to avoid disturbing the drain. Where the bed of the trench is

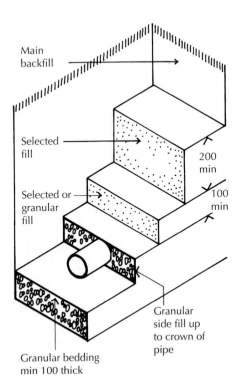

Figure 12.77 Bedding for flexible pipes.

narrow, it may be necessary to cut the sides of the trench above the level of the bed of the drain, in the form of a V, sloping out from the centre for ease of access.

Bedding rigid pipes

Where a drain trench is excavated in cohesive soil such as clay for laying rigid pipelines of clayware, for example, it is possible to lay the pipes directly on the trench bottom which has been finished to the required pipe gradient by hand trimming by shovel. The pipe lengths are lowered into the trench and soil from the trench bottom is scooped out under each socket end of the pipe so that the barrel of the pipes bears on the trench bottom and the collars keep the drain line in place. Figure 12.78 illustrates this type of bedding. For this operation to be successful, the trench needs to be sufficiently wide for a man to stand in the trench astraddle the pipeline to scoop soil out from below the pipe collars. Once the pipeline is in place, a cover of selected fill from the excavation, free from large stones or lumps of clay, is spread in the trench to a depth of 150 mm above the crown of the pipes, and the trench is backfilled to the surface level. This bedding system is suited to the use of socket and spigot and clay pipes with one of the flexible joints that can be used in all weather conditions.

When the bed of drain trenches cannot be trimmed to the pipe gradient, a bed of granular material is spread in the bed of the trench and levelled to the pipe gradient to a thickness of 100 mm, as illustrated in Figure 12.79. The rigid pipes are lowered into the trench and a layer of selected fill is carefully spread around the pipes and then filled to a level of 150 mm above the crown of the pipes. The trench is then backfilled to the surface level. This type of bedding is suited to plain clay pipes joined with flexible sleeves.

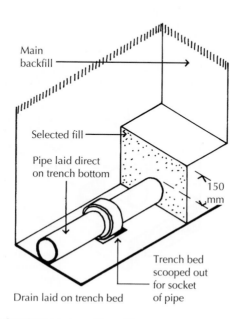

Main backfill

Selected fill

Pipe laid direct on trench bottom

150 mm

Trench bed scooped out for socket of pipe

Drain laid on trench bed

Figure 12.78 Bedding for rigid pipes – Class D.

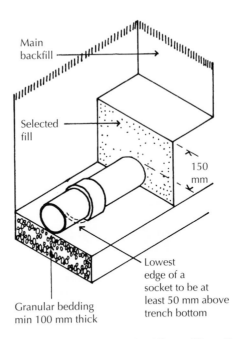

Main
backfill

Selected
fill

150
mm

Lowest
edge of a
socket to be at
least 50 mm above
trench bottom

Granular bedding
min 100 mm thick

Figure 12.79 Bedding for rigid pipes: granular bedding – Class N.

As an alternative, the granular bedding is spread 100 mm thick in the bed of the trench, the pipes are lowered into the trench and granular bedding is scooped out under the collar ends of pipes. The granular material is then spread around the pipes, up to half the outside diameter of pipe as shown in Figure 12.80. Selected fill is then spread in the trench to a depth of 150 mm above the crown of pipes, and the trench is backfilled to the surface. This system of bedding has the advantage that the granular bedding each side of the pipes maintains them in position as the trench is filled. It may well be that the most economical depth of flexible drainpipes below the surface will provide less depth of cover than recommended. To avoid unnecessary excavation, it is acceptable to lay flexible pipes with less cover when they are laid under fields or gardens. The pipes are laid on a bed of granular material 100 mm thick in the bed of the trench and then surrounded and covered with granular material up to a level of 75 mm above the crown of pipes, as illustrated in Figure 12.81. Concrete paving slabs are then laid over the granular material, bearing on offsets in the trench walls. The trench is then backfilled to the surface level with selected fill from the excavation. The protection of concrete slabs and the granular material bed and surround of pipes will provide adequate protection against deformation of the pipes.

Rigid drainpipes that are laid at a depth that provides less than the recommended cover below the surface, should be provided with protection from damage by an encasement of concrete, with flexible movement joints at each socket or sleeve joint of the pipeline. The drain trench is filled with concrete, to a depth of 100 mm, in sections along the length of the trench equal to the pipe length. Between each section, a 13 mm thick compressible board, holed for pipes, is set to the width of the trench. Before the concrete is hard, it is

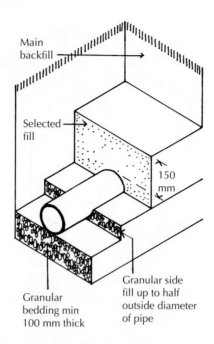

Main backfill

Selected fill

150 mm

Granular bedding min 100 mm thick

Granular side fill up to half outside diameter of pipe

Figure 12.80 Bedding for rigid pipes: granular bedding – Class B.

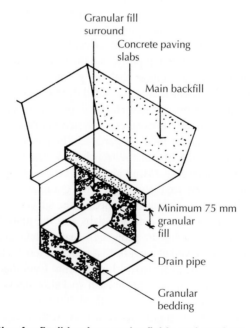

Granular fill surround

Concrete paving slabs

Main backfill

Minimum 75 mm granular fill

Drain pipe

Granular bedding

Figure 12.81 Protection for flexible pipes under fields and gardens.

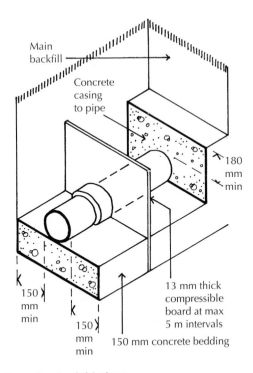

Main
backfill

Concrete
casing
to pipe

180
mm
min

150
mm
min

150
mm
min

13 mm thick
compressible
board at max
5 m intervals

150 mm concrete bedding

Figure 12.82   Concrete casing to rigid pipes.

scooped out for the socket ends of the pipe against one side of the board. A pipe is set in position up to the board and the next section of concrete is laid up to another flexible board in the trench. The concrete is scooped out for the next pipe, which is pushed into the flexible joint of the first pipe, and so on along the length of the trench, as illustrated in Figure 12.82. Once the drain line is laid, it is tested and then concrete encasement is spread around the pipes, between the flexible boards, to provide a cover of 100 mm all round the drain. The trench is then backfilled to the surface.

Drains under buildings and walls

When a drain is laid under a building and the crown of the pipes is at all points 300 mm or more below the underside of the ground floor slab, the drains should have flexible joints and be surrounded by granular material at least 100 mm thick all round the pipes. Where the crown of the drain is within 300 mm of the underside of the ground floor slab, it should be encased in concrete as an integral part of the slab.

There is a possibility that slight settlement of a wall might fracture the drain where it is laid to run through the wall under a building. There are two methods of avoiding the possibility of damage. The first method is to provide a minimum 50 mm clearance between the wall and the drain. The wall is built with a small lintel over the opening in the wall through which the pipe is to run, to provide at least 50 mm of clearance all round the drain.

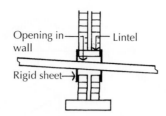

Figure 12.83 Drainpipe through an opening in the wall.

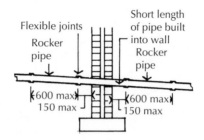

Figure 12.84 Drainpipe built into the wall.

Where the drain is laid to run through the wall, two rigid sheets, holed for the pipe, are fitted to the pipe and secured in place by screws and plugs to the wall, as illustrated in Figure 12.83, to exclude vermin or fill. The disadvantage of this system is that it is difficult to provide a close fit of a rigid board to a pipe and to brickwork and to make a watertight joint between the external board and the wall.

The other method of providing protection to a drain run through a wall is to provide for any slight settlement of the wall by means of rocker pipes. A short length of pipe is built into the wall, projecting no more than 150 mm each side of the wall. A length of pipe, at most 600 mm long, is connected to each end of the built-in pipe and connected to it with flexible joints. These rocker pipes are likewise connected to the drain line with flexible joints so that any slight movement of the pipe built into the wall is accommodated by the flexible joints of the rocker pipes. This system, illustrated in Figure 12.84, is best executed with plain clay pipes with flexible sleeve joints.

Drains close to buildings

On narrow building sites where there is only a narrow strip of land on each side of a building, it may be necessary to run a drain parallel to a flank wall of the building.

To provide the necessary gradient or fall, the drain may be at or below the level of the wall foundation. To avoid the possibility of the loads on the wall foundation imposing undue pressure on the drain, it may be necessary to provide additional protection. Where the drain trench bottom is less than 1 m from the foundation, the distance indicated by the letter A in Figure 12.85, the drain should be laid on a bed of concrete and then covered

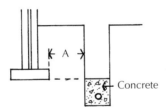

Figure 12.85   Drains less than 1 m from foundations.

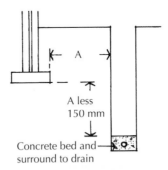

Figure 12.86   Drains 1 m or more from foundations.

with concrete up to the level of the underside of the wall foundation. A drain laid 1 m or more from the wall foundation, the distance indicated by A in Figure 12.86, and appreciably below the foundation of the adjacent wall, should be bedded on concrete and then covered with concrete up to a level equal to the distance from the wall of A minus 150 mm. This additional protection, which is expensive, may not be wholly satisfactory as it provides rigid encasement that makes no allowance for possible differential settlement along the length of the drain run, which might fracture the encasement and pipe run.

Access points to drains

There should be adequate access points to drains to provide means of clearing blockages by rodding through drains. Rodding is the operation of pushing flexible, sectional rods that can be screwed together, down drain lines to clear blockages. The three types of access point in use are:

❑ Rodding eyes, which are capped extensions of drainpipes
❑ Access fittings, which are small chambers in, or as an extension of, drainpipes with no open channel
❑ Inspection chambers, which are large chambers with an open channel. Small inspection chambers have no working space at drain level; large inspection chambers (sometimes called 'manholes') do provide working space at drain level.

Table 12.5 Minimum dimensions for access fittings and chambers

Type	Depth to invert (m)	Internal sizes Length × width (mm × mm)	Internal sizes Circular (mm)	Cover sizes Length × width (mm × mm)	Cover sizes Circular (mm)
Rodding eye	–	As drain but min. 100		–	
Access fitting		150 × 100	150	150 × 100	150
Small	0.6 or less				
Large		225 × 100	–	225 × 100	–
Inspection	0.6 or less	–	190*	–	190*
chamber	1.0 or less	450 × 450	450	450 × 450	450†
Manhole	1.5 or less	1200 × 750	1050	600 × 600	600
	over 1.5	1200 × 750	1200	600 × 600	600
	over 2.7	1200 × 840	1200	600 × 600	600
Shaft	over 2.7	900 × 840	900	600 × 600	600

*Drains up to 150 mm.
†For clayware or plastics may be reduced to 430 mm in order to provide support for cover and frame.

Table 12.6 Maximum spacing of access points in metres

From	To	Access fitting Small	Access fitting Large	Junction	Inspection chamber	Manhole
Start of external drain*		12	12	–	22	45
Rodding eye		22	22	22	45	45
Access fitting						
Small 150 dia		–	–	12	22	22
150 × 100		–	–	22	45	45
Inspection chamber		22	45	22	45	45
Manhole		22	45	45	45	90

*Connection from ground floor appliances or stack.

Access points should be provided on long drain runs and at or near the head of each drain run, at a bend or change of gradient, a change of pipe size and at junctions where each drain run to the junction cannot be cleared from an access point. The limits of the depth and minimum dimensions for access points are set out in Table 12.5 and the minimum spacing of access points in Table 12.6.

Inspection chamber

An inspection chamber is usually constructed as a brick-lined pit at the junction of drain branches and at changes of direction and gradient, to facilitate inspection, testing and clearing obstructions. An inspection chamber is formed on a 150 mm concrete bed, on which brick walls are raised. In the bed of the chamber, a half-round channel or invert takes the discharge from the branch drains, as illustrated in Figure 12.87. The walls of the

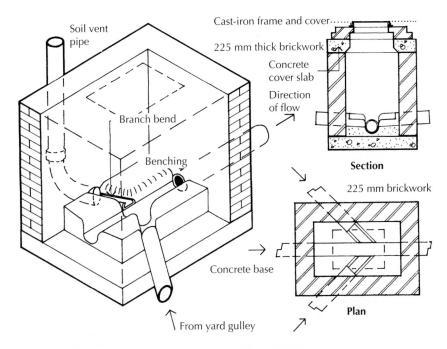

Soil vent pipe

Cast-iron frame and cover

225 mm thick brickwork

Concrete cover slab

Direction of flow

Branch bend

Benching

Section

225 mm brickwork

Concrete base

From yard gulley

Plan

Figure 12.87   Cut-away sectional view of inspection chamber.

chamber may be of dense engineering bricks. If less dense bricks are used, the chamber is lined with cement rendering to facilitate cleaning, and sometimes it is rendered outside to prevent the infiltration of groundwater. The chamber is completed with a cast iron cover and frame (Photograph 12.2).

The word invert is used to describe the lowest level of the inside of a channel in an inspection chamber, or the lowest point of the inside of a drainpipe, and measurements to the invert of a drain are used to determine the gradient of that drain (Figure 12.88). In the bed of the chamber the three-quarter section branch drains discharge over the channel in the direction of flow, and fine concrete and cement rendering termed benching is formed around the branches to encourage flow in the direction of the fall of the drain, as shown in Figure 12.87.

Backdrop inspection chamber

Where a branch drain is to be connected to a main drain or a sewer at a lower level, it is often economical to construct a backdrop inspection chamber to avoid deep excavation of a drain line. The backdrop chamber is constructed of brick on a concrete bed and the higher branch drain is connected to a vertical or drop drain that discharges to the channel in the backdrop chamber, as illustrated in Figure 12.89 (the near side and end walls are omitted for clarity). The drop drain is run in cast iron drains, supported by brackets screwed to plugs in the wall.

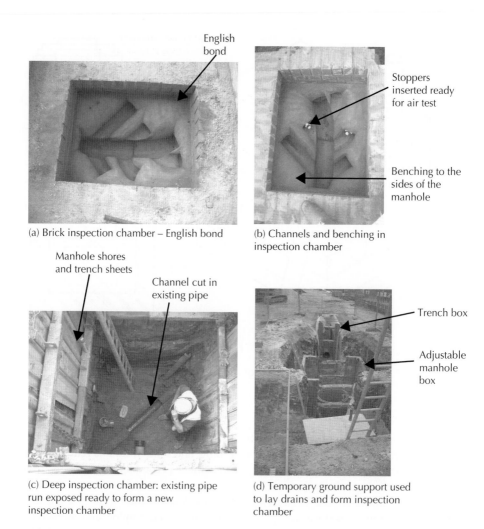

English bond

Stoppers inserted ready for air test

Benching to the sides of the manhole

(a) Brick inspection chamber – English bond

(b) Channels and benching in inspection chamber

Manhole shores and trench sheets

Channel cut in existing pipe

Trench box

Adjustable manhole box

(c) Deep inspection chamber: existing pipe run exposed ready to form a new inspection chamber

(d) Temporary ground support used to lay drains and form inspection chamber

Photograph 12.2   Laying inspection chambers.

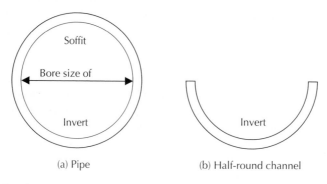

Soffit

Bore size of

Invert

Invert

(a) Pipe

(b) Half-round channel

Figure 12.88   Section through drainage showing the invert level.

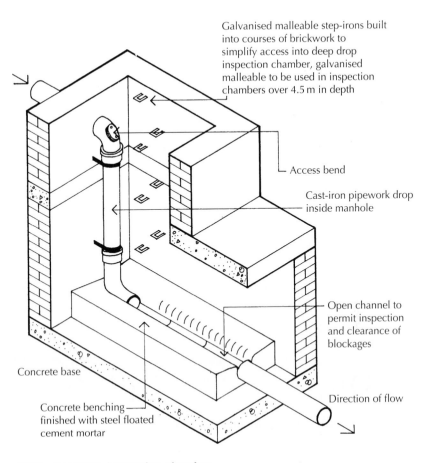

Galvanised malleable step-irons built into courses of brickwork to simplify access into deep drop inspection chamber, galvanised malleable to be used in inspection chambers over 4.5 m in depth

Access bend

Cast-iron pipework drop inside manhole

Open channel to permit inspection and clearance of blockages

Direction of flow

Concrete base

Concrete benching finished with steel floated cement mortar

Figure 12.89   Backdrop inspection chamber.

Access bowl drain system

Where uPVC drains are used for small buildings such as houses, a system of PVC access bowls or chambers may be used. The hemispherical access bowl (Figure 12.90) is supplied with one hole for the outlet drain, and up to five additional holes can be cut on site to suit inlet branches. Purpose-made inlet and outlet PVC connectors are provided for solvent welding to the bowl for outlet and inlet connections to drains. One access bowl is sited close to the building for the discharge from the soil pipe and ground floor foul water and wastewater fittings. The access bowl is extended to the ground level with PVC cylindrical extension sections, one of which is illustrated in Figure 12.90. The access bowl is set on a concrete bed and, after the drain runs have been connected, it is surrounded with a minimum of 100 mm of concrete for protection and stability. A cast iron frame and cover is set on top of the access bowl at the ground level. These access bowls may be used at changes of direction in drain lines and at the site boundary, before the connection to the sewer, for the purpose of rodding to clear blockages and inspection of flow. Access bowls can effect some cost saving compared with a brick inspection chamber.

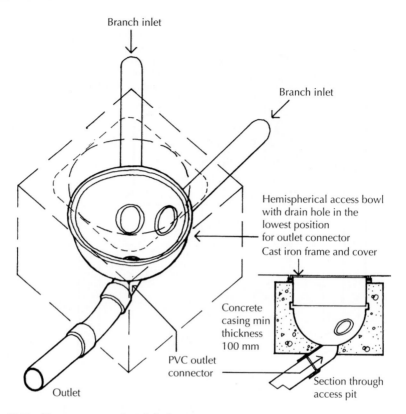

Figure 12.90   Mascar access bowl drain system.

Rodding point drain system

This sealed underground drain system combines the advantages of the long lengths of pipe and simplicity of jointing uPVC drain lines, with rodding points and one or more inspection chambers for access for inspection, testing and clearing blockages. Figure 12.91 shows a typical layout. The rodding points are used as a continuation of straight drain runs extended to the ground level with an access cap. It is possible to rod through each drain run to clear a blockage and this arrangement dispenses with the need for inspection chambers at all junctions and bends with appreciable saving in cost.

Soil stack pipe to connection drains

In the single-stack system, a single pipe serves to carry both soilwater and wastewater discharges directly down to the drains, so that the soil pipe or stack may serve as a ventilation stack pipe to the drains. In this system, the water trap seals to each of the sanitary appliances serve as a barrier to drain smells entering the building. At the base of the soil pipe is a large radius or easy sweep bend to facilitate clearing blockages. Whether the soil pipe is run inside or outside the building, there should be a large radius bend at its connection to an inspection chamber. Where the drain connection to a soil or waste pipe passes

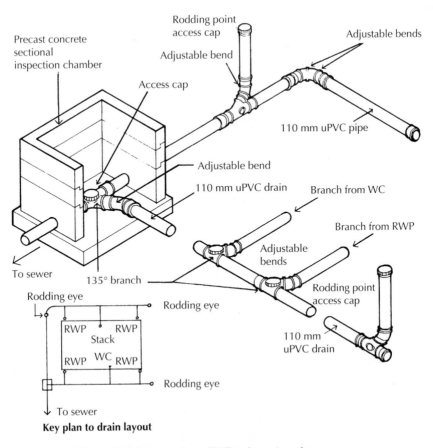

Figure 12.91   Rodding point drain system. RWP, rain water pipe.

through the wall of the building, there should be at least 50 mm of clearance all round the drain to allow for any settlement of the wall that might otherwise fracture the pipe if it were built into the wall. The opening in the wall around the pipe is supported by small lintels or brick arches and covered with rigid sheet on both sides, as illustrated in Figure 12.92.

Gullies

Where there is a combined sewer that takes soil and wastewater discharges and rainwater from roofs and paved areas, it is necessary to use a trapped gully at the foot of rainwater downpipes so that the water seal in the gully serves as a barrier to foul gases rising from the drains. The trapped gully illustrated in Figure 12.93 has a back inlet connection for the rainwater pipe and a grating that serves as an access to clear blockages and to take water running off paved areas. These gullies are made with either back or side inlet connections for rainwater pipes.

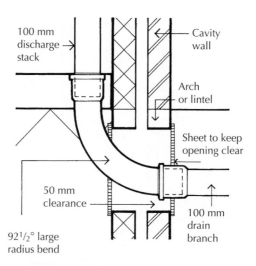

Figure 12.92 Soil stack connection to the drain.

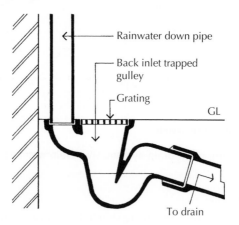

Figure 12.93 Trapped gully.

Where there are separate drain and sewer systems for foul water and rain and surface water, the gullies that collect rainwater and surface water can be connected directly to the surface or storm water drain. The fitting (Figure 12.94) is described as a rainwater 'shoe' to differentiate it from a gully that has a water seal. The shoe has a grating that fits loosely into a tray to provide access.

As there is less likelihood of blockages in surface water drains than in soilwater drains, it is not considered necessary to form inspection chambers at all junctions, bends and changes of gradient to the drain as is the case with foul water drains. Rodding eyes at salient points to facilitate clearing drains are generally considered adequate for the purpose. Because of the wide range of fittings available, it is common to use clayware gullies and

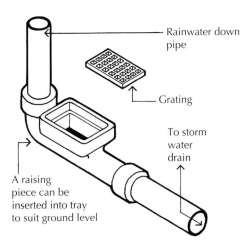

Figure 12.94   Rainwater shoe.

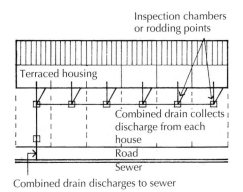

Figure 12.95   Common drain.

rainwater shoes for most drainpipe materials. Gullies and shoes are usually bedded on a small concrete base to provide a solid base for connection to stack pipes and drains.

Private sewers, common drain

There is no exact distinction between the words drain and sewer, but the most generally accepted definition is that pipelines under privately owned land, laid and maintained by the owner, are called drains, and pipelines laid and maintained by the local authority under roads are called sewers. A private sewer, also termed a combined drain, is a system of drains laid for the use of two or more buildings, paid for and maintained by the owners of the buildings and making one connection to the public sewers. A common drain or sewer to a terrace of houses such as that illustrated in Figure 12.95 requires one drain connection to the public sewer. Where properties do not directly front onto public roads and sewers,

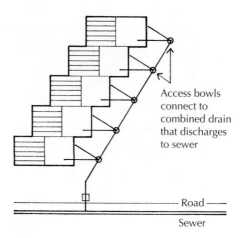

Access bowls
connect to
combined drain
that discharges
to sewer

Road

Sewer

Figure 12.96   Combined drains (private sewers).

as shown in Figure 12.96, it is convenient to lay a private sewer (sometimes called a combined drain). This allows a connection from each house to the private sewer and only one connection to the public sewer. With this type of drainage layout it is necessary to apportion the cost of maintenance and repair for the private sewer to the properties connecting to it.

Connections to sewers

Connections to public sewers are generally made by the local authority or made by the building owner's contractor under the supervision of the local authority. In both cases, the connection is paid for by the building owner. In new developments, a branch connection is constructed in a new sewer, which is then connected to an existing sewer.

Drain testing

All newly laid drain lines should be tested for watertightness after jointing and laying and again after backfilling and consolidation of trenches. Existing drains may also need to be tested. The methods available are as follows.

Water test

The drain is tested by water pressure, applied by stopping the low end of the drain and filling it with water to a minimum head of water, as illustrated in Figure 12.97. The head of water should be 1.5 m above the crown of the high end of the pipeline under test. Where long drain lines are to be tested, and the head of water would exceed 6 m due to the length and gradient of the drain, it is necessary to test in two or more sections along the line. The loss of water over a period of 30 minutes should be measured by adding water from a measuring vessel at regular intervals of 10 minutes, and noting the quantity required to maintain the original water level. The average quantity added should not exceed 1 litre per hour per linear metre, per metre of nominal internal diameter. The water test is a test of

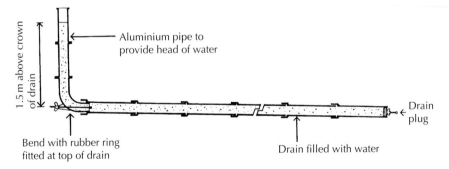

Figure 12.97   Water test.

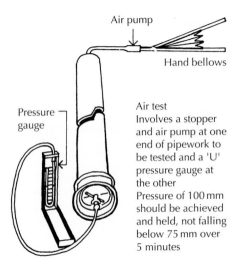

Figure 12.98   Air test.

watertightness under pressure, a condition that a freely flowing drain will never suffer, and is thus a test of watertightness far more rigorous than the drain line is designed for.

Air test

The air test is generally accepted as less rigorous test than the water test. The drain line is stoppered at both ends and air pressure is provided by a pump, the pressure being measured by a graduated U-tube or manometer, as shown in Figure 12.98. Expanding drain stoppers suited to take the tube from the pump at one end of the drain line and the tube to the pressure gauge or manometer at the other, are fitted to the ends of the drain line to be tested. Air pressure is applied to the drain through a hand- or foot-operated pump until a pressure of 100 mmHg is recorded on the graduated U-section pressure gauge and the valve to the gauge is shut. An initial period of 5 minutes is allowed for temperature stabilisation

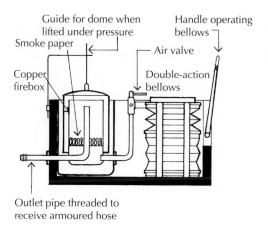

Figure 12.99   Eclipse smoke-testing machine.

and the pressure is then adjusted to 100 mmHg. The pressure recorded by the U-gauge or manometer should then not fall below 75 mm over a further 5 minutes for a satisfactory test. Where trapped gullies and ground floor appliances are connected to the drain line being tested, a pressure of 50 mmHg is adopted as the measure.

Smoke test
The smoke test has been used for old drains where the water or the air test is too rigorous. The drain is stoppered at suitable intervals and smoke is introduced under pressure from a smoke capsule or smoke machine. The purpose of the tests is to discover leaks by the escape of smoke, either when the line has been uncovered or is underground. An escape of smoke will find its way to the surface through a considerable depth of soil and all but the most dense concrete cover. Figure 12.99 illustrates a smoke-test machine. The reason for using a smoke test, particularly on old drains, is to give some indication of the whereabouts of likely leaks before any excavation to expose drains has been undertaken. This may be of use on long runs of drain and where drains run under buildings. The appearance and smell of smoke may give a useful indication as to where excavation to expose drains should take place. Water, air and smoke tests act on the whole of the internal surface of drains. These tests may indicate leaks due to cracks in the crown of the drain. As drains never run full bore, a crack in the crown of a drain may not cause any significant leakage and may not warrant repair.

Closed Circuit Television (CCT) surveys
Modern technology provides a means of making visual inspection of the inside of drain runs by the use of a small camera that is inserted into and run down the line of a drain to provide a moving picture record on a monitor of the view of the inside of the drain. This is usually recorded on a computer disc and can be used to obtain an extent of work required before excavation begins.

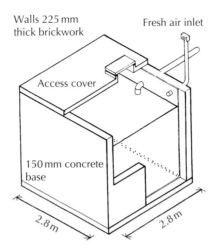

Walls 225 mm
thick brickwork

Fresh air inlet

Access cover

150 mm concrete
base

2.8 m

2.8 m

Figure 12.100   Cesspool.

Sewage collection and treatment

In areas where connection to a main sewer is not possible, it is necessary to use a cesspool, septic tank or sewage treatment plant. It is necessary to make sure that the installation conforms to current pollution prevention requirements and is appropriate for the intended purpose. Advice is available from local authority environmental control officers.

Cesspool

A cesspool is an underground chamber or container used to collect and store all foul and wastewater from buildings. They are emptied by pumping the contents to a tanker. Cesspools are used only in outlying areas where there is no ready access to a sewer and where ground is waterlogged or the slope of the site does not allow the use of a more compact and convenient septic tank. The traditional cesspool was a brick-lined pit, such as that illustrated in Figure 12.100, with engineering brick walls 220 mm thick, lined inside with cement and sand rendering on a concrete base. A reinforced concrete cover over the cesspool would be formed at, or just above, ground level with some form of access cover for emptying and a fresh air inlet (FAI). A range of prefabricated cesspools is available today. These are made from glass reinforced fibre plastic (GRP) in the form of sealed containers with capacities of 7500–54,000 L. Larger capacities of up to 240,000 L can be made. The ribbed cylindrical cesspool, with access point, inlet and FAI, is delivered ready to lower into an excavated pit with the cylinder lying on its long axis. The cesspool is laid on a bed of concrete and surrounded with a lean mix of concrete, with the cover to the access point at or just above the ground level.

Septic tank

A septic tank is designed to take the outflow of soil and wastewater, retain some solid organic matter for partial purification, and discharge the liquid sewage through a system

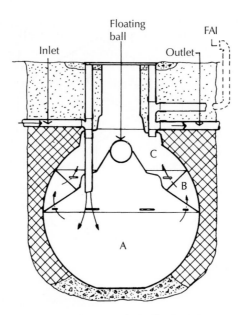

Figure 12.101   Prefabricated septic tank.

of land drains to the surrounding ground to complete the process of purification. The usual capacity of septic tanks is from 2800 to 6000 L for 4–22 people, respectively. A range of moulded, high-grade polyethylene, bulb-shaped tanks is manufactured, with access turret, inflow and outflow connections and an FAI (Figure 12.101).

The tank is bedded on concrete and surrounded with lean mix concrete with the access cover at or just above ground level. The septic tank is divided into three chambers. The outflow of the foul and wastewater drains is discharged to the lower chamber (A in Figure 12.101) in which the larger solid particles of sewage settle to the bottom of the chamber. As more sewage flows down the inlet, liquid rises through the slots in the bell-shaped division over the lower chamber and enters the middle chamber (B). In this chamber fine particles of solid matter settle on top of the bell-shaped division and settle through the slots into the lower chamber. As the tank fills, liquid now comparatively free of solid sewage matter rises through the slots in the inverted bell-shaped division to the third and upper chamber (C). As this upper chamber fills, the now partially purified liquid rises through the outflow to a system of perforated land drains.

The land or leaching drains, laid in trenches surrounded by granular material, spread the liquid sewage over an area sufficiently large to encourage further purification of the liquid sewage by aeration and the action of microorganisms in sewage. The accumulated sludge of solid sewage that settles in the base of the tank should be removed at intervals of not more than 12 months to a tanker. A hose from the tanker first empties the upper chamber, by suction. As the liquid level falls, the ball that seals the division between the upper and middle chambers falls away. This provides access to remove the sludge from the lower chamber.

Sewage treatment plant

The preferred and more expensive system for sewage outflow, where no sewer is readily available, is the installation of a sewage treatment plant that will produce a purified liquid outflow that can be discharged to nearby ditches and streams without causing pollution of water supplies. A form of sedimentation tank, similar to a septic tank, causes solid sewage to settle. The resultant liquid outflow is further purified by exposure to air to accelerate the natural effect of microorganisms, native to sewage, combining with oxygen to purify the sewage. The three systems used to speed the exposure of the liquid sewage to air are:

(1) By spreading over a filter bed
(2) By spreading over rotating discs
(3) By pumping in air to combine with the liquid (aeration)

The traditional sewage treatment plant, such as that illustrated in Figure 12.102, comprises a settlement tank which acts in the same way as a septic tank to allow solid matter to sink to the bed of the chamber either naturally or assisted by baffles. A range of prefabricated treatments is available from specialist suppliers.

Natural systems

In rural areas, it may be possible to discharge the foul water to specially designed and constructed reed beds (ponds), which naturally filter the waste. Advice should, in the first instance, be sought from the appropriate local authority, and then experts in this specialist area.

12.7 Roof drainage

The primary function of roof drainage is to get rainwater from the roof to a suitable discharge as quickly and economically as possible, thus helping to prevent water penetration to the inside of the building. Rainwater running off both pitched and flat roofs is usually collected by gutters and outlets and discharged by rainwater downpipes to drains, sewers or soakaways. More recently, emphasis has been placed on recycling the rainwater, via storage tanks, to flush WCs using grey water.

Rainwater gutters and downpipes are a prime source of dampness in walls if poorly designed and maintained. Gutters that are blocked, cracked or have sunk may cause persistent saturation of isolated areas of the wall, leading to damp staining and possible dry rot in timbers. Gutters are not readily visible to building owners and are generally difficult to access for regular clearing of leaves and other accumulated debris.

The size of gutters and downpipes is determined by the estimated volume of rainwater that will fall directly on a roof during periods of intense rainfall. It is practice to use gutters and rainwater downpipes large enough to collect, contain and discharge water that falls during short periods of intense rainfall that occur during storms.

Drainage to pitched roofs

To determine the amount of water that may fall on a pitched roof, it is necessary to make allowance for the pitch of the roof and for wind-driven rain that may fall obliquely on a

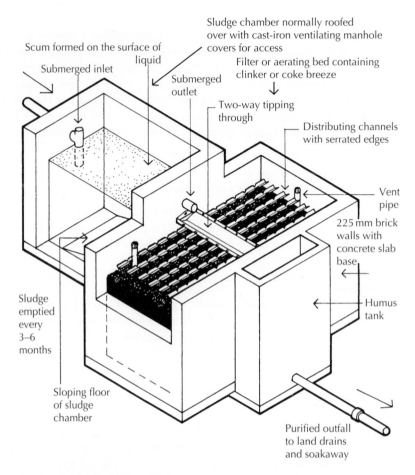

Figure 12.102 **Sewage treatment plant.**

slope facing the wind. A larger allowance for wind-driven rain is made for rain falling on steeply sloping roofs than on roofs with a shallow slope. The effective area of roof to be drained is derived from the plan (horizontal) area of a roof multiplied by a factor allowing for the pitch or slope of the roof, as set out in Table 12.7. Having determined the effective area of a roof slope to be drained, it is necessary to select a size of gutter capable of collecting and discharging the volume of water assumed to fall during storms. For the purpose of choosing a gutter size, it is assumed that the roof drains to a half-round gutter up to 8 m long with a sharp-edged outlet at one end only and laid level.

Most gutters are laid with a slight fall to encourage flow, and the majority of plastic eaves gutters have round-cornered outlets to encourage discharge, whereas square-cornered gutter outlets, such as those in cast iron, impede flow, as illustrated in Figure 12.103. In the calculation of rainwater downpipe sizes, some reduction of pipe size may be effected by the use of round-edged gutter outlets.

Table 12.7   Calculation of area drained

Type of surface	Design area (m²)
Flat roof	Plan area of relevant portion
Pitched roof at 30°	Plan area of portion × 1.15
Pitched roof at 45°	Plan area of portion × 1.40
Pitched roof at 60°	Plan area of portion × 2.00
Pitched roof over 70° or any wall	Elevational area × 0.5

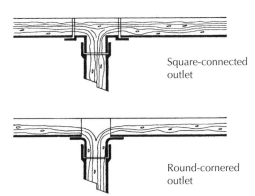

Square-connected outlet

Round-cornered outlet

Figure 12.103   Rainwater outlets.

To make the best use of the gutter, it is usual to fix it to fall towards each side of an outlet to economise on gutter size and the number of rainwater pipes. In the examples shown in Figure 12.104, it would be possible to use two rainwater pipes, one at each end of the roof, with the gutter falling each way from the centre of the roof. Each half of the length of gutter would have to collect more rainwater than any one length of fall shown in Figure 12.104, and a large gutter would be required. The length of a gutter between outlets to rainwater downpipes will depend on the position and number of rainwater pipes related to economy and convenience in drain runs (and position of windows and doors, etc.), which in turn will determine the area of roof that drains to a particular gutter length or lengths.

Table 12.8 gives an indication of the required diameter of half-round gutter sizes related to the maximum effective area of roof that drains to that gutter, where the gutter is at a fixed level and the outlets are square-edged. The figure given for flow capacity in gutters is given in litres per second, which is derived from rain falling on the relevant effective area of roof in millimetres per second where a cubic metre of water equals 1000 L. This figure may be used to check gutter sizes chosen against manufacturers' recommended flow capacities of gutters. The gutter outlet sizes will determine the size of rainwater downpipe to be used. For the majority of small buildings, such as single houses and small bungalows, a gutter with a diameter of 75 or 100 mm is adequate for the small flows between outlets.

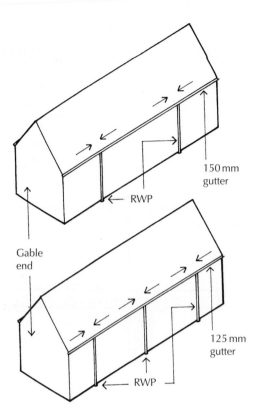

Figure 12.104   Gutters and downpipes.

Table 12.8   Gutter sizes and outlet sizes

Max roof area (m²)	Gutter size (mm dia)	Outlet size (mm dia)	Flow capacity (L/s)
6.0	–	–	–
18.0	75	50	0.38
37.0	100	63	0.78
53.0	115	63	1.11
65.0	125	75	1.37
103.0	150	89	2.16

Refers to half-round eaves gutters laid level with outlet at one end sharp-edged. Round-edged outlets allow smaller downpipe sizes.

Eaves gutters to hipped roofs are fixed to collect rainwater from all four slopes with angle fittings at corners, as shown in Figure 12.105, to allow water to run from the hipped-end slope to the outlets to main slopes. In Figure 12.105, two downpipes are shown to each main roof slope, with the gutter to the hipped ends draining each way to the gutters to the main roof slopes. A square angle in a gutter, within 4 m of an outlet, somewhat impedes

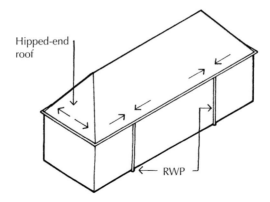

Hipped-end roof

RWP

Figure 12.105   Hipped-end roof.

flow, and allowance is made for this in the calculation of flow in gutters. Flooding and overflow of eaves gutters to hip-ended roofs most commonly occur at angles where obstruction to the flow is greatest, blockages are most likely to occur and the angle gutter fittings may sink out of alignment. Eaves gutters are usually fixed to a shallow fall of 1:360 towards outlets. This shallow fall avoids too large a gap between the edge of the roof covering and the low point of the gutter, yet is sufficient to encourage flow in the gutter and make allowance for any slight bow or settlement of the gutter.

Gutter sections

The section of gutter most commonly used for pitched roofs is the half-round gutter. Other sections are available, such as the segmental and ogee or OG gutter illustrated in Figure 12.106. The segmental may be used to collect rainwater from small shallow-pitched roofs. Gutter lengths and their associated fittings have spigot and socket ends so that the plain (spigot) end of the gutter fits to the shaped (socket) end of the gutter to provide a level bed of gutter. The necessary fittings to gutters are running outlets, stop-end outlets, angles, both internal and external, and stop ends. These fittings have socket ends to fit the spigot ends of the section of gutter.

Figure 12.107 shows typical plastic gutter fittings for half-round gutters. The angle fitting shown is an external angle fitting. Internal angle fittings are used where a wing or part of a building butts to another part at 90°. uPVC gutter sections with spigot and socket ends and angles, outlet and stop-end fittings, are commonly used to drain the majority of pitched roofs. uPVC is a lightweight material that needs no protective coating, is moderately rigid and has a smooth surface that encourages flow. This material is used for its comparatively low cost, ease of handling and freedom from maintenance.

Figure 12.108 shows uPVC gutters. Flexible seals are bedded in the socket ends of both gutters and fittings. These seals are watertight and at the same time are sufficiently flexible to accommodate the appreciable thermal expansion and contraction that is characteristic of this material. Without the flexibility of the seals, long lengths of gutter might otherwise

Segmental gutter

Half-round gutter

Ogee or og gutter

Figure 12.106   Rainwater gutter cross sections.

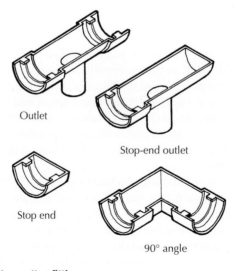

Outlet

Stop-end outlet

Stop end

90° angle

Figure 12.107   Rainwater gutter fittings.

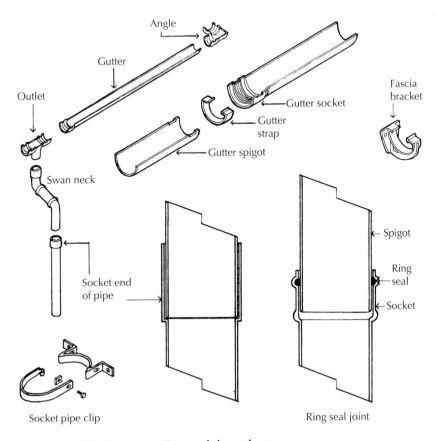

Figure 12.108 uPVC rainwater gutters and downpipes.

deform. The joint between spigot and socket ends is secured with a gutter strap clipped around the gutter.

Gutter lengths are supported by plastic fascia brackets that are screwed to fascia boards. The gutter is pressed into the fascia brackets so that the lips of the bracket clip over the edges of the gutter to keep it in place. The spacing of the brackets is determined by the section of gutter used. Rainwater downpipes are moulded with plain spigot and shaped socket ends. Socket ends are shaped to make the close-fitting joint to spigot ends, or moulded to take a flexible ring seal as illustrated in Figure 12.108. The former makes a reasonably close fit and the latter a more positive watertight fit. The shaped fitting or fittings from the gutter outlet to the downpipe are formed to make allowance for overhanging eaves so that water discharge is directed from the gutter towards the downpipe fixed to walls.

The swan-neck fittings are either moulded in one piece to accommodate particular dimensions of eaves overhang, or consist of three fittings to allow for various eaves overhangs. The swan neck shown in Figure 12.108 comprises three units: two bends and a short straight length used to allow for a particular eaves overhang. Rainwater downpipes are secured to walls with plastic pipe clips screwed to walls and the two-piece pipe clip facilitates fixing plugs and screws to the wall.

Drainage to flat roofs

Flat roofs are often surrounded by parapet walls raised above the level of the roof finish. The run-off of rainwater is encouraged by a fall or slope of the roof to outlets formed in the parapet wall. These outlets are lined with lead sheet shaped to fit in the bed and side of the outlet and are dressed down over a rainwater head, illustrated in Figure 12.109. The lead chute is dressed under the flat roof covering. Because the run-off of rainwater is comparatively slow, fairly close spacing of outlets is necessary for drainage.

Rainwater heads are usually of galvanised or galvanised plastic-coated pressed steel with an outlet to the rainwater downpipe. The traditional cast iron rainwater head is not frequently used because of its cost and the need for regular coating or painting to avoid unsightly rusting. Where a flat roof is carried over boundary walls, it discharges rainwater to an eaves or boundary wall gutter.

Large expanses of flat roof should drain to outlets in the roof as well as boundary outlets to avoid extensive lengths of fall or slope to flat roofs. Outlets are formed in the roof in some position where the necessary rainwater downpipe may be fixed to part of the supporting structure, such as a column. These outlets are shaped so that the roof covering may finish to the outlet with a watertight joint. Figure 12.110 is an illustration of an outlet to a flat roof covered with asphalt, with a lift-up grating to facilitate clearing of blockages.

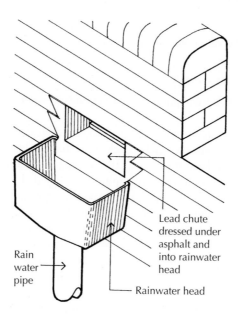

Rain water pipe

Lead chute dressed under asphalt and into rainwater head

Rainwater head

Figure 12.109   Parapet rainwater outlet.

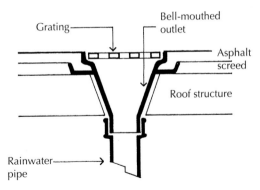

Figure 12.110   Rainwater outlet.

12.8 Surface water drainage and sustainable drainage systems (SuDS)

The underlying principle behind SuDS is to better manage surface water and ground water in a sustainable way. The idea is to try and mimic nature and manage rainfall close to where it falls. The aim is to:

❏ Reduce flood risk (quantity of water)
❏ Minimise water pollution (quality of water)
❏ Restore and maintain natural flow in water courses
❏ Improve water resources
❏ Enhance amenity and biodiversity

One of the impacts of urbanisation is the negative effect on rainwater catchment. Natural catchment would normally slow surface water run-off by infiltration of water into the ground (recharging groundwater) and by uptake of water from vegetation before it reaches natural water courses. Once a site has been developed the natural permeability of the ground is usually compromised, resulting in increased surface water run-off (and reduced groundwater recharge) and reduced uptake by vegetation. Water run-off reaches water courses much quicker, which may result in an increased risk of flooding and an increased risk of water pollution from the run-off. By following the SuDS guidance it is possible to reduce the impact on natural water drainage. Simple measures can be taken, such as minimising the use of impermeable materials and including water features, such as ponds, to slow the rate of run-off. Comprehensive guidance is available to designers and local planning authorities will provide guidance on new development in an attempt to better manage surface water drainage.

External surface water drainage

To comply with SuDS it is necessary to limit the amount of impermeable materials around buildings. Permeable materials that allow water to drain into the ground should be used

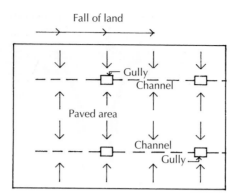

Figure 12.111   Paved area drainage.

for hard standing areas, such as patios, drives and pathways. In areas prone to flooding there will be stringent measures in place that must be complied with as part of the town planning permission.

External surfaces that are paved with impermeable materials, such as concrete, tarmac, paving slabs, bricks or granite sets, should be laid with slight slopes or falls to gullies or channels. They should also be laid to fall away from the external walls of buildings to prevent the possibility of water collecting at the base of a wall. The fall helps to discharge rainwater reasonably quickly for the convenience of people and to prevent ponding of water that would accelerate deterioration of paving materials by saturation and the effect of frost on water lying in fissures in the material. A minimum fall of 1:60 is generally recommended for paved areas on flat ground. To economise in drain runs, paved areas should drain in two or more directions to yard gullies, with drainage channels between gullies in large paved areas to effect further economies. The drainage of the paved area in Figure 12.111 is arranged by each way falls to channels that, in turn, drain to gullies. In areas prone to flooding it may be necessary to discharge the water to an underground holding tank or to a pond to reduce the rate of surface water run-off and hence reduce the quantity of water entering water courses.

Channels

Channels between gullies may be level and depend on natural run-off or may be laid to a slight fall by a gentle sinking towards a gully. A small square area of paving may be laid to fall to one central gully, with channels formed from each corner to the gully by the intersection of slopes. Forming this slope at intersections of falls, termed a current, may involve oblique cutting of paving slabs or bricks to provide a level surface. Yard or surface water gullies are set on a concrete bed to finish just below the paved surface. The gullies generally have a 100 mm square inlet and are either trapped with a water seal when they discharge to combined drains or are untrapped for discharge to a separate storm water drain. Figure 12.112 shows a trapped clay yard gully with loose cast iron grating for access to clear blockages. The graph in Figure 12.113 indicates the area of paving that can be drained relative to the estimated flow and size of the drain and to the gradient or fall.

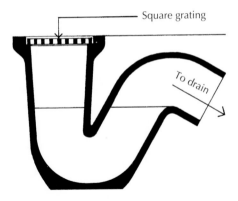

Figure 12.112 Clay yard gully.

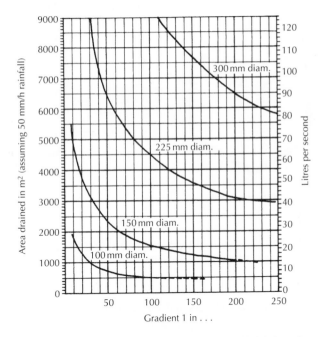

Figure 12.113 Graph for determining the diameter and gradient of surface water drains.

Drainage systems

Where there are separate drainage systems, roof and surface water drainage will connect to a separate system of underground drains discharging to the mains sewer. For the rain and surface water drains, it will generally be satisfactory to have a system of either clay or uPVC drains; with rodding eye access fittings as necessary and an inspection chamber near

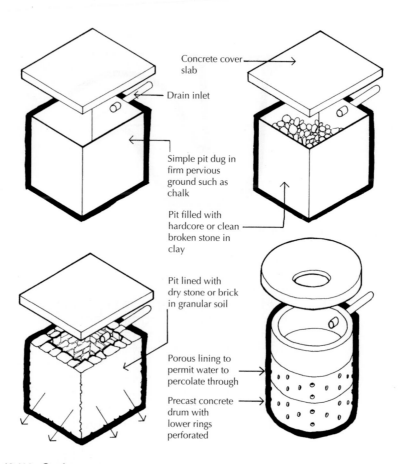

Concrete cover slab

Drain inlet

Simple pit dug in firm pervious ground such as chalk

Pit filled with hardcore or clean broken stone in clay

Pit lined with dry stone or brick in granular soil

Porous lining to permit water to percolate through

Precast concrete drum with lower rings perforated

Figure 12.114   Soakaways.

the boundary. The description of pipe laying, bedding and gradient for foul water drains will apply equally to roof and surface water drains for which untrapped gullies are used.

Soakaways

A soakaway is a pit into which roof and surface water is drained and from which the water slowly seeps into the surrounding ground (Figure 12.114). Soakaways are used where a combined sewer may not be capable of taking more water, where a separate rain and surface water sewer is distant. It can also be used as an alternative means of discharge. Outside the more densely built-up urban areas, there is often an adequate area of land surrounding the building to use soakaways. If the soil is waterlogged and the water table or natural level of sub-soilwater is near the surface, it is pointless to construct a soakaway. In this situation, the surface water will have to be discharged by pipe to the nearest sewer, stream, river or pond. Soakaways should be positioned at least 3 m away from buildings so that the soaka-way water does not affect the buildings' foundations, and they should also be on slopes down from buildings rather than towards buildings, to avoid overflowing and flooding.

12.9 Refuse storage and recycling

Waste is a result of our inefficient use of resources. As consumers, we generate a lot of waste, much of which can be recycled. Environmental pressures and legislation from Europe and the UK have resulted in increasingly stringent targets being placed on local authorities and businesses to recycle more and hence reduce waste to landfill sites. New markets are emerging based on recycling glass, paper, plastics, wood and organic materials. Indeed, there has been a steady rise in the number of building product manufacturers who produce products from recycled material. Local authorities usually provide households with various containers to encourage recycling and separation of different types of waste. Typically these containers are used for:

❏   General household waste
❏   Kitchen food waste (for composting)
❏   Glass, metal, paper and cardboard
❏   Garden waste (for composting)

Domestic waste is usually stored in sacks placed in a plastic refuse (garbage) bin and put out for collection on the prescribed collection day. Some councils provide 'wheelybins' that have larger capacity for storage of domestic waste, recycling and garden waste. Householders can also take their recycling (e.g. empty glass bottles and paper) to recycling sites.

Refuse containers

Refuse containers are large metal containers in which refuse, both domestic and trade, is stored. The limit to the size of these containers is the capacity of a collection vehicle to lift

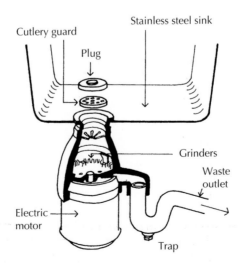

Figure 12.115   Waste disposal unit.

and carry or tow away the container, often described as a Paladin container. These galvanised steel containers are wheeled for manhandling to the collection vehicle, which is designed to lift, upturn and empty the contents into the rear of the collection vehicle. A wide variety of large, purpose-constructed, galvanised steel containers are available. Paladins should be positioned so as to be easily accessible for filling and emptying, but so as to be out of view of the public.

Sink waste disposal units

Kitchen waste is fed through the sink waste to a disposal unit in which a grinder, powered by a small electric motor, reduces the refuse to small particles that are washed down with the wastewater from the sink. These units are designed to dispose of such kitchen refuse as food remains, which rot and cause disagreeable odours in bins. They are not suited to the disposal of larger bulky lightweight refuse. Figure 12.115 is an illustration of one of these units.

13 Heat Loss and Calculations

Increased levels of thermal comfort, combined with the desire to reduce the energy consumed by buildings over their lifetime has resulted in ever more stringent legislation and the need to construct buildings that are thermally efficient. This has been addressed in many of the preceding chapters from the perspective of the various building elements, such as windows and roofs. In this chapter the emphasis is on the principles of heat loss, simple heat loss calculations and the realities of constructing thermally efficient buildings.

13.1 Principles and performance requirements

All elements of a building, from the foundations to the roof need to play their part in helping to reduce the energy demands of a building as a whole. In the preceding chapters the main elements of buildings have been discussed and in many cases explained in terms of their thermal performance.

Resistance to the passage of heat

The building interior is heated by the transfer of heat from heaters and radiators to air (conduction), the circulation of heated air (convection) and the radiation of energy from heaters and radiators to surrounding colder surfaces (radiation). This internal heat is transferred through colder enclosing walls, roofs and floors by conduction, convection and radiation to colder outside air.

Conduction

The rate at which heat is conducted through a material depends mainly on the density of the material. Dense metals conduct heat more rapidly than less dense gases. Metals have high conductivity and gases low conductivity. Conductivity is the amount of heat per unit area, conducted in unit time through a material of unit thickness, per degree of temperature

Barry's Introduction to Construction of Buildings, Third Edition. Stephen Emmitt and Christopher A. Gorse.
© 2014 John Wiley & Sons, Ltd. Published 2014 by John Wiley & Sons, Ltd.

difference. Conductivity is expressed in watts per metre of thickness of material per degree Kelvin (W/mK) and is usually denoted by the Greek letter λ (lambda).

Convection

The density of air that is heated falls, the heated air rises and is replaced by cooler air. This in turn is heated and rises so that there is a continuing movement of air as heated air loses heat to surrounding cooler air and the cooler surfaces of ceilings, walls and floors. It is not possible to quantify heat transfer by convection, so the usual practice is to make an assumption of likely total air changes per hour or volume (litres) per second and then calculate the heat required to raise the temperature of the incoming cooler air introduced by ventilation.

Radiation

Radiant energy from a body, radiating equally in all directions, is partly reflected and partly absorbed by another body and converted to heat. The rate of emission and absorption of radiant energy depends on the temperature and the nature of the surface of the radiating and receiving bodies. The heat transfer by low-temperature radiation from heaters and radiators is small, whereas the very considerable radiant energy from the sun that may penetrate glass and that from high levels of artificial illumination is converted to appreciable heat inside buildings. An estimate of solar heat gain and heat gain from artificial illumination may be assumed as part of the heat input to buildings.

Transmission of heat

Because of the complexity of the combined modes of heat transfer through the fabric of buildings, it is convenient to use a coefficient of heat transmission as a comparative measure of transfer through the external fabric of buildings. This air-to-air heat transmittance coefficient, the U-value, takes account of the transfer of heat by conduction through the solid materials and gases, convection of air in cavities and across inside and outside surfaces, and radiation to and from surfaces. The U-value is the rate of heat transfer in watts through $1\,m^2$ of a material or structure, when the combined radiant and air temperatures on each side of the material or structure differ by one Kelvin (1°C). A high rate of heat transfer is indicated by a high U-value, a low rate of heat transfer by a low U-value. Methods of compliance are given in Approved Document Part L1 and Part L2.

Ventilation

The sensation of comfort is highly subjective and depends on the age, activity and to a large extent on the expectations of the subject. For comfort, good health and well-being in buildings it is necessary to provide a controlled means of ventilation through windows and/or ventilators (see Chapter 7). For general guidance a number of air changes per hour is recommended, depending on the activity common to rooms or spaces. One air change each hour for dwellings and more frequent air changes for kitchen and sanitary accommodation is recommended to minimise condensation of moisture-laden, warm air on cold internal surfaces in those rooms.

Thermal insulation

The figures given for the thermal conductivity of materials (lambda value λ) are based on measurements in controlled environments (laboratories), not on their actual performance in the field (a real building in a particular environment). Standard lambda values assume good quality control and defect-free construction on site. Unfortunately, quality control on site has been shown to be less than satisfactory and with the best will in the world it is harder to assemble details on site than in a laboratory. Thus U-values are a reasonable guide to the thermal performance of the building fabric, and some degree of tolerance should be introduced to allow for any deviation in material quality and/or assembly on the site and any unwitting thermal bridging and/or thermal bypass (see further). It is important to recognise that the actual thermal performance of the built fabric may be less than that designed and calculated. With this in mind we would urge readers to try to move away from an attitude of trying to meet the minimum requirements as stated in the Approved Documents and try to better the thermal performance requirements in their designs and construction practices. Methods of calculating U-values are illustrated below and may also be found in the Approved Documents. Avoiding thermal bridging and ensuring airtightness are also essential in achieving a good thermally efficient building.

13.2 Heat loss calculations

Heat is lost from buildings in two ways. It is lost through the fabric of the building (the building envelope) by radiation, convection and conduction exchanges. This is known as *fabric heat loss*. Buildings are not completely airtight and heat is also lost by heated air leaving the building through gaps in the building fabric and being replaced by colder air that needs to be heated. This is known as *ventilation heat loss*.

So:

Total heat loss (Q) = Fabric heat loss (Q_f) + Ventilation heat loss (Q_v)

Fabric heat loss

The amount of heat that is lost through this route (Q_f) depends on three things:

- The difference between the inside design temperature and the outside temperature (ΔT)
- The area of the different building elements exposed to the temperature differential (A)
- The rate at which heat flows through the different building elements exposed to the temperature differential, known as the U-value (U)

This can be represented as an equation:

$$Q_f = \Sigma(U \cdot A \cdot \Delta T) \quad \text{measured in watts}$$

Σ simply means 'the sum of'

The temperature differential and area of the building elements are straightforward, but the U-value is an important concept and needs some explanation.

U-values

A measure called a U-value (also called the *thermal transmittance* coefficient) is the conventional way of expressing the rate at which heat flows through a building element such as an external wall, window, ground floor or roof. Its formal definition is:

> The rate at which heat flows, in watts, through $1\,m^2$ of a building element when the air temperature on either side differs by 1° (K or °C)

The units in which the **U-value** measure is expressed are therefore:

$$W/m^2K$$

> watts per square metre per degree Kelvin (the formal SI unit of temperature: a change in temperature of 1 K is the same as a change in temperature of 1°C)

The U-value is a measure of the rate at which a building element *transmits* heat (hence its alternative name – thermal transmittance coefficient). The higher the U-value, the more heat is transmitted, or lost, through the building element. From an energy conservation point of view, therefore, the lower the U-value of a building element, the better. The Building Regulations specify maximum U-values that should not be exceeded for different building elements. Approved Document L Part 1A (2010 edition) sets the following limiting values to reduce the risk of condensation (Table 13.1). The heat losses are used in the Standard Assessment Procedure (SAP) and the Reduced Data Standard Assessment Procedure (RdSAP) is used as a method of assessing building energy performance for compliance with Approved Document Part L.

Table 13.1   Limiting fabric parameters compared with nearly zero fabric conventions

	Limiting fabric parameter (maximum) W/m^2K	Notional target U-value 2013 W/m^2K	Nearly zero and Passiv standards W/m^2K
Walls	0.30	0.18	0.08–0.15
Floor	0.25	0.13	0.08–0.15
Roof	0.20	0.13	0.08–0.15
Party walls	0	0	0
Windows, roof lights and glazed doors	2.00	1.4	0.08–0.15 0.08–0.85
Air permeability	$10\,m^3/hr.m^2$ @ 50 Pa	$5\,m^3/hr.m^2$ @ 50 Pa	$0.6\,m^3/hr.m^2$ @ 50 Pa
Linear thermal transmittance		0.05	<0.01 W/mK

In practice, compliance with the requirements of the Building Regulations to conserve heat and power means that design U-values need to be much lower than these 'long stop' values. This is due to the challenges of achieving the required performance standards on a construction site, a point discussed in more detail as follows.

Calculating U-values

Although there are many software programmes that can calculate a U-value, it is important that readers understand the principles of how U-values are calculated. Thus readers can work out their own U-values for a proposed method of construction and also, if necessary, check U-values provided by others.

A U-value is a measure of thermal *transmittance*. Heat flow through a material is usually expressed in terms of thermal *resistance* (R). Transmittance is the inverse of resistance and can therefore be expressed as the reciprocal of resistance:

A building element is composed of a number of materials, each of which has a resistance to the flow of heat. The U-value of a building element can therefore be calculated by adding together the thermal resistances of the components of the building element (ΣR) and dividing the result by one (taking its reciprocal). The shorthand expression of that, and the formula for calculating a U-value (U), is:

$$U = \frac{1}{\Sigma R} \quad \text{Units: W/m}^2\text{K}$$

Total U-value for a building element is:

$$U = {}^1/Rsi_{(\text{inside surface resistance})} + R_{(\text{sum of components resistance})} + Rso_{(\text{outside surface resistance})}$$

Thermal resistance

The amount of resistance that a material offers to the flow heat through it depends on the thermal properties of the material and its thickness. It can be calculated from the following formula:

$$R = \frac{d}{\lambda}$$

❑ R is the thermal resistance of the material (m^2K/W)
❑ d is the thickness of the material (in metres)
❑ λ is the thermal conductivity of the material (W/mK)

Surface and air space resistances

In a building element it is not only materials that provide a resistance to the transfer of heat; surfaces and air spaces or empty cavities also offer resistances that need to be taken into account in calculating U-values. These are usually standard values that can be found from tables such as those published by the Chartered Institution of Building Services Engineers (CIBSE) (http://www.cibse.org).

Thermal conductivity

The thermal conductivity (λ) of a material is the heat flow in watts across a thickness of 1 m of material with a surface area of 1 m² when the air temperature on either side differs by 1°.

Materials which have a high thermal conductivity, such as copper, are good conductors of heat and therefore poor thermal insulators. Conversely, materials that have a low thermal conductivity, such as expanded polystyrene, are poor conductors of heat and therefore good thermal insulators.

Table 13.2 shows typical thermal conductivity values for a range of construction materials.

Table 13.2   Typical thermal conductivity of building materials

Typical thermal conductivity of building materials: Structural and finishing materials (always check manufacturer's details – variation will occur depending on the product and nature of materials)	Thermal conductivity (W/mK)
Acoustic plasterboard	0.25
Aerated concrete slab (500 kg/m³)	0.16
Aluminium	237
Asphalt (1700 kg/m³)	0.50
Bitumen-impregnated fibreboard	0.05
Brickwork (outer leaf 1700 kg/m³)	0.84
Brickwork (inner leaf 1700 kg/m³)	0.62
Dense aggregate concrete block 1800 kg/m³ (exposed)	1.21
Dense aggregate concrete block 1800 kg/m³ (protected)	1.13
Calcium silicate board (600 kg/m³)	0.17
Concrete aerated	0.14
Concrete general	1.28
Concrete (reinforced approx. 1% steel 2300 kg/m³)	2.3
Concrete (reinforced approx. 2% steel 2400 kg/m³)	2.5
Cast concrete (heavyweight 2300 kg/m³)	1.63
Cast concrete (dense 2100 kg/m³ typical floor)	1.40
Cast concrete (dense 2000 kg/m³ typical floor)	1.13
Cast concrete (medium 1400 kg/m³)	0.51
Cast concrete (lightweight 1200 kg/m³)	0.38
Cast concrete (lightweight 600 kg/m³)	0.19
Concrete slab (aerated 500 kg/m³)	0.16
Copper	390
External render sand/cement finish	1.00
External render (1300 kg/m³)	0.50
Felt. bitumen layers (1700 kg/m³)	0.50

Table 13.2   (Continued)

Typical thermal conductivity of building materials: Structural and finishing materials (always check manufacturer's details – variation will occur depending on the product and nature of materials)	Thermal conductivity (W/mK)
Fibreboard (300 kg/m^3)	0.06
Fibreboard (400 kg/m^3)	0.10
Glass	0.93
Granite	1.7–4
Gypsum (600 kg/m^3)	0.18
Gypsum (900 kg/m^3)	0.30
Gypsum (1200 kg/m^3)	0.43
Marble	3
Metal tray used in wriggly tin concrete floors (7800 kg/m^3)	50.00
Mortar (cement mortar – 1750 kg/m^3)	0.80
Mortar – exposed	0.88
Mortar – protected	0.94
Oriented strand board	0.13
Outer leaf brick	0.77
Plasterboard	0.21
Plaster dense (1300 kg/m^3)	0.50
Plaster lightweight (600 kg/m^3)	0.16
Plywood (950 kg/m^3)	0.16
Prefabricated timber wall panels (check manufacturer)	0.12
Screed (1200 kg/m^3)	0.41
Stone chippings (1800 kg/m^3)	0.96
Tile hanging (1900 kg/m^3)	0.84
Timber (650 kg/m^3)	0.14
Timber flooring (650 kg/m^3)	0.14
Timber rafters	0.13
Timber roof or floor joists	0.13
Roof tile (1900 kg/m^3)	0.84
Timber blocks (650 kg/m^3)	0.14
Web of I stud timber	0.15
Wood wool slab (500 kg/m^3)	0.10
Insulation materials and thermal conductivity	
Cellular glass	0.038–0.051
Expanded polystyrene	0.030–0.037
Expanded polystyrene slab (25 kg/m^3)	0.035
Extruded polystyrene	0.029–0.039
Glass mineral wool	0.031–0.044
Mineral quilt (12 kg/m^3)	0.040
Mineral wool slab (25 kg/m^3)	0.035
Phenolic foam	0.021–0.024
Polyisocyanurate	0.022–0.028
Polyurethane	0.022–0.028
Rigid polyurethane	0.022–0.028
Rock mineral wool	0.034–0.042
Air (0–20°C)	0.0204–0.0257
Argon (0°C)	0.016

Table 13.3 Calculation of thermal resistance

	Thickness (in metres) (d)	Thermal conductivity (W/mK) (λ)	Thermal resistance (m²K/W) R = d/λ
Internal surface	–	–	0.120
Plaster	0.013	0.50	0.026
Blockwork	0.100	0.11	0.909
Insulation	0.080	0.02	4.000
Air gap	–	–	0.180
Brickwork	0.102	0.84	0.121
External surface	–	–	0.060

Total resistance (ΣR)   5.416 U-value = 1/Σ R = 1/5.416 = 0.18W/m²K

The thermal conductivity figures are based on measurements in controlled environments and make standard assumptions about, for example, moisture content. In practice, the moisture content of materials may be higher than assumed. If a material such as mineral wool insulation becomes wet, then its thermal conductivity will be higher (because the thermal conductivity of water is greater than that of air) than that shown in the following figure.

Calculating a U-value: worked example

Table 13.3 shows the calculation required to derive the U-value for a partially filled brick and block masonry cavity wall. The structure of the wall is shown in Figure 13.1, along with information concerning the thermal conductivities of the materials and standard values for surface and air gap resistances. From this information, the thermal resistances of the components of the wall are calculated. These are then added to the resistances of the surfaces and air gap. To determine the U-value the number 1 is divided by the total resistance, as shown in Table 13.3.

Thermal conductivities	**(W/mK)**
Dense plaster	0.50
Lightweight blockwork	0.11
Brickwork	0.84
Phenolic foam insulation	0.02

Surface and air gap resistances	**(m²K/W)**
Internal surface	0.12
External surface	0.06
Air gap	0.18

Thermal bridging

Standard U-value calculations assume that the materials in the building element are homogenous, in other words, that the thermal resistance is constant in the plane of the

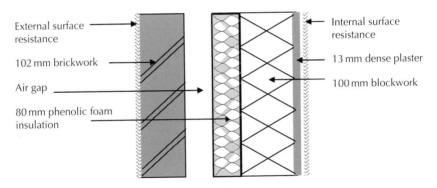

External surface
resistance

102 mm brickwork

Air gap

80 mm phenolic foam
insulation

Internal surface
resistance

13 mm dense plaster

100 mm blockwork

Figure 13.1 Calculation of heat loss through a cavity wall.

material. In practice, that is not always the case. There may be gaps in the layer of material, or the material may be 'bridged' in some way. For example in a timber frame external wall construction, the timber studs interrupt, or bridge, the insulation layer in the framed wall. In a cold roof, the ceiling joists interrupt the layer of insulation placed between the joists. In a cavity wall, a combined lintel bridges the cavity. These are all examples of *thermal bridging*. A thermal bridge is a point, length or area of building fabric which has a higher thermal transmission than the surrounding fabric. The 'bridge' is created when materials, or a build-up of materials (such as junctions or features) create an area that results in heat flowing through the created bridge. Thermal bridging has become a significant source of heat loss as insulation standards have improved and this has to be allowed for in estimating the heat loss from buildings.

Thermal bridges tend to take the form of:

❑ Geometric thermal bridges created by the morphology of the joint: the shape and geometry of, for example, a corner junction means there is an increased area of material exposed to the external environment, and hence a higher concentration of heat flux at the junction.
❑ Penetrations through the fabric: typically penetrations consist of windows, doors, air vents and services (pipes, ducts and cables).
❑ Poor detailing and/or construction of junctions and intersections, which results in discontinuities of thermal barriers

When calculating heat losses through thermal bridges a y-value is used for the heat loss coefficient. The y-value is built-up using psi (ψ W/mK), which is the rate of heat flow per degree temperature difference (ΔK or ΔC) per unit length (m). ψ-values (W/mK) are used to calculate the heat flow through geometrical and non-repeating thermal bridges and are added to the heat loss through the plane elements.

Ventilation heat loss

Building fabric has improved over time to meet more stringent Building Regulation requirements, U-values have therefore declined and building fabric heat loss has decreased. There has been less regulatory concern with airtightness with the result that ventilation heat loss can now account for up to one-third of the total heat loss in a new dwelling.

The ventilation heat losses in a building are determined by:

❑   The volume of air passing through the building that requires heating to achieve the desired internal temperature (calculated by the number of air changes per hour multiplied by the volume of the building)
❑   The amount of heat energy required to raise the air temperature by 1° (the specific heat capacity of air)
❑   The difference between the temperature to which the inside of the building is designed to be heated and the temperature outside the building

Ventilation heat loss can therefore be estimated using the following equation:

$$Q_v = \frac{C_v \cdot N \cdot V \cdot \Delta T}{3600}$$

Q_v is the ventilation heat loss (watts)
C_v is the specific heat capacity of air – usually given as 1210 J/m³K
N is the number of air changes per hour (ach): divided by 3600 gives air changes per second, which is the correct unit as Q_v is measured in watts, or joules per second
V is the volume of the building that is conditioned, or heated
ΔT is the difference in temperature between the inside and outside of the building

The recommended ventilation requirements for UK dwellings are between 0.5 ach (air changes per hour) and 1.0 ach.

Calculation of total heat loss from a building

The following worked example demonstrates how to calculate the expected heat loss from a building based on certain assumptions about the geometry of the building and its thermal design characteristics.

Design data

The detached house measures 8.5 × 12.5 m in the plan. The house has a hipped roof and the height of the walls is 4.0 m. The roof is insulated at ceiling level. Glazing accounts for 20% of the total external wall area. The house has a front and a back external door, each with an area of 2.0 m². The U-values of the building elements of the dwelling are, in W/m²K:

External walls	0.25
Windows	1.50
Doors	0.80
Roof	0.20
Floor	0.15

Table 13.4 Fabric heat loss calculation

Element	U-value U	Area (m²) A	Temperature difference ΔT	Heat loss (W) U · A · ΔT
Walls*	0.25	130.40	20	652
Windows	1.50	33.60	20	1008
Doors	0.80	4.00	20	64
Roof†	0.20	106.25	20	425
Floor‡	0.15	106.25	20	319
Total fabric heat loss				2468

*Wall area is the *opaque* area, i.e. the gross wall area less the area of the windows and doors.
†The relevant area is that exposed to unheated space. The roof construction is a cold roof, with insulation at ceiling level. The area of the roof for the purpose of this calculation is the same as the area of the floor.
‡The floor heat loss calculation is, in practice, complex. The ground temperature needs to be taken into consideration as well as the outside temperature.

Assume the specific heat capacity of air is $1200\,\text{J/m}^3\text{K}$ and that the air is changed once every 2 hours.

The design internal temperature is 20°C and an external temperature of 0°C can be assumed.

Fabric heat loss calculation: $Q_f = \Sigma(U \cdot A \cdot \Delta T)$

Table 13.4 provides an example of the fabric heat loss through a building.

Ventilation heat loss calculation: $Q_v = C_v \cdot N \cdot V \cdot \Delta T / 3600$

Putting the design data into this equation gives:

$$\text{Ventilation heat loss } Q_v = \frac{1200 \times 0.5 \times 425 \times 20}{3600} = 1417$$

In the above, N is equal to 0.5 because an air change once every 2 hours is equivalent to 0.5 ach. The volume of the building (i.e. heated) is equal to the area on plan ($106.25\,\text{m}^2$) multiplied by the height ($4.0\,\text{m}$), giving a volume of $425\,\text{m}^3$.

Total heat loss $= 2468 + 1417 = 3885\ \text{W}$

Heat loss: theory and practice

The above-mentioned equations are only an approximation of what is called 'steady state' heat transfer, in which variables such as temperature differential do not change. Such formulations cannot provide a dynamic picture of heat transfer over time, which would be necessary to assess, for example, whether summer overheating is likely to occur. To do this, much more complicated dynamic models are required.

Thermal models tend to underestimate the actual heat loss from a building. Research has shown that a significant discrepancy exists between the energy performance of a dwelling as designed and that realised in practice, typically around 20% higher than predicted by modelling (http://www.leedsmet.ac.uk/teaching/vsite/low_carbon_housing/index.htm). The difference between measured and predicted performance can be accounted for by factors such as:

❏ Thermal bypasses. A thermal bypass is set up whenever air movement is able to take place in such a way as to reduce the effectiveness of an insulation layer, for example, via the party wall cavity or if insulation boards in a cavity wall are not butted firmly up to the blockwork.
❏ Higher than predicted thermal bridging, for example, as a result of the timber fraction in a timber frame wall being significantly higher than the nominal value.
❏ Real fabric U-values higher than nominal, for example, installing windows whose 'centre pane' U-values equate to the design U-value but whose 'whole window' U-value is significantly higher. Another example would be where components such as insulation become damp, resulting in higher heat loss than predicted using manufacturer's data on thermal conductivity in laboratory conditions.
❏ As-built differing from design intent. This can happen in relation to the fabric in many ways, for example, the omission of a perimeter insulation detail. It can happen in many ways in relation to airtightness of the building and associated ventilation heat loss, for example, service penetrations not being properly sealed.

Thermal bridging

With increased levels of thermal insulation being introduced into the building fabric the problem of thermal bridging has become a major design consideration. Thermal bridging (known as a 'cold bridge' in cold climates) is caused by appreciably greater thermal conductivity through one part of a wall than the rest of the wall, which can cause condensation and encourage mould growth. Openings in walls are areas where thermal bridging can occur if not detailed correctly (Figure 13.2a and b). The greater the level of thermal insulation, the greater the effect of any thermal bridges, a point to be borne in mind when improving the thermal insulation of existing buildings that are likely to have existing thermal bridges. Thermal bridging can be designed out, or the effect minimised, through careful detailing and careful assembly on site to ensure continuity of insulation. Care must be taken, however, to consider constructability because if the detail is difficult to insulate on site, it is highly probable that it will not be constructed as detailed, thus defeating the object of the exercise.

Resistance to the passage of heat

To maintain reasonable and economical conditions of thermal comfort in buildings, walls should provide adequate insulation against excessive loss or gain of heat, have adequate thermal storage capacity, and the internal face of walls should be at a reasonable temperature. Condensation is likely to form on cold internal surfaces (Figure 13.3).

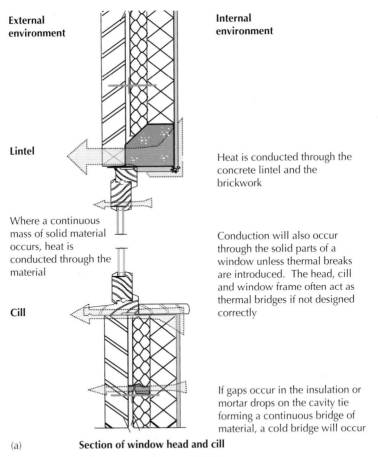

External environment

Internal environment

Lintel

Heat is conducted through the concrete lintel and the brickwork

Where a continuous mass of solid material occurs, heat is conducted through the material

Conduction will also occur through the solid parts of a window unless thermal breaks are introduced. The head, cill and window frame often act as thermal bridges if not designed correctly

Cill

If gaps occur in the insulation or mortar drops on the cavity tie forming a continuous bridge of material, a cold bridge will occur

(a)   **Section of window head and cill**

Potential thermal bridging through window and wall jamb

Where there is no insulation and the blockwork returns to meet the brickwork, at the reveal, heat energy will be conducted out of the building

Typical thermal bridges

- Single glazing

- Metal window frames

- Solid masonry reveals

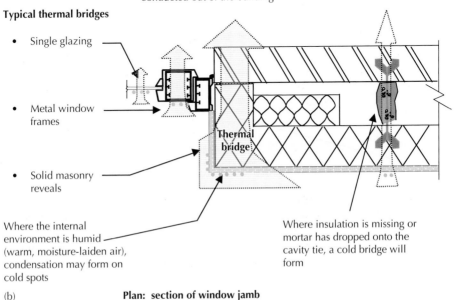

Thermal bridge

Where the internal environment is humid (warm, moisture-laiden air), condensation may form on cold spots

Where insulation is missing or mortar has dropped onto the cavity tie, a cold bridge will form

(b)   **Plan: section of window jamb**

Figure 13.2   (a) Thermal bridging: section of window head and cill. (b) Thermal bridging: plan of cavity wall return and window.

Solid walls and some uninsulated cavity walls are
susceptible to interstitial and surface condensation

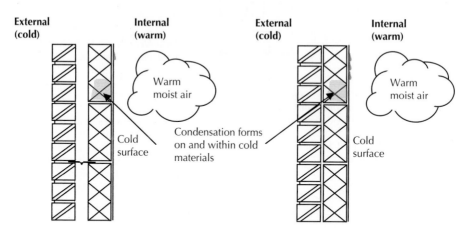

The warm internal air can hold a considerable
amount of moisture, which is generated by
people, gas burners and kettles. If the air becomes
colder it has to give up the water it contains. As the
warm air hits cold objects water will be deposited
on them

As the warm, moist air comes into contact
with the cold face of the wall, condensation
forms on the surface of the wall. If the wall
is permeable, condensation will also form
inside the cold wall (interstitial
condensation)

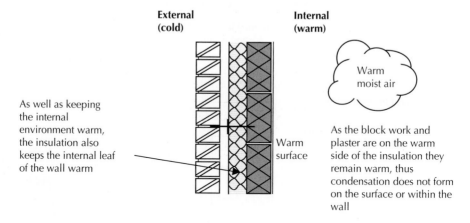

As well as keeping
the internal
environment warm,
the insulation also
keeps the internal leaf
of the wall warm

As the block work and
plaster are on the warm
side of the insulation they
remain warm, thus
condensation does not form
on the surface or within the
wall

Figure 13.3   Condensation and walls.

For insulation against loss of heat, lightweight materials with low conductivity are more
effective than dense materials with high conductivity, whereas dense materials have better
thermal storage capacity than lightweight materials (Figure 13.4). The materials that are
most effective in resisting heat transfer are those of a fibrous or cellular nature in which
very many small pockets of air are trapped to act as insulation against the transfer of heat.

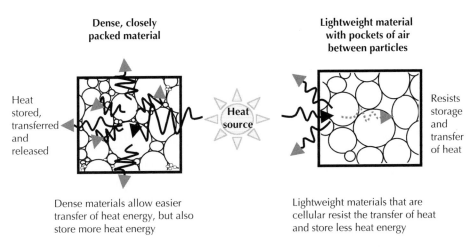

Dense, closely packed material

Heat stored, transferred and released

Heat source

Lightweight material with pockets of air between particles

Resists storage and transfer of heat

Dense materials allow easier transfer of heat energy, but also store more heat energy

Lightweight materials that are cellular resist the transfer of heat and store less heat energy

Figure 13.4   Heat transfer and storage in lightweight and dense materials.

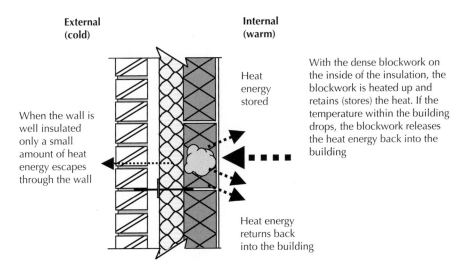

External (cold)

Internal (warm)

Heat energy stored

When the wall is well insulated only a small amount of heat energy escapes through the wall

With the dense blockwork on the inside of the insulation, the blockwork is heated up and retains (stores) the heat. If the temperature within the building drops, the blockwork releases the heat energy back into the building

Heat energy returns back into the building

Figure 13.5   Thermal storage in walls.

Because of their lightweight nature these materials do not have sufficient strength to serve as part of the structure of a wall by themselves. Lightweight insulating materials are either sandwiched between materials that have strength or behind those that resist penetration of wind and rain, or serve as internal wall finishes.

Where a building is heated it is advantageous to use the thermal storage capacity of a dense material on the inside face of the wall with the insulating properties of a lightweight material behind it (Figure 13.5). Here the combination of a brick or dense block inner leaf, a cavity filled with some lightweight insulating material and an outer leaf of brick against penetration of rain, is of advantage. The internal blockwork can act as a heat store. As the building is heated the dense blocks will become warm; once the heating is turned off the dense blocks will release the warmth back into the building. This helps to level out temperature changes when heating cycles are intermittent.

13.3 Testing and monitoring thermal performance

It is not unreasonable for clients to expect their new buildings to perform as expected and not require additional energy to achieve designed operation standards. However, few buildings are tested to establish their thermal performance either during the construction period or after completion. The SAP that is used in the UK checks the design for compliance with the Approved Documents, specifically addressing the legislative requirement for conservation of fuel and power. Information from SAP or RdSAP is used to help rate the building's performance from A through to G, within an Energy Performance Certificate (EPC). EPCs are comparative tools offering a simple 'expected' energy performance; they offer a mechanism to compare one building's expected energy efficiency to another but do not show the building's actual efficiency. Display Energy Certificates (DECs) are required for public buildings greater than $1000\,m^2$, and these do show actual energy usage. While this improves transparency, it does not provide information on the performance of the building fabric or the services.

With regard to buildings that have been constructed to earlier building codes and standards the challenge relates to thermal upgrading of the fabric. And in order to do this effectively and efficiently it is necessary to carry out a detailed investigation of the fabric to identify its thermal performance. The RdSAP assessments, used for thermal upgrades, is also used to determine a building's environmental impact, although it is based on scoring and rating systems working from design and basic inspection data.

Effectiveness of the fabric and efficiency of the services are fundamental to the overall performance of the building. Any underperformance of fabric or services will result in more energy being consumed than predicted to maintain the desired levels of comfort. The focus of this chapter is on the performance of the building fabric, and the methods outlined as follows provide an insight to the data that can be collected.

Measuring actual building performance

Actual building performance can be assessed through scientific (forensic) field tests such as those used in Technology Strategy Board (TSB) Building Performance Evaluation Programmes, and knowledge can be gleaned from the published findings of related research projects. To understand why a building underperforms, information is required on each element, component and junctions that form the whole building. Some of the approaches used are described further.

Forensic investigations are a diagnostic tool to identify the cause and effects of building failures or underperformance. Investigation of buildings is critical when trying to understand why certain results have been obtained. Whereas other tests provide context and identify anomalous results, it is the forensic investigation that identifies the actual defects that are responsible for any deviations found.

Building performance modelling

Computer modelling software may be used to determine the projected performance of construction details and whole buildings, with calculations based on the performance of individual materials, usage and exposure. This is particularly useful when assessing the thermal performance of an existing property and individual elements in different locations, under different conditions and different climates. The impact of specific building defects, alterations and improvements can also be modelled under a variety of conditions. Computer modelling offers the potential to inform and optimise physical test procedures. By understanding the interactions on a theoretical scale, features of testing such as sensor location and equipment positioning can be optimised to ensure the highest level of accuracy.

Testing during construction

Regular inspections should be made during the construction process and observations recorded with digital photography (still and video) and thermography at key stages. Timeline images of the construction process can be particularly revealing when attempting to determine the causes of poor performance of the building fabric after practical completion. By regularly visiting properties to be tested during the construction phase the impact of design changes, material modification and quality of workmanship can be recorded. Sometimes this can result in an intervention to prevent the construction of inadequate details and the correction of inappropriate work. For example, the use of thermography during the construction of buildings has helped to identify thermal bridging, which has then been corrected to prevent the error being repeated, a cheaper alternative compared to trying to correct the error after the completion of the building.

Thermal imaging

Thermal imaging (thermography) provides a powerful visual tool for identifying and representing heat loss. This is an established method for detecting and mapping a number of factors, from the amount of moisture in the building fabric to thermal bridging, discontinuity of insulation and air leakage. An infrared camera captures thermal information by using infrared sensors. By measuring the intensity of the infrared radiation from the surface of the building (or object) the camera provides a two-dimensional image of heat distribution. Thus thermal imaging identifies the warmer and cooler regions of a surface, and so assists in the understanding of thermal losses from a building. For example, thermal imaging can highlight temperature differences of areas of an external wall when viewed from the inside (see Figure 13.6). The colder areas (represented by blue colours) would suggest the presence of a thermal bypass, which may be caused by a number of factors, such as insufficient insulation and variations in moisture content of the fabric. Further tests, such as applying heat flux sensors, air flow sensors or hygrometers to the region under

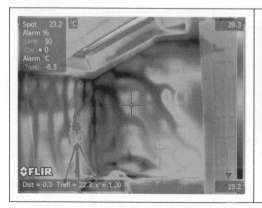

Figure 13.6   **Thermography showing cold air movement and cold spots under depressurised conditions (images courtesy of Leeds Sustainability Institute, 2012). Please refer to the book's web site, www.wiley.com/go/barrysintroduction, for the colour version of this figure.**

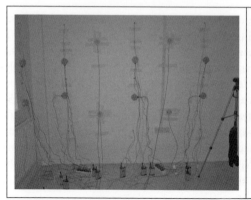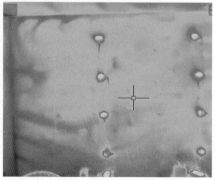

Figure 13.7   **Equipment for wall forensics: Differential pressure sensors, heat flux plates, surface and internal thermocouples, air flow transducer (images courtesy of Leeds Sustainability Institute, 2012). Please refer to the book's web site, www.wiley.com/go/ barrysintroduction, for the colour version of this figure.**

investigation can provide additional information to help inform decisions about the extent of the remedial action required.

Heat flow through building elements

Interpreting the energy performance of a building element based on the stated U-values of building elements is often unreliable because the stated U-values come from measurements made under standard laboratory conditions. In reality, the U-value within an element may vary from that expected. In situ measurements can be made using heat flux sensors in combination with temperature sensors (Figure 13.7). Heat flux sensors are used to measure the rate at which heat passes through a material or a building element. This data, together with data for the difference between internal and external temperature, can then be used to calculate an apparent in situ U-value. By calculating in situ U-values it is possible not only to ascertain the actual performance of building elements but also to identify areas which are not performing as anticipated (designed).

Coheating test

The coheating test determines the actual heat loss through the building envelope. This is achieved by heating the internal environment to an elevated temperature and maintaining the temperature. As the external temperature changes, the power input into the dwelling responds to maintain a stable temperature. When the outside temperature drops more energy is required to heat the dwelling, and when the outside temperature rises less energy is required. By monitoring the power input against the temperature differential between internal and external environments, the heat transfer through the building can be calculated. Losses due to ventilation, heat gains from solar and variations due to the wind are also considered within the calculations.

Airtightness testing

Air leakage and air infiltration are a major factor in heat loss from buildings. Controlled ventilation is necessary to maintain a comfortable and healthy internal environment. In comparison, air leakage and air infiltration refer to the uncontrolled flow of air through gaps and cracks in the building fabric. This results in draughts and associated heat loss (or heat gain in hot weather). Unfortunately, the gaps and cracks are often obscured from view by internal and external finishes, and therefore it is necessary to use smoke to help identify the cause of the air leakage/infiltration. Handheld smoke puffers may be used, or if the building is unoccupied it can be filled with smoke and pressurised to see where the smoke is emitted. The leakage may be observed using handheld smoke puffers or filling the building with smoke and pressurising the interior. Thermography may also be used. Air loses may be quantified using anemometers and differential manometers.

In-use monitoring and building user surveys

In-use monitoring is often the final stage of a full scale testing programme, obtaining data from buildings while they are being used. Monitoring can be particularly revealing with regards to occupant behaviour and the use of appliances, lighting and heating. Monitoring allows the performance of technologies such as mechanical ventilation heat recovery (MVHR) devices and photovoltaic panels to be analysed for their effectiveness. In-use monitoring studies can be combined with building fabric performance studies, helping to provide realistic performance data of the fabric under occupied conditions.

Appendix A: Websites

Electrical Safety Council	http://www.esc.org.uk
GreenSpec	http://www.greenspec.co.uk
HSE	http://www.hse.gov.uk
Leeds Sustainability Institute	http://www.leedsmet.ac.uk/research
Low Carbon Housing Learning Zone	http://www.leedsmet.ac.uk/teaching/vsite
Planning Portal	http://www.planningportal.gov.uk
The Sustainable Building Association	http://www.aecb.net

Barry's Introduction to Construction of Buildings, Third Edition. Stephen Emmitt and Christopher A. Gorse.
© 2014 John Wiley & Sons, Ltd. Published 2014 by John Wiley & Sons, Ltd.

Appendix B: References

Chapter 1

For guidance on regulations see Bett, G., Hoehnke, F. and Robinson, J. (2003) *The Scottish Building Regulations Explained and Illustrated (4th edn)*, Oxford, Blackwell Publishing, and Billington, M.J., Bright, K.T. and Waters, J.R. (2007) *The Building Regulations Explained and Illustrated (13th edn)*, Oxford, Blackwell Publishing.

For additional information about architectural technology and how to detail buildings from sustainable principles see Emmitt, S. (2012) *Architectural Technology – Second Edition*, Wiley Blackwell, Chichester, and Emmitt, S., Olie, J. and Schmid, P. (2004) *Principles of Architectural Detailing*, Oxford, Blackwell Publishing.

Some of the latest research and guidance around low carbon and sustainable construction, thermal upgrades and the testing of domestic dwellings are available on the Leeds Sustainability Institute web page, http://www.leedsmet.ac.uk/research.

Chapter 2

BRE (1995) Site investigation for low-rise building: direct investigations, *Digest* 411, CI/SfB (A3s)

Fryer, B., Egbu, C., Ellis, R. and Gorse, C.A. (2004) *The Practice of Construction Management: People and Business Performance (4th edn)*, Oxford, Blackwell Publishing

Chapter 5

BSi (2003) Recommendations for the preservation of timber, BS8417: 2003, London, BS

Greenbuildingstore (2013) Available from www.greenbuildingstore.co.uk

HSE (2001) A guide to good practice and the safe use of wood preservatives, London, HMSO

PCA (2008) Remedial timber treatment: Code of practice, Property Care Association, Huntingdon, available from http://www.property-care.org

Sutton, A., Black, D. and Walker, P. (2011) Hemp Lime: An Introduction to Low Impact Building Materials, Information Paper IP 14/11 BRE Trust

Barry's Introduction to Construction of Buildings, Third Edition. Stephen Emmitt and Christopher A. Gorse.
© 2014 John Wiley & Sons, Ltd. Published 2014 by John Wiley & Sons, Ltd.

Chapter 6

NHBC (2002) NHBC Standards, National House Building Council, Milton Keynes, Bucks
NHBC (2003) NHBC Standards, National House Building Council, Milton Keynes, Bucks

Chapter 7

BRECSU (1995) *Energy Efficiency in New Housing: Detailing for Designers and Building Professionals, Good Practice Guide 95: External Cavity Walls*, London, HMSO
Gorse, C.A. and Thomas, F. (2013) *Windows Types, Design and Good Practice*. LSi Technical Paper 2013-TP-0001. Leeds, Leeds Sustainability Institute
Gorse, C.A., Wilson, P. and Thomas, F. (2013) *Nearly Zero Fabric Details for Foundations, Design and Good Practice*. LSi Technical Paper 2013-TP-0002. Leeds, Leeds Sustainability Institute

Chapter 11

Scaddan, B. (2003) *Electrical Wiring Domestic (12th edn)*, London, Newnes

Chapter 13

Emmitt, S. (Ed.) (2013) *Architectural Technology: Research & Practice*, Chichester, Wiley-Blackwell
International Energy Institute (2013) Annex 5

Index

Barry's Introduction to Construction of Buildings, Third Edition. Stephen Emmitt and Christopher A. Gorse.
© 2014 John Wiley & Sons, Ltd. Published 2014 by John Wiley & Sons, Ltd.